The Art of Modeling in Science and Engineering with Mathematica®

Second Edition

The Art of Modeling in Science and Engineering with Mathematica®

Second Edition

Diran Basmadjian and Ramin Farnood

CRC Press is an imprint of the
Taylor & Francis Group, an **informa** business
A CHAPMAN & HALL BOOK

CRC Press
Taylor & Francis Group
6000 Broken Sound Parkway NW, Suite 300
Boca Raton, FL 33487-2742

First issued in paperback 2019

ISBN-13: 978-1-58488-460-6 (hbk)
ISBN-13: 978-0-367-39046-4 (pbk)

Visit the Taylor & Francis Web site at
http://www.taylorandfrancis.com

and the CRC Press Web site at
http://www.crcpress.com

Preface to Second Edition

In this thoroughly reorganized and updated version of the first edition, the authors have taken account of their own recent teaching experience, as well as suggestions received from both students and instructors. They have taken note of the lingering misconception of modeling either as sterile and useless, or as an all-powerful tool for describing and predicting events in the physical world. It can, in fact, be either, depending on how wisely it is applied, and we use Chapter 1 to give the reader a first glimpse of what modeling can accomplish and the mathematics that underline it, and to provide some guidelines for kick-starting the modeling process. Here, as elsewhere throughout the text, we use commentaries at the end of an illustration to highlight important results and draw attention to unexpected or startling outcomes that defy conventional wisdom.

Chapter 2 combines Chapter 4 and Chapter 5 of the first edition to provide an updated summary of the important analytical methods for solving (ODEs). This is followed by the new Chapter 3, which presents a synopsis of the Mathematica package and its use in obtaining numerical solutions of the model equations. The package is used in subsequent chapters and throughout the text to solve a variety of practical problems. Mathematica programs used in this book are prepared using Mathematica version 5.0 and are available for download at http://www.crcpress.com/e_products/downloads/default.asp

In composing Chapter 4, we culled the most useful applied problems from Chapter 1 to Chapter 6 of the first edition and supplemented them with a host of new illustrations designed to engage a wider audience. They range from the classical model of Euler for the buckling of a strut to contemporary attempts to model the progress of an HIV infection. The common feature of these applications, both old and new, is their foundation on the conservation laws of mass, energy, momentum (Newton), and electrical charge.

Partial differential equations (PDEs) and their applications and solution methods are the topics of Chapter 5 to Chapter 8. In addition to including new illustrations and practice problems, we took pains to ease the treatment of topics that students found hard to grapple with. We hope that the famous Green's functions, which have lately begun to pervade modern applied mathematics, will be more palatable and hence, more useful. Superposition methods, whose scope is often not fully appreciated or exploited, have been segregated into a separate section to emphasize their importance. The sections on integral transforms and separation of variables, on the other hand, were easily understood by students and remain largely unchanged except for the addition of new problems and the use of the Mathematica package.

Although PDEs occupy a considerable portion of the text, we remind the reader of a saying that we coined for the first edition: "PDEs if necessary, but not necessarily PDEs." In other words, the reader is urged to pay equal attention to Chapters 1 through 4, which contain models based on ODEs and algebraic equations. In fact, those chapters may well be used in a senior-level undergraduate course in modeling, whereas PDEs are best taken up in a graduate course. The intended audience remains, as before, the mature student in applied science and engineering.

It remains for us to express our thanks to Arlene Fillatre, who undertook the task of transcribing the handwritten text to readable print, and to Linda Staats, University of Toronto Press, who miraculously converted rough sketches into professional drawings. Last but not least, we wish to specially express our appreciation to our families for their great patience and understanding.

Diran Basmadjian

Ramin Farnood

Toronto, July 2006

Abstract

The text is a thoroughly revised and updated successor to the first edition of *The Art of Modeling in Science and Engineering*. As before, its purpose is to acquaint the reader with the mathematical tools and procedures used in modeling events and processes based on the laws of conservation of mass, energy, momentum, and electrical charge. The authors have culled and consolidated the best from the first edition, and expanded the range of illustrations to reach a wider audience. The text proceeds, in measured steps, from simple models of real-world problems at the algebraic and ODE levels, to more sophisticated models requiring partial differential equations. The analytical solution methods have now been supplemented with the Mathematica package, which is used throughout the text to arrive at numerical solutions of the model equations.

The text is enlivened with a host of illustrations and practice problems drawn from classical and contemporary sources. They range from Thomson's famous experiment to determine e/m and Euler's model for the buckling of a strut, to an analysis of the propagation of emissions and the performance of wind turbines. The mathematical tools required are first explained in separate chapters and then carried along throughout the text to solve and analyze the models. Commentaries at the end of each illustration draw attention to the pitfalls avoided and the difficulties overcome, and, perhaps most important of all, alert the reader to unexpected results that defy conventional wisdom.

The first four chapters can be used as a senior-level undergraduate course in modeling, while the remaining chapters dealing with PDE models are best taken up at the graduate level. The intended audience is, as before, the mature student in the applied sciences and engineering.

Biographies

Diran Basmadjian is a graduate of the Swiss Federal Institute of Technology, Zurich, and received his M.A.Sc. and Ph.D. degrees from the University of Toronto. He has been a professor of chemical engineering at Toronto since 1980. He has combined his research interests in the separation sciences, biomedical engineering, and applied mathematics with a keen interest in the craft of teaching. His current activities include writing, consulting, acting as advisor in plant design projects, and performing science experiments for kindergarten children. He has authored four books and about fifty scientific publications.

Ramin Farnood is an assistant professor at the Department of Chemical Engineering and Applied Chemistry and Associate Director of the Pulp & Paper Centre at the University of Toronto. He received his B.A.Sc. and M.A.Sc. from Sharif University of Technology, Tehran, Iran, and his Ph.D. in chemical engineering from the University of Toronto. He has nearly six years of experience in industrial research, where he used mathematical modeling as a tool to solve engineering problems. Professor Farnood's current research in the areas of material science and environmental engineering relies on the modeling of complex systems.

Table of Contents

1 A First Look at Modeling

Our opening remarks in this introductory chapter are intended to acquaint the reader with some general features of the mathematical models that we shall be encountering. A knowledge of these features is an essential prerequisite for due rational construction of a mathematical model. Without it, the process of modeling is fraught with uncertainties and often comes to a complete halt for lack of a convenient starting point. The question one is most often confronted with in these endeavors is, "Where do I start?," and it is this initial hurdle that is often the most difficult one to overcome. The reader will find that the actual solution of the model equations is frequently the easiest part of the task, and it is not uncommon to spend the preponderant part of one's efforts in composing the model and formulating appropriate simplifying assumptions.

To make this task easy, we address a number of questions and provide answers that help in overcoming these initial difficulties:

- What are the underlying laws and relations on which the model is based?
- What type of equations result from the application of these laws and relations?
- What is the role of time, distance, and geometry in the formulation of the model?
- Is there a relation between the type of physical process considered and the equations that result?
- What are some of the simplifying assumptions that can be employed?
- What type of information can be derived from the model solutions?

Connected with the last question is a process we term *solution analysis*. This step deals with scrutinizing the solution for unusual features, something more than the humdrum of a conventional answer or the confirmation of established wisdom. The analyst is asked to seek out the unexpected that will raise the answer to a new plane. Of course, this is not always possible but should nevertheless be undertaken. This text will provide a number of instances in which the answer is far from what was anticipated.

The underlying physical laws considered here are the basic conservation laws and are few in number: conservation of mass, energy, momentum, and electrical charge. The principal additional relations, which we term *auxiliary relations*, number perhaps two or three dozen. Equations are generally limited to three types: algebraic equations (AEs), ordinary differential equations (ODEs), and partial differential equations (PDEs). In these equations, time and distance usually enter as independent variables, and geometry as either a differential element or an entity of finite size.

A relation can also be established between the nature and geometry of a process and the type of equation that results. Thus, if the process takes place in a well-stirred tank, or *compartment* as it is also referred to, application of the conservation laws results in either an algebraic equation, or a first-order ordinary differential equation in time. Similarly, those that take place in unidirectional transport under steady conditions result in either algebraic or ordinary differential equations. When transport in these devices, which we will later refer to as a *one-dimensional pipe*, takes place in more than one direction, or is unsteady, the result is a partial differential equation.

Some general statements can also be made about the principal pieces of information that one can derive from these models. They are, first and foremost, distributions in time or distance of the state variables, i.e., temperature, concentration, velocity, electrical charge or current, and various types of forces. Second, one can infer from these models the size (length, volume, and area) of the equipment required to achieve a desired result. Finally, various system parameters such as transport coefficients, reaction rate constants, or structural and electrical properties can also be extracted from the model solution. Thus we can, without setting up the model equations or proceeding with their solution, make some fairly precise statements about the tools we shall require, the mathematical nature of the model equations, and the uses to which the solutions can be put.

We now proceed to a more detailed consideration of these items.

1.1 THE PHYSICAL LAWS

The physical relations underlying the models considered here are, as indicated, conveniently broken up into two categories, the so-called *basic laws* that consist of the relevant conservation laws, and additional expressions that we term *auxiliary relations*. Together, these two sets of physical laws and expressions provide us with the tools for establishing a mathematical model.

1.1.1 Conservation Laws

The conservation laws we consider are four in number. For systems that involve transport and chemical or nuclear reactions, the required conservation laws are those of mass, energy, and momentum, or Newton's law. We add to these the law of conservation of electrical charge, which is expressed through Kirchhoff's first rule.

Use of these laws, of which a summary appears in Table 1.1, is widespread in various engineering disciplines. Fluid mechanics draws heavily on the law of conservation of mass, known there as the *continuity equation*, and the law of conservation of momentum, which in its most general form leads to the celebrated Navier–Stokes equations. In nuclear processes, conservation of mass is applied to neutrons and includes diffusive transport, as well as a form of reaction when these particles are produced by nuclear fission or are absorbed in the reactor matrix. The law of conservation of energy appears in various forms in the description of thermal systems making their appearance in various mechanical, chemical, metallurgical,

TABLE 1.1
Basic Conservation Laws and the Resulting Balances

1.	Mass balance	Rate of mass in − Rate of mass out = Rate of change of mass content	
2.	Energy balance	Rate of energy in − Rate of energy out = Rate of change of energy contents	
3.	Electrical energy balance (Kirchhoff's second rule)	$\Sigma(\mathrm{V})_{\mathrm{loop}} = 0$	(1.1a)
4.	Momentum or force balance (Newton's law)	$\mathbf{F} = m\dfrac{d\mathbf{v}}{dt} = \dfrac{d\mathbf{P}}{dt}$	(1.1b)
		$\Sigma\,\mathbf{F} = 0$ (Static System)	(1.1c)
5.	Moment balance	$\Sigma\,\mathbf{r} \times \mathbf{F} = 0$ (Static System)	(1.1d)
6.	Electrical charge balance (Kirchhoff's first rule)	$\Sigma(\mathrm{i})_{\mathrm{junction}} = 0$	(1.1e)

and nuclear processes. In electrical systems, conservation of energy is expressed through Kirchhoff's second rule, which states that the sum of voltage drops in a loop equals zero. Force balances are prevalent in structural analysis, in the description of mechanical systems, and as we had noted earlier, in the analysis of fluid flow.

Application of the laws we have chosen to a system or process under consideration leads to equations that are termed *balances*. Thus, the law of conservation of mass leads to the mass balance of a particular species, e.g., a water balance or a neutron balance. Energy balances arise from the law of conservation of energy and are termed *heat balances*, or *mechanical energy balances* when consideration is restricted to thermal or mechanical energy forms. When no flow is involved, energy balances are often referred to as the *first law of thermodynamics*. The electrical energy balance is another special case that is enshrined in Kirchhoff's second rule.

Momentum balances, drawn from the corresponding conservation law, have a dual nature. The rate of change of momentum **P** is equivalent to a force **F**, hence they are also referred to as *force balances*. Moment balances usually appear in conjunction with force balances and are used extensively in statics.

The fourth conservation law, that of electrical charge, leads to what we term a *charge balance*. That balance is enshrined in Kirchhoff's first rule, which states that the algebraic sum of currents entering and leaving a junction equals zero.

1.1.2 Auxiliary Relations

Once the basic balances have been established, it is necessary to express the primary quantities they contain, such as the rates of transport or reaction, in terms of more convenient secondary state variables and parameters. Thus, an energy term that originally appears as an enthalpy H is usually converted to temperature T and specific heat C_p using the appropriate thermodynamic relation. Chemical reaction rates, which appear in the balances symbolically as r, are converted to concentration C and rate constant k_r, using appropriate empirical rate laws. Similar considerations apply to the transport of mass, energy, momentum, and electricity; all of which initially appear in the balances equations as mere symbols, such as q for the rate of heat flow, and i for the flow of electrical current. This is done using what we call *auxiliary relations*, which serve as adjuncts to the primary conservation laws. These relations are drawn from subdisciplines such as thermodynamics, chemical reaction kinetics, and transport theory.

Some of the more commonly encountered auxiliary relations have been grouped together and are displayed in Table 1.2. They include the important transport rates, chemical reaction rates, and various relations drawn from the fields of fluid flow and physical chemistry. A number of these, such as the transport and chemical

TABLE 1.2
Important Auxiliary Relations

1. Mass transport

   By diffusion (Fick's law)

   $$N = -DA\frac{dC}{dx} \tag{1.2a}$$

   By convection between phases

   $$N = k_C A \Delta C \tag{1.2b}$$

2. Heat transport

   By conduction (Fourier's law)

   $$q = -kA\frac{dT}{dx} \tag{1.2c}$$

   By convection between phases

   $$q = hA\Delta T \tag{1.2d}$$

3. Momentum transport (Newton's viscosity law)

   $$\tau = -\mu\frac{dv}{dx} \tag{1.2e}$$

4. Flow of electricity (Ohm's law)

   $$i = v/R \tag{1.2f}$$

5. Viscous drag on a sphere (Stokes' law)

   $$F_D = 3\pi d\mu v \tag{1.2g}$$

6. Viscous flow in a cylindrical pipe (Poiseuille's law)

   $$Q = \frac{1}{8}\frac{R^4}{\mu}\frac{\Delta p}{L} \tag{1.2h}$$

7. Chemical reaction rates

   First order

   $$r = k_r C \tag{1.2i}$$

   Second order

   $$r = k_r C^2 \tag{1.2j}$$

8. Phase equilibrium (Henry's law)

   $$p = Hx \tag{1.2k}$$

9. Thermodynamics (Enthalpy)

   $$\Delta H = C_p\,\Delta T = C_V - p/\rho \tag{1.2l}$$

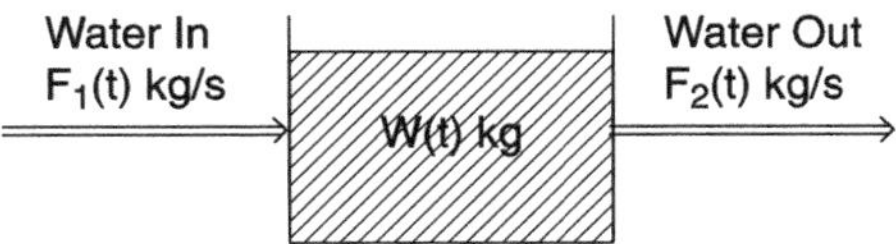

FIGURE 1.1 Diagram of a surge tank with time-varying inflow, outflow, and contents.

reaction rates, are empirical in nature and often contain empirical coefficients such as mass and heat transfer coefficients (h and k_c), and chemical reaction rate constants (k_r). In other instances, the coefficients and parameters can be derived from appropriate theory and are themselves based on conservation laws. Stokes' law for the drag force F_D on a sphere in a viscous flow field, for example, is derived in this fashion, as is Poiseuille's law (Equation 1.2g and Equation 1.2h).

1.1.3 The Balance Space and Its Geometry

We now present two simple examples that serve to introduce the reader to the craft of setting up balances. The first example requires a mass balance to be performed over an entity of finite volume (a tank) in which both mass and flow rates vary with time. This leads to a first-order differential equation in time. In the second example, a heat balance has to be performed, but the geometry here is an incremental volume that ultimately shrinks to an infinitesimally small entity. The result is again a first-order differential equation, but this time the independent variable is distance (or volume).

Illustration 1.1 The Surge Tank

In this simple device, $F_1(t)$ and $F_2(t)$, the time-varying flows of water, enter and leave a tank whose contents W(t) also vary with time (Figure 1.1). To ensure proper setting up of the balance, we start by considering a finite time interval Δt, and then proceed to the limit $\Delta t \rightarrow 0$. This is a somewhat elaborate procedure that the skilled practitioner will ultimately be able to dispense with. It pinpoints, however, some pitfalls that are best illustrated in this way.

Elementary application of the law of conservation of mass leads to the following relation:

Mass in over Δt – Mass out over Δt = Change in contents over Δt

$$[F_1(t)]_{avg}\ \Delta T - [F_2(t)]_{avg}\ \Delta t = \Delta W(t) \tag{1.3a}$$

We note that because we chose a finite time interval Δt, the flow rates by necessity have to be average quantities. They can be represented, if one wishes to do so, by time-averaged integral quantities of the type $\int_{t_1}^{t_2} F(t)dt \,/\, \Delta t$. Evidently, this leads to a complex integral equation that would be difficult to solve. We circumvent this by dividing by Δt and letting $\Delta t \rightarrow 0$. The flow rates then become instantaneous values, and we obtain:

Instantaneous rate in – Instantaneous rate out = Rate of change in contents

$$F_1(t) - F_2(t) = \frac{dW}{dt} \tag{1.3b}$$

It is important to note that the flow rates have now become algebraic quantities but are instantaneous at a point t in time, whereas the contents are represented by a rate of change derivative.

Suppose now, that in addition to the flows entering and leaving, we have convective evaporation at a rate N from the tank to the surroundings. One must recognize that this is a process instantaneous in time, identical in behavior to the outflow of water, and must therefore be placed in the "rate out" category. We obtain:

Rate of mass in – Rate of mass out = Rate of change in mass contents

$$F_1(t) - [F_2(t) + N] = \frac{dW}{dt} \tag{1.3c}$$

Let us next consider the case in which the incoming flow contains a species that undergoes a reaction at the rate r (mol/sec m^3). Here, it must be recognized that the reaction represents an instantaneous removal and must therefore be placed in the "rate out" column of the species mass balance.

There are evidently other modes of transport in and out of the tank, for example, by diffusion through porous walls, seepage through the bottom, etc. Whatever the mechanism of input and output, however, whether by flow, reaction, or diffusion, the form of the balance we had established in the preceding two equations will hold. We can generalize these balances over a finite space as follows:

$$\underset{(by\ various\ mechanisms)}{Rate\ in} - \underset{(by\ various\ mechanisms)}{Rate\ out} = \frac{d}{dt}(contents) \tag{1.3d}$$

This scheme is applicable not only to mass balances, but to energy, momentum, and charge balances as well. When there is no change with time of the contents, the expression reduces to an algebraic form, and the process is said to be at a *steady state*.

We now turn to an example in which the balance is taken over an increment in space.

Illustration 1.2 The Steam-Heated Tube

In this device, a cold liquid flows at a flow rate F_t and temperature T_{t1} into a tube in which it is heated by isothermally condensing steam of temperature T_s, exiting at a higher temperature T_{t2} (Figure 1.2). It is instructive again to start the balance with an increment Δz, and then generalize the results. Flow enters and leaves the difference element with an enthalpy H that increases in the direction of flow due to the heat q_{avg} transferred from the steam. We obtain, in the first instance, the following balance:

Rate of energy in – Rate of energy out = 0

$$[H_z + q_{avg}] - H|_{z+\Delta z} = 0 \tag{1.4a}$$

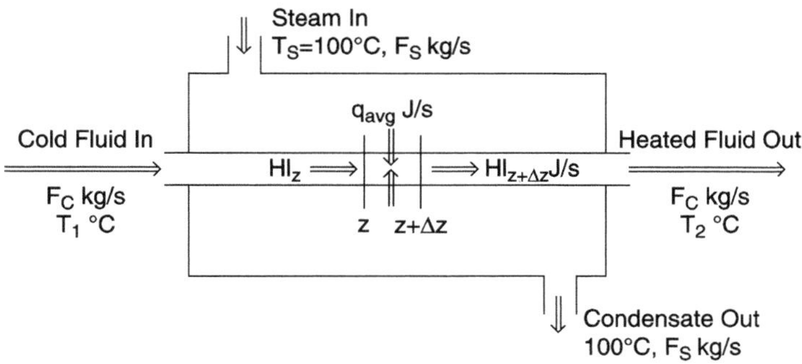

FIGURE 1.2 Energy transport in a steam-heated shell and tube heat exchanger.

where q_{avg} is the average (or mean) integral heat flow from shell to tube over the increment Δz. Before going to the limit, it will be convenient to express q_{avg} and H in terms of the independent and state variables z and T. To do this, we draw on appropriate auxiliary relations. We obtain (see Table 1.2):

$$\Delta H' \text{ (J/kg)} = C_p \text{ (J/kg K) } \Delta T \text{ (K)} \tag{1.4b}$$

which is converted to enthalpy flow H(J/sec) by multiplication with the flow rate. Thus,

$$\Delta H \text{ (J/sec)} = F_c \text{ (kg/sec) } C_p \text{ (J/kg°C) } [T - T°] \text{ (°C)} \tag{1.4c}$$

For convective heat transfer, we use the product of an overall heat transfer coefficient U, area (here, $\pi d\Delta z$), and the linear temperature driving force ΔT (here, $T_s - T_t$). Hence,

$$q_{avg} = U\pi d\Delta z(T_s - T_t)_{avg} \tag{1.4d}$$

where d = diameter of the tube and T_t = tube-side temperature. Upon dividing by Δz and letting $\Delta z \to 0$, ΔT becomes a point quantity at the distance z, much as the average flow rates to the surge tank became, upon going to the limit, point quantities in time. We obtain:

$$F_c C_p \frac{dT_t}{dz} - U\pi d(T_S - T_t) = 0 \tag{1.4e}$$

In this particular example, energy input to the incremental Δz was by flow and convective heat transfer. There are, of course, other mechanisms by which heat can enter or leave the balance space; for example, by conduction, radiation, dielectric heating, etc. Furthermore, there is no reason why the same approach cannot be used to formulate other balances over a unidirectional increment Δz. We use these arguments and results to formulate the following generalized scheme for setting up such balances. It applies to steady, time-invariant processes, and a unidirectional change in variables:

$$\underset{(by\ various\ mechanisms)}{Rate\ in\ at\ z\ and\ over\ z+\Delta z} - \underset{(by\ various\ mechanisms)}{Rate\ out\ at\ z\ and\ over\ z+\Delta z} = 0 \quad (1.4e)$$

The two generalized formulations we have presented, Equation 1.3c and Equation 1.4e provide us the basic tools for performing balances involving one independent variable, either time or distance.

To provide a visual underpinning to these results, we have, in Figure 1.3, compiled three physical configurations that are representative of the balances and the balance space encountered in modeling.

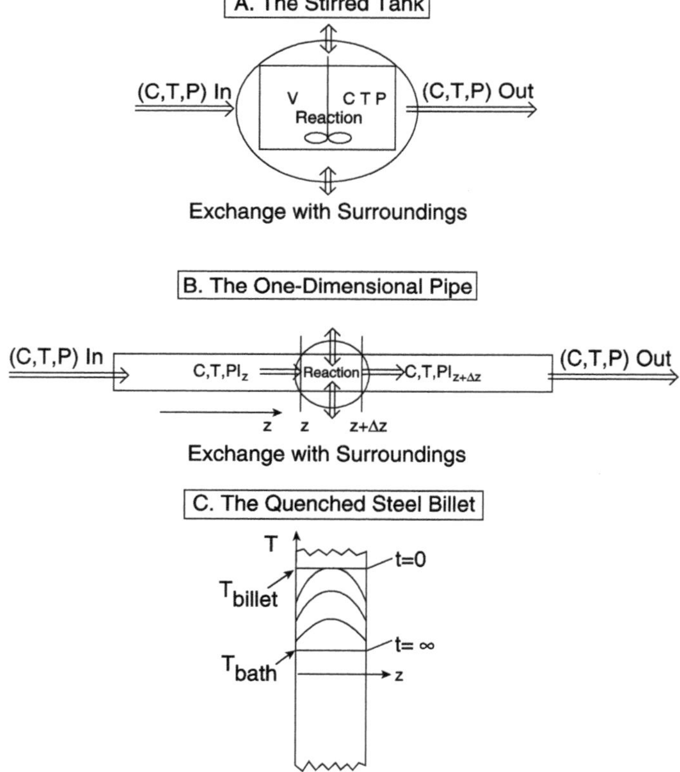

FIGURE 1.3 Diagrams of three basic physical models: (A) the stirred tank or compartment with uniform space-independent properties, (B) the one-dimensional pipe with property distribution in the longitudinal direction and at the wall, and (C) the quenched steel billet with variations of temperature in both time and space.

The device shown in Figure 1.3A describes the finite space, and is termed a *compartment* or *stirred tank*. The process variables here vary at most with time, but not with distance, i.e., they are uniformly distributed within the balance space. Input and output occurs by various mechanisms, including reaction within the compartment, and what we term "exchange with surroundings" (e.g., evaporation, radiation, leakages of various types, etc.).

Compartmental models are used extensively in describing chemical and environmental processes, as well as in biology and biomedical engineering. We shall encounter a good many of them in Chapter 4, and in different locations elsewhere in the text.

The second device shown in Figure 1.3B has an incremental balance space and is termed the *one-dimensional pipe* to convey the fact that the process variables change in one direction only. They are also time invariant, and consequently behave the exact opposite to what we saw in the compartmental model. Here again, various mechanisms of input and output can be at play, including reaction and exchange with surroundings.

Systems that can be modeled in this fashion include heat exchangers, packed columns used in separation processes, tubular membranes, and tubular reactions. They also comprise flowing systems encountered in the environment (rivers and groundwater), and in animal or human organs (blood and urine). The heat generated in an electrical conductor and dispersed to the surroundings can also be placed in this category. The range of applications of this model is thus a very broad one.

In the third device shown, termed the *quenched steel billet* (Figure 1.3C), variations occur with both time and distance. Heat given off by the hot billet to the quenching medium results in a spectrum of temperature profiles that are both time and distance dependent. Because two independent variables are involved in this instance, the heat balance required to model the process will lead to a partial differential equation (PDE). We have here a departure from the two previous single models that led to either ordinary differential equations (ODE) or algebraic equations (AE). There is, however, a way of bridging these categories that comes about as follows:

Suppose the dimensions of the billet are progressively reduced until it becomes a tiny entity, for example, a sphere of 1-cm diameter. It can then be argued that because of the high thermal conductivity of the metal, and the short distances involved, conduction will be very fast, and the temperature within the sphere will, for all practical purposes, be uniform. We will, in other words, be dealing with a compartment of "stirred tank" (although no physical stirring takes place), and the quenching process can be modeled by an ODE. Thermocouples, which consist of tiny metallic wires that respond to external temperature changes, are modeled in this fashion as well. Whether to use a PDE or ODE model is a question we shall address in Section 1.6.

1.2 THE ROLE OF THE VARIABLES: DEPENDENT AND INDEPENDENT VARIABLES

An important mathematical consideration is the distinction made between dependent and independent variables.

Dependent variables, often referred to as *state variables*, arise in a variety of forms and dimensions dictated by the particular process to be modeled. Thus, if the system involves a chemical reaction, molar concentration is the dependent variable

TABLE 1.3
Typical Variables for Various Balances

Balance	Dependent Variable	Independent Variables
Mass	Molar flux N	Time t
	Mass flux W	Coordinate distances
	Mole and mass fraction x, y	x, y, z Cartesian
	Mole and mass ratio X, Y	r, θ, z cylindrical
	Molar concentration C	r, θ, z spherical
	Partial pressure p	
Energy	Internal energy U	Time t
	Enthalpy H	Coordinate distances
	Temperature T	x, y, z Cartesian
		r, θ, z cylindrical
		r, θ, φ spherical
Momentum	Velocity $\vec{v}$	Time t
	Shear stress $\underset{\sim}{\tau}$	Coordinate distances
	Pressure p	x, y, z Cartesian
		r, θ, z cylindrical
		r, θ, φ spherical

of choice because reaction rates are usually expressed in terms of this quantity. Phase equilibria, on the other hand, call for the use of mole fractions x and y, mole or mass ratios X and Y, or partial pressures p. We had already indicated in Illustration 1.2 that temperature is the preferred variable in energy balances. Other variables of choice will be addressed as they arise. The reader is reminded that it is the dependent variables that dictate the number of equations required. Thus, if concentration and temperature occur as variables in a particular process, two equations will be required for its description, and these will usually comprise a mass and an energy balance. A convenient compilation of these factors appears in Table 1.3.

Although there is virtually an unlimited range of dependent variables at the disposal of the analyst, the basic independent variables to be dealt with here are two in number: time t and distance; it being understood that the latter can be measured in three coordinate directions; (for example, x, y, and z, or r, θ, and φ for spheres).

At a more elemental level, simple algebraic laws and balances often contain independent variables other than time and distance. In the ideal gas law, for example, PV = nRT, pressure depends on T, V, and n, but the reverse is not true. The latter are therefore to be regarded as independent variables.

There are instances in which distance, usually thought of as an independent variable, transforms into a dependent variable instead. The best-known example is given by Newton's law, which can be expressed in the three alternative forms

$$F = ma = m\frac{dv}{dt} = m\frac{d^2x}{dt^2} \tag{1.5}$$

where x, the one-dimensional distance, is now a dependent variable.

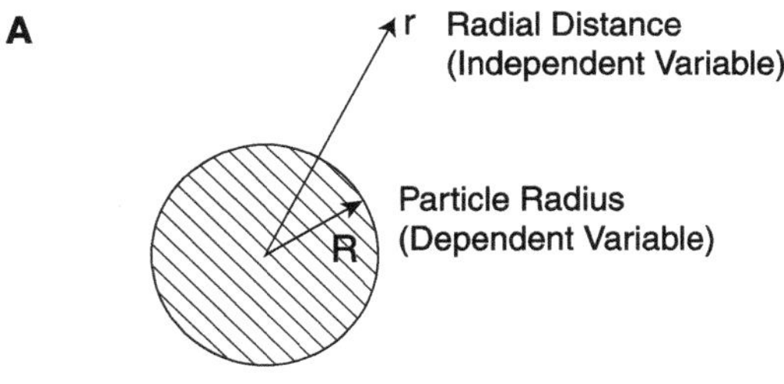

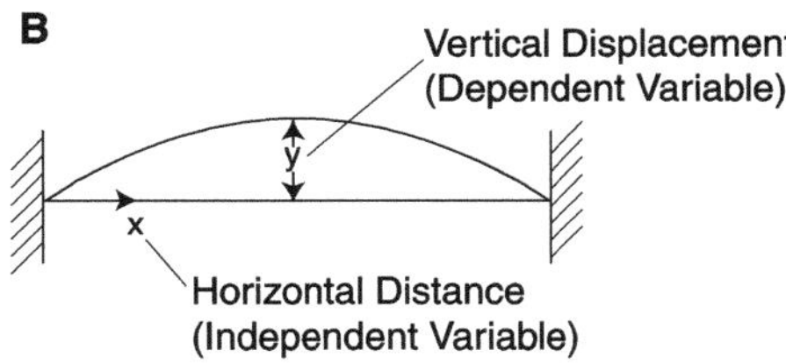

FIGURE 1.4 Distance as dependent and independent variables. (A) Reacting particle (B) Vibrating string.

Examples in which distance plays a dual role as both independent and dependent variables may also arise. In Case A of Figure 1.4, mass m of the reacting particle is usually taken to be the dependent variable, but can be easily transformed into a dependent distance variable R through the relation: mass = density × volume = $\rho(\frac{4}{3}\pi R^3)$. The general radial distance r is retained as an independent variable. A similar duality arises in the vibrating string shown as Case B. Here, the displacement distance y plays the role of the dependent variable, but is itself a function of the independent horizontal distance x.

The examples we have cited are the exceptions rather than the rules. In most situations we shall be dealing with here, both time and distance play the traditional role of independent variables.

1.3 THE ROLE OF BALANCE SPACE: DIFFERENTIAL AND INTEGRAL BALANCES

The point was made in Section 1.1 that a balance can be made over an incremental element, which generally results in a differential equation, or alternatively over a finite entity such as a tank or a column, in which case we can obtain algebraic as well as differential equations.

In the former case, we speak of "differential," "microscope," or "shell" balances, and the underlying model is often called a distributed parameter model (see Table 1.4). Such balances lead, upon solution, to distributions or "profiles" of the state variables in space, or in time and space. Thus, a one-dimensional energy balance taken over

TABLE 1.4
Categories of Balances and Resulting Equations

Names and Model Types	Equations
A. Integral or macroscopic balances	
Compartmental or lumped parameter models	
1. Steady state balance	AE
2. Unsteady state or dynamic balance	
Instantaneous in time	ODE
Cumulative in time	AE
B. Differential, microscopic, or shell balances	
Distributed parameter models	
1. Steady state one-dimensional balance	ODE
2. Unsteady state one-dimensional balance	PDE
3. Steady state multidimensional balance	PDE
4. Unsteady state multidimensional balance	PDE

an incremental element of a tube and shell heat exchanger will, upon integration, yield the longitudinal temperature profiles in both the shell and the tubes.

When the balance is taken over a finite entity, we speak of *integral* or *macroscopic* balances, and the underlying models are frequently referred to as *lumped parameter models* (see Table 1.4). Solutions of these equations usually yield relations between input to the finite space and its output. Time may or may not appear as an independent variable. This aspect is taken up in more detail in the following.

1.4 THE ROLE OF TIME: UNSTEADY STATE AND STEADY STATE BALANCES

Time considerations arise when the process is transient in nature, in which case we speak of unsteady, unsteady state, or dynamic systems and balances. Both macroscopic and microscopic balances may show time dependence. When a macroscopic balance describes an unsteady process, the result is an ordinary differential equation. Microscopic balances, which are already dependent on distance, will move up a notch to the level of partial differential equations upon the intrusion of time.

A further distinction is made between processes that are instantaneous in time, leading to differential equations, and those that are cumulative in time, usually yielding algebraic equations (Table 1.4). The rate of change of the mass in a tank being filled with water, for example, is given by the instantaneous rate of inflow and leads to a differential equation. On the other hand, the actual mass of water in the tank at a given moment equals the cumulative amount introduced to that point and yields an algebraic equation. Note that instantaneous and cumulative balances are independent entities. The difference between them is a subtle but important one and will be illustrated by examples throughout the text.

An important consideration in the analysis of transient processes is whether they are capable of attaining a steady state as $t \to \infty$. Simple physical reasoning will often provide an answer. A stirred-tank reactor, for example, which is operating at constant inflow and outflow will, after an initial start-up period, subside to a constant steady state effluent concentration in which the rate of supply of reactant is exactly balanced by its consumption. There are also mathematical methods of establishing the existence of a steady state, one of which we explore in Illustration 1.5.

When a system is dependent on both time and space, the resulting model is described by a PDE. PDEs also arise when the state variable depends on more than one dimension and are either at steady or unsteady state. Diffusion into a thin porous slab, for example, where no significant flux occurs into the edges, is described by a PDE with time and one dimension as independent variables. When the geometry is that of a cube, a PDE in three dimensions and time results. The vibrating string cited previously is another example of a model that yields a PDE in time and one dimension.

We draw the reader's attention to both Table 1.3 and Table 1.4 as useful tabulations of basic mathematical properties of the balances. Table 1.4, in particular, is designed to help in assessing the degree of mathematical difficulty to be expected, and in devising strategies for possible simplifications.

Illustration 1.3 The Countercurrent Gas Scrubber: Genesis of Steady Integral and Differential Mass Balances

Gas scrubbers are widely used devices designed to remove impurities or recover valuable substances from gases by contacting them with a suitable solvent such as water. A gas scrubber typically consists of a cylindrical shell filled with plastic or ceramic packing designed to enhance the contact area between the two phases (Figure 1.5A and Figure 1.5B). Solvent enters the column at the top and trickles down through the packing, where it contacts the gas phase that enters the scrubber at the bottom and flows upward countercurrent to the solvent stream. The purified gas leaves the column at the top, while the used solvent containing the impurity exits at the bottom. We use this example to introduce the reader to various types of steady state balances.

We start by considering the two integral balances shown in Figure 1.5A and Figure 1.5B. The balance space in these two cases is a finite one, consisting either of a part of the column (Figure 1.5A) or the entire column itself (Figure 1.5B). The corresponding solute mass balances will consist of two algebraic equations, neither of which contain the distance variable z. They can, therefore, tell us nothing about the concentration variations as a function of column height, nor help us establish the size of the column required to effect a reduction in solute content from Y_1 and Y_2.

We must, for these purposes, turn to differential balances that upon integration will yield the desired functional dependence of solute concentration on column height, i.e., $Y = f(z)$. Two such balances are sketched in Figure 1.5C and Figure 1.5D. Both are taken over the individual gas and liquid phases, and both will therefore have to contain expressions for the mass transfer rate, designated N_{avg}.

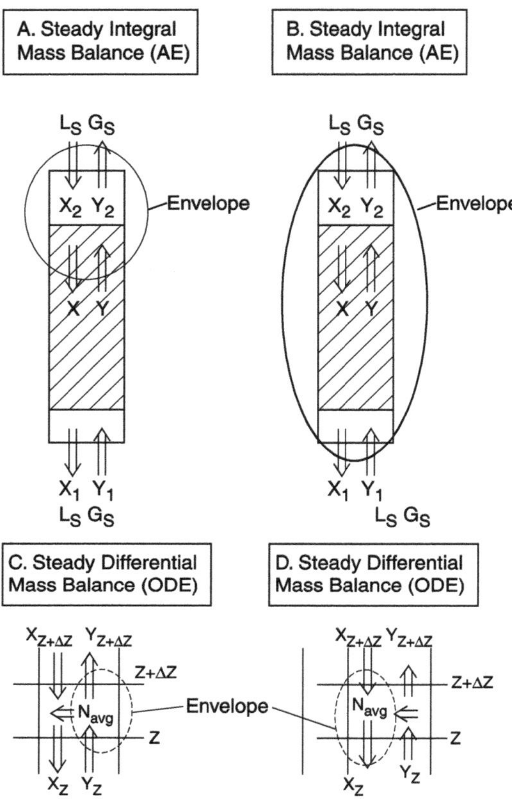

FIGURE 1.5 The packed gas absorber. Types of mass balances leading to algebraic (AE) and ordinary differential equations (ODEs).

We shall return to these balances in Chapter 4, where they will be applied for the purpose of designing a gas scrubber. We will address the problem of which balances to choose, and will demonstrate that they can be combined in different ways to achieve a specific purpose. This will lead, among other things, to the concept of a height of a transfer unit.

Illustration 1.4 Instantaneous and Cumulative Unsteady Balances: Seepage from a Land-Based Oil Spill

We use the example of the seepage from an oil spill into the underlying soil to demonstrate the simultaneous use of instantaneous and cumulative balances. A question one wishes to address in accidents of this type is the depth to which the oil will penetrate in a given time, because this will, in part, determine the strategy for subsequent clean-up operations.

It is assumed that initially an oil slick of depth h_0 has been deposited, and that it begins to penetrate the soil, with distance of penetration z being measured from

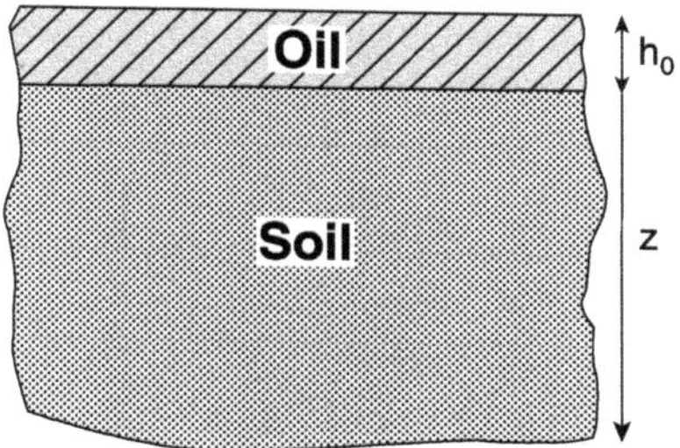

FIGURE 1.6 Ground penetration from oil spill. Use of instantaneous and cumulative balances.

the ground level down (see Figure 1.6). It is clear that some form of unsteady mass balances will be required to model the process, but it is equally evident that we will also need an expression for the velocity v at which liquids flow through porous media. That expression is given by D'Arcy's law for flow in porous media, and has the form:

$$v = \frac{K}{\mu}\frac{\Delta p}{z} \tag{1.6a}$$

where K is the so-called permeability of the soil (m^2), and μ is the viscosity of the liquid. Because the pressure gradient here is due solely to hydrostatic forces, i.e., the weight of the oil column, D'Arcy's law takes the form:

$$v = \frac{K}{\mu}\rho g \frac{z + h_0}{z} \tag{1.6b}$$

This result is used in an unsteady and instantaneous oil balance around the soil, which is formulated as follows:

Rate of oil in – Rate of oil out = Rate of change in oil content

$$\rho v A - 0 = \varepsilon \rho A \frac{dz}{dt} \tag{1.7}$$

where ε = porosity of soil.

The model thus far is composed of two equations, but still leaves us with the three dependent variables h, z, and v. A third relation is therefore needed, which may be obtained by formulating a cumulative oil balance. It reads

Initial amount of oil in the layer = Oil left in the layer + Oil contained in the soil

$$h_0 \rho A = h \rho A + z \rho \varepsilon A \tag{1.8}$$

Eliminating the variables h and v from these three equations, we obtain the final model:

$$\frac{dz}{dt} = \frac{K\rho g}{\varepsilon\mu}\left[\frac{h_0}{z} + (1-\varepsilon)\right] \tag{1.9}$$

The solution of this ODE yields the variation of the penetration distance z with time t.

Comments: It may not have escaped the attention of the reader that there are alternative ways of composing a model for this process. One can, for example, make an oil balance over the spilled layer rather than the soil, which again yields a single ODE linked to two algebraic equations. Or one can use both ODEs together with D'Arcy's law to formulate the model. This latter alternative is less desirable because it involves the solution of two simultaneous differential equations. Having learned of the existence of instantaneous and cumulative balances, we are in a position to neatly sidestep this complication.

Illustration 1.5 Response of an Electrical Network, Transient Behavior and the Ultimate Steady State

The accompanying diagram, Figure 1.7, shows an electrical circuit of some complexity that is subjected to an alternating voltage $V = V_0 \sin \omega t$. The component elements consist of an Ohmian Resistor (R), two inductors (L), and three capacitors (C) with component voltage drops:

$$V_R = iR \tag{1.10a}$$

$$V_L = L\frac{di}{dt} \tag{1.10b}$$

$$V_C = \frac{q}{C} \tag{1.10c}$$

where L = inductance and C = capacitance.

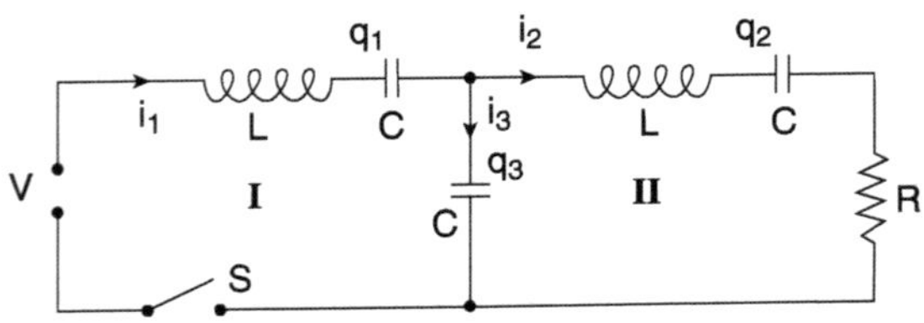

FIGURE 1.7 Two joined RLC circuits: Steady state and unsteady state behavior.

Using Kirchhoff's second rule, Equation 1.1a, which states in essence that the sum of voltage drops in a loop must equal zero, we obtain:

$$\text{For the circuit I} \quad L\frac{di_1}{dt} + \frac{1}{C}q_1 + \frac{1}{C}q_3 = V_0 \sin \omega t \tag{1.11a}$$

$$\text{For the circuit II} \quad L\frac{di_2}{dt} + \frac{1}{C}q_2 + Ri_2 + \frac{1}{C}q_3 = 0 \tag{1.11b}$$

Because current i equals the rate of change of the charge dq/dt, we can recast the model in the following equivalent form:

$$Lq_1'' + \frac{1}{C}q_1 + \frac{1}{C}q_3 = V_0 \sin \omega t \tag{1.11c}$$

$$Lq_2'' + \frac{1}{C}q_2 + Rq_2' + \frac{1}{C}q_3 = 0 \tag{1.11d}$$

These are two coupled second-order ODEs in the three dependent variables q_1, q_2, and q_3. A third relation is therefore required, which is provided by a charge balance embodied in Kirchhoff's first rule, Equation 1.1e. We have:

$$i_1 = i_2 + i_3 \tag{1.12a}$$

or equivalently:

$$q_1' = q_2' + q_3' \tag{1.12b}$$

The model is now complete and consists of Equations 1.11c, Equation 1.11d, and Equation 1.12b. The solution of such sets of linear ODEs is usually implemented, or at least attempted, by means of the Laplace transform method, to be taken up in the next chapter. It requires setting appropriate initial conditions, and when this is done and the procedure initiated, it is found that it quickly leads to an intractable cubic algebraic equation. The proceedings have thus reached a seeming impasse.

All is not lost, however. In Section 1.7, which deals with initiating the modeling process, the reader is advised that one sometimes has to be content with partial answers, or with upper and lower bounds to a solution. Such a partial answer to our problem lurks in the model in the form of the ultimate steady state attained by the system. That steady state can be established from the model by simply setting the time derivatives q'' and q' equal to zero. We obtain:

$$(q_2)_{ss} = (q_3)_{ss} \tag{1.13a}$$

and

$$(q_1)_{ss} - (q_2)_{ss} = CV_0 \sin \omega t \tag{1.13b}$$

Introducing the Kirchhoff relation Equation 1.12b into Equation 1.13b gives:

$$(q'_3)_{ss} = (i_3)_{ss} = V_0 C\omega \cos \omega t \tag{1.13c}$$

and from it

$$(i_2)_{ss} = V_0 C\omega \cos \omega t \tag{1.13d}$$

and

$$(i_1)_{ss} = 2V_0 C\omega \cos \omega t \tag{1.13e}$$

Although both currents still oscillate, they do so with a constant amplitude and frequency. This too can be considered to be a steady state.

Comments: Although we have failed in our attempt to obtain a general solution to the model, the partial answer arrived at is of considerable significance. In fact, the ultimate steady state behavior of the circuit can be of greater interest to the analyst than the transients that precede it. Partial answers are often not trivial and can be more revealing than a global solution. Numerical methods could of course have been used to obtain a more complete answer, but can evidently be applied only to specific circuits, with specified numerical values for the various parameters.

1.5 INFORMATION DERIVED FROM MODEL SOLUTIONS

A first general rule that applies to all model solutions, and should be borne in mind above all others, is the following:

Any of the variables that appear in a solution may, in principle, be assigned the role of an unknown being sought or the information to be derived. This unknown variable could be time, which appears in an unsteady balance; it could be distance or length; or it could be flow rate, a transport coefficient, voltage, electrical current, or any of the multitude of variables and parameters that appear in a solution.

Suppose, for example, that the model under consideration is one that described the response of a thermocouple to a change in the ambient temperature, a problem that is taken up in Illustration 4.18. The variables that one expects to appear in the solution are: time, mass, or dimension of the thermocouple, the heat transfer coefficient between the external medium and the device, specific heat of the thermocouple, and the magnitude of the change in external temperature. All of these variables represent information to be derived from the model, and any one of them may be assigned the role of the unknown being sought, although some of them are more likely to fall in this category than others. For example, one may wish to calculate the response time of a thermocouple of known dimensions to a prescribed change in temperature, or, conversely, derive the thermocouple dimension that will yield a prescribed response time. One can also derive the value of the heat transfer coefficient from experimental response data, or calculate the temperature change that will result in a given response time. Thus, any of these may serve as the desired unknown, although the last two items are less likely to be candidates than the first two.

One way of systematizing the wealth of information to be derived from a model solution is to distinguish between what we call *primary information* and *derived information*. The former represents the first raw result of the model solution, whereas the latter comes about by extracting lurking items, or mathematically transforming the solution into a more useful form.

Primary, information consists basically of distributions or "profiles" in time or distance, or both, of the state variables of the system. Typical of these distributions are the temperature profile in a heat exchanger, the velocity distribution in a flow field, the time response of an electrical circuit, the deflection profile of a loaded beam, or the concentration variations with time in a batch reactor. At the PDE level, two classical examples are the three-dimensional velocity and pressure distributions in a flow field, and the atmospheric concentration distributions that result from pollutant emissions. When neither time nor distance appear in the model, the primary information consists of the relation between input and output; for example, the feed concentration to a reactor and its exit concentration.

The derived information consists of a number of broad categories, some of which are more important than others. They comprise:

- Design length, height, or volume of a piece of equipment. This includes the calculation of the length of a heat exchanger that yields a prescribed exit temperature, the height of a gas scrubber that results in the reduction of the feed concentration to a desired level, or the volume of a batch reactor that will yield a prescribed conversion to product. These items are commonly known as *design problems*.
- Parameter evaluation from experimental distribution in either time or distance of the state variables. This includes the derivation of heat and mass transfer coefficients from measured changes in temperature or concentration, the extraction of reaction rate constraints from experimental conversions, or the calculation of structural parameters from the response of an applied load. Exercises of this type are referred to as *parameter estimation*.
- Equipment performance for prescribed flow rates, feed conditions, and size of equipment. Here, one is dealing with an existing piece of equipment of given dimensions, and the task is to predict its performance under a prescribed set of conditions. We term this as *performance analysis*.
- Sensitivity to disturbances. This falls in the realm of process control and seeks to establish the response of a system that is initially at steady state, to disturbances principally in feed concentration, temperature, or flow rate. The purpose of the exercise is to design and establish controller modes that will attenuate the disturbance to some desired level.

Finally, we draw the reader's attention to items that result from the mathematical transformation of the primary solution into more convenient and useful quantities. The velocity and pressure distributions in a flow field, for example, although interesting in themselves, are often not the results the analyst is seeking. Of greater usefulness, particularly in engineering, is the relation of flow rate to the existing pressure distribution. Other useful items of information can be obtained by differentiating, rather than integrating, the primary distributions obtained from the model. We summarize some items of this type in the following:

- Quantities obtained by the differentiation of profiles: heat flux from temperature distributions, mass flux from concentration distributions, and shear stress and viscous drag from velocity distributions.

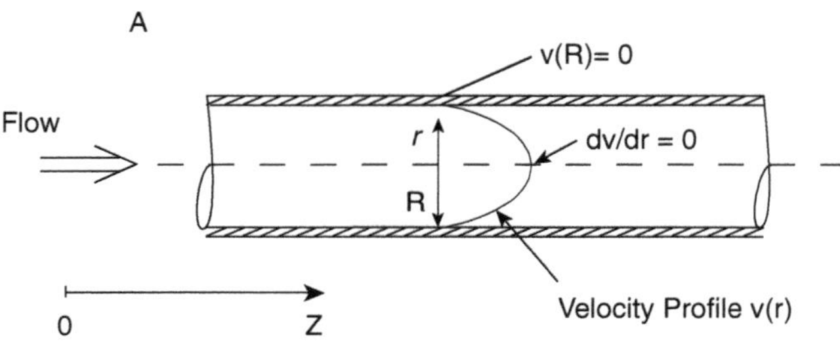

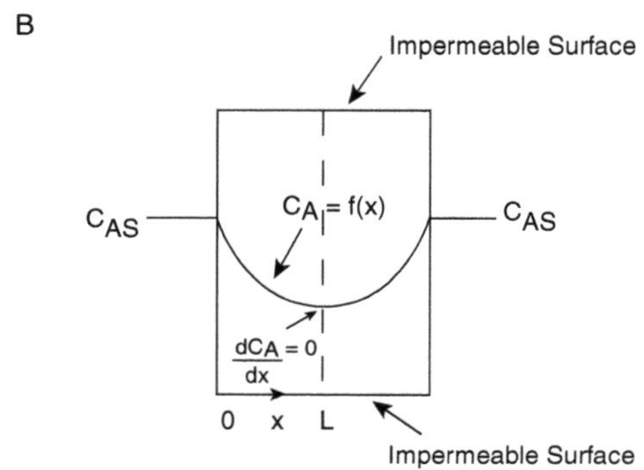

FIGURE 1.8 Two primary distributions obtained by balances. (A) Radial velocity profile in laminar pipe flow (force balance). (B) Concentration profile in a catalyst pellet (mass balance).

- Quantities obtained by the integration of profiles: flow rate from velocity distributions, cumulative energy flux from temperature profiles, accumulated or depleted mass or energy over a time interval, and pressure drag from pressure distributions.

Illustration 1.6 Transformation of the Primary Information by Integration and Differentiation

We consider here two examples in which the primary information, obtained in the form of distributions, is transformed into more useful quantities by a process of integration and differentiation. The initial profiles to be dealt with are the velocity distribution in viscous pipe flow, and the concentration profile in a catalyst pellet (Figure 1.8).

1. *Poiseuille's Law for Laminar Pipe Flow*

The primary information that results from the application of a force balance to laminar pipe flow is the radial velocity distribution in the fluid and takes the parabolic form (Figure 1.8A):

$$v(r) = \frac{R^2 \Delta p}{4\mu \Delta z}\left[1 - \left(\frac{r}{R}\right)^2\right] \tag{1.14}$$

For engineering purposes, it is more appropriate and useful to relate the pressure drop $\Delta p/\Delta z$ to the volumetric flow rate Q (m^3/sec) rather than the local velocity v(r). This results from the requirement, which is common in applications of this type, to calculate the pressure drop for a prescribed flow rate; or conversely, to deduce the flow that results from a given applied pressure drop.

The required conversion is carried out in two steps: we first obtain the average velocity v_{avg} by integrating v(r) over the cross-section, and then multiply by the cross sectional area to arrive at the flow rate Q. We obtain:

$$v_{avg} = \frac{\Delta p R^2}{4\mu \Delta z} \frac{\int_0^R \left[1-\left(\frac{r}{R}\right)^2\right] 2\pi r dr}{\pi R^2} \tag{1.15a}$$

$$= \frac{\Delta p}{2\mu \Delta z}\left[\frac{r^2}{2}\Bigg|_0^R - \frac{r^4}{4R^2}\Bigg|_0^R\right] \tag{1.15b}$$

$$v_{avg} = \frac{R^2}{8\mu}\frac{\Delta p}{\Delta z} \tag{1.15c}$$

The volumetric flow rate is consequently given by:

$$Q = v_{avg}\pi R^2 = \frac{\pi R^4}{8\mu}\frac{\Delta p}{\Delta z} \tag{1.15d}$$

This is the celebrated Poiseuille, or Hagen Poiseuille equation for viscous flow through a cylindrical tube.

2. *The Catalyst Pellet Effectiveness Factor*

The events taking place in a catalyst can be modeled by means of a mass balance that yields, in the first instance, the reactant profile within the pellet. The profile here extends from the constant value C_{As} at the surface to a lower value $C_A(L)$ at the center of the pellet, brought about by the finite rate of diffusion in the particle. Diffusion can be viewed as a resistance that slows the timely replenishment of reactants consumed by the reactions.

Here again, it is not the primary concentration distribution that is of direct practical interest, but rather the overall performance of the entire pellet. To obtain a sense of this quantity, we defined the so-called effectiveness factor E, which is given by

$$E = \frac{\text{Overall reaction rate with diffusional resistance}}{\text{Overall reaction rate without diffusional resistance}} \tag{1.16}$$

By introducing this concept, we establish a simple and useful criterion for the efficiency of the pellet. High efficiencies result when diffusional transport can keep up with reactant consumption. This happens when the pellet is small or the reaction rate is sluggish compared to the transport rate. The pellet effectiveness factor, i.e., its efficiency, will then tend to unity, $E \to 1$. Conversely, if the pellet is large, i.e., transport is over wider distances or lags behind the reaction rate, efficiency will be low, $E \to 0$.

It can be shown that for an isothermal first-order reaction $r = k_r C_A$ taking place in a catalyst pellet (Figure 1.8B), the reactant concentration profile is given by:

$$\frac{C_A}{C_{As}} = \frac{\cosh[\varphi(1-z)]}{\cosh \varphi} \tag{1.17a}$$

where $\varphi = L(k_r/D)^{1/2}$, $z = x/L$, and D = diffusivity.

It will be recalled that cosh x is the hyperbolic cosine defined by:

$$\cosh x = [\exp(x) + \exp(-x)]/2 \tag{1.17b}$$

which has the derivative:

$$\frac{d}{dx} \cosh x = \sinh x \tag{1.17c}$$

and stands in the ratio:

$$\sinh x / \cosh x = \tanh x \tag{1.17d}$$

We can go about deriving E in two ways. We can average the reaction rate over the entire pellet by integration, or we can obtain the overall reaction rate from the equivalent diffusional flow into the pellet, which requires differentiation of the concentration profile Equation 1.17a. We choose the latter and easier route and obtain:

$$E = \frac{-DA(dC_A dx)_{x=0}}{LAk_r C_{As}} \tag{1.18a}$$

Evaluation of the derivative and cancellation of terms results in:

$$E = -\frac{1}{\varphi^2} \left[\frac{-\sinh \varphi(1-z)}{\cosh \varphi} \right]_{z=0} \tag{1.18b}$$

or, alternatively, in view of Equation 1.17d:

$$E = (\tanh \varphi)/\varphi \tag{1.18c}$$

A derivation of E for a cylindrical pellet, leading to Bessel function, will be given in Chapter 2 (Illustration 2.12).

1.6 CHOOSING A MODEL

The reader will have noted that so far in our development we have failed to provide any quantitative criteria for the validity of the two simpler models, the compartment and the one-dimensional pipe (Section 1). A limited number of such criteria exist, two of which are taken up in the illustrations that follow. More often, however, one uses these simplifications to establish outer limits, i.e., maxima or minima, of what can be accomplished. Compartmental models, for example, often provide an estimate of the minimum time required to attain a given change in concentration or temperature. In similar fashion, the one-dimensional pipe, which ignores or approximates internal transport resistance, tends to maximize concentration and temperature change. Such answers can be immensely beneficial, particularly in the preliminary stages of modeling. Not infrequently, they provide all that is needed for a preliminary satisfactory analysis.

Illustration 1.7 PDE or ODE? The Biot Number Criteria

The brief analysis we had given of a thermocouple responding to an external temperature change relied on the intuitive notion that its small size and high thermal conductivity would make it an ideal candidate for compartmental modeling at the ODE level. The steel billet, on the other hand, having much more substantial dimensions, was deemed to have significant internal temperature gradients. This would call for the use of a PDE model.

These intuitive notions can be quantified by extracting appropriate criteria from the solution of the rigorous PDE model. What we are looking for is a relative measure of the external and internal transport resistances that would allow us to gauge whether internal gradients can be neglected (ODE model) or must be allowed for (PDE). Such a criterion is provided by the dimensionless Biot number (Bi) that resides in the model equations. It is defined as:

$$\text{For heat transfer } Bi_h = \frac{hl}{k} \tag{1.19a}$$

$$\text{and for mass transfer } Bi_m = \frac{k_C l}{D} \tag{1.19b}$$

where h and k_C are the external transport coefficients, l is a characteristic dimension (diameter or thickness), and k and D are the thermal conductivity and (mass) diffusivity. The number can be regarded as a ratio of internal to external resistance, with high values indicating a need to use the full PDE model. Values of Bi less than 0.1 are usually taken as a sanction for the use of the simpler ODE model.

Let us demonstrate the validity of this criterion with a numerical example. For the thermocouple cited previously, a typical heat transfer coefficient for external turbulent air flow is on the order h ≅ 0.1 kJ/m sec K, thermal conductivity k ≅ 0.1 kJ/m^2 sec K, with diameter typically in the range 0.1 to 1 mm. Hence:

$$Bi = \frac{hl}{k} = \frac{0.1 \times 10^{-3}}{0.1} to \frac{0.1 \times 10^{-2}}{0.1}$$

i.e., well below the cut-off value of Bi = 0.1. If, on the other hand, the dimension l is raised above 10 cm, which approaches the thickness of a billet, the criterion will clearly be violated.

Illustration 1.8 PDE or ODE? The Quasi-Steady State Assumption

It frequently happens that a process that is purely time dependent is linked to another process or system that is distributed in both time and distance. Two examples of such combinations are shown in Figure 1.9. Figure 1.9A shows a stirred tank linked to an

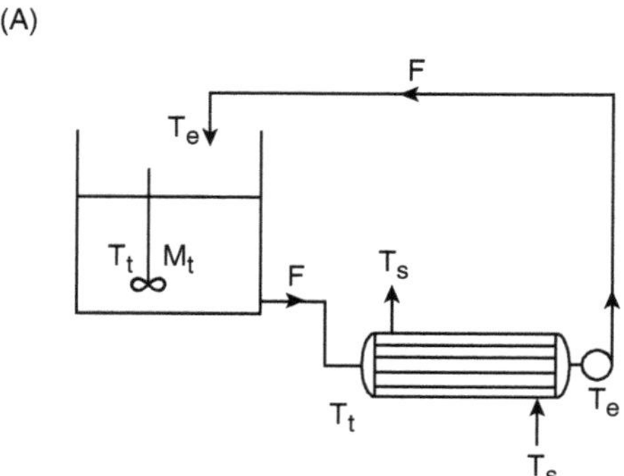

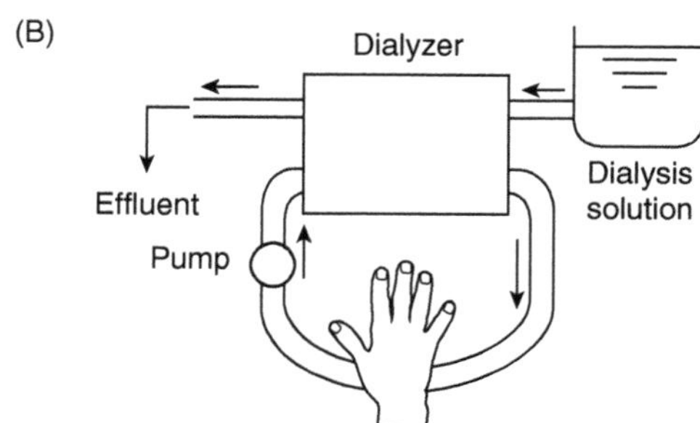

FIGURE 1.9 Linked processed distributed in time and distance. (A) Tank with external heat exchanger. (B) The hemodialyzer (artificial kidney).

external heat exchanger. The contents of the tank are continuously withdrawn and passed through the exchanger to be heated or cooled and subsequently returned to the tank. The process is continued until a desired temperature is attained in the tank. The second example (Figure 1.9B) involves the medical procedure known as *dialysis*, in which the blood of a patient is continuously pumped through a bundle of hollow permeable fibers (known as a *dialyzer*) that removes toxins into a surrounding bath and returns the cleansed blood to the patient. In both these cases, a rigorous model would call for the use of an ODE for the tank or blood compartment (assumed to be well mixed), whereas either of the two external devices would require a PDE.

An examination of these systems reveals that there is a marked difference in the timescales of the external and internal events. Within the compartments, concentration and temperature changes are very slow, on the order of many minutes or even hours. Externally, on the other hand, passage through the tubular exchangers is rapid, and the attendant changes in concentration and temperature often take place in the span of a few seconds. This entitles us to assume that the inlet conditions to the exchangers are constant, or very nearly constant, for the duration of the exchange, and remain so for a reasonable length of time. The devices can therefore be taken to operate under what is termed quasi-steady state or pseudo-steady state conditions. They are still distributed in distance, but not in time, and can consequently be modeled at the ODE level.

A reasonable ratio of timescales to attain this state can be set by the analyst. A ratio of 100:1 is a commonly used measure, but less fastidious persons may want to lower the values to 20:1 or even 10:1. In blood dialysis, for example, the duration of the treatment is 2 to 3 h, although passage through the dialyzer takes place in well under a minute. The ratio of 100:1 is thus more than fully satisfied.

Comments: The quasi- or pseudo-steady state assumption is extensively used in modeling and has proven to be enormously fruitful in reducing complex models to manageable proportions. We will, in this text, address a number of processes that fall into the class known as *moving-boundary problems*. These systems are comprised of a central core, such as a reacting particle or a melting solid, which is supplied with reactant or heat through an external stagnant layer. The front of the core (the "moving boundary") usually recedes at a relatively slow pace, whereas passage through the surrounding layer takes place at a comparably faster rate. It will be left to the exercises, and to further illustrations, to demonstrate how the quasi-steady state assumption can be brought to bear on these problems (see Illustration 2.4, Illustration 4.20, and Practice Problem 2.4).

1.7 KICK-STARTING THE MODELING PROCESS

Students who have struggled with problem solving going back to their high-school days will recall the dread on being confronted with what are commonly termed "word problems." They will recall, in particular, the difficulty of "getting started," of overcoming the first hurdle of defining underlying principles and relations that will serve as building blocks for the model. The process is often an art that requires physical insight, as well as an understanding of physical principles and the skill of making suitable simplifying assumptions.

Being at least part art, the initial steps of modeling do not easily yield to fixed recipes with a guaranteed successful outcome. There are, however, a number of useful steps the novice can take that may serve to ease the initial pain. It is these initial steps that we wish to address here, and we present them in the following in the form of a short compilation:

1. Make a sketch of the system under consideration, inserting where possible the important given variables and the desired result. This seemingly obvious step is often omitted by the novice and is replaced instead by a mental scrutiny of the problem. This may be appropriate in certain simple situations, but rapidly loses its effectiveness with increasing complexity of the system. A case can be made for using it under all circumstances.
2. If a conservation law is to be invoked, draw an envelope around the segment of the sketch to which it is to be applied, indicating entering and exiting quantities. This is not always the easy matter that it appears to be. The envelope may have to be drawn around a differential segment, or alternatively, the geometry is a finite one such as a tank. In the case of systems composed of several phases, more than one balance is usually called for and one has to choose from among several possibilities.
3. Establish whether the process is at steady state, or can be assumed to be a quasi-steady state, or whether the variations with time are such that an unsteady balance is called for.
4. Investigate the possibility of modeling the process, or part of it, as a compartment or a one-dimensional pipe. These two simple devices, previously shown in Figure 1.6, must be regarded as mainstays in any early attempts at modeling.
5. Determine whether a differential or integral balance is called for. Stirred tanks always require integral balances, but in the case of the one-dimensional pipe, both differential and integral balances can be implemented. Which of these is to be chosen is often only revealed in the course of the solution. Several trials may then become necessary, a not unusual feature of modeling.
6. Start with the simplest conservation laws, those of mass or electrical charge. Remember that it is possible to make instantaneous and cumulative balances in time. Introduce additional balances and auxiliary relations until the number of equations equals the number of unknown state variables. The model can then be considered complete.
7. Carefully consider whether the stirred tank or one-dimensional pipe has to be replaced by a PDE model. Avoid PDEs if possible, but face up to them when they become necessary. They are not always the ogres they are made out to be (see Chapters 5 et seq.). In other words, "PDEs if necessary but not necessarily PDEs."
8. Use Table 1.2 as an initial guide as to which auxiliary relations may be required.
9. Remember that the primary information often comes in the form of distributions in time or space of the state variable that may have to be

processed further to arrive at quantities of engineering usefulness. This is often done by differentiation or integration of the primary profiles.

10. Establish whether an analytical solution is possible. Table 2.4 provides a useful compilation of analytical solutions to ordinary differential equations.
11. Where a complete solution is impossible, or attainable only by numerical means, consider whether a partial answer, or asymptotic solution, will satisfy at least some of the requirements or "bracket" the solution. Illustration 1.5 is a striking example of the usefulness of this approach.
12. Borrow or steal from other disciplines, but be a jewel thief.

1.8 SOLUTION ANALYSIS

With this term, we wish to introduce the reader to the notion that the results of a model should be carefully scrutinized before being put aside. In particular, the following questions should be addressed:

1. Is the order of magnitude of the result "reasonable," or is it outlandishly low or high? There is nothing more embarrassing in modeling than to present results that are patently absurd. To avoid this, one should check for conceptual and numerical errors and for dimensional consistency. On occasion, unexpectedly high or low results may be obtained, which are in fact the correct answer. An example of this kind is found in Illustration 4.14 dealing with the evaporation of a pollutant into the atmosphere. The numerical calculations made there yield the surprising result that 87% of the mercury pollutant contained in a body of water will be transferred to the atmosphere with the evaporation of only 0.01% of the solution. This result can be justified on physical grounds because mercury has an extremely low solubility in water and consequently exhibits an inordinately high fugacity or "escaping tendency." This is the reason for its high rate of transfer into the atmosphere. The lesson for the reader here is that results that appear to be completely out of the ordinary should not call for automatic disbelief and dismissal. Instead, a careful review of the solution should be undertaken, and, if the result persists, the underlying physics of the process should be closely scrutinized. If no immediate rationale can be found, one should, with some caution, accept the result and hope that further investigation, perhaps in a changed context, will confirm the answer obtained. One of the thrills of modeling lies in the discovery of the unexpected, and its ultimate explanation and acceptance by the peer community. The currently fashionable topic of chaos owes its discovery in part to unexpectedly erratic behavior found in the solution of nonlinear algebraic and ordinary differential equations. The heroes here are the individuals who did not dismiss the results as numerical aberrations, but rather persisted in their quest for an explanation. From this sprang a host of new theories that became the foundations of the topic of chaos.
2. What is the behavior of the result for large or small values of the variables or parameters? This is also termed *asymptotic analysis* and often provides

upper and lower bounds to the solution being sought. A special case of this procedure is to let the independent variables, time and distance, go to infinity. In the former case, this leads to what is termed the *steady state solution* of the system, which can also be obtained from a steady state algebraic integral balance. Comparison of the two results provides a valuable check of the validity of the unsteady solution.

3. Does the result show unusual behavior for certain values of the variables or parameters? Suppose, for example, that the solution has the form:

$$f(x) \propto \frac{1}{b-x} \tag{1.20}$$

and we let x approach b. f(x) will then experience an enormous increase as x nears b, and will go to infinity when the two values are equal. This is sometimes referred to as "runaway" and may signal a beneficial or catastrophic event; for example, an unbounded rise in temperature. In general, whenever the result in made up of fractions that contain differences in the denominator, an immediate scrutiny is called for to establish conditions under which the solution goes to infinity.

The appearance of the exponential term e^{bx}, where b is a positive number, similarly signals an unbounded increase in the dependent variable. An example of this appears in Practice Problem 2.8, dealing with the onset of AIDS. Trigonometric terms such as sin bt, on the other hand, imply oscillatory behavior of the solution. This, too, will be of interest to the analyst.

4. Is the solution highly sensitive to small changes in one of the parameters or variables? As an example, let us revert to an equation of the form of Equation 1.20 and write:

$$f(x) \propto \frac{1}{x-1} \tag{1.21}$$

Suppose that initially x = 1.005, and its value is raised by a small increment, 0.005, to x = 1.01. This seemingly insignificant increase in x of 0.5% will increase f(x) by 100%, i.e., double it. This is again a highly interesting result that should be sought out by the analyst.

Another example of extreme sensitivity to parameter values occurs in the important contemporary subject of chaos, as alluded to before. Here, the parameters in question are the initial conditions of a set of nonlinear differential equations. Miniscule changes in these conditions lead to enormous changes in the solution and ultimately result in chaotic behavior. The phenomenon was first identified by E.N. Lorenz in a 1963 paper titled "Deterministic Nonperiodic Flows."

5. Does the solution contain maxima or minima (extrema) besides the obvious ones at the end points of the range being considered? There are two important ways in which this can be established, one by physical reasoning, the other by inspection of the analytical solution of the quantity, both of which we employ in this text. If the extrema in question contain a desirable result (minimum cost, maximum efficiency, etc.), the parameter

values within the solution that lead to that result are termed *optimum parameters*. We explore these features in the following illustration.

Illustration 1.9 Seeking Maxima or Minima: Optimum Performance of an Artillery Piece

An important consideration in the performance of an artillery piece is its horizontal range and, in particular, its maximum horizontal range. Increasing muzzle velocity, i.e., the driving charge of the shell, is one way of extending range, but is also limited by structural and other considerations. The maximum attainable by this means must perforce lie at the upper end of the muzzle velocity range. Another way of manipulating range is through the adjustment of the angle of elevation of the artillery piece. The argument here runs as follows: A completely vertical stance will result in zero horizontal range, whereas a piece aimed horizontally will result in early impact with the ground. It is therefore likely that the maximum range will be attained at some intermediate angle of elevation $90° > \alpha > 0°$. It will be shown in Illustration 4.6 that this optimum angle is 45° for frictionless flight, and falls below that value when drag is included.

Comments: In both the cases cited, we used physical reasoning to establish the existence of the maxima and the optimum parameters involved. The reasoning may be regarded as tenuous by some, and we shall use Illustration 4.6 to establish the same results in a more satisfactory fashion by analytical and numerical means.

Illustration 1.10 Process Control: Analysis of a Solution

Process control is a procedure designed to counteract disturbances in a system (for example, n unintended changes in flow rate) and to maintain its performance within the prescribed bounds. Its presence is virtually universal and inescapable in all industrial plants. Refineries, for example, would be unable to maintain product specifications without the extensive use of control devices.

Process control is also a discipline that makes use of modeling with uncommon frequency. Three solutions, which are often encountered in elementary models of controlled systems, are the following:

$$y = A + B \cos \omega t \tag{1.22a}$$

$$y = A + B \cos \omega t \exp(-kt) \tag{1.22b}$$

$$y = A + Bt \cos \omega t \tag{1.22c}$$

Here, the variable y is the quantity being controlled (for example, temperature, concentration, or the position of an object), and the equations express the reaction or response of the controlled system to a given disturbance. What we wish to do here is to analyze the results for any unusual feature they may contain.

Equation 1.22a. The controlled variable here shows oscillations about a mean value A of an indefinite duration, a feature that is generally deemed undesirable. If, however, the amplitude B is sufficiently small, the disturbance may be acceptable. For example, with a mean temperature value of 100°C, and an amplitude B of less

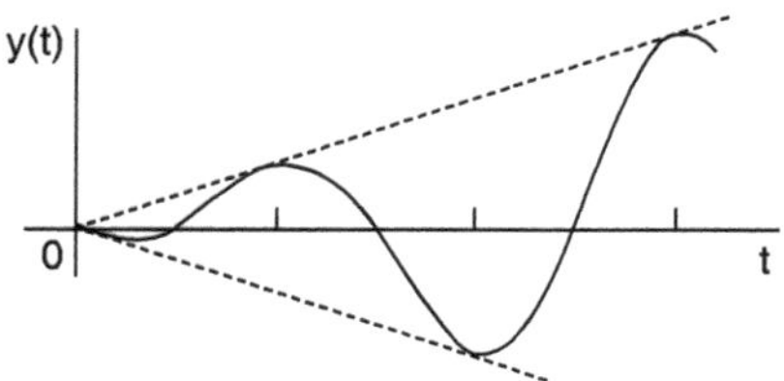

FIGURE 1.10 Oscillating response with increasing (runaway) amplitude.

than 0.1°C, the oscillations would usually not be considered a serious impediment to proper functioning.

Equation 1.22b. Although this expression contains the same oscillatory term as the preceding equation, the exponential term that follows it ensures that the oscillations will ultimately die out, returning the system to its previous steady state. This is the most desirable of all forms of responses.

Equation 1.22c. The oscillatory term here is preceded by a linear term in t that causes the amplitude of the disturbance to increase indefinitely. This response, shown in Figure 1.10, is clearly an unsatisfactory state, calling for a radical redesign of the control system.

Comments: The behavior of the analytical expressions shown in Equation 1.22 would of course also appear in any numerical simulation of the system response. Such results, however, should be treated with some reserve because they could be the consequence of numerical aberrations. The advantage of the analytical expressions lies in their ability to provide unequivocal answers.

PRACTICE PROBLEMS

1.1 To Model or Not to Model

A question that often has to be settled at the outset is whether any useful purpose will be served by modeling a system or process. If modeling will merely confirm the obvious or otherwise yield trivial results, the exercise is best avoided. Similarly, if the present-day state of knowledge does not allow for an analytical or numerical solution of the problem, or requires immense effort, modeling should be set aside in favor of experimentation. Time constraint is another factor that has to be considered in deciding how to proceed. Consider the following cases and determine what is the best course to follow:

1. *Freezing of a water pipeline.* A pipeline carrying water may be exposed to overnight subzero temperatures due to pump failure. Should it be insulated or built of heavy-gauge steel to avoid rupture?
2. The same pipeline is exposed to freezing temperatures over the course of several days.
3. *Litigation concerning the cause of an explosion.* Welding repairs carried out on the cover of a storage tank containing inert hydraulic oil led to an unexpected explosion. The suppliers were accused of having delivered oil

contaminated by gasoline left in the truck from previous deliveries. In the ensuing court case, it had to be established whether this was in fact the case.

4. Heat transfer coefficients in turbulent flow. The coefficient is to be established for heat transfer from a fluid in turbulent flow to the conduit wall kept at a constant temperature.
5. Depletion of the ozone layer. The idea of a progressive depletion of the ozone-layer by chlorofluorocarbons (CFCs) was first advanced in a 1985 landmark paper in the journal *Nature* by J. Farman, J. Shanklin, and B. Gardiner of the British Antarctic Survey in Cambridge. It led, in a remarkably short time, to the Montreal Protocol, which put controls on CFCs and was signed by 24 countries in 1987. It has resulted in the reduction of emissions from 1 Mt to less than 100,000 t/yr. Speculate on the scientific tools involved in the development of these events.

1.2 Fermi's Estimate of the Yield of the First Atomic Bomb

Shortly after the explosion of the first atomic bomb in New Mexico in 1945, Nobel laureate Enrico Fermi was seen outside the protective dugouts, holding a clutch of paper strips. When the shock wave from the explosion arrived, he released the strips and allowed them to flutter to the ground. He then paced off the distance from the point of release to where the strips had come to rest. After some quick calculations, he announced the yield of the bomb (10,000 t of TNT), which turned out to be remarkably close to the actual value (18,600 t of TNT).

Speculate as to the model, or model components, on which Fermi's calculations were based.

1.3 Critical Mass

One crucial element in the development of the atomic bomb was the minimum amount, or the *critical mass*, of fissionable material required to bring about an explosion. Calculation of this quantity has to take account of three factors: (1) neutron absorption by impurities (a net loss of neutrons), (2) the excess number of neutrons produced during fission leading to a chain reaction, and (3) the rate at which neutrons escape from the assembly (again, a net loss). Transport of the neutrons is by a diffusional mechanism and their interaction with a nucleus is expressed in terms of a *capture cross-section*, typically on the order of 10^{24} cm^2, which is the counterpart of concentration in a chemical reaction. The greater the cross section, the higher the rate of fission or absorption.

Early measurements of this quantity led to a tentative estimate of the critical mass for U^{235} ranging from 2 to 100 kg. The amount actually used in the Hiroshima bomb came to 42 kg — 2.8 times the critical mass.

1. Give a physical justification for the concept of critical mass.
2. Outline the features of a model that would yield this quantity.

Hint: Assume a spherical geometry, with a neutron source at the center.

1.4 Dependent and Independent Variables

Identify the dependent and independent variables used to describe the following system or processes:

1. *RLC circuits:* For this example, consider the two linked RLC circuits shown in Illustration 1.5.
2. *Batch distillation:* The system here is a conventional laboratory distillation apparatus consisting of a still, an overhead condenser, and a receiver for the distillate. Assume that a two-component mixture is being distilled.
3. *The loaded beam:* The behavior to be analyzed is of a horizontal beam subjected to a point force at its free end.
4. *The fluidized particle and the geosynchronous satellite:* A solid or liquid particle can be maintained in a state of suspension by subjecting it to a vertical airstream of a certain velocity. When this occurs, the upward drag and buoyancy forces acting on the particle are exactly balanced by the downward gravity force. Identify the corresponding forces and variables that determine the height to which a geosynchronous satellite has to be lifted. Such satellites are synchronous with the revolving earth, i.e., they maintain a constant position above it.
5. *Laminar flow in a rectangular duct:* The flow here is assumed to be a steady, time-invariant one, but the reader is asked to consider both the regular case of an impermeable duct and that of a porous, permeable wall. The latter situation is encountered in water desalination plants that use reverse osmosis (RO).
6. *The burning fuel particle:* The reader is asked to use Figure 1.4 and Illustration 1.6 as a guide for obtaining an answer.

1.5 Conservation Laws and Auxiliary Relations

For the examples cited in Practice Problem 1.4, establish the conservation laws, and where possible, the auxiliary relation needed to model these systems and processes. Speculate and persevere, even if the physical situations are unfamiliar.

1.6 Steady and Unsteady State: Drying of Porous Material

Consider the drying of a slab of wet porous material by passing warm air over its surface. Experiments have shown that after a brief initial adjustment period, the temperature of the surface moisture assumed a constant value (the so-called wet bulb temperature). After it has all been evaporated, the drying process proceeds into the interior of the material and continues until the slab is "bone dry."

Analyze the various stages of the drying process as to whether they are at a steady or unsteady state, and indicate as well to what extent the concentration of moisture varies with distance.

1.7 The Quasi Steady State: Concentration Fluctuations Across a Biological Membrane

A biological membrane of thickness 1 μm is exposed to a diffusing substance of concentration 10^{-9} mol/L on one side and 10^{-12} mol/L on the other. Diffusivity within the membrane is 10^{-10} m²/sec. At steady state, the concentration gradient across the membrane is linear, and the flux is:

$$N/A = D\frac{\Delta C}{\Delta x} = 10^{-10}\frac{(10^{-6} - 10^{-9})}{10^{-6}} = 10^{-10}\, mol/m^2 s$$

If the external concentration doubles over a period of 60 sec, is it reasonable to assume that the concentration gradient will remain linear over this interval, i.e., that the system is at a quasi steady state?

1.8 The Biot Number for Mass Transfer

Figure 1.11 exemplifies a mass transfer process in which a substance, for example, water vapor, is transferred from a flowing gas stream through a stagnant gas film into a stationary porous particle (for example, a desiccant), where it is adsorbed and retained. Mass transfer coefficients through the gas film are on the order 10^{-3} to 10^{-2} m/sec, whereas the diffusivity of vapors in porous particles is at the most 10^{-6} m²/sec. How small does the particle have to be before the internal gradients vanish and it starts acting like a "stirred tank"? What do you conclude from this?

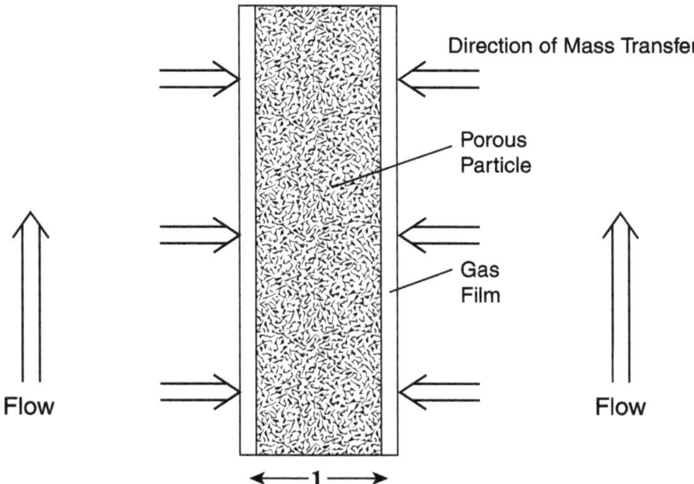

FIGURE 1.11 Mass transfer from a flowing gas stream into a porous particle.

1.9 Being Content with Less

Suppose the time course of events in a compartment or "stirred tank" is described by the two ODEs:

$$\frac{dx}{dt} = k_1 x^2 y$$

$$\frac{dy}{dx} = k_2 x y^2$$

Here, x and y can be viewed as the concentrations of two species taking part in a chemical reaction.

A solution of this system usually calls for the use of a numerical procedure. Can this be avoided by seeking a partial answer?

1.10 Information Obtained from Model Solutions

Consider again the case of laminar flow in a rectangular duct with an impermeable wall. What is the form of the model (PDE, ODE, or AE)? What is the primary information contained in the solution? Qualitatively indicate how the primary information can be converted into quantities of greater engineering usefulness? (Hint: Consult Illustration 1.4.)

1.11 Locating Maxima and Minima

1. Present an argument showing that there is an optimum pipe diameter that will minimize the combined capital and operating costs of a pipeline, whatever its configuration.
2. The combined effect of attractive and repulsive forces between two molecules can be expressed through the celebrated "6–12" Lennard-Jones potential:

$$E = -A\left(\frac{1}{r^6} - \frac{1}{r^{12}}\right)$$

where r is the distance between the molecules. Attractive forces vary inversely with the sixth power of distance, whereas the r^{12} dependence expresses the effect of the repulsive forces. E is the potential between the two particles, i.e., the work required to move one molecule over the distance (∞, r). Explore the possible existence of maxima and minima, both conceptually and by conventional calculus.

1.12 Analyzing the Solution

The text had made the point that a solution should always be scrutinized for unusual features before being set aside. Consider the following solutions, and indicate any noteworthy features:

(1) $$y = ax^3 + bx^2 + cx + d$$

(2) $$y = \frac{K}{ax^2 + bx + c}$$

(3) $$y = 0.01 \exp(10^{-5} t) + 10^5 \exp(-3t)$$

(4) $$y = \sin \frac{Kt}{a^2 - b^2}$$

The parameters a, b, and c are to be considered flexible, which can assume a range of values.

2 Analytical Tools: The Solution of Ordinary Differential Equations

Differential equations arise, as we have seen, whenever a state variable such as temperature, concentration, velocity, or electrical current, varies with time or distance, or both. The state variables usually assume the role of the dependent variable, whereas time and distance become the independent variables. As we have seen in Chapter 1, some exceptions to this rule may occur. For example, distance may be used as a dimension to describe a time-varying mass or volume, or the deflection of a physical structure. These exceptions, however, are relatively few in number, and the original definitions given earlier apply in general.

A distinction is also made between ordinary differential equations (ODEs), which contain only one independent variable, and partial differential equations (PDEs), which have two or more such variables. It is the former that we shall be addressing here; the topic of PDEs will be taken up after Chapter 4.

This chapter starts with a classification of ODEs as to order, linearity, homogeneity, and other properties. The boundary and initial conditions, which are indispensable parts of a differential equation, are examined and related to the underlying physics of the ODEs. We take up some classical analytical solution methods for both linear and nonlinear ODEs, followed by a brief survey of nonlinear analysis. A short introduction to the Laplace transformation, of which more will be seen at the PDE level, rounds off the discussion. Numerical solutions using the Mathematica package are taken up in the next chapter.

2.1 DEFINITIONS AND CLASSIFICATIONS

2.1.1 ORDER OF AN ODE

The order of a differential equation is that of its highest derivative. Some simple examples describing physical systems drawn from various disciplines will serve to demonstrate the principle:

The steam-heated tube: We have seen in Illustration 1.2 that the model for this device is given by Equation 1.4e:

$$FCp\frac{dT}{dx} - U\pi d(T_s - T) = 0 \tag{1.4e}$$

Here, T is the temperature of a fluid in tubular flow, jacketed with steam condensing at a constant temperature T_s. The ODE is first order in the fluid temperature, and its solution yields the temperature profile in the direction of flow, x.

Population growth:

Malthus' law:

$$\frac{dp}{dt} = ap \tag{2.1a}$$

Logistic law:

$$\frac{dp}{dt} = ap - bp^2 \tag{2.1b}$$

Both of these expressions are first-order ODEs in the population p, the logistic law being an extension and refinement of Malthus' law. They are calibrated using past data and are used to project future population growth, p = f(t).

The catalyst pellet

Platelet:

$$\frac{d^2C_A}{dx^2} - k_r C_A / D_e = 0 \tag{2.1c}$$

Sphere:

$$\frac{d^2C_A}{dr^2} + \frac{2}{r}\frac{dC_A}{dr} - k_r C_A / D_e = 0 \tag{2.1d}$$

The ODEs here are both second order in the concentration C_A of a substance diffusing into a porous catalyst pellet and undergoing a first-order reaction. The solutions describe the resulting concentration profiles that are often converted into the so-called pellet effectiveness factor (see Illustration 1.4 (2)). Note that although pellet geometry affects the form of the ODE, it does not change its order.

Bending of beams

$$EI\frac{d^2y}{dx^2} = M \tag{2.1e}$$

This expression, encountered in Chapter 4, describes the reaction of a beam to an applied moment M, and is a second-order ODE in the deflection y. It applies as well to a loaded strut (vertical beam) and can be used to predict the onset of buckling.

The task of solving these equations is considerably eased by the fact that most ODEs that arise in engineering and the physical sciences are of either first or second order. In particular:

- All unsteady integral balances and all steady state differential balances that arise from the simple compartmental and one-dimensional pipe models shown in Figure 1.3 lead to first-order ODEs. Similarly, electrical circuits that are subjected to transient or oscillatory inputs yield first-order ODEs.
- Steady state balances that involve molecular or diffusive transport described by Fick's and Fourier's laws lead to second-order ODEs. In general, whenever the auxiliary relations already contain an nth-order derivative, the order of the resulting differential balance will be increased by n.
- Second- and higher-order ODEs also arise when combining several first-order ODEs. For example, if we replace steam in the simple heat exchanger described by Equation 1.4e with a nonisothermal heating medium that cools in the direction of flow, a second energy balance will be required to describe the attendant temperature change. We now have:

$$F_h C_{ph} \frac{dT_h}{dx} - U\pi d(T_h - T_c) = 0 \quad (2.3a)$$

$$F_c C_{pc} \frac{dT_c}{dx} - U\pi d(T_h - T_c) = 0 \quad (2.3b)$$

where we have used the subscripts h and c to denote the heating medium and the cold fluid being heated.

Solving Equation 2.3a for T_h and substituting the result into Equation 2.3b enables us to reduce the system to a single, but higher-order ODE:

$$\frac{d^2T_c}{dx^2} + K\frac{dT_c}{dx} = 0 \quad (2.3c)$$

We note in this connection that whenever an analytical solution is being sought, combining lower-order ODEs in this fashion is a fruitful approach. In numerical work, on the other hand, the reverse procedure is often preferred, i.e., one decomposes higher-order equations to a set of equivalent first-order equations. This is done to take advantage of standard ODE solver packages (e.g., Runge–Kutta routines) that are specifically designed to solve sets of first-order ODEs.
- ODEs of order higher than 2, although less common, also arise in certain areas of fluid and solid mechanics. Some examples of these are taken up in the subsequent chapters.

2.1.2 Linear and Nonlinear ODEs

This distinction and categorization is of great importance in determining the method of solution, and indeed the ease of solution, of an ODE by analytical means. Thus, a host of methods exist and can be applied without undue difficulty to solve linear ODEs. This is primarily because of the fact that one can make use of the important superposition principle, which in essence states that the general solution of a linear ODE can be composed of the sum of all independent particular solutions. In systems of nonlinear ODEs, this important principle is lost, and one must resort to *ad hoc* methods that lack generality and are relatively few in number.

An informal definition of linear ODEs is that all dependent variables and their derivatives must appear in linear form, i.e., they are not multiplied or divided by each other, or raised to a power other than 1. A more formal definition consists of the requirement that the ODE must satisfy the following two conditions.

An ODE $f(y^{(n)} \ldots y^1, y, x) = 0$ and two particular solutions y_1 and y_2 are given. Then, if

$$f[y_1(x) + y_2(x)] = f(y_1) + f(y_2) \tag{2.4a}$$

and

$$f(ky) = kf(y) \tag{2.4b}$$

the ODE is said to be linear.

Note that these definitions do not require the independent variable to be linear. In fact, the latter can be as complex as one likes without violating the superposition principle. Some examples:

$$\text{Linear ODE} \quad \frac{d^2y}{dx^2} + e^x \frac{dy}{dx} + y = x^3$$

$$\text{Nonlinear ODE} \quad y\frac{d^2y}{dx^2} + \frac{1}{x}\frac{dy}{dx} + y = 0$$

Set of linear ODEs

$$\frac{dy}{dx} = x + z$$

$$\frac{dz}{dx} = y - z$$

Set of nonlinear ODEs

$$\frac{dy}{dx} + y/z = 0$$

$$\frac{dz}{dx} x - y^2$$

TABLE 2.1
Linear and Nonlinear Auxiliary Relations

Linear Expressions	
1. First-order reaction	$r = k_r C$
2. Newton's law of cooling	$q/A = h\Delta T$
3. Mass-transfer rate	$N/A = k_C \Delta C$
4. Fick's law of diffusion	$N/A = -D\dfrac{dC}{dx}$
5. Fourier's law of conduction	$q/A = -k\dfrac{dT}{dx}$
6. Drag on sphere in laminar flow field (Stokes' law)	$F = 3\pi\mu dv$
7. Linear phase equilibrium (Henry's law)	$D_p = Hx$
8. Alternating voltage	$V = V_0 \sin \omega t$
9. Linear spring (Hooke's law)	$F_s = k_s x$
Nonlinear Expressions	
10. nth-order reaction	$r = k_r C^n$
11. Bimolecular reaction	$r = kC_A C_B$
12. Radiation heat transfer	$q \propto \Delta(T^4)$
13. Drag on sphere in turbulent flow field	$F_D \propto v^2$
14. Nonlinear phase equilibrium (Antoine equation)	$\ln p = A - B/(T + C)$
15. Nonlinear spring	$F_s = kx^2$

In physical systems, nonlinearities are most often brought into the model by nonlinear auxiliary relations and physical properties. We have summarized those of most common occurrence and their sources in Table 2.1, and contrasted them against their linear counterparts as well.

It is to be noted in this table that linearity usually refers to a state or dependent variable, such as concentration, temperature, or velocity, but can equally well involve linear derivatives or differences of these variables (see items 2 to 5 of Table 2.1). The relations become nonlinear when the dependent variables or their derivations are raised to some power other than 1 (see items 10, 12, 13, and 15), expressed as some other nonlinear function (see item 14), or when they occur in combination with other dependent variables (see item 11). We emphasize again that nonlinearities in the independent variable, even severe ones, do not by themselves cause an expression to become nonlinear (see sin ωt in item 8).

Whereas most analytical solution methods are confined to linear ODEs, numerical methods can handle both linear and nonlinear equations with almost equal ease. They are, however, more susceptible in the nonlinear case to instabilities and other aberrations.

2.1.3 ODEs with Variable Coefficients

This classification denotes differential equations in which the coefficients of the derivatives are functions of the independent variable, i.e., not constant. The classification is usually only applied to linear ODEs of order greater than one.

The reason for making a distinction between ODEs with constant and variable coefficients lies in the difference in analytical solution techniques that have to be applied. In the former case, the classical D-Operator Method, the Laplace transformation, or the method of eigenvalues are the tools of choice, and the solutions are usually expressed in terms of simple trigonometric or exponential functions. In the case of ODEs with variable coefficients, these methods often become inconvenient or inapplicable. One then resorts to a solution in infinite power series that give rise to new classes of functions, such as the Bessel and Legendre functions.

Variable coefficient differential equations most commonly arise in mass or energy balances involving molecular or diffusive transport through a variable area, e.g., radially in a cylinder, sphere, or circle. Both area A and the gradient du/dr have to be differentiated in this case, yielding coefficients that vary with radial distance r. Thus, for diffusion and reaction in a spherical catalyst pellet, the result was previously given by the ODE:

$$\frac{d^2C_A}{dr^2} + \frac{1}{r}\frac{dC_A}{dr} - k_r C_A / D_e = 0 \tag{2.1d}$$

Note that this equation is still linear because the nonlinear term 1/r is a function of the independent, not the dependent variable. However, if the order of the reaction rate k_rC_A is changed, e.g., to $k_rC_A^2$, the equation becomes nonlinear.

2.1.4 Homogeneous and Nonhomogeneous ODEs

A linear ODE that does not contain an isolated function of the independent variable f(x) or an isolated constant is termed *homogeneous*. When this is not the case, the equation is said to be *nonhomogeneous*. Thus,

$$\frac{dy}{dt} + ky = 0 \tag{2.5a}$$

and

$$\frac{d^2y}{dx^2} + K\frac{dy}{dx} + xy = 0 \tag{2.5b}$$

are homogeneous ODEs.

Examples of nonhomogeneous equations are:

$$\frac{dy}{dt} + K_1 y = K_2 \tag{2.5c}$$

$$\frac{dy}{dt} + K_1 y = K_2 \sin K_3 t \tag{2.5d}$$

and more generally

$$\frac{d^2y}{dt^2} - K_1 \frac{dy}{dt} - K_2 y = f(t) \tag{2.5e}$$

The nonhomogeneous terms appearing on the right side of these equations are commonly referred to as *forcing functions*, and their functional form has a direct impact on the form of the solution. They are usually associated with time-dependent models and, hence, appear extensively in process control theory and other areas dealing with dynamic systems.

The preferred solution method of nonhomogeneous, as well as homogeneous time-dependent equations, is the Laplace transformation. Nonhomogeneous ODEs can also be solved by adding a particular integral to the solution of the homogeneous ODE. Both of these methods are taken up in subsequent sections.

2.1.5 Autonomous ODEs

These equations can be linear or nonlinear and are characterized by an absence of terms in the independent variable other than the derivatives themselves. When this is not the case, they are said to be *nonautonomous*. Examples of both classes, which are usually converted to first-order systems, appear below:

$$\text{Autonomous ODEs:} \quad \frac{dy}{dx} = f(y) \tag{2.6a}$$

$$\text{or for a set:} \quad \dot{\underset{\sim}{y}} = \underset{\sim}{f}(\underset{\sim}{y}) \tag{2.6b}$$

$$\text{Nonautonomous ODEs:} \quad \frac{dy}{dx} = f(x, y) \tag{2.6c}$$

$$\text{or for a set:} \quad \dot{\underset{\sim}{y}} = \underset{\sim}{f}(x, \underset{\sim}{y}) \tag{2.6d}$$

where we use a vector-matrix notation to generalize the classes to sets of simultaneous equations.

What prompts this classification is again a marked difference in the form of the solutions, as well as in the analytical solution methods. Thus, single autonomous ODEs are immediately integrable by separation of variables and, if linear, by the Laplace transformation as well. Nonautonomous equations are solved by special analytical techniques, unless separable, or by numerical techniques.

Illustration 2.1 Classification of Model ODEs

We undertake here the classification of model ODEs that were encountered in previous illustrations. We establish their order, and distinguish between linear/nonlinear equations, homogeneous and nonhomogeneous forms, second-order ODEs with constant and variable coefficients, and autonomous/nonautonomous behavior.

The Land-Based Oil Spill

$$\frac{dz}{dt} = \frac{K\rho g}{\varepsilon\mu}\left[\frac{h_0}{z} + (1-\varepsilon)\right] \tag{1.9a}$$

This is a first-order ODE that is also nonlinear by virtue of the term h_0/z. It is autonomous because no term in time t appears in it other than the derivative.

Coupled RLC Circuits

$$Lq_1'' + \frac{1}{C}q_1 + \frac{1}{C}q_3 = V_o \sin wt \tag{1.11c}$$

$$Lq_2'' + \frac{1}{C}q_2 + Rq_2' + \frac{1}{C}q_3 = 0 \tag{1.11d}$$

$$q_1' = q_2' + q_3' \tag{1.12b}$$

Here we are dealing with a mixed set of first- and second-order linear ODEs in the state variables q_1, q_2, and q_3. The forcing function $V_0 \sin \omega t$ makes the system nonhomogeneous.

The Steam-Heated Tube

$$FC_p\frac{dT}{dx} - U\pi d(T_s - T) = 0 \tag{2.1a}$$

This ODE, describing the transfer of heat from condensing steam to a colder fluid in tubular flow, is first order and linear in the temperature T. It is autonomous, but the term T_s makes it nonhomogeneous. Note that by introducing the new variable $y = T_s - T$, the ODE transforms into a simpler, homogeneous equation.

The Spherical Catalyst Pellet

$$\frac{d^2C_A}{dr^2} + \frac{2}{r}\frac{dC_A}{dr} - k_rC_A/D_e = 0 \tag{2.1d}$$

The second derivative makes this a second-order ODE in the reactant concentration C_A. It has a variable coefficient 2/r brought about by the spherical geometry, but is still linear and contains no nonhomogeneous terms.

2.2 BOUNDARY AND INITIAL CONDITIONS

Boundary and initial conditions (BCs and ICs), usually expressed as equations, are needed to evaluate integration constants and to provide starting values for numerical integration procedures. The number of such conditions required equals the order of the ODE.

For first-order ODEs, these conditions are usually identified from the prevailing values of the state variables at position $z = 0$ or time $t = 0$. Some typical examples of the information used to obtain these conditions are the following:

- Concentration, pressure, and temperature of the feed to a tubular reactor
- Initial level, concentration, or temperature of a compartment or tank
- Initial velocity and position of a falling or rising particle
- Initial velocity and position of an oscillating system (pendulum, mass on spring, etc.)
- Inlet temperature to a steam-heated exchanger
- Initial charge in an electrical circuit

When all the conditions required for a set of first-order ODEs are given at the same point in time or space, one speaks of the problem as being an *initial value problem* (IVP). All of the examples cited here are illustrations of IVPs. Also in this category are multiphase contacting devices in co-current flow, such as co-current heat exchangers. When flow is counter-current, the boundary conditions are generally given at opposite ends of the device. In these cases, i.e., when the required boundary conditions are only known at different locations, one speaks of a *boundary value problem* (BVP). Such problems arise in a good many other physical processes. Analytical solution methods do not make a special distinction between IVPs and BVPs and can be applied with equal ease to either case. In numerical work, one generally has to know the value of all state variables at $z = 0$ or $t = 0$ to initiate the integration procedure. Hence, most standard ODE solver packages are designed to handle IVPs only.

For second-order ODEs, one requires two boundary conditions, and these are frequently given at different locations, resulting in a BVP. A typical example is the ODE that describes diffusion and reaction in a catalyst pellet (Equation 2.1d). Figure 2.1A shows the boundary conditions for a spherical pellet with no external film resistance. A boundary condition (BC) is immediately obtained at the pellet surface where $C_A(R) = C_{Ab}$, where C_{Ab} is the prescribed or known surface or bulk fluid concentration. For the second BC, one argues that because the concentration profile must be symmetrical about the center, the derivative dC_A/dr at that point will be zero. Alternatively, the same condition may be deduced from the fact that the mass flux at the center is zero; hence, $(dC_A/dr)_{r=0} = 0$.

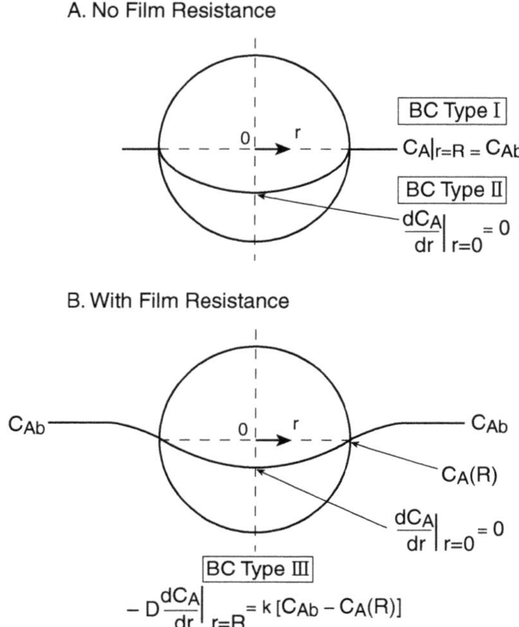

FIGURE 2.1 Boundary conditions for diffusion and reaction in a spherical catalyst pellet.

The type of BCs used has a significant impact on the form of the solution, and this has led to a formal classification of boundary conditions depending on whether they contain the state variable, its derivative alone, or a combination of the two. The resulting classes of BCs, their nomenclature, and their occurrence are tabulated in Table 2.2.

Using this table and the example of the catalyst pellet shown in Figure 2.1A, we see that it has a Type I boundary condition at the surface and a Type II condition at the center. For the catalyst pellet with film resistance, shown in Figure 2.1B, the center boundary condition is retained, whereas the surface BC now becomes a

TABLE 2.2
Types of Boundary Conditions

Type	Nomenclature	Property	Occurrence
I	BC of Type I	Contains y only	y specified at a particular location
II	BC of Type II	Contains dy/dx only	Profile symmetrical Constant or zero diffusional flux
III	BC of Type III or mixed type	Contains dy/dx and y	Diffusional flux with film resistance

mixed-type or Type III condition. Note that the latter is obtained by equating convective transport of reactant through the film to diffusive transport away from the surface and into the interior. Thus,

Type III BC:

$$k_C 4\pi R^2 (C_A - C_A(R)) = D_e 4\pi R^2 \left. \frac{dC_A}{dr} \right|_{r=R} \tag{2.7}$$

Rate of transport through film — Rate of diffusion into interior

2.2.1 Some Useful Hints on Boundary Conditions

It is frequently sufficient to formulate a boundary condition as "y is finite at z = L" or "y is bounded at y = L," rather than specifying actual values of the state variable or its derivatives at that position. For example, in the frequently encountered solution for a bounded state variable y:

$$y = C_1 \exp(kx) + C_2 \exp(-kx) \tag{2.8a}$$

the integration constant C_1 can be easily evaluated by invoking the boundedness condition:

$$y \big|_{x \to \infty} = \text{finite} \tag{2.8b}$$

i.e., C_1 must perforce be zero.

The boundary condition dy/dx = 0 arises whenever a profile is symmetrical or when the diffusive flux is zero. There are numerous physical situations in which one or the other of these conditions applies, including:

- $dv/dr|_{r=0} = 0$ in viscous flow of a fluid in a pipe (symmetry)
- $dC/dr|_{r=0} = 0$ in a cylindrical or spherical catalyst pellet (symmetry)
- $dT/dr|_{r=0} = 0$ in a radial conduction through a cylinder or sphere (symmetry)
- $dT/dz|_{z=L} = 0$ at an insulated surface (zero flux)
- $dP/dz|_{z=L} = 0$ in a porous duct with one end sealed (zero flux)
- $dC/dx|_{x=L}$ in diffusion into a sealed tube

Application of boundary conditions of Type III leads to awkward and lengthy expressions involving both state variables and their derivatives. This can be avoided by "bracketing" the exact solution with a Type I BC representing zero film resistance and a Type II condition dy/dx = 0 representative of infinite resistance. The technique is often invoked when the only solutions available are those for Type I and Type II conditions. Boundary conditions involving higher-order derivatives arise in certain problems of solid mechanics.

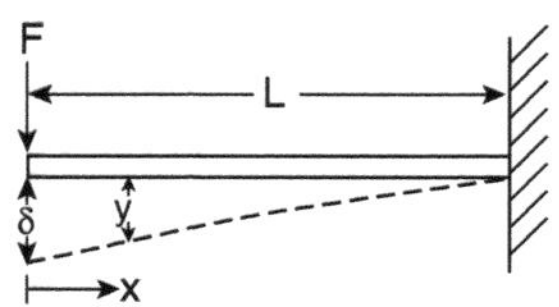

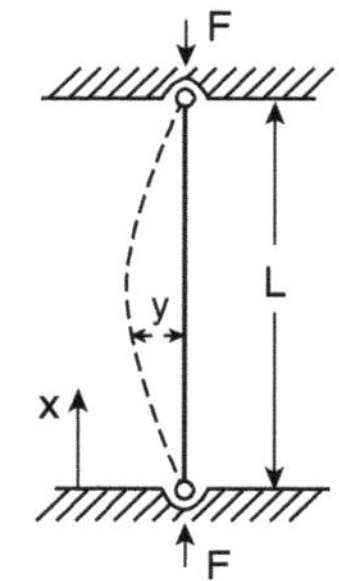

FIGURE 2.2 Bending of beams caused by lateral and axial loads. (A) Cantilever with a lateral end load. (B) Hinged strut.

Illustration 2.2 Boundary Conditions for Loaded Structures: IVPs and BVPs

Figure 2.2 shows the behavior of two classical structures in response to a point load or force F. The deflection y that results from these loads is described by the second-order ODE

$$EI \frac{d^2y}{dx^2} = M \tag{2.9}$$

where M is the moment because of the applied force, E is the modulus of elasticity (a property of the material), and I its second area moment (a structural parameter).

The application of Equation 2.9 to the buckling of a strut will be taken up in Illustration 2.6. Our intent here is to establish the boundary conditions required to solve structural problems of this type.

Because Equation 2.9 is a second-order ODE, two boundary conditions have to be provided: For the cantilever (Figure 2.2A), both of these conditions are given at the clamped end, which makes this an IVP. The first, and most obvious condition, is that of zero deflection at this location. We obtain the following BC of Type I:

$$y(0) = 0 \tag{2.10a}$$

The second condition, which is slightly less self-evident, is that the derivative of y also vanishes at this point. This follows from the clamped condition of the beam. We have

$$y'(0) = 0 \tag{2.10b}$$

which will be recognized as a boundary condition of Type II.

For the hinged strut, Figure 2.2B, which is anchored at both ends, the necessary and sufficient conditions are that both deflections at these locations vanish:

$$y(0) = 0 \tag{2.10c}$$

$$y(L) = 0 \tag{2.10d}$$

These are two BCs of Type I but they are situated at two different locations, resulting in a BVP.

2.3 ANALYTICAL SOLUTIONS OF ODEs

Although numerical methods have by now become the standard tool for the solution of ODEs, particularly of nonlinear sets, analytical techniques are far from being obsolete and continue to maintain a strong presence and hold in the field. There are several reasons for this. Foremost among them is the fact that analytical methods are unsurpassed in providing a general and precise sense of solution behavior and in linking it to the physics of a process. An important example is the solution of second-order linear and nonautonomous (i.e., forced) systems. Here, the analysis and solution of a single ODE gives us a clear picture of the transition of exponentially decaying solutions to oscillatory behavior, and provide precise criteria for the occurrence, usually undesirable, of unbounded growth in the state variable (see Illustration 1.8). Such *linear responses*, as they are often called, are the logical starting points for the analysis of more complex nonlinear phenomena. In nonlinear analysis, to be taken up in Section 2.4, one dispenses with precise solutions of the ODEs and attempts instead to define domains in which certain types of behavior occur. This is oftentimes of greater interest and benefit than precise numerical response data for a specific set of initial or boundary conditions.

Analytical solutions and criteria are also indispensable in numerical work. Here they provide precise and proven expressions against which the numerical solution to a particular problem can be tested. This is a sound practice, given that numerical methods can, in spite of all refinements and safeguards, lead to unstable and other aberrant behavior.

We commence this section by providing the reader with a list, shown in Table 2.3, of the more important classical ODEs that arise in various fields of science and engineering and are associated with the names of famous mathematicians. They have been the subject of detailed analyses, and a good deal of their general behavior is well known. In Table 2.4, we have added a summary of classical analytical solutions of various types of single ODEs. A distinction is made between major methods that

TABLE 2.3
Important ODEs

Name	Form	Occurrence
	A. Linear ODEs	
1. Airy equation	$y'' + xy = 0$	Quantum mechanics diffraction of waves
2. Cauchy–Euler equation	$x^2y'' + bxy + cy = 0$	Solution of PDEs
3. Bessel equation	$y'' + \frac{1}{x}y' + \left(1 + \frac{n^2}{x^2}\right)y = 0$	Radial diffusive transport in circle or cylinder, vibrations of circular membranes
4. Legendre equation	$(1 - x^2)y'' - 2xy' + n = 0$	Diffusive transport in sphere
5. Laguerre equation	$xy'' + (1 - x)y' + cy = 0$	Solution of PDEs
6. Hermite equation	$y'' - 2xy' + \lambda y = 0$	Wave mechanics
	B. Nonlinear ODEs	
7. Bernoulli equation	$y' + f(x)y - g(x)y^n = 0$	Nonlinear electrical circuits
8. Riccati equation	$y' + f(x)y + g(x)y^2 = h(x)$	Intermediate result in various engineering problems
9. Duffing equation	$\ddot{y} + a\dot{y} + by + cy^3 = d \cos \omega t$	Electrical oscillations
10. van der Pol equation	$\ddot{y} - \lambda(1 - y^2)\dot{y} + y = 0$	Electrical and biological oscillations
11. Lotka–Volterra (predator–prey) model	$\dot{x} = x(a - by)$ $\dot{y} = y(-c + dx)$	Population growth
12. Lorenz attractor	$\dot{x} = 10(y - x)$ $\dot{y} = -xz + 28x - y$ $\dot{z} = xy - (8/3)x$	Free convection flow, chaos

are encountered with great frequency in the sciences and engineering, and other methods of less frequent occurrence. Among the major methods, Laplace transformation has been singled out for separate treatment, both in this chapter and in subsequent treatments of PDEs.

2.3.1 Separation of Variables

This is a powerful and sometimes underrated method for solving both linear and nonlinear first-order ODEs. The following points and recommendations are drawn to the attention of the reader:

- Separation of variables is the preferred method for solving single ODEs and is always to be tried first.

TABLE 2.4
Analytical Solutions of ODEs

System	Solution
A. Major Methods	
1. Separable ODEs	By separation of variables
$y' - f(x)/g(y) = 0$	$\int g(y)dy = \int f(x)dx + C$
$y'' - f(x)/g(y') = 0$	$\int g(y')dy' = \int f(x)dx + C$
2. Linear homogeneous second-order ODEs with constant coefficients	By D-operator method
$y'' + ay' + by = 0$	$y = C_1 \exp(D_1x) + C_2 \exp(D_2x)$
3. Linear nonhomogeneous second-order ODEs with constant coefficients	Solution of homogeneous ODE + particular integral y_P (see Table 2.7 for listing of y_P)
$ay'' + by' + cy = f(x)$	
4. Linear homogeneous second-order ODEs with variable coefficients	By power series
$a(x)y'' + b(x)y' + c(x)y = 0$	$y = C_1\sum_1^\infty a_n x^{n+k_1} + C_2\sum_1^\infty b_n x^{n+k_2}$
5. Sets of linear first-order initial value ODEs with constant coefficients	By Laplace transformation
B. Other Methods	
6. Linear first-order nonhomogeneous ODE with variable coefficients	Directly given by formula
$y' + f(x)y = g(x)$	$y = \exp\int -f(x)dx$ $\left[\int g(x)\exp\int f(x)dxdx + C\right]$
7. General first ODEs.	Various solutions arise, depending on the method used:
$y' + \frac{f(x,y)}{g(x,y)} = 0$	
(a) by substitution $y = v\,x$	If substitution yields ODE in v, x only, the solution is
	$\ln x = C - \int \frac{g(1,v)dv}{f(1,v) + vg(1,v)}$
(b) ODE is exact, i.e.,	Directly given by formula
$\frac{\partial f}{\partial y} = \frac{\partial g}{\partial x}$	$\int f(x,c)dx + \int g(c,y)\frac{\partial}{\partial y}\int f(x,c)dxdy = C$

(continued)

TABLE 2.4 (Continued)
Analytical Solutions of ODEs

System	Solution
8. Nonlinear second-order ODE with first derivative and terms in x missing. $y''+f(y)=0$	Multiply equation by 2(dy/dx)dx = 2dy to obtain $y'=[2\int f(y)dy+C]^{1/2}$, then apply separation of variables
9. Nonlinear second-order ODE with missing dependent variable $f(y'', y', x)=0$	Reduce to first-order ODE by "p-Substitution" $p=\frac{dy}{dx}$, and attempt integration by one of the preceding methods, followed by a second integration.
10. Nonlinear second-order ODE with missing independent variable $f(y''', y', y)=0$	As under 9. Note that second derivative becomes $\frac{d^2y}{dx^2}=\frac{dp}{dx}=\frac{dp}{dy}\frac{dy}{dx}=p\frac{dp}{dy}$

- Single second-order ODEs may be amenable to solution by separation of variables, provided one of the boundary conditions is given in terms of the first derivative (see item 1 in Table 2.4).
- ODEs that appear to defy separation of the variable at first glance may, by proper manipulation, be reduced to a separable form. An example of this is given in the following Illustration.

Illustration 2.3 Solution of Complex ODEs by Separation of Variables

Consider the following first-order ODE:

$$x\sin y\frac{dy}{dx}+\frac{ye^x}{1-y}=e^x(y^3+y) \tag{2.11a}$$

This equation is highly nonlinear, not because of the exponential terms that are in x, but by virtue of the y^3 term as well as the expression sin y(dy/dx). The equation does not at first sight appear to be separable. The situation is improved, however, by factoring out e^x and collecting terms. We obtain:

$$e^x\left(y^3+y-\frac{y}{1-y}\right)=x\sin y\frac{dy}{dx} \tag{2.11b}$$

which is clearly separable. Formal integration leads to the result:

$$\int\frac{e^x}{x}dx=\int\frac{\sin y\, dy}{y^3+y-y/(1-y)}+C \tag{2.11c}$$

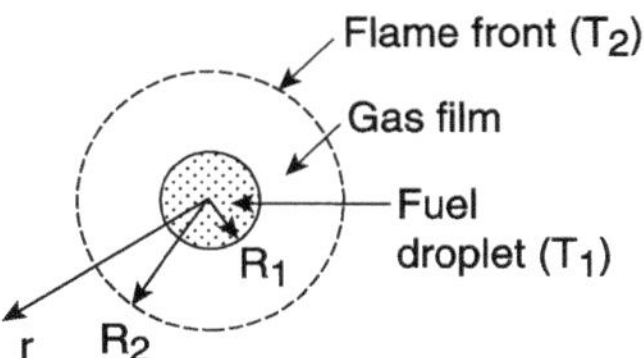

FIGURE 2.3 Configuration of a burning liquid fuel droplet.

Analytical evaluation of the left side is easily accomplished; that of the right side is more problematical. One has to resort to numerical methods, and the question then arises whether these are not better applied at the source, i.e., at the ODE level. The answer here is no. The integrated form (Equation 2.11c), although not fully evaluated, provides a better picture of the solution. In particular, it enables us to identify conditions on y that leads to a divergence of the integral. These may occur when $y^3 + y - y/1 - y$ goes to zero. For these reasons, it is preferable to deal with integrated expressions, if attainable, rather than the ODE itself.

Illustration 2.4 Repeated Separation of Variables: The Burning Fuel Droplet as a Moving-Boundary Problem

In the combustion of liquid fuels, the rate-determining step is frequently taken to be the rate of heat transfer through the stagnant gas film. Heat is conducted through this film to the drop surface, where it vaporizes the fuel. Fuel vapor in turn diffuses to the flame front where the actual combustion takes place. The fuel droplet is assumed to be at its boiling point T_1, and the flame front is assumed to be at a constant and known temperature T_2 (see Figure 2.3).

The droplet size variation with time has been established experimentally to be of the form:

$$R_1^{\,2} = \left(R_1^{\,0}\right)^2 - \alpha t \tag{2.12a}$$

where R_1^0 is the initial droplet radius.

The model solution is to be used to verify this expression and to relate the constant α to the physical parameters of the system.

An outline of the preferred procedure for solving moving boundary problems was given in Illustration 1.8. One typically starts with unsteady mass and energy balance about the core, which is assumed to have uniform properties, and proceeds outward into the film, which is assumed to be at quasi-steady state conditions. We obtain:

Core Mass Balance:

Rate of fuel in − Rate of fuel out = Rate of change of fuel content

$$0 - F = \frac{d}{dt}m = \rho_L 4\pi R_1^2 \frac{dR_1}{dt} \tag{2.12b}$$

Core Energy Balance:

Rate of energy in – Rate of energy out = Rate of change of energy contents

$$q - FH_v = \frac{d}{dt}H_L \tag{2.12c}$$

where F = rate of vapor formed (kg/sec), and $H_{V,L}$ = vapor and liquid fuel enthalpies.

Taking the liquid enthalpy H_L as the reference state and using the appropriate auxiliary relation for q, we obtain from Equation 2.12c:

$$k_V 4\pi R_1^2 \left(\frac{dT}{dr}\right)_{R_1} - F\Delta H_v = 0 \tag{2.12d}$$

Quasi-Steady-State Differential Energy Balance in Gas Film:

Rate of energy in – Rate of energy out = 0

$$\left[H\big|_r + q\big|_{r+\Delta r}\right] - \left[H\big|_{r+\Delta r} + q_r\right] = 0$$

or equivalently

$$dq - dH = 0 \tag{2.12e}$$

which upon introduction of the relevant auxiliary relations yields

$$d\left(k_V 4\pi r^2 \frac{dT}{dr}\right) - d[FC_{pV}(T - T_0)] = 0 \tag{2.12f}$$

Note that conduction takes a positive sign because it takes place in the negative direction.

This expression can be directly integrated from the surface of the droplet (R_1, T_1) to some arbitrary position (r, T). We obtain:

$$k_V 4\pi r^2 \frac{dT}{dr} - k_V 4\pi R_1^2 \left(\frac{dT}{dr}\right)_{R_1} = FC_{pV}(T - T_1) \tag{2.12g}$$

A second integration is performed, this time by separation of variables. This yields the result

$$\int_{R_1}^{R_2} \frac{dr}{r^2} = \int_{T_1}^{T_2} \frac{dT}{K(T - T_1) + R_1^2 \left(\frac{dT}{dr}\right)_{R_1}} \tag{2.12h}$$

and

$$\frac{1}{R_1} - \frac{1}{R_2} = \frac{1}{K} \ln\left[\frac{T_2 - T_1}{(R_1^2/K)\left(\frac{dT}{dr}\right)_{R_1}} + 1\right] \tag{2.12i}$$

where $K = FC_{pV}/4\pi k_V$. Note that in this integration both R_1 and $(dT/dr)_{R1}$ are held constant by virtue of the steady state assumption.

We pause at this point for a brief inventory. The model equations on hand are now three in number: the core mass balance Equation 2.12b, the core energy balance Equation 2.12d, and the integrated steady state gas film energy balance Equation 2.12i. The associated unknown state variables are F, $(dT/dr)_{R_1}$, and R_1, of which we wish to retain only R_1. Elimination of F and $(dT/dr)_{R_1}$ from these three equations and a formal second integration by separation of variables lead us to:

$$\ln\left[1 + \frac{C_{pV}}{\Delta H_v}(T_2 - T_1)\right]\int_0^t dt = -\frac{C_{pV}\rho_L}{k_v}\int_{R_1^0}^{R_1} R_1(1 - R_1/R_2)dR_1 \tag{2.12j}$$

One notes that to satisfy the experimental finding, Equation 2.12a, the right-side integral must yield the form R_1^2, i.e., the ratio of inner to outer radius R_1/R_2 has to be a constant. This is an acceptable assumption, given that the flame front recedes in proportion to the shrinking fuel core.

We now finalize the result by carrying out the indicated integration and obtain the following after rearrangement:

$$R_1^2 = \left(R_1^0\right)^2 - \alpha t = \left(R_1^0\right)^2 - \frac{2k_V}{\rho_L C_{pV}(1 - R_1/R_2)} \ln\left[1 + \frac{C_{pV}}{\Delta H_v}(T_2 - T_1)\right] t \tag{2.12k}$$

where k_{pV}, C_{pV} = thermal conductivity and heat capacity of the vapor, ΔH_v = latent heat of vaporization, and T_1, T_2 = boiling point of fuel and temperature of the flame front, respectively. Most of these physical parameters are readily available. The unknown flame temperature T_2 is arrived at by equating the heat of vaporization and sensible heat of fuel vapor to the heat of combustion, i.e., by performing an integral energy balance. This leaves the ratio of radii R_1/R_2, which has to be obtained by fitting at least one set of experimental data to the Equation 2.12k. This has in one case yielded a value of $R_1/R_2 = 0.48$, an acceptable number for the geometry in question.

Comments: We have succeeded, in this example, in modeling a process which, to the uninitiated at least, is one of considerable complexity. Let us summarize the features that led to the successful solution of the problem.

1. The principal simplifying step was the use of the combination of a uniform shrinking core tied to an external vapor film taken to be at a quasi steady state. This resulted in a decoupling of the process into three ODEs, two of them in time t (Equation 2.12b and Equation 2.12c) and one in distance R_1 (Equation 2.12f).
2. The second step was the recognition that the ODEs could be solved in succession and are independent of each other, and that this could be done by a double application of the method of separation of variables. Note that the inner and outer radii (R_1 and R_2) were quasi-steady state variables in this process, and that the initial radius R_1^0 was only brought in at the last integration step (Equation 2.12j).
3. The solution was aided considerably by adopting a systematic procedure that started with balances around the core and then moved outward into the gas film. Along the way we kept a running account of the number of dependent variables and the number of equations. When the two were equal, we stopped adding new equations and decided which variables to eliminate. This process led to the final result (Equation 2.12k).
4. The solution we obtained (Equation 2.12k) can be adapted to other fuel systems as well, using the relevant physical parameters and flame temperatures. Although the ratio R_1/R_2 may differ among systems, the change is not expected to be major, so that the value of ~0.5 can be used as a good first approximation.

2.3.2 The D-Operator Method: Solution of Linear n-th-Order ODEs with Constant Coefficients

We start this section with an example that we use to introduce the reader to the concept of characteristic roots or eigenvalues of a linear ODE, and to the important superposition principle. Consider the equation:

$$\frac{d^2y}{dx^2} - y = 0 \tag{2.13a}$$

A relative novice to the field might attempt a solution by substituting trial functions into the ODE and seeing whether the ODE is satisfied. It might further be argued that because the equation is a second-order one, two boundary conditions will have to be satisfied; hence, two integration constants will have to be evaluated. These integration constants must be associated with two independent functions, for if they are not, the two constants would coalesce into a single one, and we would be unable to satisfy the two boundary conditions.

Let us attempt a solution with some simple trial functions. Neither sin x nor cos x satisfy Equation 2.13a. However, both e^x and e^{-x} do, and, furthermore, they are independent of each other. Therefore, one might formulate a general solution of the form:

$$y = C_1 e^x + C_2 e^{-x} \tag{2.13b}$$

This sum also satisfies the ODE. Thus, we have, somewhat inadvertently, discovered the superposition principle, at least as it applies to this example. This principle in essence states that the general solution to an n-th-order linear ODE is composed of the sum of n independent functions. Uniqueness of the solution is guaranteed by uniqueness theorems that are described in most texts dealing with ODEs.

If one were to conduct extensive trials of this type, one would discover that the solution of any linear homogeneous ODE with constant coefficients, of whatever order, is always composed of the sum of exponential functions with either real or imaginary arguments. Early workers in the field were well aware of this fact. They had also noted a precise connection between the (constant) coefficients of the ODE and those of the arguments of the exponential functions. This led to the development of a formalism known as the *D-operator method.* In it, the operational part of a derivative, i.e., d/dx, is replaced by the operator symbol D, and that symbol is treated as an algebraic entity, subject to the usual rules of algebra. Equation 2.13a can then be written in the form:

$$(D^2 - 1)y = 0 \tag{2.13c}$$

Equivalently,

$$D^2 - 1 = 0 \tag{2.13d}$$

with the solutions:

$$D_1 = 1,\ D_2 = -1 \tag{2.13e}$$

The Equation 2.13d is termed the *characteristic equation of the ODE*, and its solution is termed its *characteristic roots*. These roots are identical to the coefficients of the arguments of the exponential functions in Equation 2.13b.

Table 2.5 lists a compilation of characteristic roots and the corresponding solutions for the most frequently encountered case of a second-order ODE. For real and distinct roots, the solution is the sum of the corresponding exponential functions that can also be expressed in terms of equivalent hyperbolic functions, listed for

TABLE 2.5
Solutions of the Second-Order ODE ay″ + by′ + cy = 0

Characteristic Roots or Eigenvalues	Solution
1. Distinct and real: $D_{1,2}$	$y = C_1 e^{D_1 x} + C_2 e^{D_2 x}$
	if $D_1 = -D_2$: $y = C_1' \sinh D_1 x + C_2' \cosh D_2 x$
2. Identical and real: $D_1 = D_2 = D$	$y = C_1 e^{Dx} + C_2 x e^{Dx}$
3. Imaginary: $D_{1,2} = \pm$ bi	$y = C_1 e^{bix} + C_2 e^{-bix}$
	or $y = C_1' \cos bx + C_2' \sin bx$
4. Complex conjugate: $D_{1,2} = a \pm bi$	$y = C_1 e^{(a+bi)x} + C_2 e^{(a-bi)x}$
	or $y = (C_1' \cos bx + C_2' \sin bx)e^{ax}$

convenience in Table 2.6. When the roots are identical, one of the exponential functions is premultiplied by the independent variable. Exponential functions with imaginary arguments that result from complex conjugate characteristic roots are converted to trigonometric functions with real arguments using the Euler formula given in the table.

These characteristic roots can also be obtained by matrix methods. To accomplish this, we decompose the n-th-order equation into an equivalent set of n first-order ODEs and evaluate the eigenvalues λ_i of the coefficient matrix. For Equation 2.13a, we obtain the equivalent set:

$$\frac{dy}{dx} = 0 + z$$

$$\frac{dz}{dx} = y + 0 \tag{2.13f}$$

for which the coefficient matrix is given by:

$$\underset{\sim}{A} = \begin{bmatrix} 0 & 1 \\ 1 & 0 \end{bmatrix} \tag{2.13g}$$

The eigenvalues follow from the relation:

$$\det\,(\underset{\sim}{A} - \lambda \underset{\sim}{I}) = 0 \tag{2.13h}$$

TABLE 2.6
Table of Hyperbolic Functions

1. Hyperbolic sine of u	$\sinh u = \frac{1}{2}(e^u - e^{-u}) = \frac{1}{\text{csch } u}$
2. Hyperbolic cosine of u	$\cosh u = \frac{1}{2}(e^u + e^{-u}) = \frac{1}{\text{sech } u}$
3. Hyperbolic tangent of u	$\tanh u = (e^u - e^{-u})/(e^u + e^{-u})$
4. Hyperbolic cotangent of u	$\coth u = (e^u + e^{-u})/(e^u - e^{-u})$
Relation to Trigonometric Functions	
5. $\sinh iu = i \sin u$	8. $\sinh u = -i \sin iu$
6. $\cosh iu = \cos u$	9. $\cosh u = \cos iu$
7. $\tanh iu = i \tan u$	10. $\tanh u = i \tan iu$
Derivatives	
11. $\frac{d}{dx}\sinh x = \cosh x$	13. $\frac{d}{dx}\tanh x = \text{sech}^2\, x$
12. $\frac{d}{dx}\cosh x = \sinh x$	14. $\frac{d}{dx}\coth x = \text{csch}^2\, x$
Integrals	
15. $\int \sinh x\, dx = \cosh x$	17. $\int \tanh x\, dx = \ln(\cosh x)$
16. $\int \cosh x\, dx = \sinh x$	18. $\int \coth x\, dx = \ln(\sinh x)$
Other Relations	
19. $\cosh^2 u - \sinh^2 u = 1$	
20. $\sinh(-u) = -\sinh u$	
21. $\cosh(-u) = \cosh u$	
22. $\tanh(-u) = -\tanh u$	
23. $\coth(-u) = -\coth u$	
24. $e^{ix} = \cos x + i \sin x$ Euler's Formula	

or, equivalently,

$$\begin{vmatrix} 0-\lambda & 1 \\ 1 & 0-\lambda \end{vmatrix} = 0 \tag{2.13i}$$

Hence, $\lambda^2 - 1 = 0$ and $\lambda_{1,2} = \pm 1$. The eigenvalues of the coefficient matrix of the set of two first-order ODEs (Equation 2.13g) are thus seen to be identical to the characteristic roots of the corresponding second-order ODE (Equation 2.13a).

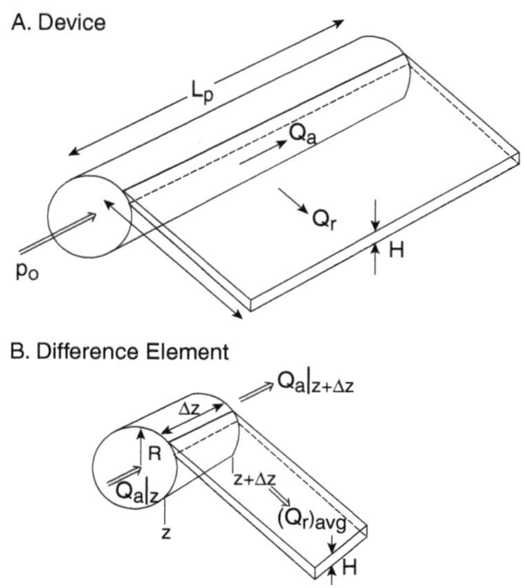

FIGURE 2.4 (A) Schematic diagram of a polymer sheet extruder, (B) Difference element for the mass balance.

Illustration 2.5 Polymer Sheet Extrusion and the Uniformity Index: An Example of a Second-Order Linear ODE

In this example, we consider the performance of a simple device used in the extrusion of polymer sheets. The molten polymer is forced, with an inlet pressure p^0, into a pipe with a lateral slit or lip extending some distance away from the pipe wall (Figure 2.4). The polymer flows into the pipe axially and exits radially through the extruder lip. A problem that arises in these devices is that the pressure driving the melt through the slit diminishes in the axial direction, causing nonuniformity in the thickness of the extruded sheet. A model is required to relate sheet thickness to the system parameters and to axial distance so that these can be properly modified to ensure high uniformity. This is done by considering the ratio of radial flow at the inlet $Q_r(0)$, and the corresponding flow $Q_r(L_p)$ at the pipe end that is sealed off. The ratio of the two quantities is known as the uniformity index $E = Q_r(L_p)/Q_r(0)$. Thus, for complete uniformity, $E = 1$, and for nonuniform sheets, $E < 1$. In analyzing the system, we shall assume that flow is Newtonian, so that standard flow-pressure drop relations may be applied.

We start by considering a mass balance over the difference element shown in Figure 2.4B. Axial flow Q_a enters and leaves the element at position z and $z + \Delta z$, whereas at the same time there is a radial outflow over the distance Δz. Flow rate Q_a can, in principle, be related to pressure drop via the Poiseuille equation derived in Chapter 1, Equation (1.15d). However, owing to the lateral outflow in the extruder, we will no longer be dealing with a constant pressure gradient $\Delta p/\Delta z$ and must instead write:

$$Q_a = -\frac{\pi R^4}{8\mu}\frac{dp}{dz} \tag{2.14a}$$

where dp/dz is now the local pressure gradient.

Flow in the radial direction proceeds through a rectangular slit rather than a circular pipe and calls for a modification of the Poiseuille equation. Viscosity and pressure drop remain in place, but the radial term R^4 has to be replaced by the product $H^3\Delta z$, where H is now the height of the slit. The full expression is given by:

$$(Q_r)_{avg} = \frac{H^3\Delta z}{12\mu}\frac{(\Delta p)_{avg}}{L_s} \tag{2.14b}$$

where Δp and Q_r are average values taken over the increment Δz. For a derivation of this expression, see Illustration 4.7.

We are now in a position to compose a mass balance for that increment and write:

Rate of flow in − Rate of flow out = 0

$$Q_a\Big|_z - \left[(Q_a\Big|_{z+\Delta z} + (Q_r)_{avg}\right] = 0$$

$$\frac{\pi R^4}{8\mu}\Delta\frac{dp}{dz} - \frac{H^3\Delta z}{12\mu}\frac{(\Delta p_r)_{avg}}{L_s} = 0 \tag{2.14c}$$

Dividing by Δz and going to the limit yields the second-order ODE:

$$\frac{d^2 p}{dx^2} - m^2 p = 0 \tag{2.14d}$$

where $m^2 = \dfrac{2}{3}\dfrac{H^3}{\pi R^4 L_s}$ and p_{ext} has been set = 0.

Solution of this equation by the D-operator method yields:

$$p = C_1 \sinh mx + C_2 \cosh mx$$

where we use the hyperbolic rather than exponential form for later convenience. Two boundary conditions are required, which are as follows:

$$\text{BC 1} \qquad \left.\frac{dp}{dz}\right|_{z=L_p} = 0 \quad \textit{(No flow at sealed end)} \tag{2.14e}$$

$$\text{BC 2} \qquad p(0) = p^0 \ \textit{(Inlet pressure above atmospheric)}$$

Using the derivatives of hyperbolic functions listed in Table 2.6, we obtain from BC 1,

$$C_1 = -\,C_2 \tanh\,(\mathrm{mL_p}) \tag{2.14f}$$

and from BC 2

$$C_2 = p^0$$

The resulting axial pressure profile is given by the relation

$$p = p^0[(\cosh mx - \tanh\,(mL_p)\sinh\,(mx)] \tag{2.14g}$$

A quick look at the uniformity index is warranted. We have

$$E = \frac{Q_r(L_p)}{Q_r(0)} = \frac{p(L_p)}{p(0)}$$

and, hence,

$$E = \cosh(mL_p) - \frac{\sinh^2(mL_p)}{\cosh(mL_p)} \tag{2.14h}$$

Using the hyperbolic relations of Items 19 and 2 of Table 2.6, this reduces to the simple expression:

$$E = \operatorname{sech}(mL_p) = \operatorname{sech}\left(\frac{2H^3L_p^{\ 2}}{3\pi L_s R^4}\right) \tag{2.14i}$$

$\mathrm{mL_p}$ is usually much less than one, so that one can use a truncated series expansion of the hyperbolic secant found in standard mathematical handbooks:

$$E = \operatorname{sech}(mL_p) \cong 1 - \frac{\left(mL_p\right)^2}{2} = 1 - \frac{H^3L_p^{\ 2}}{3\pi L_s R^4} \tag{2.14j}$$

To obtain a sense of parameter sensitivity, suppose that for a given configuration E was found to be 0.95, i.e., $H^3L_p^2/3\pi L_sR^4 = 0.05$. It is now proposed to double the sheet thickness H. How will this affect the sheet uniformity? We find $H^3_{new}/H^3_{old} = 8$, which translates into a new index value of $E_{new} = 0.40$. Thus, E has dropped from an acceptable value of 95% to a low and usually unacceptable level of 40%.

Comments: The first impression one gains from the formulation of the problem is that it calls for a PDE model. Velocities vary in a complex way both radially and axially, and it is possible that an angular component has to be contended with as well. The geometry is a discontinuous one, leading to discontinuous boundary conditions that add to the complexity of the problem. We are, thus, dealing with a fairly difficult application of the viscous flow equations. Non-Newtonian behavior would further aggravate the situation.

The principal tool in sidestepping these difficulties was the tacit assumption that the opening width of the lip is small in comparison to the circumference of the pipe and, consequently, the normal parabolic velocity profile remains essentially undisturbed. This is a reasonable simplification in view of the small thickness of normal polymer sheets. Its consequences, however, are quite considerable, because we can now lump the radial flow into the axial mass balance as a "rate-out" term that is determined solely by the local radial pressure drop and the geometry of the slit. Thus, we have reduced the number of state variables from four (three velocities and pressure) to only one, pressure, and the number of independent variables from three to one, the axial distance. Note that the model can easily accommodate non-Newtonian flow, but this requires the use of an appropriate non-Newtonian $Q - dP/dz$ relation to replace Poiseuille's law, Equation 2.14a.

There is a parallel to be found here to the catalyst pellet we had analyzed in Illustration 1.4. In both cases, the primary profiles, given here by Equation 2.14g, are converted to criteria that serve to compare actual performance to an established ideal. For the pellet, the criterion was the effectiveness that conveyed a sense of the degree of nonuniformity in reactant concentration. In a similar way, the uniformity index in polymer extrusion establishes the degree of nonuniformity of tubular pressure and, hence, sheet thickness.

The reason for choosing the hyperbolic form of solution of the ODE becomes apparent when we reach Equation 2.14i. The uniformity index E can now be expressed in compact form as the hyperbolic secant of a single dimensionless group $H^3L_p^2/3\pi L_sR^4$. Never content to stop simplifying, we reached back to elementary calculus, an often neglected area, to expand the hyperbolic secant into the simple truncated algebraic Equation 2.14j. This equation related the uniformity index in revealing fashion to the geometry of the system. Of particular note is the dependence of E on H^3/R^4, which indicates that any adverse effect on E caused by an increase in polymer sheet thickness H can easily be compensated for by an increase in pipe radius R. An increase in slit length L_s can be used to similar good effect. Thus, the simple Equation 2.14j manages to illuminate the entire problem and enables us to quickly address important design questions.

Illustration 2.6 Another Second-Order Linear ODE: Euler's Formula for the Buckling of a Strut

In Figure 2.2, we had briefly indicated the behavior of two classical structures, the cantilever and the strut, in response to an applied load. We now amplify the type further.

The principal difference between the two cases is the following. Application of the end load to a cantilever causes it to bend, the deflection δ increasing gradually with an increase in load (Figure 2.2). Ultimately, if the load is raised to the level of the tensile strength of the material, the beam will rupture or crack, but this happens only after it has gone through a gradual process of increasing deflection. In the case of the strut, there is at first no lateral movement at all with an increase in load, until a point is reached at which the resistance to bending has become so small that the slightest increase causes the strut to suddenly buckle. This is often termed a *catastrophic failure* or *instability*. Note again that the failure is sudden and occurs at a particular critical load F_C given by Euler's celebrated formula. It is the derivation of this formula we wish to address here. The gradual bending of a cantilever, which makes use of the same basic formulae but under different conditions (lateral rather than axial application of the load), is taken up in Practice Problem 2.7.

We start with Equation 2.9 and set M = –Fy, where the minus sign follows from the fact that at equilibrium the restoring moment (left side of Equation 2.9) must be equal and opposite to the externally applied moment. We obtain:

$$EI\frac{d^2y}{dx^2} = -Fy \tag{2.15a}$$

which has the solution (see Table 2.5):

$$y = A\cos\alpha x + B\sin\alpha x \tag{2.15b}$$

with the boundary conditions set by the requirement of zero deflection at both ends. Thus,

$$y(0) = y(L) = 0 \tag{2.15c}$$

Introduction of Equation 2.15c into Equation 2.15b leads to the following expressions for the integration constants:

$$A = 0 \text{ and } B\sin\alpha L = 0 \tag{2.15d}$$

Because B cannot be zero, we must have:

$$\sin\alpha L = 0 \quad \text{and} \quad \alpha L = 0, \pi, 2\pi, \ldots, n\pi \tag{2.15e}$$

where:

$$\alpha = (F/EI)^{1/2} \tag{2.15f}$$

From Equation 2.15e and Equation 2.15f, it follows that the relation between load F and the system properties E, I, and L is given by:

$$F = \frac{n^2\pi^2 EI}{L^2} \tag{2.15g}$$

Because we are only interested in the smallest load F_b that will cause buckling, we set n = 1 and obtain:

$$F_b = \frac{\pi^2 EI}{L^2} \tag{2.15h}$$

This is the buckling expression formulated by Euler.

Comments: There are two ingenious moves to be noted in this development. First, Euler modeled the system after buckling had occurred, rather than analyzing the onset of buckling, which is more complex and less fruitful. Having done so, and being confronted with an infinite set of solutions, he resisted the temptation to declare the model invalid and chose the solution corresponding to the lowest load. This too required a leap in thought.

The boundary conditions to be applied in the solution of Equation 4.19 turn out to be of importance. It was already pointed out in Figure 2.2 that beams can be mounted in different ways: clamped, hinged, or with one end free. For a strut that is clamped at both ends, for example, the buckling load turns out to be four times that of a hinged beam, a considerable difference in values.

2.3.3 Nonhomogeneous Linear Second-Order ODEs with Constant Coefficients

We consider here systems of the form

$$L(y) = ay'' + by' + cy = f(x) \tag{2.16}$$

where a, b, and c are constants, and f(x) is the nonhomogeneous term. It can be shown by the superposition principle that the solution will in this case be made up of the sum of the solution of the homogeneous form of Equation 2.16, termed the *complementary function*, and a particular integral y_p that has the same functional form as the nonhomogeneous term f(x). Thus,

General Solution = Complementary Solution + Particular Integral (2.17)

Methods for the evaluation of the complementary solution were given in Subsection 2.3.2. Evaluation of the particular integral is done by the so-called *method of undetermined coefficients*. It consists of substituting the known form of y_p, which

TABLE 2.7
Particular Integrals y_p of the Second-Order ODE $ay'' + by' + cy = f(x)$

f(x)	Form of y_p	Coefficients of y_p
a_0	K	$K = a_0/c$
$a_0 + a_1x + \ldots a_nx^n$	$A_0 + A_1x + \ldots A_nx^n$	Determined by substituting into ODE and equating coefficients
a_0e^{rx}	A_0e^{rx}	$A_0 = \frac{a_o}{ar^2 + br + c}$
$a_0 \sin nx$	$A_0 \sin nx$	$A_0 = \frac{(c - n^2a)a_o}{(c - n^2a)^2 + n^2b^2}$
$a_0 \cos nx$	$A_0 \cos nx$	$A_0 = \frac{(c - n^2a)a_o + nb}{(c - n^2a)^2 + n^2b^2}$
$a_0 \sin nx + b_0 \cos nx$	$A_0 \sin nx + B_0 \cos nx$	$A_0 = \frac{(c - n^2a)a_o + nbb_o}{(c - n^2a)^2 + n^2b^2}$
		$B_0 = \frac{(c - n^2a)b_o - nb\,a_o}{(c - n^2a)^2 + n^2b^2}$ o

is identical to that of f(x), into the ODE (Equation 2.16) and evaluating the undetermined coefficients by setting the sum of coefficients of a particular function equal to zero. This method, which also finds use in series solutions taken up in the next section, will be discussed in more detail there. For our present purposes, we content ourselves with a listing of the most frequently required particular integrals, shown in Table 2.7. These can be used directly in the formulation (Equation 2.17) to arrive at a general solution of the nonhomogeneous ODE. We demonstrate its application in the following example.

Illustration 2.7 Vibrating Spring with a Forcing Function

Vibrating systems give rise to a host of interesting solutions. We consider only the simplest of these, that of a mass suspended from a spring and vibrating under its own weight. A full analysis of such systems is deferred to the next section dealing with the Laplace transformation, which is the preferred method of solution in these cases.

Applying Newton's law to the system, we obtain, in the first instance:

$$\sum Forces = m\frac{d^2x}{dt^2}$$

or

$$F_s - F_g = m\frac{d^2x}{dt^2} \tag{2.18a}$$

where F_g and F_s are the gravity force and the restoring force of the spring, respectively. The latter varies directly with the extension x, and is expressed for linear behavior by Hooke's law

$$F_s = kx \tag{2.18b}$$

Equation 2.18a then becomes, after substitution and rearrangement:

$$\frac{d^2x}{dt^2} + (k/m)x = g \tag{2.18c}$$

where the gravitational constant g is the nonhomogeneous term.

Although other solution methods for this equation exist, including the Laplace transformation, we shall use the example to demonstrate the use of the particular integral and its superposition on the complementary solution. The latter is obtained by the D-operator method. We write:

$$(D^2 + k/ms)x = 0 \tag{2.18d}$$

which has the characteristic roots:

$$D_{1,2} = \pm\, i(k/m)^{1/2} = \pm\, bi \tag{2.18e}$$

so that the complementary solution becomes (see Table 2.5)

$$x_c = C_1 \sin bt + C_2 \cos bt \tag{2.18f}$$

The particular integral y_p is established with the aid of Table 2.7, yielding:

$$y_p = g/(k/m) \tag{2.18g}$$

Hence, the general solution is given by:

$$x = C_1 \sin bt + C_2 \cos bt + mg/k \tag{2.18h}$$

We now introduce the boundary conditions (initial conditions here) by specifying that the mass is initially extended to a position x_o and that the velocity at t = 0 is zero. Note that these conditions have to be applied to the whole of Equation 2.18h, and not just the complementary solution (Equation 2.18f). We obtain:

From IC 1 $$C_1 = 0 \tag{2.18i}$$

From IC 2 $$C_2 = x_o - mg/k$$

The general solution then takes the form:

$$x = (x_o - mg/k) \cos [(k/m)^{1/2}t] + mg/k \tag{2.18j}$$

The equation reveals that the response of the system to the forcing function g is an oscillatory one with amplitude (x_o– mg/k) and frequency $(k/m)^{1/2}$ in units of sec^1. The oscillations persist indefinitely without any decay in the amplitude. Time-dependent amplitudes arise when the forcing function is itself time dependent or when friction exercises a dampening effect. These cases that arise in the classical analysis of second-order linear systems are taken up in greater detail in the section dealing with Laplace transformation.

2.3.4 Series Solutions of Linear ODEs with Variable Coefficients

When the coefficients of a linear ODE themselves become functions of the independent variable, the D-operator method can no longer be applied. One must turn to alternative methods, which have led to the development of *series solutions*. The series solutions belong to a wider class of solution techniques in which a specific form of the solution is guessed or assumed, for example, y = a sin x + b cos x. The unknown coefficients a and b here are evaluated by substituting the solution into the ODE and setting the coefficients of like terms equal to zero. This procedure is known as the *method of undetermined coefficients*, and was encountered briefly in connection with particular integrals.

The solution form we shall assume here is a power series in x, i.e.,

$$y = a_o + a_1x^{1+k} + a_2x^{2+k} + \ldots \tag{2.19a}$$

This is not an unreasonable guess to make, because the solution to the constant coefficient case can also be expressed in terms of power series in x, for example:

$$e^{ax} = 1 + \frac{ax}{1!} + \frac{(ax)^2}{2!} + \ldots \tag{2.19b}$$

It can then be argued that variable coefficients, particularly those of a polynomial form, will merely alter the coefficients and the exponents of x but will not otherwise deviate from the power series forms. To allow for this effect, we have included an undetermined parameter k in the exponent, which will be a function of the variable coefficients contained in the ODE.

We start by demonstrating these concepts and their validity with a simple example.

Illustration 2.8 Solution of a Linear ODE with Constant Coefficients by a Power Series Expansion

We consider the first-order ODE:

$$y' + y = 0 \tag{2.20a}$$

and assume a series solution of the form:

$$y = a_o + a_1x + a_2x^2 + a_3x^3 + \ldots \tag{2.20b}$$

Note that a solution can also be arrived at by separation of variables, which we can use to validate the series solution.

Substitution of Equation 2.20b into Equation 2.20a yields:

$$(a_o + a_1) + (2a_2 + a_1)x + \ldots = 0 \tag{2.20c}$$

We now proceed to evaluate a_o, a_1, and a_2 by the method of undetermined coefficients, setting the aggregate coefficient of each power of x equal to zero. This is justified by the fact that the series expansion (Equation 2.20b) must equal zero for any arbitrary value of x. We obtain:

$$a_1 = -a_o \tag{2.20d}$$

$$a_2 = -\frac{1}{2}a_1 = \frac{1}{2}a_o$$

and the series solution becomes:

$$y = a_o\left(1 - x + \frac{1}{2}x^2 - \ldots\right) \tag{2.20e}$$

where a_o will evidently play the role of an integration constant.

Let us compare this with the solution obtained by separation of variables that has the exponential form and associated series expansion:

$$y = C\ \exp(-x) = C\left(1 - x + \frac{x^2}{2} - \ldots\right) \tag{2.20f}$$

This is proof, at least for the initial three terms, of the validity of the series solution.

We now turn to the more general case of a second-order ODE with variable coefficients:

$$a(x)y'' + b(x)y' + c(x)y = 0 \tag{2.20g}$$

where the coefficients are assumed to be of polynomial form:

$$a(x) = a_o + a_1x + a_2x^2 \ldots$$

$$b(x) = b_o + b_1x + b_2x^2 \ldots \tag{2.20h}$$

$$c(x) = c_o + c_1x + c_2x^2 \ldots$$

For this case, it can be shown that the solution takes the form:

$$y = C_1 \sum_{n=0}^{\infty} a_n x^{n+k_1} + C_2 \sum_{n=0}^{\infty} b_n x^{n+k_2} \tag{2.20i}$$

Evaluation of the coefficients is cumbersome, but can be accelerated by the so-called Method of Frobenius, which is described in standard texts on ODEs. We shall not go into the details of this procedure but will examine instead the functions that arise in this solution. The following points are of note.

- The functions that result from the series solution of linear second-order ODEs with polynomial coefficients are either finite polynomials or infinite power series in x.
- A series of new functions arise as a result, which are denoted by the name of the associated ODEs, in particular: Bessel functions, Legendre polynomials, Laguerre polynomials, and Hermite polynomials. Other classes of functions include Chebyshev polynomials and hypergeometric functions.
- These functions do not differ in their general properties from the classical exponential, circular, or hyperbolic functions. They are usually exponential or periodic in behavior, can be differentiated or integrated, and are either bounded or unbounded at the origin and at infinity.
- Extensive tabulations for most functions appear in various mathematical handbooks (see References). The reassuring fact emerges that for each value of x, a corresponding value of the function can be looked up or deduced from certain relations (recursion formulae). In this and other respects, these seemingly exotic functions with forbidding German, French, and Russian names are no different from their more conventional counterparts.

We focus our attention here on Bessel functions that arise in conduction and diffusion in circular and cylindrical geometries, particularly at the PDE level. To acquaint the reader with their general behavior, we list in Table 2.8 some of their more important properties, including their integrals and derivatives that are used to derive expressions for diffusional flux from the primary profiles. The list of derivatives also contains the so-called recursion formulae, i.e., relations between Bessel functions of different orders. We note in this connection that the order of a Bessel function, denoted by a subscript, is related to and determined by the form of the variable coefficients of the ODE, and resides in the exponent k of the series expansion (Equation 2.20i). Both fractional and integer orders can arise, each order being associated with a distinct function. Thus, a zero-order Bessel function is not identical to a first-order Bessel function, but may be similar in form, i.e., periodic or exponential. The recursion formulas serve to interrelate them.

The four Bessel functions listed in Table 2.8 are infinite power series that give rise to both periodic and exponential behavior. This is demonstrated for zero-order

TABLE 2.8
Properties of Bessel Functions

A. Types and Designation

Symbol		Designation
$J_k(x)$		Bessel function of the first kind and order k
$Y_k(x)$		Bessel function of the second kind and order k
$I_k(x)$	Modified	Bessel function of the first kind and order k
$K_k(x)$	Modified	Bessel function of the second kind and order k

B. Functional Form

Bessel Function	Functional Form
$J_k(x)$	Damped periodic
$Y_n(x)$	Damped periodic
$I_k(x)$	Exponential
$K_n(x)$	Exponential
In particular:	
$J_{1/2}(x)$	$(2/\pi x)^{1/2} \sin x$
$J_{-1/2}(x)$	$(2/\pi x)^{1/2} \cos x$
$I_{1/2}(x)$	$(2/\pi x)^{1/2} \sinh x$
$I_{-1/2}(x)$	$(2/\pi x)^{1/2} \cosh x$

C. Values of Various Functions at x = 0 and x = ∞

	x = 0	x = ∞
$J_k(x)$	0[a]	0
$I_k(x)$	0[a]	∞
$Y_k(x)$	−∞	0
$K_k(x)$	∞	0
sin (x)	0	—
cos (x)	1	—
sinh (x)	0	∞
cosh (x)	1	∞

D. Derivatives of Bessel Functions and Recursion Formulae

$$x\frac{d}{dx}J_k(\alpha x) = kJ_k(\alpha x) - \alpha x J_{k+1}(\alpha x)$$

$$= \alpha x J_{k1}(\alpha x) - kJ_k(\alpha x)$$

$$x\frac{d}{dx}Y_k(\alpha x) = kY_k(\alpha x) - \alpha x Y_{k+1}(\alpha x)$$

$$= \alpha x Y_{k-1}(\alpha x) - kY_k(\alpha x)$$

(continued)

TABLE 2.8 (Continued)
Properties of Bessel Functions

D. Derivatives of Bessel Functions and Recursion Formulae
$x\frac{d}{dx}I_k(\alpha x) = kI_k(\alpha x) + \alpha x I_{k+1}(\alpha x)$
$= \alpha x I_{k-1}(\alpha x) - kI_k(\alpha x)$
$x\frac{d}{dx}K_k(\alpha x) = kK_k(\alpha x) - \alpha x K_{k+1}(\alpha x)$
$= -\alpha x K_{k-1}(\alpha x) - kK_k(\alpha x)$

E. Integrals of Bessel Functions
$\alpha\int x^k J_{k-1}(\alpha x)dx = x^k J_k(\alpha x) + C$
$\alpha\int x^k Y_{k-1}(\alpha x)dx = x^k Y_k(\alpha x) + C$
$\alpha\int x^k I_{k-1}(\alpha x)dx = x^k I_k(\alpha x) + C$
$\alpha\int x^k K_{k-1}(\alpha x)dx = -x^k K_k(\alpha x) + C$

F. Values of Bessel Functions for Small Arguments
$J_n = I_n \cong \left(\frac{x}{2}\right)^n$

[a] However, note the special cases $J_0(0) = I_0(0) = 1$ and $J_k(0) = I_k(0) = \pm\infty$.

Bessel functions (k = 0) in the plots shown in Figure 2.5. One notes that the modified Bessel function $K_0(x)$ and $Y_0(x)$ are unbounded at the origin. This rules out their use in domains that include the origin of a radial geometry.

We now present three illustrations involving the derivation and use of such Bessel functions, as well as an example (Illustration 2.10) in which a series solution and Bessel functions can be avoided by reformulating the ODE into a separable form.

Illustration 2.9 Evaluation of a Bessel Function

Let us consider the evaluation of a second-order Bessel function of the first kind at a value of x = 5, i.e., we wish to determine the value of $I_2(5)$. We note that tabulations in handbooks usually list values only for zero-th and first-order functions. To obtain

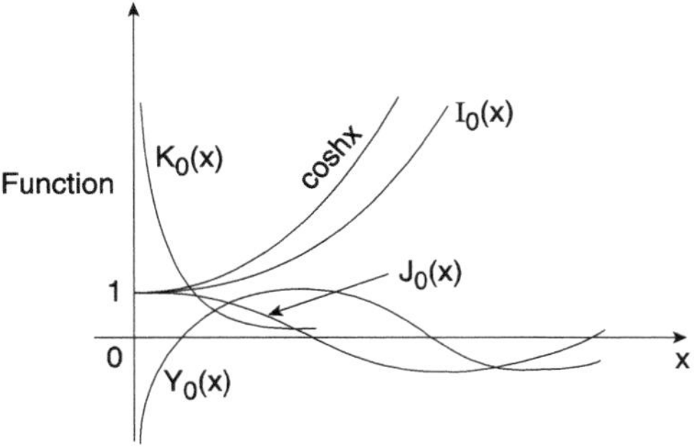

FIGURE 2.5 Graphical representation of zero-order Bessel functions. $J_0(x)$ and $Y_0(x)$ are Bessel functions of the first and second type; $I_0(x)$ and $K_0(x)$ are *modified* Bessel functions of the first and second type.

values for higher-order Bessel functions, use must be made of the recursion formulae listed under Table 2.8. Setting k = 1, we obtain:

$$I_2(x) = I_0(x) - \frac{2}{x} I_1(x)$$

or

$$I_2(5) = I_0(5) - \frac{2}{5} I_1(5)$$

Tabulations for $I_0(5)$ and $I_1(5)$ give values of 27.24 and 24.34, respectively. There results:

$$I_2(5) = 27.24 - \frac{2}{5} 24.34 = 17.50$$

Illustration 2.10 Second-Order ODEs with Variable Coefficients: Solution by Separation of Variables

The most common source of variable coefficient ODEs are diffusional processes in which the traversed area varies in the direction of the flux. Among the many geometries that fit this description, the sphere and the cylinder or circular disk are the most obvious and most frequently encountered examples. The processes in question are those of conduction and diffusion with sources described by Fourier's and Fick's law. What we wish to demonstrate here is that series solutions (and Bessel functions) can often be avoided by proper structuring of the ODE.

The difference element over which the relevant balances have to be made, and the flux entering and leaving that element, are shown in Figure 2.6. S represents the production or consumption of heat or mass because of chemical

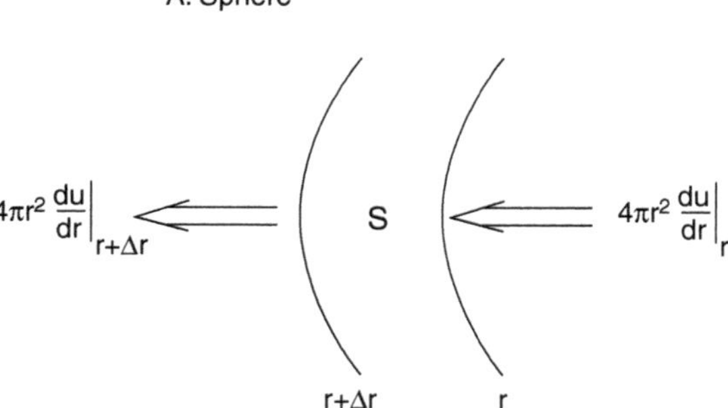

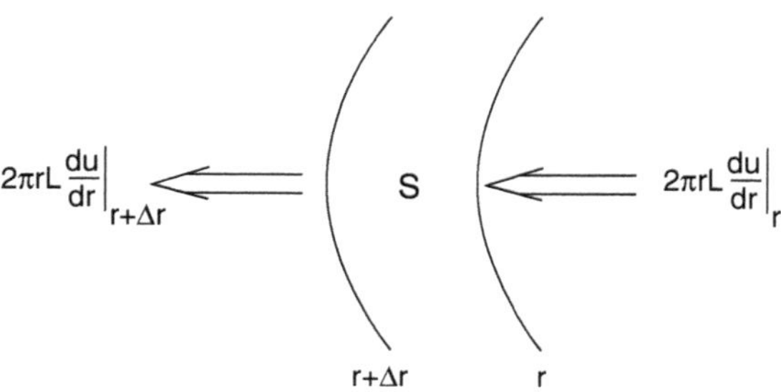

FIGURE 2.6 Genesis of second-order ODEs with variable coefficients: (A) Radial flux in a sphere. (B) Radial flux in a cylinder or circular disk. S = Source or sink term.

or nuclear reactions or brought about by internally or externally induced electrical heating.

Consider as an example the radial conduction through a cylinder, with heat being produced at the rate S (J/sec m^3). The ends of the cylinder are assumed to be insulated. We have:

$$\text{Rate of heat in} - \text{Rate of heat out} = 0$$

$$\begin{bmatrix} -k2\pi rL \left.\dfrac{dT}{dr}\right|_r \\ +S2\pi rL\Delta r \end{bmatrix} - \left[-k2\pi rL \left.\dfrac{dT}{dr}\right|_{r+\Delta r} \right] = 0 \tag{2.21a}$$

Dividing by $2\pi L\Delta r$ and letting $\Delta r \rightarrow 0$ yields:

$$-k\left(\frac{d}{dr} r \frac{dT}{dr}\right) = -k\left[r\frac{d^2T}{dr^2} + \frac{dT}{dr}\right] = Sr \tag{2.21b}$$

For the sphere, the corresponding expression is given by:

$$-k\left(\frac{d}{dr}r^2\frac{dT}{dr}\right) = Sr^2 \tag{2.21c}$$

If the source term S is constant, i.e., uniformly distributed throughout the cylinder and independent of temperature, integration by separation of variables becomes possible, provided we use the compact rather than the expanded forms of Equation 2.21b and Equation 2.21c. A first integration of Equation 2.21b, for example, would then yield:

$$-kr\frac{dT}{dr} = \frac{Sr^2}{2} + C_1 \tag{2.21d}$$

where $C_1 = 0$ by virtue of the symmetry condition $(dT/dr)_{r=0} = 0$. A second integration leads directly to the final form:

$$-kT = \frac{Sr^2}{4} + C_2 \tag{2.21e}$$

where C_2 is evaluated from the surface boundary condition T(R). If, for example, we assume a BC of Type I:

$$T(R) = T_s \tag{2.21f}$$

we obtain the following radial temperature profile for the cylinder:

$$T - T_s = \frac{SR^2}{4k}\left[1 - \left(\frac{r}{R}\right)^2\right] \tag{2.21g}$$

For an interesting application of this equation, see Illustration 4.24.

Comments:

- The success of the procedure depended entirely on the source term S being constant throughout the cylinder, and on the terse form of the ODE. If S is distributed, i.e., depends on radial distance S(r), integration by separation of variables is still possible. A temperature-dependent source presents a more complex problem. A linear variation in T can still be accommodated by analytical means, but requires a power series solution and gives rise to Bessel functions. Illustration 2.21 will address this problem in the context of diffusion of a reactant undergoing a first-order reaction in a cylindrical catalyst pellet.
- The temperature distribution Equation 2.21f, which is a parabolic one, is not often an item of importance. Of greater interest to the analyst is the

heat given off to the surroundings. This quantity follows immediately from Equation 2.21c. With $C_1 = 0$, we have:

$$q/A = -k\frac{dT}{dr}\bigg|_R = \frac{SR}{2} \tag{2.21g}$$

Note that this result is independent of the boundary condition prevailing at the surface. The same result is obtained in the presented of a surface film resistance (BC of Type III).

Illustration 2.11 Second-Order ODEs with Variable Coefficients: Solution by Power Series

Procedures for obtaining power series solutions of ODEs are fairly lengthy and cumbersome. Fortunately, these procedures may be avoided in most problems by making use of the following generalized formula. The accompanying illustration will demonstrate its use:

The differential equation

$$x^2\frac{d^2y}{dx^2} + x(a+2bx^r)\frac{dy}{dx} + [c + dx^{2s} - b(1-a-r)x^r + b^2x^{2r}]y = 0 \tag{2.22a}$$

has the generalized solution:

$$y = x^{(1-a)/2} \exp(-bx^r/r)\,[C_1Z_k(d^{1/2}x^s/s) + C_2Z_{-k}(d^{1/2}x^s/s)] \tag{2.22b}$$

where $$k = \frac{1}{s}\left[\left(\frac{1-a}{2}\right)^2 - c\right]^{1/2}.$$

Z_k denotes one of the Bessel functions. If $\sqrt{d/s}$ is real and k is not zero or an integer, Z_k denotes J_k; if k is zero or an integer n, Z_k denotes J_n, Z_{-k} denotes Y_n. If $\sqrt{d/s}$ is imaginary and k is not zero or an integer, Z_k denotes I_k, Z_{-k} denotes I_{-k}; I and K then assume real arguments. If k is zero or an integer n, Z_k denotes I_n, and Z_{-k} denotes K_n.

Suppose the solution of the following ODE is to be established, using Equation 2.22a and Equation 2.22b:

$$x^2\frac{d^2y}{dx^2} + x(1-2\beta)\frac{dy}{dx} + \beta^2x^{2\beta}y = 0 \tag{2.22c}$$

When comparing with the generalized equation given previously, the following will be observed:

1. $1 - 2\beta$ will equal $a + 2bx^r$ if $b = 0$ and $a = 1 - 2\beta$.
2. $\beta^2x^{2\beta}$ will equal $c + dx^{2s} - b(1 - a - r)x^r + b^2x^{2r}$ if the preceding condition is granted and $c = 0$, $d = \beta^2$, and $s = \beta$.

Consequently, $a = 1 - 2\beta, b = 0, c = 0, d = \beta^2, s = \beta$. Then $k = (1/\beta)\sqrt{\beta^2} = 1$, $\sqrt{d}/s = 1$. The solution is then:

$$y = x^{\beta}[C_1 J_1(x^{\beta}) + C_2 Y_1(x^{\beta})] \tag{2.22d}$$

Illustration 2.12 Concentration Profile and Effectiveness Factor of a Cylindrical Catalyst Pellet

We return here to the problem of diffusion and reaction in a catalyst particle. This time the process is assumed to take place in a cylindrical pellet of sufficient length L that only radial diffusion needs to be considered, i.e., the flux through the end of the cylinder is neglected. The reaction is assumed to be first order. A mass balance over a radial difference element then yields the expression:

Rate of reactant in − Rate of reactant out = 0

$$D_{eff} 2\pi r L \left.\frac{dC}{dr}\right|_{r+\Delta r} - \left[\begin{array}{c} D_{eff} 2\pi r L \left.\frac{dC_A}{dr}\right|_r \\ + k_r (C_A)_{avg} 2\pi r L \Delta r \end{array} \right] = 0 \tag{2.23a}$$

Upon dividing by $2\pi L\Delta r$ and letting $\Delta r \to 0$ we obtain

$$\frac{d}{dr}\left(r\frac{dC}{dr} \right) - (k_r/D_{eff}) r C_A = 0 \tag{2.23b}$$

or equivalently,

$$r^2 \frac{d^2 C_A}{dr^2} + r\frac{dC_A}{dr} - \alpha^2 r^2 C_A = 0 \tag{2.23c}$$

where $\alpha^2 = k_r/D_{eff}$.

Neither of these equations can be solved by separation of variables, and we must instead resort to a power series solution.

Comparison with the generalized solution of Illustration 2.11 yields the following parameter values:

$$a = 1, \quad b = c = 0, \quad d = \alpha^2, \quad s = 1$$

The solution is then immediately given as

$$C = A\, I_0(\alpha r) + B\, K_0(\alpha r) \tag{2.23d}$$

with BC 1: $\frac{dC}{dr}(0) = 0$ (symmetry) or $C(0)$ = finite

BC 2: $C(R) = C_s$ (surface concentration of reactant)

Because $K_0(0) = \infty$ (Table 2.8C), it follows from BC 1 that the integration constant B must be zero. Together with BC 2, this yields the final result:

$$\frac{C}{C_s} = \frac{I_0(\alpha r)}{I_0(\alpha R)} \tag{2.23e}$$

The plot in Figure 2.5 shows that $I_0(r)$ increases exponentially with r and has a zero slope at the origin. This is in qualitative agreement with the concentration profile one would expect to see.

The effectiveness factor E can be determined from the expression:

$$E = \frac{\int_0^R k_r C_s 2\pi r L I_0(\alpha r)dr}{I_0(\alpha R)k_r C_s \pi R^2 L} \tag{2.23f}$$

Use of the Table 2.8E of integrals of Bessel functions then yields, with k set = 1:

$$\int_0^R r I_0(\alpha r)dr = (r/\alpha)I_1(\alpha r)\Big|_0^R = (R/\alpha)I_1(\alpha R) \tag{2.23g}$$

and the final effectiveness factor becomes, after cancellation of terms:

$$E = \frac{2I_1(\alpha R)}{\alpha R I_0(\alpha R)} \tag{2.23h}$$

Comments:

- The appearance of Bessel functions in the final expression (Equation 2.23h) need not deter us from evaluating E. As noted previously, convenient tabulations of zero and first-order Bessel functions are available in handbooks of mathematical functions.
- For small values of the argument αR, the effectiveness factor E should approach unity. It can be shown that this is indeed the case. Using the tabulations of Table 2.8 (item F), we find $I_1(\alpha R) \rightarrow (\alpha R/2)$ and $I_0(\alpha R) \rightarrow 1$, so that $2I_1(\alpha R)/\alpha R I_0(\alpha R) \rightarrow 1$, as required.

2.3.5 Other Methods

Methods in this category, though less sweeping in scope than those discussed in the previous sections, nevertheless find their use in the solution of a host of special problems. These techniques, listed in Table 2.4, are capable of solving both linear and nonlinear first- and second-order ODEs. We mention in particular the use of various transformations and the p-substitution. The following two examples will serve as illustrations.

Illustration 2.13 Product Distributions in Reactions in Series: Use of the Substitution y = vx

The following consecutive reactions give a good representation of many organic reactions, such as the chlorination, nitration, and sulfonation of aromatics:

$$A + B \xrightarrow{k_1} R \tag{2.24a}$$

$$R + B \xrightarrow{k_2} S$$

The calculation of the variation of these species with time or distance (stirred tank or plug flow reactors) is usually achieved numerically, but the important product distributions $R/A_0 = f(A)$ and $S/A_0 = g(A)$ can, for a batch reactor, be obtained in an easy fashion analytically from the rate laws and the stoichiometry of the reactions. If the intermediate R is the desired product, the distributions can be used to calculate optimum conversion of A to achieve a maximum yield in R.

We assume the rate laws to correspond to the stoichiometry of the reaction. The process is assumed to be isothermal and carried out in a batch reactor. We obtain mass balances or rate laws:

$$-k_1 C_A C_B = dC_A/dt \tag{2.24b}$$

$$k_1 C_A C_B - k_2 C_R C_B = dC_R/dt \tag{2.24c}$$

Division of the two equations eliminates dt (our favourite trick!), and we obtain:

$$\frac{dC_R}{dC_A} + \frac{k_1 C_A - k_2 C_R}{k_1 C_A} = 0 \tag{2.24d}$$

We try the substitution $C_R = vC_A$ (item 7a in Table 2.4) and obtain:

$$(v + C_A(dv/dC_A) + (1 - k_2/k_1)v = 0 \tag{2.24e}$$

Thus, the substitution has successfully reduced the ODE to one in v, C_A only, which has the solution (see Table 2.4, item 7a):

$$\ln C_A = C - \int \frac{g(1, C_R/C_A)d(C_R/C_A)}{f(1, C_R/C_A) + (C_B/C_A)g(1, C_R/C_A)}$$

with $g = k_1$ and $f = k_1 - k_2\ (C_R/C_A)$.

Evaluation of the integral leads to the expression

$$C_A = C[1+(1-(k_2/k_1)(C_R/C_A))]^{k_1/(k_2-k_1)} \tag{2.24f}$$

where C = integration constant. It is evaluated from the initial condition $C_R(0) = 0$ and $C_A(0) = C_A{}^0$ and yields:

$$C = C_{AO}$$

The solution then becomes

$$C_A/C_{A0} = [1+(1-(k_2/k_1)(C_R/C_A))]^{k_1/(k_2-k_1)} \tag{2.24g}$$

Alternatively, solving for C_R, one obtains:

$$C_R/C_{A0} = \frac{1}{1-k_2/k_1}[(C_A/C_{A0})^{k_2/k_1} - (C_A/C_{A0})] \tag{2.24h}$$

which is the distribution of the intermediate R.

If R is the desired product, it is useful to know at which point the reaction should be stopped before C_R begins to decline. We find this by maximizing the yield of C_R, i.e., we write, using Equation 2.24h:

$$d(C_R/C_{A0})/d(C_A/C_{A0}) = 0 = \frac{1}{1-k_2/k_1}\left[\frac{k_2}{k_1}\left(\frac{C_A}{C_{A0}}\right)^{(k_2/k_2)-1} - 1\right] \tag{2.24i}$$

and hence,

$$(C_A/C_{A0}) = (k_1/k_2)^{k_1/(k_2-k_1)} \tag{2.24j}$$

This means that to maximize the yield of C_R, the reaction should be stopped when the conversion X of C_A has reached the value:

$$X = 1-(C_A/C_A{}^0) = 1-(k_1/k_2)^{k_1/(k_2-k_1)} \tag{2.24k}$$

Illustration 2.14 Path of Pursuit

In this example, involving geometry as well as the geometrical aspects of calculus, we consider a predator B, moving at a constant speed b, in pursuit of a prey A,

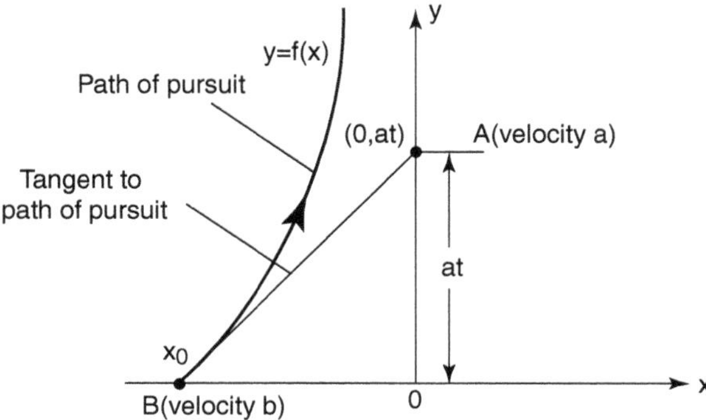

FIGURE 2.7 Path of pursuit.

moving in a straight line at a constant speed a (Figure 2.7). The task will be to find the path of pursuit, y = f(x), which the predator has to follow so as to intercept the prey in minimum time.

To accommodate this situation, we position the prey A on the y-axis, which is also the straight-line path taken by it. The predator B is initially placed on the x-axis, from which it moves away as the pursuit develops. Because the predator will keep the prey in constant sight, a "line of sight" can be drawn connecting the two, which will perforce be tangential to the moving predator and, hence, to the path of pursuit. This is shown in Figure 2.7. Note that the variable position y of the prey has been related to elapsed time and velocity of the prey through the term "at." This is the first of a series of ingenious moves to quantify the problem.

We start the modeling process by writing out the equation for the tangent. This serves to introduce the derivative of the function we ultimately seek. We obtain:

$$\frac{dy}{dx} = \frac{y - y_1}{x - x_1} = \frac{y - at}{x - 0} \tag{2.25a}$$

or, alternatively,

$$xy' - y = -at \tag{2.25b}$$

The task is now to eliminate time t as a variable by expressing it in terms of the system coordinates x, y. This is done in a series of clever steps, which we outline below:

We start by differentiating Equation 2.25b with respect to x. This leads to the result

$$xy'' = -a\frac{dt}{dx} \tag{2.25c}$$

Next, we decompose dt/dx along the arc s of the path, obtaining

$$(dt/dx) = (dt/dx)(ds/dx) \tag{2.25d}$$

This appears to be a retrograde step, because we have merely managed to introduce an additional variable, s. Closer examination shows, however, that the derivatives appearing on the right side of Equation 2.25d can be expressed in terms of the system parameters and coordinates, devoid of either time t or arc length s. We first recognize dt/ds as the inverse of the speed of the predator. Thus:

$$dt/dx = 1/b \tag{2.25e}$$

The second term in Equation 2.25d, ds/dx, will be recognized as the directional derivatives that can be related to dy/dx by the standard formula known from elementary calculus. We have:

$$ds/dx = [1 + (y')^2]^{1/2} \tag{2.25f}$$

We have thus managed, at one and the same time, to eliminate the undesired variables t and s. It remains to introduce the Equation 2.25d and Equation 2.25e into Equation 2.25c. This gives the result:

$$xy'' = -\frac{a}{b}[1+(y')^2]^{1/2}$$

or, equivalently,

$$1+\left(\frac{dy}{dx}\right)^2 = \frac{b^2}{a^2}x^2\left(\frac{d^2y}{dx^2}\right)^2 \tag{2.25g}$$

where the boundary conditions are given by:

$$dy/dx = 0 \quad \text{at } x = x_0 \tag{2.25h}$$

$$y = 0 \quad \text{at } x = x_0 \tag{2.25i}$$

We recognize this as a second-order nonlinear ODE with missing terms in y, which calls for the use of the p-substitution (see item 9 in Table 2.4). After substitution of p and separation of variables, we obtain:

$$\int \frac{kdp}{(1+p^2)^{1/2}} = C_1 \int \frac{dx}{x^2} \tag{2.25j}$$

where the integral on the left is found in mathematical tables in the form:

$$\int \frac{dx}{(a^2+x^2)^{1/2}} = \ln[x+(a^2+x^2)^{1/2}] \tag{2.25k}$$

A first integration is performed, yielding:

$$[p+(1+p^2)^{1/2}]^k = \frac{x}{C_1} \tag{2.25l}$$

which, upon solving for p, gives the result:

$$\frac{dy}{dx} = p = \frac{1}{2}\left[\left(\frac{C_1}{x}\right)^{-1/k} - \left(\frac{x}{C_1}\right)^{-1/k}\right] \tag{2.25m}$$

Introduction of the boundary condition Equation 2.25h yields

$$C_1 = x_0 \tag{2.25n}$$

so that Equation 2.25m now becomes

$$\frac{dy}{dx} = p = \frac{1}{2}\left[\left(\frac{x_0}{x}\right)^{-1/k} - \left(\frac{x}{x_0}\right)^{-1/k}\right] \tag{2.25o}$$

A final integration then leads to the result:

$$y = \frac{1}{2}x_0\left[\frac{k}{1-k}\left(\frac{x}{x_0}\right)^{(k-1)/k} + \frac{k}{k+1}\left(\frac{x}{x_0}\right)^{(k+1)/k}\right] + C_2 \tag{2.25p}$$

The second integration constant C_2 is evaluated using the boundary condition (Equation 2.25i) and gives:

$$C_2 = x_0\frac{k}{k^2-1} \tag{2.25q}$$

The final result then becomes:

$$y = \frac{1}{2}x_0\left[\frac{k}{1-k}\left(\frac{x}{x_0}\right)^{(k-1)/k} + \frac{k}{k+1}\left(\frac{x}{x_0}\right)^{(k+1)/k} + \frac{2k}{k^2-1}\right] \tag{2.25r}$$

Comments:

- One will note the ingenious use of several concepts and relations, drawn mainly from calculus, to arrive at the final model, Equation 2.25g. To many, the procedure will appear to have been circuitous, starting with the expression for the location of the prey in terms of velocity and elapsed time, the latter of which then had to be eliminated by introducing the directional derivatives ds/dx. The temptation here would have been to denote the position of the prey as y, but this would have brought the proceedings to a halt. The next step might have been to try and express prey location in terms of x, but this is clearly not possible because x = 0 along the path. This leaves, as the only alternative, an expression in t, and here it takes a good sense of the tools available from calculus to realize that a conversion of t to x, or dy/dx, is not only possible, but can be accomplished with relative ease.
- The reader will have noted that the path presented here is only one of several alternative trajectories that present themselves. The result obtained from Equation 2.25g is clearly an optimum, because interception is achieved in minimum time by continually adjusting the trajectory. A more straightforward but less efficient method is to use a path of constant slope, i.e., to proceed along a straight line. Finally, interception may be complicated by evasive action taken by the prey. This is clearly a more complex problem, which is not addressed here. The model equations we have established here, however, provide some guidelines for the solution of that more complicated case.

2.4 NONLINEAR ANALYSIS

The term *analysis*, as it is applied here, refers to the methodology of obtaining an understanding of the qualitative nature of the solution and identifying regions of unusual behavior without actually solving the equations, or by solving them only partially. Thus, we do not seek to derive a solution for a particular set of parameter values and boundary conditions, but instead ask ourselves these questions: Under what conditions or set of values will the solution be periodic, and when will it be exponentially decaying? When does it become unstable or unbounded? Can there be more than one solution?

An analytical tool of sorts for linear systems has already been provided by the so-called characteristic roots of the D-operator method described in Section 2.3.2. Although this was not demonstrated explicitly, we were able to predict the form of the solution by a mere inspection of the characteristic roots without actually solving the ODE. Thus, when the roots were real, exponential solutions would result; when the roots were imaginary, they were purely oscillatory; and when they were complex, they became periodic with exponentially rising or decaying amplitudes (see Table 2.5). A more detailed examination of these phenomena and of linear analysis in general is deferred to Section 2.5, which deals with the Laplace transformation.

This transformation is not only a highly useful solution method for linear initial-value ODEs, but it also provides a convenient vehicle for an analysis of such systems.

What we wish to do in this section is to address systems of nonlinear ODEs and the associated subject of nonlinear analysis. This is a vast topic, of which we will address a particular aspect. We hope, nevertheless, to provide the reader with an understanding of the principal tools used in nonlinear analysis and to open the door for a glimpse of the exotic phenomena that can arise in nonlinear systems. These include multiplicities giving rise to catastrophe, bifurcations (including the Hopf bifurcations), period doubling, and chaos. Such phenomena arise only in nonlinear systems and were addressed early on in the development of mathematical analysis. Euler's calculation of the load that will buckle a strut is one of the first examples of catastrophe theory. The full development of the theories of nonlinear phenomena had to await the advent of the computer which aided immensely in the discovery of various forms of exotic nonlinear behavior. We confine ourselves to the presentation of one of the simpler tools, the phase plane analysis.

2.4.1 Phase Plane Analysis: Critical Points

Phase plane analysis refers to the examination of the interrelation of the dependent variables of a system. This interrelation is established through a favorite algebraic trick of ours, the division of two first-order ODEs to eliminate the independent variable, usually time t, resulting in the relation $y_1 = f(y_2)$. Although extensions to sets of equations exist — one then speaks of phase space analysis — the two-equation system is best suited for an illustration of the power of the method and the insight it provides.

We use, as an introduction to the method, the behavior of a pendulum considered to consist of a point mass m fixed to a rigid rod of length l. Both gravitational and frictional forces are involved. The latter is assumed to be proportional to the instantaneous velocity of the mass, or equivalently, the rate of change of the angle of deflection dθ/dt. A force balance then leads to the expression:

$$ml\frac{d^2\theta}{dt^2} \quad + \quad kl\frac{d\theta}{dt} \quad + \quad mg\sin\theta \quad = \quad 0 \tag{2.26a}$$

Acceleration Friction Gravity

or, equivalently,

$$a_o\ddot{\theta} + a_1\dot{\theta} + a_2\sin\theta = 0$$

This is a second-order nonlinear ODE in θ, which can be decomposed into an equivalent set of two first-order ODEs by defining the velocity dθ/dt as a new variable

θ_2, and designating the angle of deflection as θ_1. We obtain the set:

$$\frac{d\theta_1}{dt} = \theta_2 = f_1(\theta_1, \theta_2) \tag{2.26b}$$

and

$$\frac{d\theta_2}{dt} = -\frac{a_1}{a_0}\theta_2 - \frac{a_2}{a_0}\sin\theta_1 = f_2(\theta_1, \theta_2)$$

Division of the two equations eliminates dt and yields the result:

$$\frac{d\theta_2}{d\theta_1} = \frac{a_1\theta_2 + a_2\sin\theta_1}{a_0\theta_2} \tag{2.26c}$$

The solution of this equation, usually done numerically, leads to a family of curves $\theta_2 = f(\theta_1)$, which can be plotted in the phase plane θ_2 vs. θ_1. The curves, a typical example of which is shown in Figure 2.8, are "trajectories" or "pathways" of the pendulum, each point representing its velocity θ_2 at a particular position or deflection θ_1. Each pair of initial conditions $\theta_1(0)$ and $\theta_2(0)$ is associated with a particular curve, leading to an infinite number of trajectories in (θ_1, θ_2) space.

The various points at which the trajectories converge are called "critical points," "equilibrium points," or "stationary points." They correspond to the steady states of the pendulum attained as $t \to \infty$, and can be calculated by setting the time derivatives in Equation 2.26b equal to zero. There is an infinite set of such points that occur at $\sin\theta_1 = 0$, i.e., at $\theta_1 = 0, \pi, 2\pi, \ldots$, as well as negative values of same.

A first set of critical points are those at $-\pi, \pi, 3\pi, \ldots$, etc., which are unstable and are referred to as *saddle points*. They correspond to the condition of the pendulum that has come to rest in a vertically upward position. That position is clearly unstable, because a slight deviation of θ_1 from it will cause the pendulum to resume swinging. The situation is depicted by arrows pointing away from the equilibrium point.

A second set of stationary points are located at $-2\pi, 0, 2\pi, 4\pi$, etc., and carry the designation focus. They correspond to the condition of the pendulum that has come to rest in a vertically downward position. That state is clearly stable, and is indicated by arrows pointing toward it.

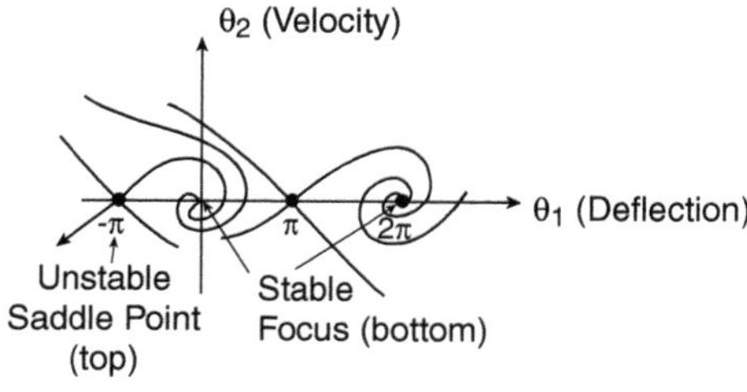

FIGURE 2.8 Phase plane representative of the trajectories of a pendulum with friction.

Stationary points are always calculated by setting time derivatives of the ODEs equal to zero and solving the set of nonlinear algebraic equations that result, given here by:

$$f_1(\theta_1, \theta_2) = 0$$

$$f_2(\theta_1, \theta_2) = 0 \tag{2.27a}$$

A host of other interesting critical points may arise, depending on the form of the ODEs involved. To seek out these points, and to characterize them in relation to the functional form of the model, use is made of the so-called Jacobian J of the set of Equation 2.27a. J is a matrix whose components here are the four partial derivatives of the algebraic set of Equation 2.27a. Thus,

$$\underset{\sim}{J}(\theta_1, \theta_2) = \begin{pmatrix} \partial f_1/\partial\theta_1 & \partial f_2/\partial\theta_1 \\ \partial f_1/\partial\theta_2 & \partial f_2/\partial\theta_2 \end{pmatrix} \tag{2.27b}$$

The order of the Jacobian matrix equals the number of ODEs involved.

Characterization of the critical points is achieved by determining the eigenvalues μ of that matrix, and examining their properties. That is to say we set:

$$\begin{vmatrix} \dfrac{\partial f_1}{\partial\theta_1} - \mu & \dfrac{\partial f_1}{\partial\theta_2} \\ \dfrac{\partial f_2}{\partial\theta_1} & \dfrac{\partial f_2}{\partial\theta_2} - \mu \end{vmatrix} = 0 \tag{2.27c}$$

and examine the roots $\mu_{1,2}$ of the resulting quadratic equation in μ. Using our pendulum as an example, the entire procedure of critical point analysis is then made up of the following steps:

1. Solve the set of Equation 2.27a to establish the coordinates θ_1^e, θ_2^e of the stationary or equilibrium points.
2. Derive expressions for the partial derivative in Equation 2.27c from the set Equation 2.27a.
3. With Step 2 having been established, expand the determinant Equation 2.27c and solve the resulting quadratic in μ. This yields the roots $\mu_{1,2}$ as a function of θ_1 and θ_2.
4. Substitute θ_1^e and θ_2^e obtained in Step 1, into the expressions for $\mu_{1,2}$.
5. Examine the nature of $\mu_{1,2}$ thus obtained.

It is the nature of these roots, i.e., whether they are positive or negative, real or complex, which determines the type of critical points involved. We have summarized

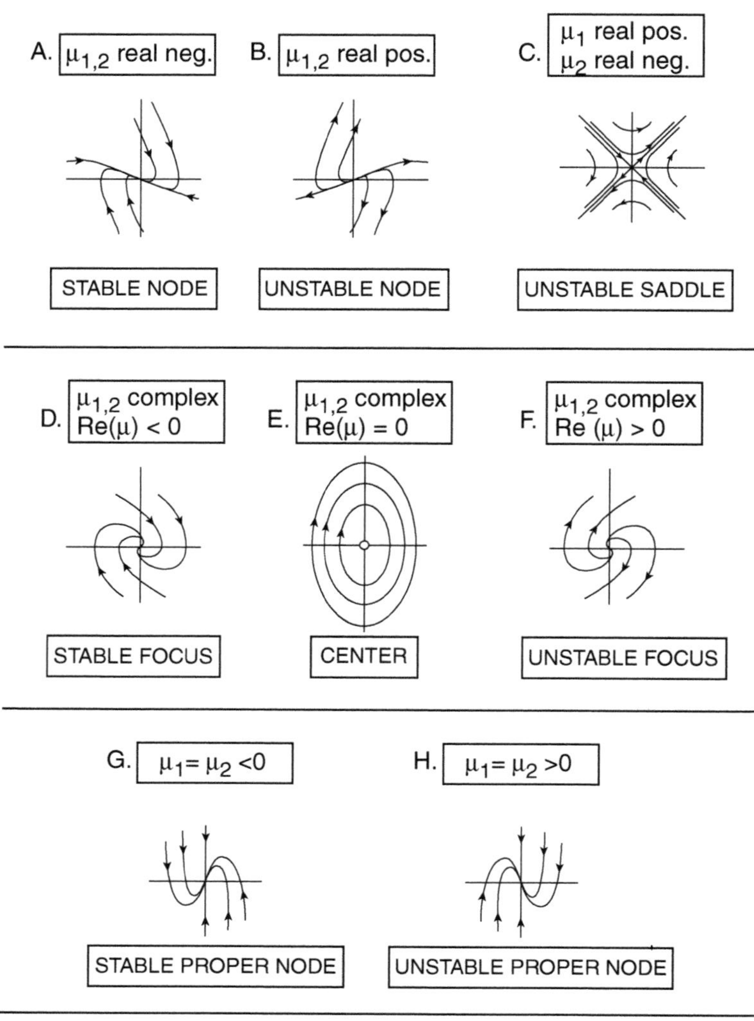

FIGURE 2.9 Phase plane representation of critical points for various eigenvalues $\mu_{1,2}$.

the principal characteristic points in Figure 2.9 that reveal the existence of eight types of such points, depending on the properties of $\mu_{1,2}$. Thus, when $\mu_{1,2}$ are complex with a negative real part Re(μ), the critical point will be a stable focus (Figure 2.9D) with all trajectories converging at that critical point, but none emerging from it. This corresponds to a pendulum coming to rest in the bottom position. When one of the roots is real positive and the other real negative, the critical point will be an unstable saddle (Figure 2.9C), with trajectories leading into it, as well as away from it. This corresponds to a pendulum having come to a precarious rest in the top position. We now proceed to a more complete examination of this device.

Illustration 2.14 Analysis of the Pendulum

The operative equations in the state variables θ_1 (angle of deflection) and θ_2 (time derivative of θ_1) are given by the expression Equation 2.26b. We start by establishing the stationary values of θ_1 and θ_2. To obtain these, we set:

$$f_1(\theta_1, \theta_2) = f_2(\theta_1, \theta_2) = 0 \tag{2.28a}$$

so that

$$\begin{gathered} \theta_2 = 0 \\ \sin\theta_1 = 0 \end{gathered} \tag{2.28b}$$

with roots $\theta_1 = \pm n\pi$, n = 0, 1, 2, ...

Next, we evaluate the components of the Jacobian matrix (Equation 2.27b), obtaining:

$$\begin{aligned} &\partial f_1/\partial\theta_1 = 0 \qquad \partial f_2/\partial\theta_1 = \frac{a_2}{a_0}\cos\theta_1 \\ &\partial f_1/\partial\theta_2 = 0 \qquad \partial f_2/\partial\theta_2 = -\frac{a_1}{a_0} \end{aligned} \tag{2.28c}$$

and from Equation 2.27c

$$\begin{vmatrix} 0-\mu & \dfrac{a_2}{a_0}\cos\theta_1 \\ 1 & -\dfrac{a_1}{a_0}-\mu \end{vmatrix} = 0 \tag{2.28d}$$

Let us consider the two cases of the pendulum with and without friction.

Frictionless pendulum: For this case, we set $a_1 = 0$ and obtain from Equation (2.28d)

$$\mu_{1,2} = \pm\left(-\frac{a_2}{a_0}\cos\theta_1\right)^{1/2} \tag{2.28e}$$

For the stationary points $\theta_1 = 0, \pm 2\pi, \pm 4\pi, \ldots$ (Equation 2.28b), we have $\cos\theta_1 = 1$, and the eigenvalues are pure imaginary. Thus,

$$\mu_{1,2} = \pm\frac{a_2}{a_0}i = \pm bi \tag{2.28f}$$

This corresponds to Case E in Figure 2.9, i.e., the critical point is a center, and the pendulum swings indefinitely with a constant amplitude given by the abscissa intercepts.

The remaining stationary points in Equation 2.28b, i.e., $\theta_1 = \pm\,\pi, \pm\,3\pi, \pm\,5\pi, \ldots$ lead to real values of $\mu_{1,2}$, because we now have $\cos\,\theta_1 = 1$. The characteristic roots become:

$$\mu_{1,2} = \pm\frac{a_2}{a_0} \tag{2.28g}$$

that is, they are real positive and real negative, respectively. This corresponds to Case C in Figure 2.9 and leads to an unstable saddle as the critical point. We recognize this as the situation in which the pendulum points vertically upward.

Pendulum with friction: Here, the expansion of the full determinant, Equation 2.28d, leads to the following quadratic in μ:

$$\mu^2 + \frac{a_1}{a_o}\mu + \frac{a_2}{a_o}\cos\,\theta_1 = 0 \tag{2.28h}$$

with roots

$$\mu_{1,2} = \frac{-a_1/a_0 \pm [(a_1/a_0)^2 - 4(a_2/a_0)\cos\,\theta_1]^{1/2}}{2} \tag{2.28i}$$

One recognizes immediately that for $\cos\,\theta_1 = -1$, one always obtains a real positive and a real negative root, respectively, leading to the same unstable saddle point we have seen previously.

For $\cos\,\theta_1 = 1$, we distinguish three cases, depending on the value of the discriminant (bracketed term in Equation 2.28i).

1. $4a_2/a_0 > (a_1/a_0)^2$: Here, the roots become complex, with a real negative part:

$$\mu_{1,2} = \frac{-a_1/2a_0 \pm i[4\,a_2/a_0 - (a_1/a_0)]^2}{2} \tag{2.28j}$$

   This corresponds to Case D of Figure 2.9, i.e., the critical point is a stable focus, with the pendulum swinging repeatedly with decreasing amplitude until the friction brings it to a stop at the bottom of its trajectory.

2. $4a_2/a_0 = (a_1/a_0)^2$: For this case, we have:

$$\mu_1 = \mu_2 = -a_1/2a_0 < 0 \tag{2.28k}$$

   that is, we are dealing with a stable proper node, Case H of Figure 2.9. The pendulum is critically damped and comes to rest with no oscillations.

3. $4a_2/a_0 < (a_1/a_0)^2$: This condition always yields two real negative roots so that Case A of Figure 2.9 applies. The pendulum is overdamped, and comes to rest with no oscillations.

All of these types of behavior — undamped oscillations, damped and critically damped oscillations, stable and unstable states, etc. — will be seen again.

2.5 LAPLACE TRANSFORMATION

The Laplace transformation (or Laplace transform for short) belongs to a broader class of integral transform operations in which a function f(t) is multiplied by a "Kernel" K(s,t) and integrated between the limits a and b. Thus,

$$T\{F(t)\} = \int_a^b F(t)K(s,t)dt = f(s) \tag{2.29a}$$

where T = operational symbol for the transformation and f(s) = transform of F(t).

F(t), the function operated on, is quite arbitrary in form and can be an ordinary function in t, such as sin at, a derivative d^2F/dt^2, or even an integral. The kernel and the integration limits (a,b) define the type of transform. Apart from Laplace transforms, there are Fourier transforms of various types, those involving Bessel functions, and several others, all of which have their own special kernels and integration limits. These are taken up in more detail at the PDE level. For the Laplace transform, the kernel K(s,t) is the function e^{-st}, and the integration limits (a,b) are from zero to infinity. Thus, the Laplace transform of f(t) takes the form:

$$L\{F(t)\} = \int_0^\infty F(t)\exp(-st)dt = f(s) \tag{2.29b}$$

where L is its operational symbol and L^{-1} is its inverse, i.e.,

$$L^{-1}\{f(s)\} = F(t) \tag{2.29c}$$

This transformation, when applied to each term of an ordinary differential equation, has the effect of eliminating the independent variable t, thus reducing the ODE to an algebraic equation in the transformed state variable f(s). After solving for f(s), the AE is then translated back into the solution space by means of appropriate "dictionaries," the Laplace transform tables. The reader will note that this procedure bears a resemblance to the D-operator method that likewise transforms an ODE to an AE, and after solving the latter, translates the result back into the solution domain by means of an appropriate dictionary, Table 2.10.

The Laplace transform is generally used to solve linear initial value ODEs or PDE's with constant coefficients. Although it can, in principle, be applied to boundary value problems, the procedure is somewhat cumbersome because the missing initial conditions have to be evaluated by substitution of given boundary conditions into the solution, much like integration. It is also of little help in solving variable coefficient ODEs and is generally inapplicable to nonlinear ODEs. Once these limitations are accepted, however, it becomes an extremely powerful tool both for the solution and the analysis of linear ODEs, as well as PDEs. It can directly solve many nonhomogeneous initial value problems without first finding the fundamental solution of the corresponding homogeneous problems. Second, it can handle a wide array of nonhomogeneous terms, including the important class of discontinuous or impulsive forcing functions. Third, the initial conditions appear automatically in the solution, thus obviating the need to evaluate integration constants.

2.5.1 General Properties of the Laplace Transform

In Table 2.9, we have tabulated some general properties of the Laplace transform, and supplement this, in Table 2.10, with tabulations of transforms of some common functions. Our comments here are with respect to Table 2.9.

To begin with, we note that the table is akin to a table of integrals in that one can move from left to right or from right to left, depending on the information sought. In this instance, one moves from right to left to obtain the transform f(s) of a function F(t), and from left to right to obtain what is termed the *inverse transform* of f(s), i.e., the function F(t) itself. The latter process is commonly referred to as an *inversion*. The formula given for the general inversion, item 2, involves a line integral in the complex plane and requires some background knowledge of complex variable theory. It is used only sparingly, and we shall limit ourselves here to making use of the more convenient items.

Item 5, the transform of derivatives, requires special mention. Its main feature is that it converts the derivatives into algebraic expressions in s. In this, it resembles the D-operator method and, as we shall see, also leads to a set of characteristic values termed *poles*. It has, however, the additional advantage of incorporating the initial conditions in the resulting algebraic expression. This was not the case in the D-operator method, which required the somewhat cumbersome evaluation of integration constants.

Two additional items call for special mention. One is the convolution integral, item 7, which allows the inversion of the product of two arbitrary functions, f(s) and g(s). It is frequently used to carry a general and unspecified function, say F(t), into the transformation process and return it upon inversion as an integrand of the convolution integral. The second item of special interest is the Heaviside expansion, item 8. To use it in the inversion of a ratio of polynomials, one first has to evaluate the roots of the denominator q(s), which are then substituted into the inversion formula given on the right side. That formula applies only to distinct roots. Extensions to repeated roots are available. Alternatively, the ratio can be decomposed into partial fractions, and each fraction thus obtained can be inverted on an individual basis. We demonstrate the use of these formulas with some sample illustrations.

TABLE 2.9
General Properties of the Laplace Transform

f(s)	F(t)
1. Laplace transform	
$\int_0^\infty F(t)e^{-st}dt$	F(t)
2. Inverse transform	
f(s)	$\frac{1}{2\pi i} \lim_{\beta \to \infty} \int_{\gamma-i\beta}^{\gamma+i\beta} e^{tz} f(z)dz$
3. Transform of a constant	
C/s	C
4. Transform of a sum	
af(s) + bg(s)	aF(t) + bG(t)
5. Transform of derivatives	
sf(s) − F(0)	F′(t)
s^2f(s) − sF(0) − F′(0)	F″(t)
$s^n f(s) - s^{n-1}F(0) - s^{n-2}F'(0) \ldots F^{(n-1)}(0)$	$F^{(n)}(t)$
6. Transform of an integral	
$\frac{1}{s}f(s)$	$\int_0^t F(\tau)d\tau$
7. Inverse of a product: the convolution theorem	
f(s) g(s)	$\int_0^t F(\tau)G(t-\tau)d\tau = F(t) * G(t)$
8. Inverse of a ratio of polynomials (heaviside expansion)	
$\frac{p(s)}{q(s)}$ (order of $p(s)<q(s)$)	$\sum_{n=1}^{\infty} \frac{(s-a_n)p(a_n)}{q(a_n)} e^{a_n t} = \sum_{n=1}^{\infty} \frac{p(a_n)}{q'(a_n)} e^{a_n t}$ where a_n = roots of q(s)
9. Inverse of derivatives	
f′(s)	−t F(t)
$f^{(n)}$ (s)	$(-1)^n t^n$ F(t)
10. Inverse of an integral	
$\int_0^\infty f(x)dx$	$\frac{1}{t}F(t)$

(continued)

TABLE 2.9 (Continued)
General Properties of the Laplace Transform

f(s)	F(t)
11. Translation or shifting properties	
(a) $f(s - a)$	$e^{at} F(t)$
(b) $e^{-as} f(s)$	$\begin{cases} F(t-a) & t > a \\ 0 & t < a \end{cases}$
12. Initial value theorem	
$\lim_{s \to \infty} sf(s)$	$F(0)$
13. Final value theorem	
$\lim_{s \to \infty} sf(s)$	$\lim_{t \to \infty} F(t)$

TABLE 2.10
Laplace Transforms of Some Functions

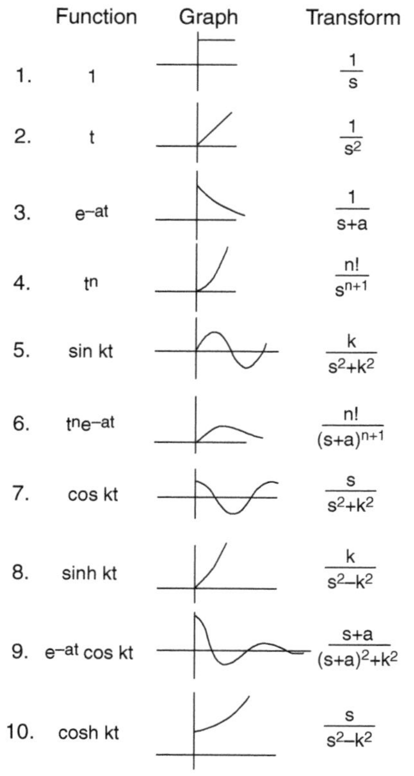

	Function	Graph	Transform
1.	1		$\frac{1}{s}$
2.	t		$\frac{1}{s^2}$
3.	e^{-at}		$\frac{1}{s+a}$
4.	t^n		$\frac{n!}{s^{n+1}}$
5.	$\sin kt$		$\frac{k}{s^2+k^2}$
6.	$t^n e^{-at}$		$\frac{n!}{(s+a)^{n+1}}$
7.	$\cos kt$		$\frac{s}{s^2+k^2}$
8.	$\sinh kt$		$\frac{k}{s^2-k^2}$
9.	$e^{-at} \cos kt$		$\frac{s+a}{(s+a)^2+k^2}$
10.	$\cosh kt$		$\frac{s}{s^2-k^2}$

Illustration 2.15 Inversion of Various Transforms

1. Obtain the Inverse $L^{-1}\left\{\dfrac{1}{(s^2+k^2)^2}\right\}$

   Because the roots of the denominator are both repeated, one cannot use the Heaviside expansion shown in item 8 of Table 2.9. We note, however, from Table 2.10, item 5, that:

$$L^{-1}\left\{\frac{k}{(s^2+k^2)}\right\}=\sin\,kt \tag{2.30a}$$

   so that we are in a position to apply the convolution theorem. We obtain:

$$L^{-1}\left\{\frac{1}{(s^2+k^2)^2}\right\}=\frac{1}{k^2}\sin\,kt * \sin\,kt$$

$$=\frac{1}{k^2}\int_0^t \sin\,k\tau \sin\,k(t-\tau)d\tau \tag{2.30b}$$

   Expanding sin k(t – τ) gives:

$$\sin\,(kt-k\tau)=\sin\,kt\,\cos\,(-k\tau)+\cos\,kt\,\sin\,(-k\tau) \tag{2.30c}$$

   and using the following formulae obtained from tables of integrals:

$$\int_0^t \sin^2\,k\tau d\tau=\frac{1}{2k}\left[t-\frac{1}{2}\sin^2\,kt\right] \tag{2.30d}$$

$$\int_0^t \sin\,k\tau\,\cos\,k\tau\,d\tau=\frac{1}{2k}\sin^2\,kt$$

   as well as,

$$\sin\,2\,kt=2\,\sin\,kt\,\cos\,kt \tag{2.30e}$$

$$\sin^2\,kt+\cos^2\,kt=1$$

   there finally results:

$$L^{-1}\left\{\frac{1}{(s^2+k^2)^2}\right\}=\frac{1}{2k^3}\{\sin\,kt-kt\,\cos\,kt\} \tag{2.30f}$$

2. Find $L^{-1}\left\{\dfrac{e^{-5s}}{(s-2)^4}\right\}$

Here, some thought will reveal that application of the two shifting properties 11a and 11b in Table 2.9 to the polynomial and exponential terms in succession will accomplish the desired inversion. We first apply 11a, using item 4 of Table 2.10 as well, and obtain

$$L^{-1}\left\{\frac{1}{(s-2)^4}\right\} = e^{2t}\,L^{-1}\left\{\frac{1}{s^4}\right\} = e^{2t}\,\frac{1}{6}t^3 \tag{2.30g}$$

Knowing the inversion of the polynomial part, we then apply the shifting property 11b to the full expression. This yields:

$$L^{-1}\left\{\frac{e^{-5s}}{(s-2)^4}\right\} = \begin{cases} \frac{1}{6}(t-5)^3 e^{2(t-5)} & t>5 \\ 0 & t<5 \end{cases} \tag{2.30h}$$

3. Invert $\left\{\frac{3s+1}{(s-1)(s^2+1)}\right\}$

We recognize this as a ratio of polynomials p(s)/q(s) where p(s) = 3s + 1, q(s) = (s − 1)(s^2 + 1) = $s^3 - s^2 + s - 1$, and the roots of q(s) are $a_1 = 1$, $a_2 = i$, $a_3 = -i$. Because the roots are not repeated, and the order of p(s) is less than that of q(s), the Heaviside expansion, item 8 of Table 2.9, may be applied.
We have:

$$q'(s) = 3s^2 - 2s + 1 \tag{2.30i}$$

Expanding the sum of the Heaviside expression, we obtain:

$$\frac{p(1)}{q'(1)}e^t + \frac{p(i)}{q'(i)}e^{it} + \frac{p(-i)}{q'(-i)}e^{-it} \tag{2.30j}$$

$$= 2e^t + \left(-1-\frac{1}{2}i\right)(\cos t + i\sin t) + \left(-1+\frac{1}{2}i\right)(\cos t - i\sin t)$$

Evaluation of the products reduces Equation 2.30j to the expression:

$$L^{-1}\left\{\frac{3s+1}{(s-1)(s^2+1)}\right\} = 2e^t - 2\cos t + \sin t \tag{2.30k}$$

2.5.2 Application to Differential Equations

Among the attractive features of the Laplace transformation we had listed in the introduction, the one with the greatest appeal to the novice is the ease with which it can be applied to the solution of ODEs. There are three steps to the solution procedure, which can be implemented in near-automatic fashion:

1. Apply the Laplace transform in turn to each term of the ODE in Y(t). If that term is a derivative, one obtains a composite of the transform of the unknown function, y(s) and its initial values (see item 5 of Table 2.9). For example, dY/dt becomes, when transformed, y(s) – Y(0). Terms directly containing the unknown state function become the transform of that variable. Thus, L{kY(t)} will become ky(s). Finally, the nonhomogeneous terms or forcing functions are directly reduced to explicit functions of s. For example, the forcing function e^{at} is transformed directly into the explicit form 1/(s – a).
2. Solve the algebraic equation in y(s) that resulted from step 1. This yields expressions of the form:

$$y(s) = G(s) \tag{2.31a}$$

where y(s) is the transformed state variable Y(t), and G(s) is an explicit function of s containing, among other things, the transforms of the forcing functions and the initial conditions. The following is an example.

$$y(s) = G(s) = \frac{Y_0}{(s-1)}$$

3. Invert the expression Equation 2.31a, i.e., apply the operator L^{-1} to each side. For y(s), this automatically yields the desired state variable Y(t). Inversion of the right-side term G(s) is accomplished by means of tables, such as Table 2.10, and by one or more of the procedures listed in Table 2.9, such as the use of the convolution integral or the Heaviside expansion. Together, these two tools yield the desired solution in the final form:

$$Y(t) = f(t, \text{Initial Conditions}) \tag{2.31b}$$

Illustration 2.16 The Mass-Spring System Revisited: Resonance

We consider here the system previously seen in Illustration 2.7, that of a vibrating spring with a forcing function. The forcing function used there was the weight of the mass m attached to the spring so that gravity and the restoring force kx of the spring were the forces to be considered. We obtained, by application of Newton's law:

$$\frac{d^2x}{dt^2} + (k/m)x = g \tag{2.32a}$$

where the gravitational constant g represented the forcing function. We now generalize this equation to apply to an arbitrary forcing function F(t) and specify more general initial conditions, so that:

$$\frac{d^2X}{dt^2} + \omega_0^{\ 2}X = F(t)/m \tag{2.32b}$$

where $X(0) = X_0$ (initial position), $X'(0) = v_0$ (initial velocity), and $\omega_0 = (k/m)^{1/2}$.

We carry the forcing function in unspecified form into the Laplace transformation process and obtain:

$$s^2\, x(s) - sX_0 - v_0 + \omega_0^2 x(s) = f(s)/m \tag{2.32c}$$

Solving for x(s), we obtain:

$$x(s) = \frac{X_0 s + v_0}{s^2 + \omega_0^{\ 2}} + \frac{f(s)}{m}\frac{1}{s^2 + \omega_0^{\ 2}} \tag{2.32d}$$

This is the expression that now has to be inverted. Inversion of the first fraction is by items 5 and 7 of Table 2.10, and that of the second term by means of the convolution integral, item 7 in Table 2.9. We obtain:

$$X(t) = X_0 \cos \omega_0 t + \frac{v_0}{\omega_0} \sin \omega_0 t + \frac{1}{m\omega_0}\int_0^t \sin (\omega_0 \tau)\ F(t-\tau)d\tau \tag{2.32e}$$

Note the appearance of the initial conditions X_0 and v_0 in this expression, and that the forcing function F(t) remains unspecified under the convolution integral. The advantage here of the convolution formulation is its ability to yield closed-form solutions without the need to specify certain terms of the original ODE.

We consider the following two cases for F(t).

1. $F(t) = constant = mg$, *and* $v_o = 0$. Under these conditions, the sine term drops out and the integral becomes:

$$\frac{1}{m\omega_0}\int_0^t \sin (\omega_o t)F(t-\tau)d\tau = -\frac{g}{\omega_0^{\ 2}}\cos \omega_o t \tag{2.32f}$$

We, thus, recover the solution given in Illustration 2.7, i.e.:

$$X(t) = \left(X_0 - \frac{mg}{k}\right)\cos (k/m)^{1/2}\, t \tag{2.32g}$$

or in equivalent form,

$$X(t) = (X_0 - g/\omega_0^2)\cos\omega_0 t \tag{2.32h}$$

2. $F(t) = F_0 \sin\omega t$. Here, the expression to be inverted becomes:

$$x(s) = \frac{X_0 s + v_0}{s^2 + w_0^2} + \frac{F_0}{m}\frac{\omega}{(s^2+\omega_o^2)(s^2+\omega^2)} \tag{2.32i}$$

The first term on the right is inverted via items 5 and 7 of Table 2.10, the second term either by partial fractions or the Heaviside expansion. We obtain:

$$X(t) = X_0 \cos\omega_o t + \frac{1}{\omega_o}\left[v_0 + \frac{F_0\omega}{m\left(\omega^2-\omega_0^2\right)}\right]\sin\omega_o t - \frac{F_0}{m\left(\omega^2-\omega_0^2\right)}\sin\omega t \tag{2.32j}$$

where, as before, $\omega_0 = (k/m)^{1/2}$.

Here, ω_0 is referred to as the *natural vibration frequency* and ω as the *forced vibration frequency* of the system. Let us now ask ourselves what happens when the two frequencies are identical. One could intuitively argue that this superposition of two vibrations with the same frequency might lead to an escalation in the amplitude of the vibrations of the system. This is indeed the case, as shown in the following.

3. $F = F_0 \sin\omega_0 t$. *Resonance* . The expression to be inverted now becomes

$$x(s) = \frac{X_0 s + v_0}{s^2 + \omega_0^2} + \frac{F_o\omega_0}{m(s^2+\omega_0)^2} \tag{2.32k}$$

One notes here the appearance of a repeated term in the denominator of the second fraction. Its inversion was demonstrated in Illustration 2.15 and led to the appearance of a term in t as kt cos kt, i.e., a time-dependent amplitude. For the case in hand, we obtain, in similar fashion:

$$X(t) = X_0\cos\omega_0 t + \frac{1}{\omega_0^2}\left[v_0\omega_0 + \frac{F_0}{2m}\right]\sin\omega_0 t - \frac{F_0}{2m\omega_0} t\cos\omega_0 t \tag{2.32l}$$

In view of the last term, the amplitude of the oscillations increases indefinitely, and one speaks of the forcing function F(t) as being *in resonance* with the system frequency. In particular, for $X_0 = 0$ and $v_0 = -F_0/(2m\omega_0)$, one obtains the expression:

$$X(t) = -\frac{F}{2m\omega_0} t\cos\omega_0 t \tag{2.32m}$$

A. Vibrating Mass

F(t)
k_1 m k_2
0 x
Dashpot

B. Oscillating Electrical Circuit

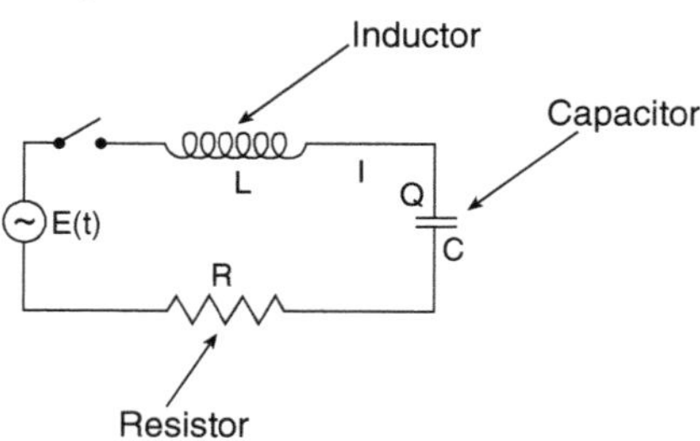

FIGURE 2.10 Two different physical systems that give rise to identical forms of second-order ODEs.

This equation, which we previously encountered in Illustration 1.8 and graphed in Figure 1.8, clearly shows a linear increase in amplitude with t and would, in time, lead to a rupture of the mass-spring system.

Illustration 2.17 Equivalence of Mechanical Systems and Electrical Circuits

We wish to demonstrate here the equivalence of the equations that describe mechanical vibrations on the one hand, and the oscillations that arise in certain electrical circuits on the other.

We do this by first deriving a more general equation for forced mechanical vibrations that now includes a damper or dashpot in the system, as illustrated in Figure 2.10A. The damper has the effect of producing a countervailing force that is proportional to the velocity of the vibrating mass. This, in turn, leads to the appearance of a first-order derivative $k_1 dX/dt$ in the constitutive equation. We now obtain:

$$m\frac{d^2X}{dt^2} \quad + \quad k_1\frac{dX}{dt} \quad + \quad k_2X \quad = \quad F(t) \tag{2.33a}$$

Acceleration   Damper   Spring   Forcing function

where the forcing function is again kept in a general and unspecified form.

We next turn to the consideration of the simple electrical circuit shown in Figure 2.10B. It is made up of a coil with inductance L, a resistor with resistance R, a capacitor with capacitance C, and a voltage supply E(t). Previously in Chapter 1, we have given the expressions for the voltage drop for these elements that we repeat here.

$$\text{Inductor } V_I = L \, di/dt \tag{1.10}$$

$$\text{Resistor } V_R = iR$$

$$\text{Capacitor } V_C = q/C$$

where i = current, q = charge.

The sum of three voltage drops must equal the imposed voltage E(t) so that:

$$L\frac{di}{dt} + \frac{q}{C} + iR = E(t) \tag{2.33b}$$

Because current equals the rate of flow of charge q at any time, we have i = dq/dt. Introducing this relation into Equation 2.33b, we obtain the second-order ODE:

$$L\frac{d^2q}{dt^2} \quad + \quad R\frac{dq}{dt} \quad + \quad q(C) \quad = \quad E(t) \tag{2.33c}$$

Inductor Resistor Capacitor Imposed voltage

The analogy to the vibrating mass and spring, Equation 2.33a, is immediately evident. We note in particular that the inductor corresponds to the acceleration term, i.e., accelerates the charge, the resistor has a "damping" effect on charge flow, and the capacitor, likewise, resists charge flow much like the action of the spring.

The response of these systems to various forcing functions gives rise to some interesting phenomena. We have already given a partial analysis of this type in Illustration 2.16. In the following two illustrations, we shall first, by way of preamble, address the response of first-order systems and ODEs (Illustration 2.18), and follow this up, in Illustration 2.19, with a more general analysis of second-order systems of the type given earlier. We will show in particular, how these responses are related to the coefficients of the constituent equations, and how they are affected by the form of the forcing function.

Illustration 2.18 Response of First-Order Systems

We use here, as an example of a first-order system, a thermocouple or a thermometer that is exposed to a change in the ambient temperature T_a. That change represents a forcing function F(t). We have referred to this case repeatedly in Chapter 1, where a step change in the ambient temperature was considered. Here, we generalize the treatment to accommodate arbitrary forcing functions. A number of new features in both the form of the ODE and its transform are introduced that help in the generalization of the results.

The model is arrived at by performing an energy balance over the thermocouple.

$$\text{Rate of energy in} - \text{Rate of energy out} = \frac{d}{dt}\text{energy contents}$$

$$hA(T_a - T) - 0 = \frac{d}{dt}H = \rho V C_p \frac{dT}{dt} \tag{2.34a}$$

where ρ, V, C_p are the density, volume, and specific heat of the thermocouple, respectively, and A its surface area.
Note that at steady state, the relation becomes:

$$hA(T_{as} - T_s) = 0 \tag{2.34b}$$

where the subscript s denotes the (previous) steady state. Subtraction of the two equations yields:

$$hA[(T_a - T_{as}) - (T - T_s)] = mC_p \frac{d(T - T_s)}{dt} \tag{2.34c}$$

or

$$[X - Y] = \tau \frac{dY}{dt} \tag{2.34d}$$

where X and Y are the so-called *deviation variables*, defined as

$X = T_a - T_{as}$, the forcing function or input

and $Y = T - T_s$, the response function or output

$\tau = \rho V C_p / hA$ is the so-called *time constant*, with units of time.

Noting that Y(0) = 0, the Laplace transform of Equation 2.34d yields the expression

$$x(s) - y(s) = \tau s y(s) \tag{2.34e}$$

or in rearranged form,

$$\frac{\textit{Transform of output}}{\textit{Transform of input}} = \frac{y(s)}{x(s)} = \frac{1}{\tau s + 1} \tag{2.34f}$$

The term on the right is referred to as the *transfer function* of the system, and is denoted by the symbol G(s). It can also be viewed as the ratio of the transforms of the output to that of the input. Thus,

$$G(s) = \frac{y(s)}{x(s)} = \frac{1}{\tau s + 1} \tag{2.34g}$$

We now consider various inputs X to the system, and examine the resulting response Y.

1. Step input A — Here, we have, from item 3 of Table 2.9,

$$x(s) = A/s \tag{2.35a}$$

and the relation to be inverted becomes:

$$y(s) = \frac{A}{s} \frac{1}{\tau s + 1} \tag{2.35b}$$

Inversion can be accomplished by either the Heaviside expansion or by partial fractions and yields:

$$Y(t) = A(1 - e^{t/\tau}) \tag{2.35c}$$

This expression depicts a smooth exponential rise to the new steady state value A, as shown in Figure 2.11.

2. Unit impulse input — The input here consists of a pulse of magnitude 1 applied over a infinitesimally small time interval. The transform of such

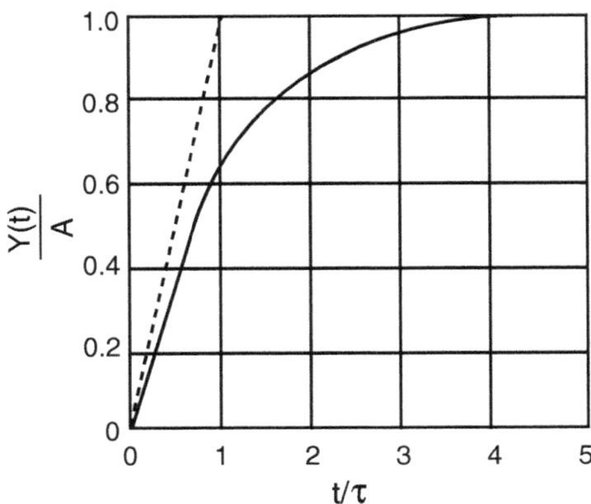

FIGURE 2.11 Response Y(t)/A of a first-order ODE to a step change in input.

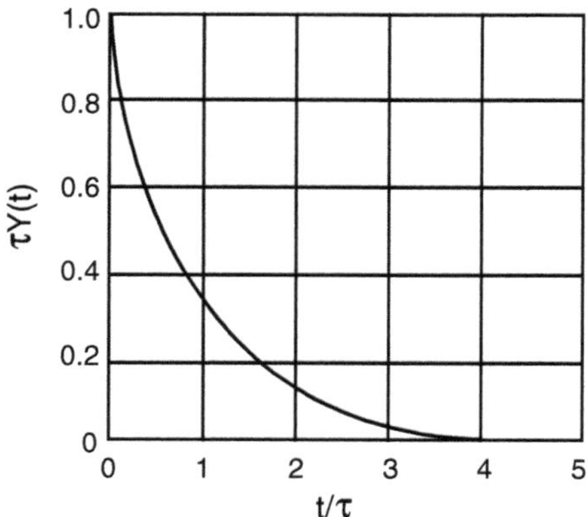

FIGURE 2.12 Response Y(t) of a first-order system to an impulse input.

a pulse, also known as the *Dirac delta function*, equals 1. The transform of the ODE is then given by

$$y(s) = \frac{1}{\tau s + 1} \tag{2.36a}$$

with an inverse (see item 3 of Table 2.10)

$$Y(t) = \frac{1}{\tau} e^{-t/\tau} \tag{2.36b}$$

This expression is plotted in Figure 2.12.

One notices that the response rises immediately to 1.0 and then decays exponentially. Such an abrupt rise is not physically possible, but is closely approached by the response of the thermocouple to a pulse of narrow width.

3. Sinusoidal input — This is an interesting case in which the forcing function or input is of the form:

$$X(t) = A \; sin \; \omega \mathrm{t} \tag{2.37a}$$

The Laplace transform is now given by:

$$y(s) = \frac{A\omega}{s^2 + \omega^2} \frac{1}{\tau s + 1} \tag{2.37b}$$

which upon inversion by partial fraction or the Heaviside expansion yields:

$$Y(t) = \frac{A\omega\tau e^{-t/\tau}}{\tau^2\omega^2 + 1} - \frac{A\omega\tau}{\tau^2\omega^2 + 1}\cos \omega t + \frac{A}{\tau^2\omega^2 + 1}\sin \omega t \tag{2.37c}$$

This expression can be further consolidated by combining the two trigonometric terms using the identity:

$$p \cos A + q \sin A = r \sin(A + \phi) \tag{2.37d}$$

$$r = (p^2 + q^2)^{1/2}, \ \tan\phi = p/q$$

Application to Equation 2.37c leads to the transformed expression:

$$Y(t) = \frac{A\omega\tau}{\tau^2\omega^2 + 1} e^{-t/\tau} + \frac{A}{(\tau^2\omega^2 + 1)^{1/2}} \sin(\omega t + \phi) \tag{2.37e}$$

with $\phi = \tan^{-1}(-\omega t)$.
Several features of this expression are worth noting.

1. The solution is made up of an exponentially decaying term and a periodic term. This latter term, called the *ultimate periodic solution*, constitutes the response as $t \to \infty$, i.e.,

$$\underset{t \to \infty}{Y(t)} = \frac{A}{(\tau^2\omega^2 + 1)^{1/2}} \sin(\omega t + \phi) \tag{2.37f}$$

2. The output given by Equation 2.37f has the same frequency ω as the forcing function but lags behind it by an angle $|\phi|$.
3. The ratio of output amplitude to input amplitude of the ultimate periodic solution is $(\tau^2\omega^2 + 1)^{1/2}$, which is always smaller than 1. The output signal is said to be attenuated with respect to the input amplitude.

We note that first-order systems do not oscillate on their own and hence do not have a natural frequency ω_0 as do second-order systems. Thus, resonance does not arise here. These and other features are discussed in the following illustration.

Illustration 2.19 Response of Second-Order Systems

The genesis of second-order ODEs in oscillating mechanical and electrical systems that yield to analysis and solution by the Laplace transformation was briefly discussed in Illustration 2.17. The ODEs involved were of the form:

$$K_1 \frac{d^2Y}{dt^2} + K_2 \frac{dY}{dt} + K_3 Y = F(t) \tag{2.38a}$$

For the purpose of analyzing the solution behavior, it is convenient to recast the expression into the following form:

$$\tau^2 \frac{d^2Y}{dt^2} + 2\lambda\tau \frac{dY}{dt} + Y = X(t) \tag{2.38b}$$

where $\tau = (K_1/K_3)^{1/2}$ time constant (sec)

$$\lambda = \frac{K_2}{2}(K_1K_3)^{-1/2} \text{ characteristic parameter, dimensionless}$$

And

$$X(t) = F(t)/K_3$$

Assuming initial conditions $Y(0) = Y'(0) = 0$, the transform of Equation 2.38b becomes:

$$\frac{y(s)}{x(s)} = \frac{1}{\tau^2 s^2 + 2\lambda\tau s + 1} \quad (2.38c)$$

We now examine this expression and its inverse for various types of forcing functions or inputs, as we did in the case of first-order systems. In particular, we shall make use of the roots of the denominator, $s_{1,2}$, which are known as the *poles* of the transfer function.

Unit step input — This input leads to the transform:

$$y(s) = \frac{1}{s}\frac{1}{(\tau^2 s^2 + 2\lambda\tau s + 1)} \quad (2.39a)$$

whose inverse will depend on the magnitude of the characteristic parameter λ. We distinguish three cases:

Step response for $\lambda > 1$ — Here, the roots are real and distinct, and the Heaviside expansion yields a sum of exponentials given by:

$$Y(t) = 1 - e^{-\lambda t/\tau}\left[\frac{\lambda}{(\lambda^2 - 1)^{1/2}}\sinh(\lambda^2 - 1)^{1/2} t/\tau + \cosh(\lambda^2 - 1)^{1/2} t/\tau\right] \quad (2.39b)$$

Plots of this relation for $\lambda > 1$ appear in Figure 2.13. The response in all cases is seen to be a smooth exponential rise to the new steady state at $Y(t) = 1$. That rise, however, becomes increasingly sluggish as λ is raised, i.e., it takes increasingly longer times to reach the new steady state. One speaks of the system as being overdamped. In a mass and spring system with dashpot, this would correspond to the displaced mass coming to rest in a single half-swing with no oscillations because the strong damping effect of the dashpot.

Step response for $\lambda < 1$ — This case is the one most frequently encountered in practice and involves oscillations whose amplitude decay with time, corresponding to a vibrating mass or an oscillating pendulum coming to rest after a few swings. The roots of the quadratic term in Equation 2.39b are now complex conjugate with a real negative part, the latter accounting for the exponential decay of the amplitude. Inversion is by partial fractions or by the Heaviside expansion. Using in addition

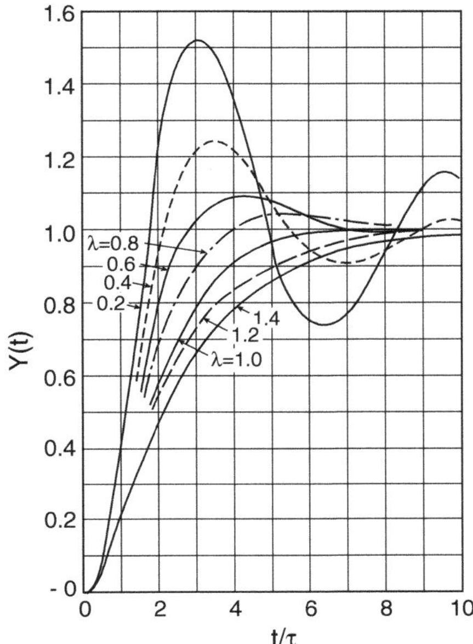

FIGURE 2.13 Response of a second-order system to a step change in input. System is underdamped for λ > 1, overdamped for λ < 1, and critically damped at λ = 1.

the trigonometric transformation expressed by Equation 2.37d, one obtains, after some manipulation, the inverted form:

$$Y(t) = 1 - \frac{1}{(1-\lambda^2)^{1/2}} e^{-\lambda t/\tau} \sin[(1-\lambda^2)^{1/2} t/\tau + \phi] \qquad (2.39c)$$

where $\phi = \tan^{-1} (1 - \lambda^2)^{1/2}/\lambda$

Plots of this oscillatory response also appear in Figure 2.13 for various values of λ. One notes that the amplitudes of the oscillations increase with decreasing values of λ, but all curves ultimately converge to a value of Y(t) = 1.

We single out a response curve for a particular value of λ, shown in Figure 2.14, for more detailed scrutiny. One notes a number of characteristic features of the plot, including the rise time t_r, the period of oscillation T, and parameters A, B, and C (which are related to the degree of overshoot) and the decay ratio. We define these as follows:

1. *Overshoot* is the quantity that expresses the degree by which the response exceeds the ultimate steady state. It is defined as the ratio A/B (Figure 2.14) that is related to the characteristic parameter λ by the relation:

$$\text{Overshoot} = A/B = \exp(-\pi\lambda)/(1 - \lambda^2)^{1/2} \qquad (2.40a)$$

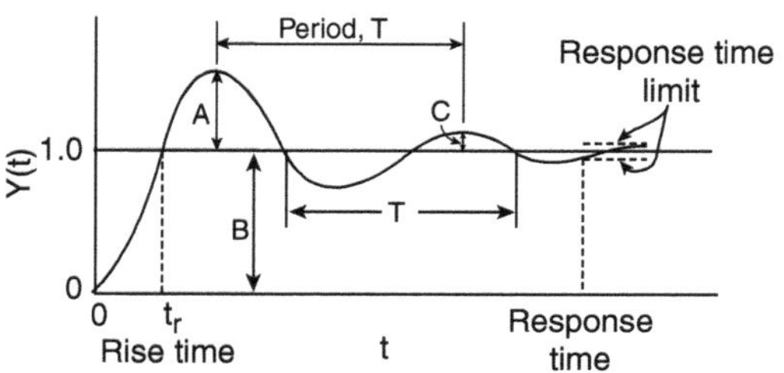

FIGURE 2.14 Characteristic parameters of an underdamped system.

2. *Decay ratio C/A* is a measure of the relative magnitude of successive peaks and is related to λ by the expression:

$$\text{Decay ratio} = C/A = \exp\left[(-2\pi\lambda)/(1-\lambda^2)^{1/2}\right] = (\text{Overshoot})^2 \qquad (2.40b)$$

3. *Rise time* is the time required for the response to reach its first steady state value. There is no explicit expression for it in terms of system parameters but it can be shown that it increases with the value of λ.
4. *Response time* is the time required for the response to come within a specified interval, usually taken as ± 5%, of the ultimate steady state (see Figure 2.14).
5. *Oscillation period and frequency* is yet another quantity that can be directly expressed in terms of the system parameters. The pertinent relation is given by:

$$T = \frac{1}{\nu} = \frac{2\pi}{\omega} = \frac{2\pi\tau}{(1-\lambda^2)^{1/2}} \qquad (2.40c)$$

   where ω = circular frequency in radians, ν = frequency in cycles/time, and the period T in time/cycle.
6. *Natural frequency* ω_0. We have seen in Illustration 2.16 that an undamped system has its own natural frequency ω_0. This frequency is obtained by removing the first derivative in Equation 2.38b, i.e., by setting λ = 0. We obtain from Equation 2.40c

$$T_0 = \frac{1}{\nu_0} = \frac{2\pi}{\omega_0} = (2\pi\tau) \qquad (2.40d)$$

Note that the phenomenon of resonance that we had seen for the undamped system with sinusoidal forcing does not arise when λ ≠ 0, i.e., when the system is damped.

The noteworthy feature of this development is that some important parameters of the system response can be directly deduced, and deduced quantitatively, from the coefficients of the underlying ODE. No solution is required, only the evaluation of the characteristic parameter λ, and in the case of Equation 2.40c, that of the time constant τ as well. This parallels to some extent features we have noted in connection with the D-operator method. The connection there was between the qualitative shape of the solution — periodic, exponential, or a combination thereof — and the coefficients of the ODE contained in the so-called characteristic roots.

Step response for $\lambda = 1$ — For this value of the characteristic parameter, the quadratic term in Equation 2.39a yields identical roots. Inversion of the equation that can be accomplished by convolution yields the expression:

$$Y(t) = 1 - \left(1 + \frac{t}{\tau}\right)e^{-t/\tau} \tag{2.40e}$$

Figure 2.13 shows this to be the borderline between overdamped and underdamped behavior, and the system is consequently referred to as being *critically damped.* Physically, it represents the condition of quickest attainment of the new steady state, a desirable response but one that is difficult to implement in practice.

Unit impulse input — Response of Equation 2.38b to a unit dirac pulse, with a transform of 1, yields:

$$y(s) = \frac{1}{\tau^2 s^2 + 2\lambda\tau s + 1} \tag{2.41a}$$

We refrain from going into the details of the inversion and merely summarize the resulting responses for the three values of the critical parameter λ. We have:

For $\lambda > 1$:

$$Y(t) = \frac{1}{\tau}\frac{1}{(\lambda^2 - 1)^{1/2}} e^{-\lambda t/\tau} \sinh(\lambda^2 - 1)^{1/2} t/\tau \tag{2.41b}$$

For $\lambda < 1$:

$$Y(t) = \frac{1}{\tau}\frac{1}{(1-\lambda^2)^{1/2}} e^{-\lambda t/\tau} \sin(1-\lambda^2)^{1/2} t/\tau \tag{2.41c}$$

For $\lambda = 1$

$$Y(t) = \frac{1}{\tau^2} t e^{-t/\tau} \tag{2.41d}$$

A graphical representation of these curves appears in Figure 2.15. The behavior is very similar to that seen in Figure 2.13 for the response to a unit step input, except

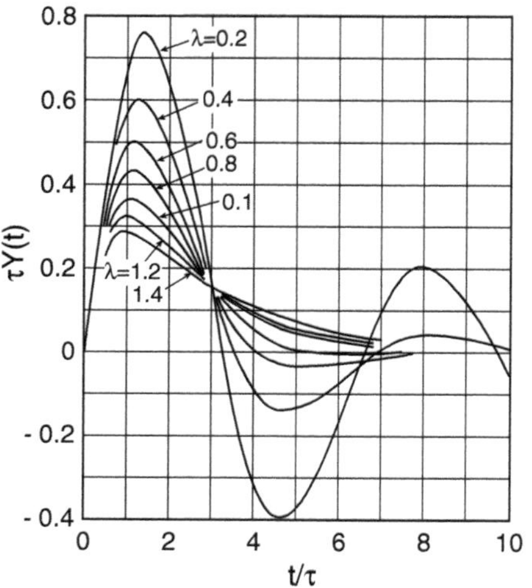

FIGURE 2.15 Response τY(t) to an impulse input.

that the system now returns to its original state. $\lambda > 1$ again corresponds to an overdamped system, $\lambda < 1$ to an underdamped system, and $\lambda = 1$ is the dividing line between the two, i.e., it corresponds to critical damping.

Sinusoidal input — The input considered here again has the form:

$$X(t) = A \sin \omega t$$

We apply it to the full Equation 2.38b, i.e., we include damping. The system consequently has no natural frequency of its own and resonance does not arise.

The transform to be inverted is of the form:

$$y(s) = \frac{A\omega}{(s^2 + \omega^2)(\tau^2 s^2 + 2\lambda\tau s + 1)} \tag{2.42a}$$

which upon application of standard inversion methods yields the response:

$$Y(t) = \frac{A}{\{[1-(\omega\tau)^2]^2 + (2\lambda\omega\tau)^2\}^{1/2}} \sin(\omega t + \phi) \tag{2.42b}$$

with the phase angle ϕ given by:

$$\phi = -\tan^{-1} \frac{2\lambda\omega\tau}{1-(\omega\tau)^2} \tag{2.42c}$$

We note the following features of Equation 2.42b:

- The frequency of the response ω is identical to that of the forcing function.
- The output lags the input by a phase angle $|\phi|$. It can be seen from Equation 2.42c that the argument of the inverse tangent approaches zero as $\omega \rightarrow \infty$, and that $|\phi|$ consequently approaches 180° asymptotically. This is in contrast to the response of first-order systems whose phase angle is at most 90°.
- The ratio of output amplitude to input amplitude is given by:

$$\{[1 - (\omega\tau)^2]^2 + (2\lambda\omega\tau)^2\}^{-1/2} \tag{2.42d}$$

and can be greater or smaller than 1, depending on the magnitude of λ and $\omega\tau$. Both amplification and attenuation are thus possible. This is again in contrast to the behavior of first-order systems, whose amplitude ratio never exceeds 1.

Comments: The preceding two illustrations have given us a fairly thorough look at the responses of first- and second-order systems to a variety of forcing functions or disturbances. Although the models for real systems are considerably more complex and of a higher order, the principal features shown by these two simple models, such as over- and underdamped behavior, resonance, etc., are observed in more complex systems. In fact, it is quite common to approximate them by the simpler second-order model and to use the results as a guide to their actual behavior.

PRACTICE PROBLEMS

2.1 Classification of ODEs

a. Classify the following ODEs:

$$y'' + y' + (\sin x)y = 0$$

$$y'' + y' + y = \sin x$$

$$y'' + y' + (\sin^2 x)y = 0$$

$$y'' + y' + \sin y = 0$$

$$y'' + y' + y = 0$$

b. Give an example each of an autonomous ODE, a nonautonomous ODE, a homogeneous ODE, and a nonhomogeneous ODE. Are all autonomous ODEs homogeneous? Are all homogeneous ODEs autonomous?

2.2 Boundary Conditions: Initial Value and Boundary Value Problems (IVP and BVP)

Specify the boundary and initial conditions for the following systems and situations, and indicate whether they lead to an IVP or BVP.

a. The velocity profile that arises in steady laminar flow through a cylindrical pipe (see Equation 1.14).
b. Diffusion of oxygen into a liquid contained in a tube open to the atmosphere at one end and sealed at the other.
c. A tubular reactor coated internally with a catalyst. The fluid containing the reactant is in unsteady laminar flow.
d. An artillery piece in which the gun elevation β and muzzle velocity v^0 are specified.
e. The same artillery piece, but with the distance L to the target ("range") specified.

2.3 Solution by Separation of Variables. A Simple Cellular Process

Cellular processes can assume a variety of forms. In one version, substances within the cell undergo a reaction, accompanied by diffusional transport through the cell wall. This is the most easily visualized process that can be described by standard compartmental models. Events may also take place within the membranes of the cell rather than its interior and are triggered by so-called receptors embedded within the cell wall. These receptors are large protein molecules capable of "communicating" chemically with the external, extracellular fluid, as well as the internal cytoplasmic domain (Figure 2.16). The process involves the binding of ligand molecules from the extracellular fluid, which enables the receptors to "sense" the environment and to signal this fact chemically to the interior of the cell. Two other major events, termed coupling and trafficking, involve changes in the receptors themselves.

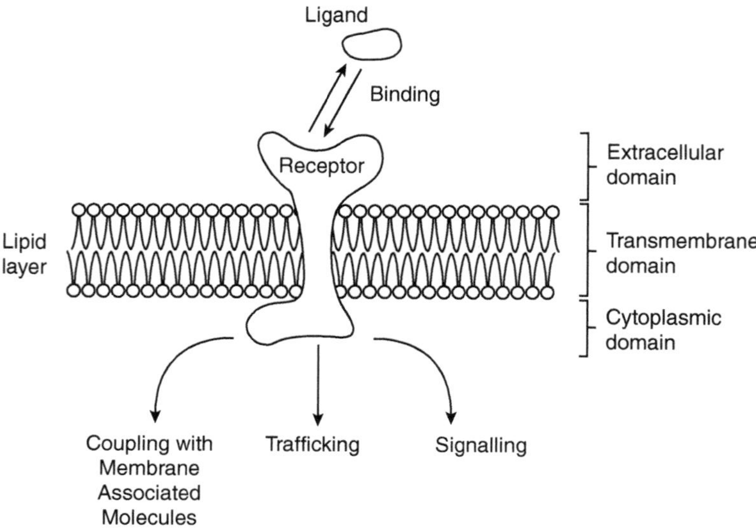

FIGURE 2.16 Processes involving receptors embedded in a cell membrane.

The process we address here is that of ligand binding, which is commonly modeled as a reversible bimolecular reaction:

$$R + L \underset{k_r}{\overset{k_f}{\leftarrow\!\!\rightarrow}} C \tag{A}$$

with the rate given by:

$$\frac{dC}{dt} = k_f RL - k_r C \tag{B}$$

where R = number of receptors per cell, C = number of ligand–receptor complexes per cell, and L = external ligand concentration in moles/L.

Because the model contains three dependent variables R, C, and L, two additional relations are required, which are provided by cumulative receptor and ligand balances. Thus:

$$R_{Tot} = R + C \tag{C}$$

and

$$L_{Tot} = L + \frac{n}{N_{av}} C \tag{D}$$

where n = number of cells and N_{av} = Avogadro's number. The ratio n/N_{av} is used to convert the units of C to the molar concentration of L.

Substitution of (C) and (D) into the expression (B) yields:

$$\frac{dC}{dt} = k_f (R_T - C)\left[L_0 - \left(\frac{n}{N_{av}}\right)C\right] - k_r C \tag{E}$$

or in equivalent compact form:

$$\frac{dC}{dt} = aC^2 + bC + d \tag{F}$$

Show that solution by separation of variables yields the result:

$$C(t) = \frac{C_2 - C_1 \dfrac{C_2 - C_0}{C_1 - C_0} \exp[-(C_1 - C_2)t]}{1 - \dfrac{C_2 - C_0}{C_1 - C_0} \exp[-(C_1 - C_2)t]} \tag{G}$$

and identify the meanings of C_0, C_1, and C_2.

2.4 The Shrinking-Core Model

Show by repeated application of the separation of variables method that the relation between core radius r_C and time of reaction for a reacting spherical particle is given by:

$$t = \frac{\rho_p R^2}{6bD_e C_{As}}\left[1-3\left(\frac{r_c}{R}\right)^2+2\left(\frac{r_c}{R}\right)^3\right]$$

where b = moles of solid reacting per mole of reactant gas A, and D_e = diffusivity through the ash layer. Assume diffusion through the ash layer is controlling. Use the systematic approach employed in Illustration 2.4.

2.5 The Countercurrent Heat Exchanger Solution by the D-Operator Method

The operation of a standard shell-and-tube heat exchanger shown in Figure 2.17 is described by the following two differential energy balances:

For the tube side:

$$F_t C_{pt}\frac{dT_t}{dz} - U\pi d(T_s - T_t) = 0 \tag{A}$$

and for the shell side:

$$F_s C_{ps}\frac{dT_s}{dz} - U\pi d(T_s - T_t) = 0 \tag{B}$$

where enthalpy H has been replaced by the product of flow rate F, specific heat C_p, and temperature difference ΔT (see Table 1.2). U is the overall heat transfer coefficient between shell and tube, and the subscripts t and s denote the tube and shell, respectively.

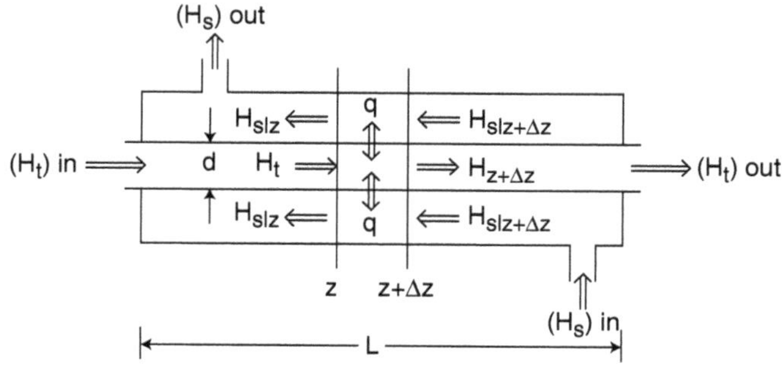

FIGURE 2.17 The countercurrent single-pass shell-and-tube heat exchanger.

Using the D-operator method, show that the tube-side temperature profile is given by:

$$\frac{T_t - (T_t)_{in}}{(T_s)_{in} - (T_t)_{in}} = \frac{F_s C_{ps}}{F_t C_{pt}} \frac{\left[\exp U\pi D\left(\frac{1}{F_s C_{ps}} - \frac{1}{F_t C_{pt}}\right)Z - 1\right]}{\left[\exp U\pi D\left(\frac{1}{F_s C_{ps}} - \frac{1}{F_t C_{pt}}\right)L - \frac{F_s C_{ps}}{F_t C_{pt}}\right]} \quad \text{(C)}$$

Hint: Solve (A) for T_t and substitute into (B).

2.6 Tubular Reactors with Axial Diffusion: The Danckwerts Boundary Conditions

A tubular reactor involving a first-order reaction at constant flow velocity v is described by the following differential mass balance:

$$v\frac{dC}{dz} + k_r C = 0 \quad \text{(A)}$$

The concentration gradient caused by the reaction can, in principle, give rise to a diffusional flux of reactant in the direction of flow. When this is taken into account, the mass balance reads:

$$-D\frac{d^2C}{dz^2} + v\frac{dC}{dz} + k_r C = 0 \quad \text{(B)}$$

a. Show that for a tubular reactor with axial dispersion, conditions at the inlet and outlet are given by the so-called Danckwerts, or close boundary conditions:

$$vC(0^-) = vC(0^+) - D\left.\frac{dC}{dz}\right|_{0^+} \quad \text{(C)}$$

$$\left.\frac{dC}{dz}\right|_{x=L} = 0$$

b. Using the D-operator method and the preceding boundary condition, show that the outlet concentration for a first-order reaction is given by:

$$C_{out} = C_{in}\frac{4s\exp(Pe/2)}{(1+s)^2\exp(s\,Pe/2) - (1-s)^2\exp(-s\,Pe/2)} \quad \text{(D)}$$

where $s = \left(1 + \frac{4k\tau}{Pe}\right)^{1/2}$, Pe = Peclet Number = vL/D.

2.7 Bending of a Beam under a Lateral Load

For the beam shown in Figure 2.2A, derive a formula for the deflection δ at the point of application of the force F.

Hint:The moment M at any point along the beam is given by –Fx.

2.8 Dynamics of the Human Immunodeficiency Virus (HIV)

The major target of HIV infection is a class of white blood cells known a $CD4^+T$ cells, or T cells for short, which secrete substances required by the immune system. The course of the disorder is sketched in Figure 2.18. Immediately after infection, the amount of virus rises dramatically, accompanied by flu-like symptoms. After a few weeks to months, these symptoms disappear, and the virus concentration subsides to a low level. An immune response to the virus has occurred, and antibodies against the virus can be detected in the blood. Persons in whom such antibodies are found are declared to be HIV positive.

After the primary infection, the virus concentration may pursue three paths. In the rarer instances, it may stay constant indefinitely or decline further. More commonly, and unless dealt with effectively, it will ultimately begin to creep up again and the decrease in T cells will resume. If in this latter stage the cell count drops below 200 cells/mm^3, the patient is classified as having acquired immuno deficiency syndrome (AIDS).

The earliest mathematical descriptions of infection by HIV were based on one-compartment models with the following postulated rate laws:

Rate of virus production $r_v = N\ k_{ei}\ C_i$ (A)

Rate of infected cell production $r_i = k_r\ C_v\ C_u^0$ (B)

Elimination rate for virus $r_{ev} = k_{ev}\ C_v$ (C)

Elimination rate for infected T cells $k_{ei}\ C_i$ (D)

where the subscripts I, v, and u denote infected cells, virus, and uninfected cells, respectively. N equals the number of viruses produced during the lifetime of the

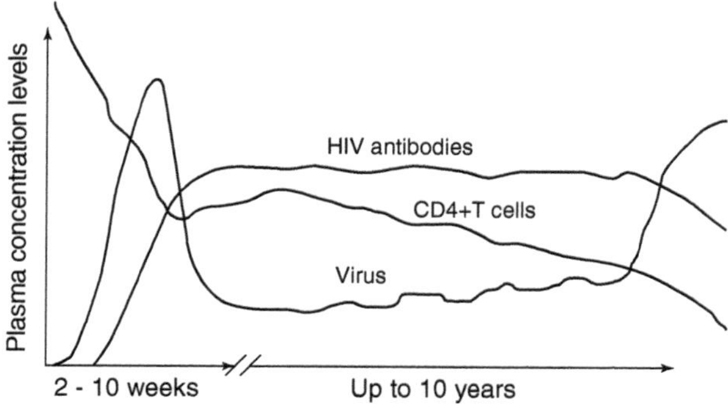

FIGURE 2.18 The course of HIV infection in a typical patient.

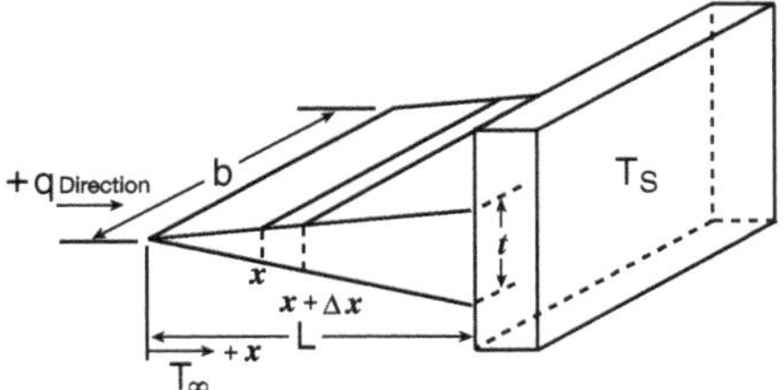

FIGURE 2.19 Geometry of a tapered heal exchanger fin.

cell, and the concentration of uninfected cells, being a large entity during the pre-AIDS period, is taken to be constant at C_u^0.

a. Set up unsteady mass balances for virus and infected cells.
b. Analyze the solution of these ODEs to establish the range of parameters that would yield the three possible viral pathways outlined earlier.

Hint: Combine the two equations into a second-order ODE.

2.9 Variable Coefficient ODEs: The Krogh Cylinder Solved by Separation of Variables

The Krogh cylinder is a model of the uptake of oxygen and nutrients from flowing blood into the surrounding tissue that metabolizes the substance at a constant rate k_0. The tissue region is taken to be a cylindrical shell with an impermeable outer surface ($r = r_0$) and an inner wall ($r = r_i$) that is exposed to the constant concentration C_0 in the blood. The mass balance for the diffusing species given in the literature has the form:

$$D_e\left(\frac{d^2C}{dr^2}+\frac{1}{r}\frac{dC}{dr}\right)-k_0=0 \tag{A}$$

where D_e = effective diffusivity in the tissue.

Transform this ODE to a form that can be integrated by separation of variables and solve for the concentration profile C:

Answer:
$$C-C_0=(k_0/D_e)\left[\frac{r^2-r_i^2}{4}-\frac{r_0^2}{2}\ln\frac{r}{r_i}\right]$$

2.10 More Variable Coefficient ODEs: The Heat Exchanger Fin and the Genesis of Bessel Functions

The rate of heat transfer in a conventional shell-and-tube heat exchanger can be increased by attaching external fins to the exchanger tubes. These fins, which are typically circular, rectangular, or tapered in shape, serve to enhance the tubular surface area, and through it, the rate of heat transfer.

A fin in the shape of a tapered wedge is shown in Figure 2.18. This particular geometry is of practical interest because it approximates the shape yielding the maximum heat flow per unit weight. In a typical application, heat will flow from the tubular wall by conduction into the fin, which loses heat by convection to an external medium held at a temperature T_∞. Ideally, if conduction were infinitely fast, the entire fin area would assume the same temperature as the tubular wall, and the total heat transfer area A_T would equal the sum of the exposed tubular and fin areas:

$$A_{Tot} = A_t + A_f \tag{A}$$

Because of the finite rate of conduction, however, a radial temperature profile will develop within the fin, which will reduce its effective contribution to the total heat transfer area below the value A_f. It is customary to express this reduction in terms of a fin efficiency E_f, defined as the ratio of the actual heat transfer rate to the ideal, maximum rate corresponding to infinitely fast conduction. Thus:

$$E = \frac{\textit{Actual rate of heat transfer}}{\textit{Maximum rate of heat transfer}} \tag{B}$$

Equation A must then be modified to read

$$A_T = A_t + E_f A_f \tag{C}$$

To derive an expression for E_f, the reader is asked to proceed as follows:

a. Derive the ODE that describes the radial temperature profile in the fin (Hint: Consult Illustration 2.12). It is of a form that cannot be solved by separation of variables as in Illustration 2.9.
b. Solve the ODE using Illustration 2.12 as a guide.
c. Derive an expression for E_f using the actual flux at the base of the fin.

Hint: Approximate the perimeter P of the fin by $P \approx 2b$.

Answer: (a) $x^2 \frac{d^2y}{dx^2} + x\frac{dy}{dx} - B^2 xy = 0$

2.11 The Bernoulli Equation

Derive a closed-form solution of the Bernoulli equation:

$$y' + f(x)y = g(x)y^n$$

using the substitution $u = y^{1-n}$.
(Hint: Use item 6 of Table 2.4).

2.12 The Pendulum

The equation of motion for a frictionless pendulum is given by:

$$\frac{d^2\theta}{dt^2} + \frac{g}{1}\sin\theta = 0$$

where θ = angle of deflection. Show that the closed-form solution of this problem is given:

$$\int \frac{d\theta}{\sqrt{g\cos\theta + C_1}} = \sqrt{\frac{2}{1}}t + C_2$$

2.13 The Duffing Equation

For a Duffing-type equation of the form:

$$\ddot{y} + \dot{y} - y + y^3 = 0$$

derive the stationary solutions and analyze their character and stability.

Answer: Critical points consist of an unstable saddle and stable focus.

2.14 The van der Pol Equation

Deduce the nature and stability of the stationary solution of the van der Pol equation:

$$\ddot{y} - \lambda(1 - y^2)\dot{y} + y = 0$$

in the parameter range

$$3 > \lambda > -3$$

Partial answer: There is a center at $y_1 = y_2 = \lambda = 0$.

2.15 Transformation and Inversion of Complex Functions

a. Show that:

$$L^{-1}\left\{\ln\left(1 + \frac{1}{s^2}\right)\right\} = \frac{2}{t}(1 - \cos t)$$

(Hint: Differentiate the transform with respect to s and decompose by partial fraction.)

b. Find the inverse of $s/(s + 1)^3$. (Hint: Decompose by partial fractions.) Answer: $te^{-1}(1 - t/2)$

c. Use a shifting theorem to evaluate $L^{-1}\ (s\ e^{-4\pi s/5})/(s^2 + 25)$.

d. Show that L{δ – a)} = e^{-st}. Use the fact that L{δ(t)} = 1{δ(t)} = Dirac function.

e. Find $L^{-1}\dfrac{2s^2 - 4}{(s + 1)(s - 2)(s - 3)}$.

Answer: $-\frac{1}{6}e^{-t} - \frac{4}{3}e^{2t} + \frac{7}{2}e^{3t}$

2.16 Solution of Simultaneous ODEs

Find the solution to the system of equations:

$$X'(t) - 2Y'(t) = F(t)$$

$$X''(t) - Y''(t) + Y(t) = 0$$

with $X(0) = X'(0) = Y(0) = Y'(0) = 0$.

(Hint: Eliminate one of the transforms, say y(s), algebraically and invert. Repeat the procedure for x(s).)

Answer:

$$X(t) = \int_0^t F(\tau)d\tau - 2\int_0^t F(\tau)\cos(t-\tau)d\tau$$

$$Y(t) = -\int_0^t F(\tau)\cos(t-\tau)d\tau$$

2.17 Radioactive Decay Series

The decay of radioactive elements is the classical example of a reaction that is accurately described by a first order rate law. Consider the series:

$$N_1 \xrightarrow{k_1} N_2 \xrightarrow{k_2} N_3 \xrightarrow{k_3} N_4$$

where N_i = number of atoms of element I and the reaction rates are represented by:

1. $dN_1/dt = -k_1N_1$
2. $dN_2/dt = -k_2N_2 + k_1N_1$
3. $dN_3/dt = -k_3N_3 + k_2N_2$
4. $dN_4/dt = k_3N_3$

Show that the number of atoms of the last and stable species is given by:

$$\frac{N_4}{N_1^0} = 1 - \frac{k_2k_3e^{-k_1t}}{(k_2-k_1)(k_3-k_1)} - \frac{k_1k_3e^{-k_2t}}{(k_1-k_2)(k_3-k_2)} - \frac{k_1k_2e^{-k_3t}}{(k_1-k_3)(k_2-k_3)}$$

Note that the ODEs are not coupled so that they can, in principle, be solved in succession. This could be done by using the D-operator method, but the repeated appearance of nonhomogeneous terms would require the evaluation of a set of particular integrals. The Laplace transformation avoids this step and is therefore easier to apply.

2.18 Oscillation of an Electrical RLC Circuit

A circuit consisting of an induction coil with inductance L, a resistor with resistance R, and a capacitor with capacitance C are connected in series and subjected to an alternating voltage $E = E_0 \sin \omega t$. Derive an expression for the current i as a function of time.

2.19 Design of a Thermocouple for Oscillating Temperature Fluctuation

A thermocouple is to be used to register sinusoidal temperature oscillations given by the expression:

$$T(t) - 50 = 100 \sin [2t(\text{sec})]$$

The dimension of the thermocouple, composed of cylindrical wires, should be such that the maximum temperature is registered within 3% of its actual value. What is the maximum permissible diameter to achieve this?

Data: $h = 100$ J/sm^{2}K, $\rho = 9000$ kg/m^3, $C_p = 0.4$ kJ/kg K.

(Hint: Use an equation of the form given by Equation 2.37f.)

2.20 More on the Response of a Second-Order System

A second-order system has a transfer function given by:

$$\frac{y(s)}{x(s)} = \frac{s-10}{5s^2+3s+1}$$

Verify that the system is underdamped, and determine its characteristics without inverting.

3 The Use of Mathematica in Modeling Physical Systems

Advances in computing technology have led to the development of various tools for mathematical modeling in recent years. These tools significantly simplify the derivation and analysis of physical models. In addition to a rich library of analytical and numerical functions, such software are capable of plotting results of a model in terms of xy or xyz graphs, which are powerful ways to visualize the behavior of the system. The tools available for mathematical modeling can be broadly divided into numerical and analytical calculations, although some software can handle both types of calculations to various degrees. The use of software significantly simplifies the modeling process, however, it does not replace the insight and creativity of the modeler. Instead, the combination of the two offers an excellent way for system modeling and analysis.

In this chapter, we provide a concise introduction to one of such tools, Mathematica (Wolfram Research Inc., IL), and demonstrate its application in the modeling of physical systems. Several examples, including a number of illustrations already discussed in Chapter 1 and Chapter 2, will be revisited with the aid of Mathematica. This is an efficient way to develop and analyze models all in the same environment and to consolidate the modeling process. However, the procedure is far from automatic, and user intervention is required to achieve the desired answer.

3.1 HANDLING ALGEBRAIC EXPRESSIONS

In the process of developing mathematical models, we always need to rearrange and simplify analytical expressions. There are several functions in Mathematica that can perform such an operation, including `Simplify`, `Factor`, `Expand`, `Apart`, `Refine`, and `FullSimplify`. The use of these functions is demonstrated in the following examples:

```
 In[1]:= (x + y)(x^2 - 2xy + y^2)
Out[1] = (x + y)(x² - 2xy + y²)
```

```
 In[2]:= Simplify[%]
Out[2] = (x - y)²(x + y)
```

```
 In[3]:= Expand[%]
Out[3] = x³ - x²y - xy² + y³
```

In[4]: = **Factor[%]**
out[4] = $(x - y)^2(x + y)$

The symbol "%" in the preceding expressions refers to the output from the previous line. Now, consider:

In[5]:= $\dfrac{\sqrt{a^2b^2}}{a + b}$

Out[5] = $\dfrac{\text{Sqrt}[a^2b^2]}{a + b}$

As a and b are unknown parameters, neither `Simplify` nor `Expand` is able to further reduce this expression. Let us try:

In[6]:= **Refine[%,{a > 0 && b < 0}]**

Out[6]= $-\left(\dfrac{ab}{a + b}\right)$

The function `Refine` gives the form of an expression under the specified assumption that a is positive and b is negative. In simplifying the preceding expression, we may use the function `PowerExpand;` however, as you can see, this function by default assumes that both a and b are positive.

In[7]:= **PowerExpand[Out[5]]**

Out[7]= $\dfrac{ab}{a + b}$

Now let us rearrange this expression:

In[8]:= **Apart[%]**

Out[8]= $a - \dfrac{a^2}{a + b}$

Suppose we need to determine the preceding expression for a specific value of a, say for $a = \dfrac{1}{2+b}$:

In[9]:= **% /. a →** $\dfrac{1}{2 + b}$

Out[9]= $\dfrac{1}{2 + b} - \dfrac{1}{(2 + b)^2\left(b + \dfrac{1}{2 + b}\right)}$

```
In[10]:= Simplify[%]
```

$$\text{Out[10]}= \frac{b}{(1 + b)^2}$$

The construct /. a −> $\frac{1}{2+b}$ assigns $\frac{1}{2+b}$ to a.

In handling and rearranging algebraic equations, we often need to use a combination of several functions to convert an expression to a desirable form.

3.2 ALGEBRAIC EQUATIONS

Algebraic equations are the most common type of expressions encountered in engineering modeling. Such an equation in Mathematica is expressed as $f(x) == 0$.

3.2.1 Analytical Solution to Algebraic Equations

The analytical solution to an algebraic equation, if available, can be obtained using the Mathematica function `Solve`.

Let us consider x^3 − 3x^2 − x − 3 == 0. Now, let us examine the behavior of this expression and its intercept with the x-axis by plotting it:

```
In[1]:= Plot[X^3 + 3X^2 - X - 3,{X,-5,5}]

out[1]= -Graphics-
```

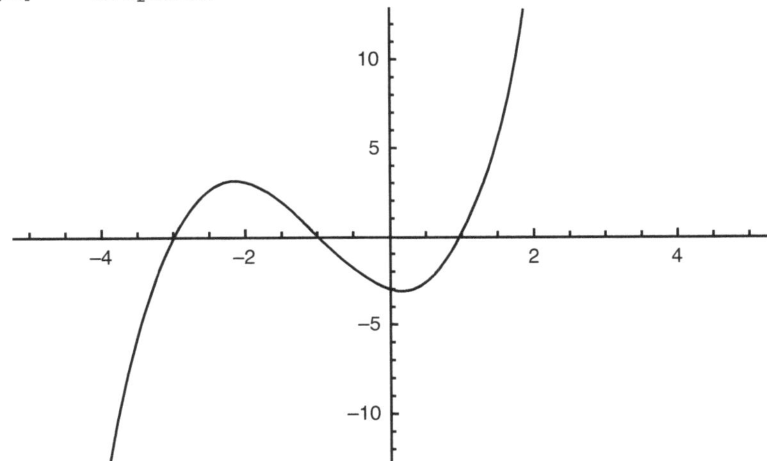

Examination of the preceding plot shows that there are three real solutions. To obtain the exact values of these solutions, we write:

```
In[2]:= Solve[x^3 + 3x^2 - x - 3 == 0, x]
Out[2]= {{x -> -3},{x -> -1},{x -> 1}}
```

Now, we can check the validity of answers. For instance:

```
In[3]:= x^3 + 3x^2 - x - 3 /. x -> -3
Out[3] = 0
```

Similarly, the function `Solve` can handle equations containing unknown parameters or a system of equations in more than one variable. For example, consider the following equations: the first one is a nonlinear equation with two unknown parameters a and b, and the second one is a nonlinear system of three variables:

In[4]:= **Solve[(1 + x^3)^(a+x) == (1 + x^3)^b, x]**

Out[4]= {(x -> 0), {x -> -a + b}}

In[5]:= **Solve[{(x - y)z^2 == 1, x^2 == 4, y^2 = = 4}, {x, y, z}] // TraditionalForm**

Out[5]//TraditionalForm=

$$\left\{\left\{x \to -2, y \to 2, z \to -\frac{1}{2}\right\}, \left\{x \to -2, y \to 2, z \to \frac{1}{2}\right\}, \left\{x \to 2, y \to -2, z \to -\frac{1}{2}\right\}, \left\{x \to 2, y \to -2, z \to \frac{1}{2}\right\}\right\}$$

If only real solutions of the preceding system of equation are desired, one can use the function `Reduce` to narrow down the results:

In[6]:= **Reduce[{(x - y) z^2 == 1, x^2 == 4, y^2 == 4}, {x, y, z}, Reals]**

Out[6]= $x == 2 \&\& y == -2 \&\& \left(z == -\left(\frac{1}{2}\right) \,||\, z == \frac{1}{2}\right)$

3.2.2 Numerical Solution to Algebraic Equations

For numerical solution of a system of polynomial equations, the function `NSolve` can be used:

```
In[7]:= NSolve[x^3.1 + x^2 - 5.822 == 0, x]
Out[7]= {{x -> -1.11407 + 1.60899 I},
        {x -> -1.11407 - 1.60899 I}, {x -> 1.50557}}
```

If the expression is not convertible to a polynomial, the function `FindRoot` should be used. The following examples demonstrate the use of this function.

In[8]:= **FindRoot[x + x^2 - 5.822 = = 0, {x, 1}]**

Out[8]= {x -> 1.49977}

In the previous examples, `FindRoot` uses the Newton method based on the initial value of 1. If two initial values are used, the secant method is applied to determine the solution:

```
In[9]:= FindRoot[x - e^x/Sin[x] ==0, {x, 1, 4}]
Out[9]= {x -> 1.76343}
```

3.3 INTEGRATION

In analyzing physical systems, we often need to determine the cumulative or average value of a distributed variable over space or time. Such a calculation requires the use of integration, and tables of integrals are commonly used to help with this operation. In Mathematica we can take advantage of the function `Integrate` to perform definite or indefinite integrals:

```
In[1]:= Integrate[x^n a^x, x] // TraditionalForm
Out[1]//TraditionalForm=
```

$$-\ x^{x+1}\Gamma(\eta + 1, -x\log(\alpha))\,(-x\log(\alpha))^{-x-1}$$

Let us consider Illustration 2.12. Starting from Equation 2.23f, we obtain the expression for catalyst effectiveness:

```
In[2]:= Integrate[2 pi kr Cs L r BesselI[0, αr], {r, 0, R}]
Out[2]= 2 Cs kr L Pi R BesselI[1, R α] / α
```

```
In[3]:= % / (BesselI[0, αR]kr Cs Pi R^2L)//TraditionalForm
Out[3]//TraditionalForm=
```

$$\frac{2I_1(R\alpha)}{R\alpha I_0(R\alpha)}$$

Now, let us consider the heating up of an unstirred tank (Illustration 4.19); the integration of Equation 4.27 is conducted here using Mathematica:

```
In[4]:= Integrate[1/(Ts - T)^(4/3), {T, Ti, Tf},
        GenerateConditions → False]
```

$$\text{Out[4]= } \frac{3}{(-Tf + Ts)^{1/3}} - \frac{3}{(-Ti + Ts)^{1/3}}$$

It is important to use the option GenerateConditions ->False in the preceding expression to narrow down the result and to obtain the relevant answer. Otherwise, depending on the relative values of Ts, Tf, and Ti, several possible answers will be obtained. Alternatively, we may achieve the same outcome by using the option Assumption:

```
In[5]:= Integrate[1/(Ts - T)^(4/3), {T, Ti, Tf},
        Assumptions → (Ts > Tf && Tf > = T && T > =
        Ti && 0 < Ti < Tf)]
```

$$\text{Out[5]=}\ \frac{3}{(-\text{Tf} + \text{Ts})^{1/3}} - \frac{3}{(-\text{Ti} + \text{Ts})^{1/3}}$$

The numerical value of this integral can be calculated if the actual values of the upper and lower limits are used in the integration:

```
In[6]:= Integrate[1/(120 - T)^(4/3), {T, 20, 100}]
```

$$\text{Out[6]=}\ \frac{3(-1 + 5^{1/3})}{10^{2/3}}$$

```
In[7]:= % // N
Out[7]= 0.458879
```

In case there is no closed-form solution to the integral, numerical integration can be performed using NIntegrate function:

```
In[8]:= NIntegrate[1/(120 - T)^(4/3), {T, 20, 100}]
Out[8]= 0.458879
```

NIntegrate uses an adaptive algorithm to divide the integration region as needed to achieve the desired precision. The default method for numerical integration in Mathematica is Gauss–Kronrod. This is a Gaussian quadrature with error estimation based on evaluation at Kronrod points.

3.4 ORDINARY DIFFERENTIAL EQUATIONS

As discussed earlier, in modeling of physical systems, we often resort to a differential approach in which conservation laws and auxiliary equations are applied over a differential element. Such models lead to differential equations with various degrees of complexity. Broadly speaking, we can divide differential equations into two categories: ordinary differential equations (ODEs) and partial differential equations (PDEs). In this section, we will illustrate the use of Mathematica to such equations.

3.4.1 Analytical Solution to ODEs

In Chapter 2, we provided a comprehensive introduction to the analytical solutions of ODEs. Separation of variables is perhaps the most widely used method in solving differential equations encountered in the modeling of physical systems. The solution in such a case involves the segregation of dependent and independent variables and the integration of the resulting equation. This requires the manipulation of algebraic equations and use of `Integrate` or `NIntegrate` functions to obtain the final solution. Let us revisit the burning fuel problem.

Illustration 3.1 The Burning Fuel Problem Revisited

Here, we take on solving Illustration 2.4 with the aid of Mathematica. We begin with the definition of a function that describes the heat transfer by conduction, q, in terms of the radial position, r:

In [1]:= **q[r_] := k_v4 r^2T′[x]**

In the preceding expression, the syntax ": =" reflects the definition of function `q[r]`. Here, T'(r) is the temperature gradient in the gas phase. Similarly, we define function H(r) that represents the head convection in the gas phase:

In[2]:= **H[r_] := Fc C_{pv}(T[x] - To)**

Let us begin from the quasi-steady-state differential energy balance in the gas film. We focus on the left-hand side of Equation 2.12e:

In[3]:= **lhs = D[q[r], r] - D[H[r], r]**

Out[3]= $-(Fc\ C_{pv}\ T'[r]) + 8\ Pi\ r\ k_v\ T'[r] + 4\ Pi\ r^2\ k_v T''[r]$

Integrating from the surface of the droplet (R_1) to some arbitrary distance (r) yields:

In[4]:= **lhs = Integrate[lhs, {r, R_1, r}]**
/ .{T[R_1]'T_1, T'[R_1] → T'_s)

Out[4]= $Fc\ C_{pv}\ (T_1 - T[r]) + 4\ Pi\ k_v(-R_1^2 T'_s) + r^2 T'[r])$

The temperature and temperature gradient at the surface of the droplet are assigned T_1 and T'_s, respectively. Now, we reorganize this expression:

In[5]:= **lhs = Collect[lhs / (4 k_vT′[r]), r]**

$$\text{Out[5]= } r^2 + \frac{-4\ Pi\ k_v\ R_1^2 T'_s + Fc\ C_{pv}\ (T_1 - T[r])}{4\ Pi\ k_v\ T'[r]}$$

We now separate the variables r and T by moving the second term of the preceding expression to the right-hand side and inverse both sides. By doing this we will arrive at Equation 2.12h:

```
In[6]:= rhs = -1/Part[lhs, 2]
```

$$\text{Out[6]=}\ \frac{-4\,\text{Pi}\,k_v\,T'[r]}{-4\,\text{Pi}\,k_v\,R_1^{\,2}\,T'_s + \text{Fc}\,C_{pv}\,(T_1 - T[r])}$$

```
In[7]:= lhs = 1/Part[lhs, 1]
```

$$\text{Out[7]=}\ r^{-2}$$

Integrating the left- and right-hand sides from the surface of the droplet (R_1) to some arbitrary distance (R_2) and on further simplification yields:

```
In[8]:= rhs = Integrate[rhs, {r, R1, R2}]
        / .{T[R1] → T1, T[R2] → T2}
        // TraditionalForm
```

$$\text{Out[8]//TraditionalForm=}\ \frac{4\pi(\log(4\pi k_v(T)\,,R_1^2 + \text{Fc}\,C_{pv}(T_2 - T_1)) - \log(4\pi k_v R_1^2(T)_s))k_v}{\text{Fc}\,C_{pv}}$$

```
In[9]:= 1/(FcCpv/(4πkv)) Log[FullSimplify[Exp[Fc Cpv/(4πkv) × %]]]
```

$$\text{Out[9]=}\ \frac{4\,\text{Pi}\,\text{Log}[1 + \frac{\text{Fc}\,C_{pv}\,(-T_1 + T_2)}{4\,\text{Pi}\,k_v\,R_1^{\,2}\,T'_s}]k_v}{\text{Fc}\,C_{pv}}$$

At this point, we introduce the time dependency of the droplet diameter:

```
In[10]: = rhs = % / . R1 -> R1[t]
```

$$\text{Out[10]=}\ \frac{4\,\text{Pi}\,\text{Log}[1 + \frac{\text{Fc}\,C_{pv}\,(-T_1 + T_2)}{4\,\text{Pi}\,k_v\,T'_s\,R_1[t]^2}]k_v}{\text{Fc}\,C_{pv}}$$

Using the core energy balance, Equation 2.12c, and assuming that $H_L = 0$, obtain an expression for the temperature gradient at the surface of the droplet and eliminate it from the right-hand side:

```
In[11]:= q[R1[t]] - Fc ΔHv/. T' [R1[t]] -> T's
```

$$\text{Out[11]=}\ -(\text{Fc}\ \Delta H_v) + 4\,\text{Pi}\,k_v\,T'_s\,R_1[t]^2$$

In[12]:= **Solve[% == 0, T'_s]**

Out[12]= $\{(T'_s \to \dfrac{Fc\ \Delta H_v}{4\ Pi\ k_v R_1[t]^2}\}\}$

In[13]:= **rhs = rhs /. T'_s -> T'_s /.%[[1, 1]]**

Out[13]= $\dfrac{4\ Pi\ Log[1 + \dfrac{C_{pv}(-T_1 + T_2)}{\Delta H_v}]k_v}{Fc\ C_{pv}}$

Similarly, left-hand side of Equation 2.12h can be integrated. Here, we need to specify that both limits of integral (R_1 and R_2) are positive numbers:

In[15]:= **lhs = Integrate[lhs, {r, R_1[t], R_2[t]}, Assumptions → (R_1[t] > 0 && R_2[t] > 0)]**

Out[15]= $\dfrac{1}{R_1[t]} - \dfrac{1}{R_2[t]}$

At this point, let us simplify the right-hand side and left-hand side and further assume that R_2 remains proportional to R_1:

In[16]:= **lhs = Simplify[lhs $\dfrac{C_{pv}\rho_L R_1[t]^2 R'_1[t]}{k_v}$ /. $R_2[t] \to R_1[t]/\beta$]**

Out[16]= $-(\dfrac{(-1 + \beta)\ C_{pv}\rho_L R_1[t]\ (R_1)'\ [t]}{k_v})$

In[17]:= **rhs = rhs $\dfrac{C_{pv}\rho_L R_1[t]^2 R_1{}'[t]}{k_v}$**

Out[17]= $-Log[1 + \dfrac{C_{pv}(-T_1 + T_2)}{\Delta H_v}]$

Assuming that only R_1 varies with time, the preceding expressions provide a way to find the time dependency of the droplet diameter by integrating both sides over t and solving for R_1

In[18]:= **lhs = Integrate[lhs, {t, 0, t}] /. {R_1[0] → $R_{1,0}$, R_1[t] -> R_1}**

Out[18]= $\dfrac{(-1 + \beta)\ C_{pv}\ \rho_L(-R_1{}^2 + R_{1,0}{}^2)}{2\ k_v}$

In[19]:= **rhs = Integrate[rhs,{t,0,t}]**

$$\text{Out[19]= } -(t\,\text{Log}[1 + \frac{C_{pv}(-T_1 + T_2)}{\Delta H_v}])$$

In[20]:= **Solve[rhs - lhs == 0, R_1]**

Out[20]=

$$\{\{R_1 \to -(\frac{\text{Sqrt}[t\,\text{Log}[1+\frac{C_{pv}(-T_1+T_2)}{\Delta H_v}] - \frac{C_{pv}\,\rho_L\,R_{1,0}^2}{2\,k_v} + \frac{\beta\,C_{pv}\,\rho_L\,R_{1,0}^2}{2\,k_v}]}{\text{Sqrt}[\frac{-(C_{pv}\,\rho_L)}{2\,k_v} + \frac{\beta\,C_{pv}\,\rho_L}{2\,k_v}]})\},$$

$$\{R_1 \to \frac{\text{Sqrt}[t\,\text{Log}[1+\frac{C_{pv}(-T_1+T_2)}{\Delta H_v}] - \frac{C_{pv}\,\rho_L\,R_{1,0}^2}{2\,k_v} + \frac{\beta\,C_{pv}\,\rho_L\,R_{1,0}^2}{2\,k_v}]}{\text{Sqrt}[\frac{-(C_{pv}\,\rho_L)}{2\,k_v} + \frac{\beta\,C_{pv}\,\rho_L}{2\,k_v}]})\}$$

Where $R_{1,0}$ is the initial droplet diameter. Finally, R_1^2 (t) will become:

In[21]:= **(R_1/.%[[1,1]])^2**

$$\text{Out[21]= } \frac{t\,\text{Log}[1+\frac{C_{pv}(-T_1+T_2)}{\Delta H_v}] - \frac{C_{pv}\,\rho_L\,R_{1,0}^2}{2\,k_v} + \frac{\beta\,C_{pv}\,\rho_L\,R_{1,0}^2}{2\,k_v}}{\frac{-(C_{pv}\,\rho_L)}{2\,k_v} + \frac{\beta\,C_{pv}\,\rho_L}{2\,k_v}}$$

In[22]:= **FullSimplify[%]**

$$\text{Out[22]= } \frac{2t\,\text{Log}[1+\frac{C_{pv}(-T_1+T_2)}{\Delta H_v}]\,k_v}{(-1+\beta)\,C_{pv}\,\rho_L} + R_{1,0}^2$$

Comments: As the reader may realize from this illustration, modeling in Mathematica is certainly not an automated process and, in fact, involves similar steps when compared to a pen-and-paper approach to modeling. The advantage of Mathematica, however, is that it simplifies algebraic manipulation, integration, and differentiation by reducing the labor involved. Moreover, consolidating the model in a Mathematica file, called a notebook, offers an excellent opportunity to explore various scenarios and analyze the behavior of a physical system.

Analytical solutions to differential equations may be obtained using `DSolve`. This function can solve linear ODEs of any order with constant coefficients, many linear equations up to second order with nonconstant coefficients, and differential algebraic equations. Linear systems of differential equations can also be solved using `DSolve`.

Illustration 3.2 Polymer Sheet Extrusion Revisited

Here, we revisit the polymer sheet extrusion problem, and specifically the solution to the differential equation, Equation 2.14d:

```
In[1]:= DSolve[p''[x] - m^2p[x] == 0, p[x], x]
```

$$\text{Out[1]= } \{\{p[x] \to E^{m\,x}C[1] + \frac{C[2]}{E^{m\,x}}\}\}$$

The integration constants, C[1] and C[2], can be obtained from the boundary conditions. Alternatively, we can introduce the boundary conditions in the DSolve:

```
In[2]:= DSolve[{p''[x]-m^2p[x]==0, p[0]==p0,
        p'[Lp] == 0}, p[x], x]
```

$$\text{Out[2]= } \{\{p[x] \to \frac{(E^{2\,Lp\,m} + E^{2\,m\,x})p0}{E^{m\,x}\,(1+E^{2\,Lp\,m})}\}\}$$

We now convert the exponentials to hyperbolic functions and simplify the result:

```
In[3]:= ExpToTrig[p[x] /.%[[1, 1]]]
Out[3]= (p0 (Cosh[m x] - Sinh[m x]) (Cosh[2Lp m]
        +Cosh[2 m x] + Sinh[2 Lp m] + Sinh[2 m x])) /
        (1 + Cosh[2 Lp m] + Sinh[2 Lp m])

In[4]:= TrigFactor[%]
Out[4]= p0 Cosh[Lp m - m x] Sech[Lp m]
```

Illustration 3.3 The Path of Pursuit Revisited

We begin from Equation 2.25g and use Mathematica to solve this differential equation with the boundary conditions, Equation 2.25h and Equation 2.25i. Hence, we consolidate all the operations in a nested fashion in such a way that only the final answer will be viewed.

```
In[1]:= Apart[
         Simplify[
          PowerExpand[
            Apart[
             Simplify[
              Simplify[
                 TrigToExp[
                                                   b^2  2
                      y[x]/. DSolve[{1+(y'[x])^2 == ---- x (y"[x])^2,y'
                                                   a^2
                      [x0]==0, y[x0]==0},y[x],x][[2,1]]
                 ] /.a -> kb
              ] /.
              {Sqrt[-2+x0^(-2*k)+x0^(2*k)]->((-1+x0^K)*
              (1+x0^k))/x0^k, x->s x0}
             ]
            ]
          ] /.k -> 1/k
         ]
        ]
```

$$\text{Out[1]= } \frac{k\,x0}{-1 + k^2} - \frac{k\,s^{1-1/k}x0}{2(-1 + k)} + \frac{k\,s^{1+1/k}x0}{2(1 + k)}$$

where s represents the ratio x/x_o.

Comment:

1. Solving Equation 2.25g using `DSolve` function results in four possible solutions. However, only one of these solutions is acceptable (the proof is left to the interested reader).
2. To develop a nested procedure requires a step-by-step approach similar to Illustration 3.1. Once the procedure is developed, one may combine the operation in a concise fashion, such as the one presented in the preceding example.

3.4.2 Numerical Solution to Ordinary Differential Equation

The numerical solution to ODEs can be obtained using the function `NDSolve`. This function by default uses an LSODA approach, switching between a nonstiff Adams method and a stiff Gear backward differentiation formula method [ref: Mathematica handbook]. However, a different method may be selected in Mathematica from a list of available techniques including several variations of the Runge–Kutta method. The solutions for the differential equation are given in terms of an interpolating function that provides approximations to the solution over the specified domain. For instance, consider:

```
In[1]:= V = NDSolve[{x''[t] == x[t]^2, x[0] == 1, x'[0]
        == 0}, x, {t, 0, 2.2}]
Out[1]= {{x -> InterpolatingFunction[{{0., 2.2}},<>]}}
```

We now plot the result:

```
In[2]:= Plot[Evaluate[ x[t] /. v],{t, 0, 2.2}, Frame->True,
        FrameLabel->("t", "x(t)", "Numerical solution to
        ODE", ""}];
```

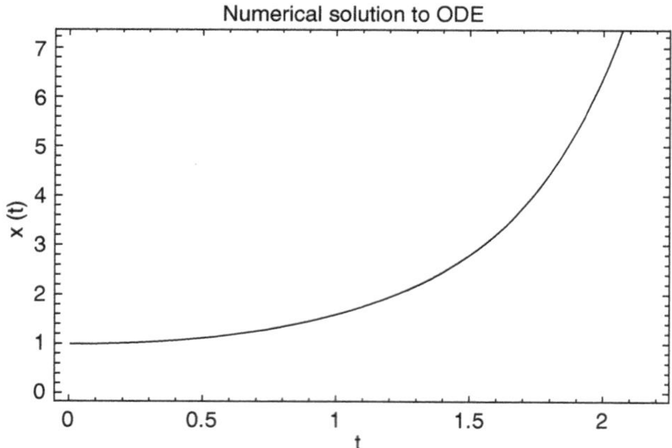

Illustration 3.4 Numerical Solution to the Path of Pursuit Problem

Here, we solve the path of pursuit problem using the Runge–Kutta method and compare the solution to the exact analytical solution presented in the previous illustration, defined as function f[x].
We assume that the initial position of the prey $x_0 = 1$, and that the speed of prey and predator are 1 and 1.5, respectively.

```
In[1]:= f[x_] := b x0/(b^2 - a^2) - b x0(x / x0)^(1-a/b)/(2(b - a)) + b x0(x / x0)^(1.a/b)/(2(a + b));
        a = 1; b = 1.5; x0 = 1;
        path = NDSolve[{(1 + (y'[x])^2)/(y''[x])^2 == b^2/a^2 x^2, y'[x0]
        == 0, y[x0] == 0}, y[x],
          {x, 1, 100}, Method → ExplicitRungeKutta, MaxSteps → 100000];
        p1 = Plot[Evaluate[Abs[y[x] /.path]], {x, 1, 100},
          PlotRange → All, Frame → True, DisplayFunction
          → Identity];
        p2 = Plot[f[x], {x, 1, 100}, PlotRange → All, Frame → True,
          DisplayFunction → Identity];
        Show[p1, p2, DisplayFunction → $DisplayFunction,
         FrameLabel → {"x", "y(x)", "Path of Pursuit", ""}]
```

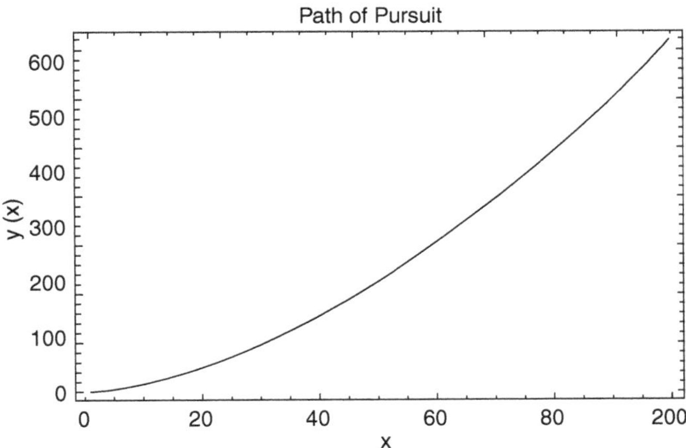

Plots p1 and p2 represent the numerical and analytical solutions of the differential equations. As seen from the preceding graph, p1 and p2 exactly overlay on top of each other, indicating that the numerical and analytical solutions lead to practically identical results.

Comment: The number of steps in the numerical solution was set at 100,000, up from the default value of 10,000, to reach the correct answer. Using the default number of steps will lead to an erroneous result.

3.5 PARTIAL DIFFERENTIAL EQUATIONS

The uses of Mathematica for solving PDEs are discussed in Chapters 5 to 8. Here, we only point out that, unlike the ODEs, the analytical solution to PDEs in Mathematica, in general, is not automated. Therefore, user should follow the standard solution methods for PDEs. In contrast, the numerical solutions to PDEs are handled using `NDSolve`.

PRACTICE PROBLEMS

3.1 Tubular Heat Exchanger

Solve Practice Problem 2.5 using Mathematica.

3.2 Flow Through a Porous Duct

A fluid flows in the positive x-direction through a long flat duct of length L, width W, and thickness B, where L >> W >> B. The duct has porous walls at y = 0 and y = B, so that a constant cross flow can be maintained (with $v_y = v_o$) everywhere. The differential equation governing the velocity profile in the x-direction for this system, v_x, is given by:

$$\rho v_0 \frac{dv_x}{dy} = \mu \frac{d^2 v_x}{dy^2} + \frac{\Delta P}{L}$$

Using Mathematica, obtain an expression for the velocity profile $v_x(y)$. For simplicity, use the following change of variables: $\eta = \frac{y}{B}$, $\Phi = \frac{v_x}{B^2 \frac{\Delta P}{\mu L}}$, and $A = \frac{B v_0 \rho}{\mu}$.

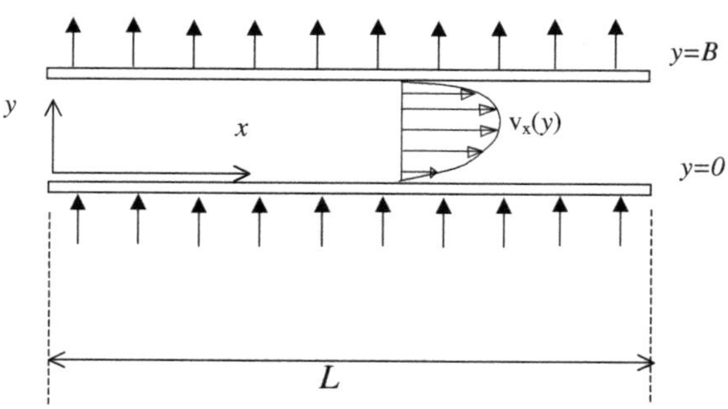

FIGURE 3.1 Flow through porous duct (Practice Problem 3.2.).

3.3 Flow in a Packed Bed

The modified Darcy's law describing the superficial velocity, V_{0z}, along the axial, z, direction through a uniform packed bed with circular cross section of radius R is given by (Brinkman correction):

$$\mu \frac{1}{r}\frac{d}{dr}\left(r\frac{dV_{0z}}{dr}\right) - \frac{\mu}{\kappa}V_{0z} - \frac{dP}{dz} = 0$$

where r is the radial position from the centerline of packed column.

For a constant applied pressure gradient ($-dP/dz = \Delta P/L$ = constant) and by using Mathematica, obtain an expression for the velocity distribution in the packed column, and find the total volumetric flow rate through the packed bed.

3.4 Concentration of Ligand within a Receptor

Cellular processes can assume a variety of forms. In one version, substances within the cell undergo a reaction, accompanied by diffusional inflow and outflow of reactants and products through the cell wall. One of such processes is the binding of ligand molecules from the extracellular fluid.

Consider the simple case that both the ligand and the receptor of radius R move freely and interact in the extracellular fluid. The steady state differential equation expressing the diffusion of ligand within a spherical receptor is:

$$\frac{d^2L}{dr^2} + \frac{2}{r}\frac{dL}{dr} = 0$$

where L is the ligand concentration at position r within the receptor.

Assuming that the concentration of ligand in the bulk of extracellular fluid ($r = \infty$) is known as L_∞, and that at the surface of the receptor the rate of diffusion of ligand is:

$$4\pi R^2 D \left.\frac{dL}{dr}\right|_{r=R} = k_r L(R)$$

where D is the diffusion coefficient of ligand in the receptor, k_r is the rate of ligand debinding, and $L(R)$ is the concentration of ligand at the surface of the receptor. Determine an expression for the concentration distribution of ligand within the spherical receptor.

3.5 Tanks in Series

Consider two tanks in series shown in the following figure. Assume that the volumetric flow rate q leaving each tank is proportional to the volume of liquid in the

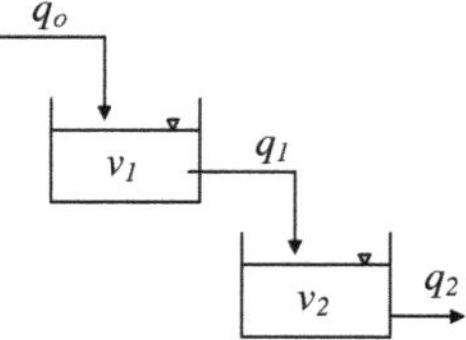

FIGURE 3.2 Tanks in series (Practice Problem 3.5).

tank v (i.e., $q_i = kv_i$, for $i = 1$ and 2). At time $t = 0$, liquid starts flowing into the first tank at a constant rate of q_o. The mass balances in the tanks are:

$$\frac{dv_1}{dt} = q_o - kv_1$$

$$\frac{dv_2}{dt} = k(v_1 - v_2)$$

The tanks are initially empty. Determine an expression for the liquid volume in tank 2 as a function of time.

4 Elementary Applications of the Conservation Laws

In the previous three chapters, we laid the groundwork required to model physical systems based on the conservation laws and presented a number of illustrations which provided the reader a first glimpse of the scope and power of mathematical modeling. The reach of this discipline is, in fact, quite considerable, perhaps even unsurpassed, within the applied sciences and engineering. When properly applied, it illuminates the topic it addresses in a way that is only matched by direct experimentation. As with experimentation, modeling pokes and prods the subject, often revealing it in a new light or uncovering the unexpected. Most major discoveries have come about through the twin efforts of experiment and theory, of which modeling is a major weapon.

The present chapter reinforces these notions by applying the conservation laws, together with appropriate auxiliary relations, to a range of physical problems. The treatment at this stage is kept at an elementary level involving algebraic and ordinary differential equations, as well as making occasional use of basic vector and matrix manipulations. The organization is in accordance with the principal conservation laws involved, rather than the category of the physical system considered. In this fashion, we progress through applications of force, mass, and energy balances, as well as combinations of them. The mathematical tools having already been presented, the main emphasis will be on the physics of the systems, the lessons provided by the modeling process, and revelations of the unexpected. In fact, we hope that readers will regard the illustrations as a way of exploring and comprehending the physical world around them.

4.1 APPLICATION OF FORCE BALANCES

Force balances, and various associated versions of Newton's law, are used with unusual frequency over a wide spectrum of the physical and engineering sciences.

The range and diversity of forces is often submerged in the conventional view which associates force with purely mechanical systems involving solids (e.g., the force acting on or produced by a moving object). The range of systems involving forces is, in fact, much wider. Gravitation, buoyancy, centrifugal force, and electrical and magnetic field forces are prominent categories. Fluid forces, both static and dynamic, are called upon to compose force balances in the domain of fluid mechanics. At the molecular level, the impact of gas molecules (which translates into pressure) and the intermolecular liquid forces that give rise to surface tension are other important contributors.

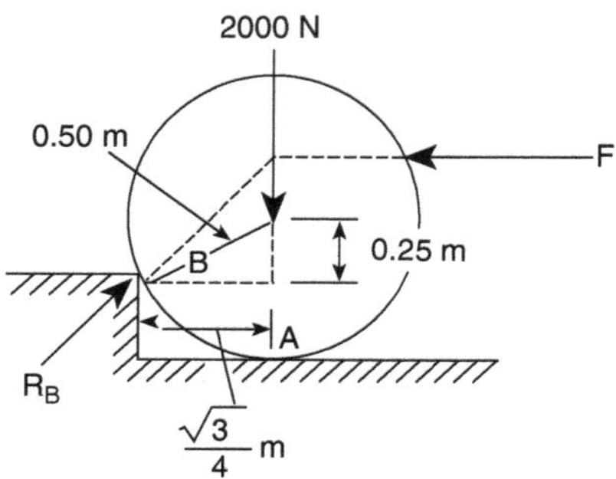

FIGURE 4.1 Rolling on a cylinder over an obstacle.

What we wish to do in this section is to illustrate the use of force balances with a number of examples which are marked by diversity of application rather than mathematical complexity. It is surprising how the simplest of expressions often yield the most interesting results. This will also be our theme in subsequent sections.

Illustration 4.1 A Simple Problem of Solid Statics: Rolling a Body over an Obstacle

A cylindrical liquid-storage tank weighing 2000 N (~ 204 kg) and of diameter 1 m is to be pushed over a change of elevation of 25 cm by a horizontal force **F** as indicated in the accompanying Figure 4.1. The task is to determine the magnitude of **F** for the tank to just lift off the ground.

As no actual movement occurs, we are dealing with a static system for which both the sum of forces and moments must vanish. The latter condition is the working equation for our case. We take moments about the point B and obtain:

$$\sum M = \sum r \times F = 0 \tag{4.1a}$$

where **r** is the vector from the moment center to any point on **F**. For the moment due to the weight of the tank, we choose the shortest distance d (or level arm) from B to **W**, so that:

$$|\boldsymbol{M}| = |\boldsymbol{W}|\,|\boldsymbol{r}| \times sm\alpha = (d)(W) \tag{4.1b}$$

where from the geometry of the system:

$$d = [(0.5)^2 - (0.25)^2]^{1/2} = \frac{\sqrt{3}}{4} \tag{4.1c}$$

The moment due to **F** is handled in a similar fashion so that the total moment is given by:

$$\sum |Moments| = -0.50|\boldsymbol{F}| + \left(\frac{\sqrt{3}}{4}\right)2000 = 0 \tag{4.1d}$$

from which

$$|\boldsymbol{F}| = 1732 \text{ N} \tag{4.1e}$$

Comments: Equation 4.1d allows a quick assessment to be made of the force required as a function of its position. The minimum requirement occurs when the force is applied at the top. We obtain:

$$F_{Min} = \frac{0.50}{0.75} \times 1732 = 1160\,N \tag{4.1f}$$

i.e., two thirds of the previous value or about one half the tank weight W. We next turn to two examples involving static fluid forces in which pressure (force per unit area) features prominently.

Illustration 4.2 Forces on Submerged Surfaces: Archimides' Law

We start by presenting the general formula for the force F acting on submerged surfaces of arbitrary shape and area A because of the so-called hydrostatic head h of the fluid. It takes the form:

$$dF = pdA \tag{4.2a}$$

where p = hydrostatic pressure. By means of a force balance, it can be established that p is related to h by the expression:

$$p = \rho gh \tag{4.2b}$$

where ρ = density of the fluid. This expression is known as the *fundamental equation of fluid statics*. It shows, for example, that at a depth of 100 m below the water surface, the pressure amounts to p = 1000 × 9.81 × 100 = 981 kPa, or approximately 10 atm.

We proceed to apply it to a number of cases of interest.

1. Flat horizontal surface:
   Substitution of Equation 4.2b into Equation 4.2a yields:

   $$dF = \rho ghdA \tag{4.2c}$$

   which, upon integration for h = constant, leads to the relation:

   $$F = \rho ghA \tag{4.2d}$$

2. Flat vertical surface:
   Equation 4.2c is applied again, yielding:

   $$dF = \rho ghdA \tag{4.2e}$$

   or

   $$F = \rho g\int_{h_1}^{h_2} hdA \tag{4.2f}$$

where the hydrostatic head h now varies over the surface area, and it is recognized that for a plate of width W, dA = Wdh. We also recognize that $\int hdA = h_{cg}A$, where h_{cg} is the centroidal distance, or the distance of the center of gravity of the plate from the fluid surface. We then obtain for the total force F the abbreviated expression:

$$F = \rho g h_{cg} A \tag{4.2g}$$

Suppose, for example, that the vertical surface is a rectangular wall of height H = 10 m and width W = 1 m, with the upper edge 10 m below the water surface. The centroid of the rectangle will be at mid-height and its distance from the water surface $h_{cg} = 10 + 5 = 15$ m. Hence, the total (horizontal) force acting on the submerged wall is $F = \rho g h_{cg} HW = (1000)(9.81)(15)(10)(1) = 1.47$ MN or approximately 15 atm

3. Arbitrary surfaces:
Integration of Equation 4.2c over curved and other surfaces of arbitrary shape can be cumbersome, even after decomposing the total force F into horizontal and vertical components F_h and F_v. The difficulty can be circumvented by applying the following two simple rules:
   a. The horizontal force component F_h equals the force on the area A_p formed by projecting the surface onto a vertical plane. Thus,

$$F_h = \rho g (h_{cg})_p A_p \tag{4.2h}$$

   where $(h_{cg})_p$ = distance of the centroid of the projected area to the surface.
   b. The vertical force component F_v equals the weight of the entire column of fluid W_f, both liquid and atmosphere, resting on the submerged surface. Thus,

$$F_v = W_f \tag{4.2i}$$

   The total force is then obtained as the square root of the sum of squares:

$$F = (F_h^2 + F_v^2)^{1/2} \tag{4.2j}$$

   Suppose, for example, that the vertical plate considered previously is now inclined with an angle $\alpha = 30°$ to the water surface. The projected area A_p is then HW sin α = (10)(1) sin 30° = 5 m^2. The centroidal distance $(h_{cg})_p$ is the same as before and we obtain $F_h = \rho p (h_{cg})_p A_p = (1000)(9.81)(15)(5) = 0.74$ MN, i.e., one half the value for the vertical plane.
4. Archimides' law:
It is now possible to give a simple explanation of Archimides' law. For a submerged body of arbitrary shape, the horizontal component $F_h = 0$,

because the projections to the left and right side vertical planes will have the same area A_p and centroidal distance h_{cg} but with associated forces of opposite sign. We are left with the vertical component F_v, which will be the difference between the fluid weights resting on the upper and lower surfaces of the body, W_1 and W_2. We therefore, obtain:

$$F = F_v = W_2 - W_1 = \text{Weight of fluid contained in the volume of the body} \quad (4.2k).$$

This in effect confirms Archimides' law, which states that the vertical force, or buoyancy, equals the weight of the displaced fluid.

Illustration 4.3 Forces Acting on a Pressurized Container: Analysis of an Aneurysm

In the design of pressure vessels, such as gas cylinders, one wishes to know the minimum wall thickness required to withstand a given internal pressure. That internal pressure would have to overcome the tensile strength or stress τ_t of the material to rupture the vessel. Tensile strength of common steel alloys are of the order of 500 MPa.

To prevent rupture, the forces due to the tensile strength and the pressure must, at a minimum, be in balance. Integration of the forces over the surface can be avoided as before by making use of the fact that the net force due to the pressure equals the force acting on the projected area of the vessel surface. This can be shown, e.g., for a cylinder, by integration of the component force pLR sin α so that:

$$F_{net} = pLR\int_{2\pi}^{0} \sin\alpha \, d\alpha = 2pRL \quad (4.3a)$$

where 2RL = dL is seen to be the projection of the cylinder surface onto a plane. Equating this force to the tensile force holding the vessel together we obtain, for a cylinder and sphere, respectively,

Cylinder $$p = \tau_t \frac{\Delta d}{d} \quad (4.3b)$$

Sphere $$p = 4\tau_t \frac{\Delta d}{d} \quad (4.3c)$$

where Δd is taken to be the wall thickness. These are the so-called *hoop stress formulae* for the cylinder and sphere, which are the most commonly encountered geometries. It can be observed that the sphere can accommodate a permissible pressure four times that of the cylinder. This fact is often taken advantage of in the design of pressure vessels. Note also that the allowable pressure for a particular vessel varies inversely with diameter. For high-pressure applications, therefore, it is

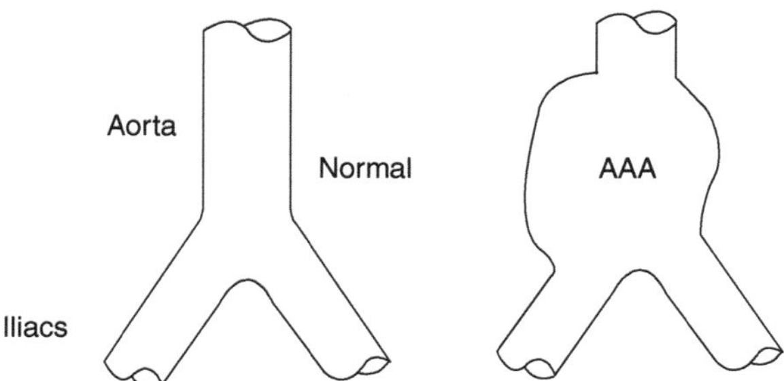

FIGURE 4.2 Diagram of abdominal aortic aneurysm (AAA).

preferable to use small diameter containers in order to avoid excessively thick walls (Δd) with an attendant increase in weight. Practice Problem 4.4 addresses the problem of calculating the minimum required wall thickness of a commercial gas cylinder.

What we wish to do here is to use the hoop stress formula to analyze the behavior of an aneurysm, a serious pathological condition that arises in the human body.

Aneurysms are balloonlike dilatations of the arterial wall, which most commonly occur in the circle of Willis, a network of arteries that supplies blood to the brain. Another common location is the abdominal aorta, which gives rise to the so-called abdominal aortic aneurysms, or AAA for short (see Figure 4.2). It was a ruptured AAA that led to the death of Albert Einstein.

Although aneurysms come in a variety of shapes and sizes, a considerable number of them can be treated as thin-walled, pressurized spherical structures that then become amenable to analysis by the hoop stress formula. The dilemma faced by a neurosurgeon, and one which an analysis can help resolve, is that the rupture potential of an aneurysm of this type is actually quite low, of the order 0.1 to 1% per year, making an operation unnecessary. Once they rupture, however, 50% of the patients die, and 50% of the survivors will be severely debilitated. It then becomes a question of balancing the risk inherent in an operation against the risk of leaving the lesion untreated.

To demonstrate the utility of an analysis in this context, consider the following example:

A detected aneurysm is found to have a diameter of 5 mm, an estimated wall thickness of 15 μm, and is under a mean blood pressure of 120 mmHg = 16,000 N/m^2. The task is to ascertain whether the critical stress τ_c, which is of the order of 5 MPa, will be substantially approached or even exceeded.

We have from Equation 4.3c:

$$\tau_c = \frac{pd}{4\Delta d} = \frac{16{,}000 \times 5 \times 10^{-3}\, m}{4 \times 15 \times 10^{-6}\, m} = 1.3\, MPa \tag{4.4}$$

which, although less than the critical stress, is of the same order of magnitude.

Comments: Although the result yields a safety factor of about 4, one may prefer, in view of the high mortality rate, to be dealing with a safety factor of at least 10. This would come about if the lesion were 60 μm thick rather than 15 μm, and highlights the importance of high resolution medical imaging to determine the wall thickness as exactly as possible. Given a safety factor of 10, and considering the inevitable risk of surgical intervention, the surgeon may then decide to simply monitor the lesion over time rather than treat it immediately. What we see in this example is that although the model does not convey a precise answer (a not uncommon occurrence in modeling), it does provide helpful guidelines for an eventual medical decision.

The derivatives of force we had invoked in the previous two examples involved dividing force by area, which gave us pressure and tensile stress as new variables. We now wish to consider a variable, surface tension, which is composed by dividing force by unit length. The genesis of this variable, and some of its applications are discussed in the following illustration.

Illustration 4.4 The Effects of Surface Tension: Laplace's Equation and Capillary Rise

A liquid, being unable to expand freely like a gas, will form an interface with a second liquid or a gas. This arises essentially because, within the liquid interior, molecules are densely packed and repel each other; whereas at the surface, with half the neighbors missing, the packing is looser and the molecules attract one another. The net effect is that the surface is under tension, the tensile attractive forces counterbalancing the repulsive forces that prevail in the interior (Figure 4.3). The quantity which characterizes this effect is the surface tension γ, with units of N/m or Nm/m^2. The latter unit reveals that surface tension is equivalent to the energy to form or eliminate a unit area of surface. Values of γ for various liquids generally range from 2×10^2 to 8×10^2 N/m; the higher values corresponding to polar liquids with stronger attractive forces due to hydrogen-bonding (e.g., 7.3×10^2 N/m for water). For liquid metals, e.g., mercury, those forces are even larger, leading to surface tensions an order of magnitude higher than normal.

As a consequence of surface tension, a liquid in contact with a solid surface will have a contact angle θ with that solid, as shown in Figure 4.3C. If $\theta < 90°$, the internal attractive forces are small and the liquid is said to wet the solid; if θ is in the range 90 to 180°, the liquid is termed *nonwetting*. Water and organic solvents are extremely wetting in contact with clean glass, with $\theta \approx 0$. Mercury, on the other hand, has a high contact angle of $\theta = 130°$ because of its abnormally large surface tension, i.e., high internal attractive forces. Surface tension and contact angle depend on the nature of the surface. Thus, water wets clean glass but not wax, which is hydrophobic. In the former case, the strong hydrophilic nature of the glass will overcome the internal attractive forces of the water and cause spreading. On hydrophobic (e.g., greasy) surfaces, water droplets maintain their integrity.

Laplace's equation — A second consequence of surface tension is that the interior pressure caused by the repulsive forces is higher than the exterior (usually

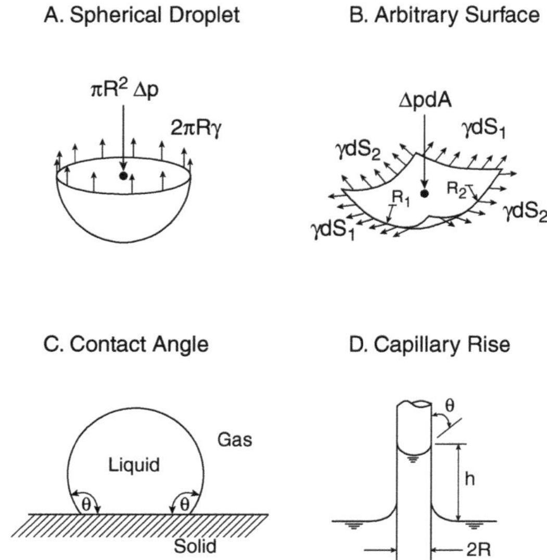

FIGURE 4.3 Aspects of surface tension: (A) Balance of forces in a spherical droplet, (B) Balance of forces in an arbitrary surface, (C) Contact angle for nonwetting liquid, and (D) Capillary rise of a wetting liquid.

atmospheric) force and is precisely balanced by the surface tensile force. Thus for a spherical droplet (see Figure 4.3A):

$$\underset{\text{Interior pressure force}}{\frac{\pi d^2}{4}\Delta p} = \underset{\text{Surface tensile force}}{\pi d \gamma} \tag{4.5a}$$

so that

$$p_{int} - p_{ext} = \Delta p = 4\gamma/d \tag{4.5b}$$

This equation can be generalized to an arbitrary curved surface and terms of its principal (and orthogonal) radii of curvature, and is then known as *Laplace's equation* (see Figure 4.3B):

$$\Delta p = \gamma\left[\frac{1}{R_1} + \frac{1}{R_2}\right] = 2\gamma\left[\frac{1}{d_1} + \frac{1}{d_2}\right] \tag{4.5c}$$

Another equation named after Laplace, a PDE with wide ramifications, will be encountered in Chapter 5.

Capillary rise — One consequence of surface tension is that the tendency to spread in contact with a solid causes liquids with $\theta < 90°$ to rise in capillary tubes. The physical situation is depicted in Figure 4.3D. A vertical force balance for this configuration leads to the equation:

$$\pi d \gamma \cos\theta \quad = \quad \rho g h \frac{\pi d^2}{4} \tag{4.5d}$$

Tensile force   Hydrostatic force

or

$$h = \frac{4\gamma \cos\theta}{\rho g d} \tag{4.5e}$$

For $\theta > 90°$, $\cos \theta < 0$; hence, h is negative and a depression results.

Capillary rise is the means by which nutrients dissolved in water are conveyed to the upper reaches of a plant or tree. Let us calculate the diameter of a capillary required to convey water to the top of a full-grown tree, h = 10 m. The water will be in contact with hydrophilic cell material, so that we can assume, as a first approximation, $\theta \approx 0$. We obtain from Equation 4.5e:

$$d = \frac{4\gamma}{\rho g h} = \frac{(4)(7.3 \times 10^{-2})}{(1000)(9.81)(10)} = 2.98 \times 10^{-6} m$$

Mercury porosimetry of nanopores — Another interesting application of Laplace's equation, and one which is not easily foreseen, involves the determination of pore size, or pore size distributions, in porous solids such as catalysts, adsorbents, and compacted soils. The pores, usually taken to be cylindrical, range in diameter from a few Angstroms (10 angstroms = 1 nanometer), to several thousand Angstroms. The method consists of forcing mercury into the pores and relating the applied pressure to the diameter of the pore. The smaller the pore, the higher the pressure p required to overcome the tensile forces and spread the mercury into the pore. p can be calculated from Laplace's equation or from the following simple force balance:

$$\pi d \gamma \cos \theta \quad = \quad p\, \pi d^2/4 \tag{4.5f}$$

Tensile force   Applied pressure Force

Hence:

$$p = (4\ \gamma \cos \theta)/d$$

For example, for a pore of 10 nm diameter, and using the pertinent parameters for mercury ($\gamma = 0.48$ N/m, $\theta = 130°$), we obtain:

$$p = [4 \times 0.48 \times (0.64)]/10^{-8}$$

$$p = 1.17 \times 10^{8} \text{ Pa or approximately 1000 atm}$$

(The negative sign indicates that work is done on the system.)

The high-pressure equipment required to carry out such experiments is available commercially.

The previous four illustrations were designed to convey the range of physical systems that are subject to static forces or derived variables. Additional examples were seen in Illustration 2.6 dealing with the bending of beams and struts. Practice Problem 4.7 applies the same principles to a nanodevice known as the *atomic force microscope*.

We now turn to dynamic systems which require the full use of Newton's law. We consider the forces which arise when an object is in relative motion to a fluid, the velocities that arise, and the trajectory of the object.

Illustration 4.5 Particle Movement in a Fluid

The behavior of particles rising or falling through a fluid, or in suspension in a flowing medium, represents yet another classical example of the application of a force balance. A distinction is made between the steady state, in which the sum of forces is zero, and the unsteady state, which requires inclusion of an acceleration term. A particle released in a medium of higher density will commence to rise with a steadily increasing velocity and an attendant increase in the drag force. A state is ultimately reached when that force is exactly in balance with the gravity and buoyancy forces. The particle is then said to have attained its terminal velocity. A body falling in a medium of higher density undergoes a similar period of acceleration and ultimate attainment of a steady velocity. In the third case considered, that of a particle in suspension, the steady state is assumed to have already been attained and the sum of forces is equal to zero.

Steady state — The forces in balance to be considered here are threefold: gravity force F_g, buoyancy force F_b, and drag force F_d. We first examine the case of a falling particle, in which the gravity force F_g acts downward, and F_b and F_d act upward. We obtain:

$$\sum Forces = 0$$

or

$$F_g - F_d - F_b = 0 \tag{4.6a}$$

and expanding these terms by appropriate auxiliary relations yields:

$$\underset{\text{Buoyancy}}{V_p \rho_p g} \quad - \quad \underset{\text{Drag}}{C_D \rho_f A_c \frac{v_t^2}{2}} \quad - \quad \underset{\text{Gravity}}{V_p \rho_f g} \quad = \quad 0 \tag{4.6b}$$

where V_p = volume of the particle.

Here the subscripts p and f denote properties of the particle and fluid, respectively, and A_c is the cross-sectional area of the particle at right angles to the direction of motion. This equation applies to a particle suspended in a flowing fluid as well.

For a rising particle, $\rho_f > \rho_s$ and the sign of the drag force is reversed. Grouping together the terms dependent on terminal velocity v_t, Equation 4.6b yields the following expressions for the various cases considered:

Falling particle and particle in suspension:

$$C_D v_t^2 = \frac{2V_p g}{A_c}\left(\frac{\rho_p - \rho_f}{\rho_f}\right) \tag{4.6c}$$

Rising particle:

$$C_D v_t^2 = \frac{2V_p g}{A_c}\left(\frac{\rho_f - \rho_p}{\rho_f}\right) \tag{4.6d}$$

Equations 4.6a to 4.6d have general validity irrespective of the flow regime or the geometry of the particle.

Let us now consider a specific case, that of a falling or suspended sphere. In the laminar regime, the drag coefficient is known to vary inversely with Reynolds' number Re: $C_D = 24/\text{Re}$ where $\text{Re} = dv\rho/\mu$, μ = viscosity. In the fully turbulent regime, it is constant, $C_D = 0.44$. We obtain, taking $g = 9.81$ m²/sec into account:

Laminar region, Re < 0.1:

$$v_t = 0.454\frac{\rho_f}{\mu}d^2\left(\frac{\rho_p - \rho_f}{\rho_f}\right) \tag{4.6e}$$

Turbulent region, $500 < \text{Re} < 2 \times 10^5$:

$$v_t = 5.45\left[d\left(\frac{\rho_p - \rho_f}{\rho_f}\right)\right]^{1/2} \tag{4.6f}$$

We first note that in both the laminar and turbulent cases, the terminal velocity or fluid velocity required to keep a particle in suspension is proportional to the diameter of the sphere, although only weakly so in the turbulent regime. This is in agreement with the conventional wisdom that larger bodies fall faster. Viscosity plays a role in the laminar region, as expected, and the quotient $(\rho_p - \rho_f)/\rho_f$ may be regarded as a dimensionless driving force for the falling particle in both laminar and turbulent reactions.

A second point to note is that the two variables of greatest interest, d and v_t, appear in combination in the Reynolds number. Thus, with only one of them specified (which is the usual case), one cannot determine the Reynolds number and, hence, the flow regime and pertinent drag coefficient C_D. In most practical cases it is, therefore, difficult to decide which of the two equations (Equation 4.6e or Equation 4.6f) is applicable, or whether one is located in the transition regime between the two cases.

One can circumvent this difficulty by introducing the lower limits of validity of the drag coefficients into rearranged forms of Equation 4.6e and Equation 4.6f and using the result to establish upper and lower bounds on the terminal velocities or diameters in each regime. For example, to establish the upper bound for the diameter of a sphere falling in water in the laminar region, we obtain, in the first instance:

$$v_t d = 0.454\frac{\rho_f}{\mu}d^3\left(\frac{\rho_p-\rho_f}{\rho_f}\right) \tag{4.6g}$$

where $v_t d$ has to satisfy the relation Re < 0.1 and, hence, using a kinematic viscosity for water $\nu = \mu/\rho_f = 10^{-6}$ m²/s, the inequality:

$$v_t d < 10^{-7} \text{ m}^2/\text{s} \tag{4.6h}$$

Combining Equation 4.6g and Equation 4.6h, it follows that the diameter has to fulfill the condition:

$$d^3 < \frac{1}{0.454}\times 10^{-13}\left(\frac{\rho_f}{\rho_p-\rho_f}\right) \tag{4.6i}$$

resulting in the following:

$$d < 6.04\times 10^{-5}\left(\frac{\rho_f}{\rho_p-\rho_f}\right)^{1/3} \tag{4.6j}$$

We have done similar calculations for d in the turbulent region, and for v in both the laminar and turbulent regions, and have summarized the results in Table 4.1.

For water, one can assume the density ratio $[\rho_f/(\rho_p-\rho_f)]^{1/3}$ to be of the order of one. It follows that for the laminar region equation (Equation 4.6e) to apply, the particle diameter has to be less than about 0.06 mm, and for the turbulent equation (Equation 4.6f) to apply, the particle diameter has to be greater than about 2 mm. Similar statements can be made for particles falling in air and for the bounds on velocity. One notes that the region between the two regimes, which is the transition region, encompasses some two orders of magnitude in both velocity and diameter. Table 4.1 serves the useful purpose of providing a range of diameters and velocities that span the transition region. If, for example, the given diameter of a sphere is 10^4 m = 0.1 mm, the table signals that the transition region applies. One can then substitute the

TABLE 4.1
Bounds on Diameter and Terminal Velocity for a Sphere Falling in Water and Air

Regime	Velocity (m/sec)	Diameter (m)
	Water	
Laminar	$v_t < 1.66\times10^{-3}\left(\frac{\rho_p-\rho_f}{\rho_f}\right)^{1/3}$	$d < 6.04\times10^{-5}\left(\frac{\rho_f}{\rho_p-\rho_f}\right)^{1/3}$
Turbulent	$v_t > 0.246\left(\frac{\rho_p-\rho_f}{\rho_f}\right)^{1/3}$	$d > 2.03\times10^{-3}\left(\frac{\rho_f}{\rho_p-\rho_f}\right)^{1/3}$
	Air	
Laminar	$v_t < 4.3\times10^{-3}\left(\frac{\rho_p-\rho_f}{\rho_f}\right)^{1/3}$	$d < 4.15\times10^{-4}\left(\frac{\rho_f}{\rho_p-\rho_f}\right)^{1/3}$
Turbulent	$v_t > 0.644\left(\frac{\rho_p-\rho_f}{\rho_f}\right)^{1/3}$	$d > 1.40\times10^{-2}\left(\frac{\rho_f}{\rho_p-\rho_f}\right)^{1/3}$

transition relation $C_D = 18.5/Re^{3/5}$ into Equation 4.6c and solve for the desired velocity.

The unsteady state and approach to steady state — It is frequently necessary in problems involving rising or falling particles to calculate distance traveled in a given time, or conversely, the time necessary to travel a given distance. In problems involving settling tanks, for example, one may wish to know the time necessary for a particle of given size to settle to the bottom, i.e., to fall a given distance (see Practice Problem 4.14). In Practice Problem 4.5, a marker particle to be used by divers is to be designed, which would rise with a given steady velocity. Here again it would be desirable to know the length of time the particle spends in the unsteady state, so that deviations from the desired design velocity can be assessed.

To model the unsteady state, full use of Newton's law must be made, i.e., an acceleration term has to be added to the previous steady state balance. Thus, for falling particles,

$$F_g - F_d - F_b = m\frac{dv}{dt} \tag{4.7a}$$

Expansion in terms of the usual auxiliary relation then yields:

$$V_p\rho_p g - C_D\rho_f A_c\frac{v^2}{2} - V_p\rho_f g = \rho_p V_p\frac{dv}{dt} \tag{4.7b}$$

which can be formally integrated to yield the time dependence of the velocity. If the particle ultimately reaches its terminal velocity in the turbulent region, integration of C_D would have to take account of the full spectrum of C_D variations with Reynolds' number, i.e., velocity. We circumvent this difficulty by considering the limiting case of laminar flow. This not only describes the behavior of particles that fall entirely in the laminar region, but also provides an upper bound to the time involved and a lower bound to the distance traveled.

Integration by separation of variables of Equation 4.7b then yields:

$$t = -\frac{1}{B}\ln\left(1 - \frac{B}{A}v\right) \tag{4.7c}$$

where

$$A = \frac{\rho_p - \rho_f}{\rho_f}g, \quad B = 18\frac{\mu}{\rho_f d^2}$$

Comparison with the laminar flow Equation (4.6e) shows that $B/A = 1/v_t$, i.e., the inverse of the steady state terminal velocity. Hence Equation (4.7c) becomes:

$$t = -\frac{\rho_f d^2}{18\mu}\ln\left(1 - \frac{v}{v_t}\right) \tag{4.7d}$$

To obtain a sense of the order of magnitude of the times involved, consider a sphere of density $\rho_p = 2000$ kg/m^3 and diameter just below the upper bound given in Table 4.1, $d = 10^{-5}$ m, falling through water. For a 95% approach to the steady state terminal velocity, we obtain:

$$t = -\frac{(2000)(10^{-5})^2}{(18)(10^{-3})}\ln(1 - 0.95) = 3.3 \times 10^{-5} \text{ sec}$$

In general, calculations for other cases, such as fall through air or when all three flow regimes are involved, show that the times required to attain the steady state terminal velocity are quite short and can often be neglected in calculations involving falling or rising particles. We make use of this fact in Practice Problem 4.5 in the design of marker particles for scuba diving.

The foregoing example provides the foundation for the next illustration, which deals with the path of a projectile fired from an artillery piece. Two of the three forces we had considered before are operative here as well: gravity and air resistance on drag. Buoyancy effects are negligible in this case and are omitted.

The reader will recognize that the solution will call for the use of Newton's law in two dimensions: the vertical distance y and the horizontal distance x. Furthermore, if both gravity and drag are to be considered, the model will consist of two nonlinear ODEs because drag, being invariably in the turbulent region, will vary with v^2 [i.e., with $(\frac{dx}{dt})^2$ and $(\frac{dy}{dt})^2$]. Solution of this full model requires the use of

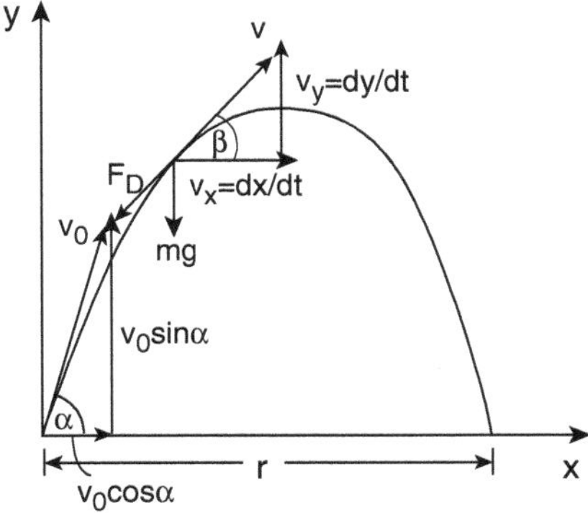

FIGURE 4.4 Path of a projectile.

numerical procedures. If on the other hand drag is omitted as a first approximation, an analytical solution becomes possible. In what follows we address a number of questions using, in the first instance, the analytical approximation, and following this up with a numerical simulation of the full nonlinear model using the Mathematica package.

Illustration 4.6 Path of a Projectile

Figure 4.4 displays a sketch of the relevant model parameters and the shell trajectory.

The questions we wish to address here are the following:

1. How high does the projectile rise vertically?
2. How far does the projectile travel horizontally; that is, what is its range r (Figure 4.4)?
3. What angle of elevation produces the maximum range? (See Illustration 1.9 in this connection.)
4. What is the equation for the trajectory y = F(x)?

Answers to questions 2 and 3 are of interest in the use of ground artillery, whereas question 1 has its importance in the use of antiaircraft guns.

With air drag neglected, the model equations to be solved are as follows:

$$0 = m\frac{d^2x}{dt^2} \tag{4.8a}$$

$$-mg = m\frac{d^2y}{dt^2} \tag{4.8b}$$

with the initial conditions (t = 0):

$$v_x = \frac{dx}{dt} = v_0 \cos\alpha$$

$$v_y = \frac{dy}{dt} = v_0 \sin\alpha \tag{4.8c}$$

$$x = 0$$

$$y = 0$$

where v_0 is the muzzle velocity.

Straightforward integration by separation of variables yields the following general solutions:

For the horizontal component:

$$x = (v_0 \cos\alpha)t \tag{4.8d}$$

For the vertical component:

$$y = -\frac{1}{2}gt^2 + (v_0 \sin\alpha)t \tag{4.8e}$$

We use these expressions to provide answers to the questions posed in the beginning.

1. To establish maximum height y_{max}, we set dy/dt = 0 and obtain:

$$\frac{dy}{dt} = -gt + v_0 \sin\alpha = 0 \tag{4.9a}$$

from which it follows:

$$t_{y_{max}} = \frac{v_0 \sin\alpha}{g} \tag{4.9b}$$

and consequently:

$$y_{max} = -\frac{1}{2}gt_{max}^2 + (v_0 \sin\alpha)t_{max} = \frac{(v_0 \sin\alpha)^2}{2g} \tag{4.9e}$$

2. To calculate the range, we set y = 0, so that from Equation 4.8e we have:

$$y = t\left[-\frac{1}{2}gt + v_0 \sin\alpha\right] = 0 \tag{4.10a}$$

and

$$t = 0 \tag{4.10b}$$

$$t = \frac{2v_0 \sin\alpha}{g} \tag{4.10c}$$

where Equation 4.10b corresponds to the instant when the shell is fired, whereas Equation 4.10c represents the time it strikes the ground. Note that the latter is twice the time t_{max} of Equation 4.9b. Substitution of Equation 4.10c into Equation 4.8d yields the desired expression for the range r:

$$r = v_0 \cos\alpha \frac{2v_0 \sin\alpha}{g} = \frac{v_0^2}{g} \sin 2\alpha \tag{4.10d}$$

3. It follows immediately from Equation 4.10d that r_{max} results when sin $2\alpha = 1$, so that

$$\alpha_{opt} = 45^\circ \qquad r_{max} = \frac{v_0^2}{g} \tag{4.11}$$

Note that both y_{max} and the range r are independent of the mass m of the shell. This goes against the intuitive notion that a lighter piece might travel further, but is a logical consequence of the cancellation of m in Equation 4.8b. When air resistance is included, that cancellation no longer occurs. In this case, the optimum angle of elevation is no longer at 45°, and the range r will, as expected, fall below the value given by Equation 4.10d. The set of differential equations that govern the trajectory of a projectile in the presence of drag are * :

$$\frac{du}{dt} = -ku\sqrt{u^2 + v^2} \tag{4.12a}$$

$$\frac{du}{dt} = -g - kv\sqrt{u^2 + v^2} \tag{4.12b}$$

$$\frac{dx}{dt} = u \tag{4.12c}$$

$$\frac{dy}{dt} = v \tag{4.12d}$$

with initial conditions

$$x(0) = 0 \tag{4.12e}$$

$$y(0) = 0 \tag{4.12f}$$

$$u(0) = v_o \cos\theta \tag{4.12g}$$

$$v(0) = v_o \sin\theta \tag{4.12h}$$

* Basmadjian, D., Mathematical Modelling of Physical Systems.

For the numerical solution of this problem, let us assume $v_o = 400$ m/sec and k=0.0001, while varying θ over 30° to 60°:

```
In[1]:= k = 0.0001;
        g = 9.81;
        Vo = 400;
        tab = Table[
             NDSolve[
              {u'[t] == -k u[t] Sqrt[u[t]^2 + v[t]^2],
              v'[t] == -g - kv[t] Sqrt[u[t]^2 + v[t]^2],
              x'[t] == u[t],
              y'[t] == v[t],
              x[0] == 0,
              y[0] == 0,
              u[0] == Vo Cos[θ π/180],
              v[0] == Vo Sin[θ π/180]},
              {u[t], v[t], x[t], y[t], {t, 0, 60}],
              {θ, 30, 60, 5}];

In[5]:= ParametricPlot[
         Evaluate[
          Table[{x[t], y[t]} /. tab[[i]], {i, 1, Length[tab]}],
          {t, 0, 60}]
          , PlotRange → All,
          Frame → True,
          TextStyle → {FontFamily → "Times", FontSize → 14},
          FrameLabel → {"x", "y", "", ""}]
```



```
Out[5]= -Graphics-
```

The examples seen so far in this chapter have either been at steady state or have varied with time. None of them showed variations with distance, although two examples in which forces or their derivatives were distributed appeared in Chapter 2 (Illustration 2.5 and Illustration 2.6). We wish to add to these examples with two problems drawn from fluid mechanics. The first deals with distributions which arise in inviscid (i.e., frictionless) steady flow and leads to the celebrated Bernoulli equation. We address this case, and some relevant applications, in Practice Problem 4.6. The second case deals with laminar flow in a parallel plane channel, a geometry we had encountered briefly before (Illustration 2.5). We examine it in greater detail in the following illustration, extrapolate the results to other geometries, and draw the reader's attention to some interesting biomedical applications.

Illustration 4.7 Viscous Flow in a Parallel Plate Channel: The Effect of Shear Stress on Cell Behavior

The forces operative in viscous or laminar flow can be defined with precision without recourse to empirical friction factors. They are the forces due to the pressure drop and that caused by the Newtonian shear stress (Equation 1.2e).

We consider the flow between two parallel plates held apart by two narrow vertical walls of height T. Because of their small dimension compared to that of the plates, frictional forces due to these walls are neglected. A force balance over a finite length L of a slit of 2x then leads to the expression:

$$(p_1 - p_2)2xW = \tau 2LW = -\mu \frac{dv}{dx} 2LW \tag{4.13a}$$

where W = width of the channel and x = distance from the center plane.

Integration by separation of variables yields in the first instance:

$$\frac{\Delta p}{\mu L}\frac{x^2}{2} = -v + C \tag{4.13b}$$

where the integration constant $C = (\Delta P/L)(x^2/2)$ is obtained from the so-called no-slip condition, which states that the velocity at the wall, i.e., at x = T/2, is zero. The final result is then given by:

$$v(x) = \frac{\Delta p}{8\mu L} T^2 \left[1 - \left(\frac{2x}{T} \right)^2 \right] \tag{4.13c}$$

This is a parabolic velocity distribution, symmetric about the center line, much like the profile obtained in viscous flow through a circular pipe, with x replacing radial distance r. The maximum velocity occurs at the centerline and is given by:

$$v_{Max} = \frac{\Delta p}{8\mu L} T^2 \tag{4.13d}$$

To obtain a flow rate–pressure drop relation, which is the quantity of greatest interest in engineering calculations, the profile (Equation 4.13c) is integrated over the cross-sectional area of flow to yield:

$$Q = \frac{\Delta p}{12\mu L} T^3 W \tag{4.13e}$$

It follows from this expression that the average velocity, equal to Q/TW, is given by:

$$v_{avg} = \frac{\Delta P}{12\mu L} T^2 = \frac{2}{3} v_{Max} \tag{4.13f}$$

Similar force balances can be performed for viscous flow in circular and annular conduits. The results are, for convenience, summarized in Table 4.2. It can be noted from these results that in all cases pressure drop varies directly with flow rate Q, length of conduit L, and viscosity μ, but is inversely proportional to the fourth power of a linear dimension of the conduit or a combination thereof. For the circular pipe, the linear dimension is the diameter d_1; for the annulus, it is the outer diameter d_o; and for the parallel plate channel, it is a combination of channel height T and its width W.

Comments: Flow in parallel plate channels finds a number of practical applications. The attachment to the polymer extruder, Figure 2.4, is one example; another is the flow through a certain type of reverse osmosis membrane termed *spiral-wound*. Although the plates here exhibit a certain curvature, this is usually neglected in calculating the performance of these devices.

An intriguing application of this flow geometry arises in a biological context, which is explained as follows:

It has long been known that animal or human cells, in particular the endothelial cells lining the walls of blood vessels, undergo significant changes in response to an applied stress. Visually it has been observed that there is an increase in cell perimeter and length, and a decrease in cell width. Internal structural changes also take place. More importantly, however, the production of a host of molecules, including those that influence growth, inflammation, and adhesion of blood components, are significantly affected by stress and changes in applied stress. This was first shown by Rosen et al., in 1974, who used a parallel plate flow chamber of the type shown in Figure 4.5. They found that endothelial cells alter the production of a specific molecule, histamine, in response to altered shear stresses. Since then, there has been an upsurge of such studies which continue to this day.

Wall shear stress τ_w, which is the correlating factor in these studies, is obtained from Equation 4.13a by setting x = T/2 and eliminating Δp/L via Equation 4.13e. The result for the parallel plate channel is:

$$\tau_w = \frac{6\mu Q}{T^2 W} \tag{4.14a}$$

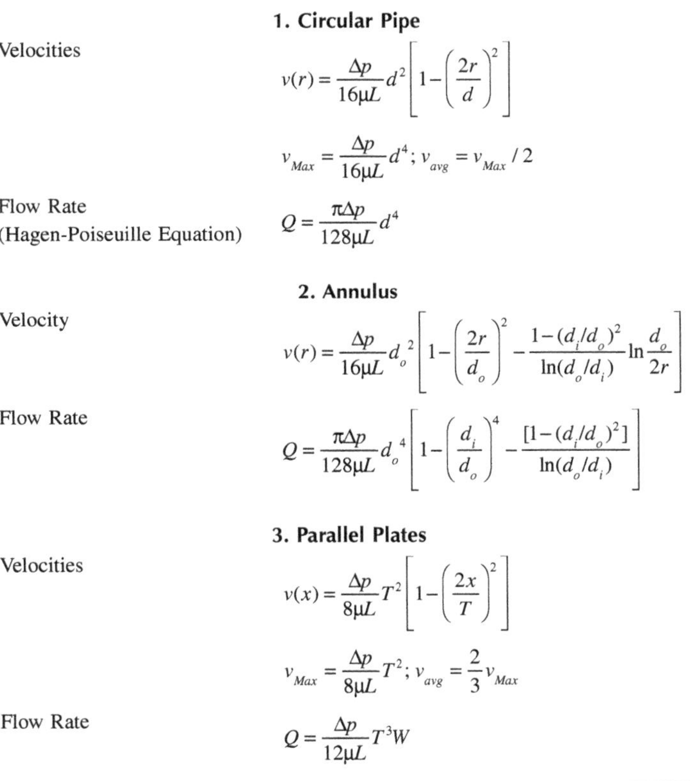

TABLE 4.2
Velocities and Flow Rates in Viscous Flow through Various Conduits

	1. Circular Pipe
Velocities	$v(r) = \frac{\Delta p}{16\mu L} d^2 \left[1 - \left(\frac{2r}{d}\right)^2\right]$
	$v_{Max} = \frac{\Delta p}{16\mu L} d^4;\ v_{avg} = v_{Max} / 2$
Flow Rate (Hagen-Poiseuille Equation)	$Q = \frac{\pi \Delta p}{128\mu L} d^4$
	2. Annulus
Velocity	$v(r) = \frac{\Delta p}{16\mu L} d_o^{\,2} \left[1 - \left(\frac{2r}{d_o}\right)^2 - \frac{1-(d_i/d_o)^2}{\ln(d_o/d_i)} \ln \frac{d_o}{2r}\right]$
Flow Rate	$Q = \frac{\pi \Delta p}{128\mu L} d_o^{\,4} \left[1 - \left(\frac{d_i}{d_o}\right)^4 - \frac{[1-(d_i/d_o)^2]}{\ln(d_o/d_i)}\right]$
	3. Parallel Plates
Velocities	$v(x) = \frac{\Delta p}{8\mu L} T^2 \left[1 - \left(\frac{2x}{T}\right)^2\right]$
	$v_{Max} = \frac{\Delta p}{8\mu L} T^2;\ v_{avg} = \frac{2}{3} v_{Max}$
Flow Rate	$Q = \frac{\Delta p}{12\mu L} T^3 W$

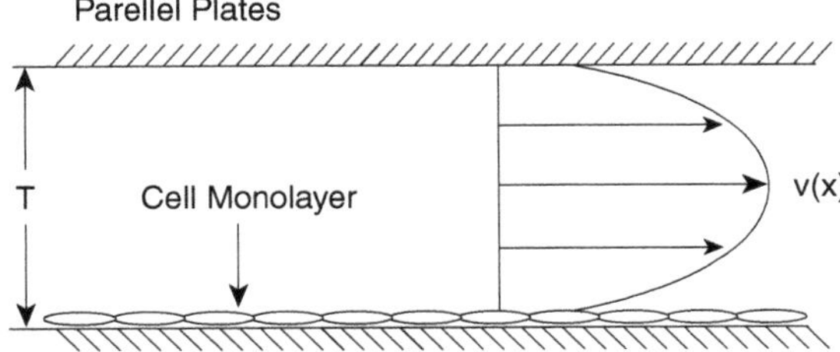

FIGURE 4.5 Flow in a parallel plate channel. The device is used to study the response of human cells to changes in flow.

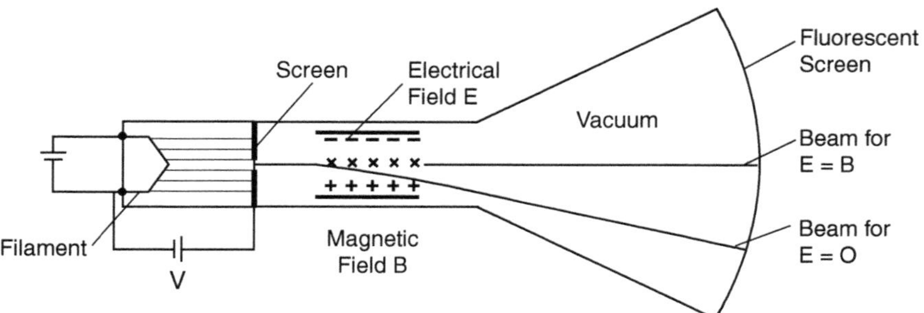

FIGURE 4.6 Thomson's determination of e/m.

The corresponding result for a cylindrical tube is given by:

Cylindrical channel $$\tau_w = \frac{32\mu Q}{\pi d^3} \tag{4.14b}$$

The latter expression is the *most frequently cited equation in vascular biology* related to endothelial cell studies of the type mentioned. The strong inverse dependence on tubular diameter allows stress to be relieved by slightly dilating the blood vessel. This is, in fact, the body's way of maintaining wall shear stress within normal bounds.

We conclude the section of force balances by drawing on an example involving electromagnetic forces. These forces played a crucial role in the pathbreaking measurement in 1897 by J.J. Thomson of the ratio of charge e to mass m of an electron. The experiment, which amounted to the discovery of the electron, consisted of an observation of its deflection in combined electrical and magnetic fields, and also considered the deflection experienced by an electron in a pure electrical field. The three items together provide the tools for deducing the ratio e/m from experimental observations.

Illustration 4.8 Thomson's Determination of e/m

If we place a test charge of magnitude q_0 into the space near a charged rod, an electrostatic force will act on the charge. We speak of an electric field E in this space and define it as the force per unit charge, that is:

$$\mathbf{E} = \frac{\mathbf{F}}{q_0} \tag{4.15a}$$

As force is a vector, the electric field, which has units of newton per coulomb (N/C), must also be a vector. Their magnitudes are related by the scalar equation:

$$E = \frac{F}{q_0} \tag{4.15b}$$

The scalar form of a magnetic field is similar, i.e., it involves a force per unit charge, but it also contains the velocity v of the charge. This arises from the fact that a magnetic field exerts a force only when the charge is in motion.

We have:

$$B = \frac{F}{q_0 v \sin(\mathbf{v}_1\mathbf{F})} \tag{4.15c}$$

where $(\mathbf{v}_1\mathbf{F})$ represents the angle between v and F vectors.

This expression reflects the following experimental observations:

- For a charge at rest (v = 0) the magnetic field force F is zero.
- For a charge moving tangentially to the magnetic field lines (as revealed by iron filling, for example), the field force is likely equal to zero as sin $(\mathbf{v}, \mathbf{F}) = 0$.
- For a charge moving at right angles to the field lines, the field force is a maximum [sin $(\mathbf{v}, \mathbf{F}) = 1$].
- Thomson's experiment also requires an expression for the deflection of a charge in a pure electrical field. The trajectory is a parabolic one, given by

$$y = \frac{1}{2}\frac{eE}{mv^2}x^2 \tag{4.15d}$$

Its derivation is left as an exercise (Practice Problem 4.9). With these expressions in place, we are in a position to address Thomson's experiment. In Figure 4.7, which is a modernized version of Thomson's apparatus, electrons are emitted from a hot filament and accelerated by a potential V while passing through a hole in a screen. They then enter a region in which they move at right angles to an electric field **E** and a magnetic field **B**. The two fields themselves are arranged in such a way that they cause deflections in opposite directions.

Thomson's threefold procedure was (a) to note the position of the deflected beam spot with **E** = **B** = 0; (b) to expose the beam to a pure electrical field and measure the attendant deflection; and (c) to apply a magnetic field and adjust its value until the beam deflection is restored to zero. For step b, Equation 4.15d applies. In this expression, x (length of deflecting plates) and E are known, and y (deflection at far edge of the field) can be calculated from the observed beam displacement and the known geometry of the apparatus. This leaves v_0 as an extra unknown. The problem is resolved by the ingenious addition of step c, for under its conditions the deflecting force **F** becomes zero, and it follows from Equation 4.15b and Equation 4.15c that:

$$eE = ev_0B\sin(v_0, B) = ev_0B \times 1 \tag{4.15e}$$

or equivalently

$$v_0 = \frac{E}{B} \tag{4.15f}$$

Note that Equation 4.15e makes use of the fact that **E** and **B** act in opposite directions. It remains to substitute Equation 4.15f into Equation 4.15d to obtain:

$$\frac{e}{m} = \frac{2yE}{B^2x^2} \tag{4.15g}$$

Thomson's value for e/m was 1.7×10^{11} C/kg, in almost exact agreement with the 1977 value of 1.758803×10^{11} C/kg.
Comments:

- It is clear that the underlying model played a crucial role in the design of Thomson's experiment. Equally important was the ingenious introduction of step c, which eliminated v_0 as an unknown.
- The determination of e (and hence the mass m of the electron) was accomplished by Robert Millikan in 1911 using tiny charged oil drops suspended in an electrical field; e was then calculated by equating the electrical field force with gravity.

4.2 APPLICATIONS OF MASS BALANCES

We had, in our previous treatment of force balances, approached the topic by making a distinction between the two broad categories of static and dynamic systems. That distinction, which is an apt one for applications of Newton's law, is not well suited for organizing mass balances. We draw attention instead to the two devices we introduced in Chapter 1: the stirred tank or compartment, and the one-dimensional pipe (see Figure 1.3). In other words, we divide the systems into those that vary at most with time but not with distance (lumped parameter), and those that vary with one dimension but are also at steady state (distributed parameter). Both cases lead to either algebraic or ordinary differential equations.

In Chapter 1, we indicated the wide range of systems and processes that fall into these two categories. This is reflected in the illustrations that follow. We draw on the environment for several examples (Illustration 4.13, Illustration 4.14, and Illustration 4.16), on biology and biomedical engineering (Illustration 4.9, Illustration 4.13, and Illustration 4.17), as well as on traditional chemical engineering operations. Together with the practice problems, these will expose the reader to the diversity of applications of mass balances, which are mostly simple and often ingeniously used.

4.2.1 Compartmental Models

Illustration 4.9 Measurement of Plasma Volume and Cardiac Output by the Dye Dilution Method

One standard and historical method of measuring the amount of certain body fluids present in the system is the use of a measured amount of dye that is preferentially

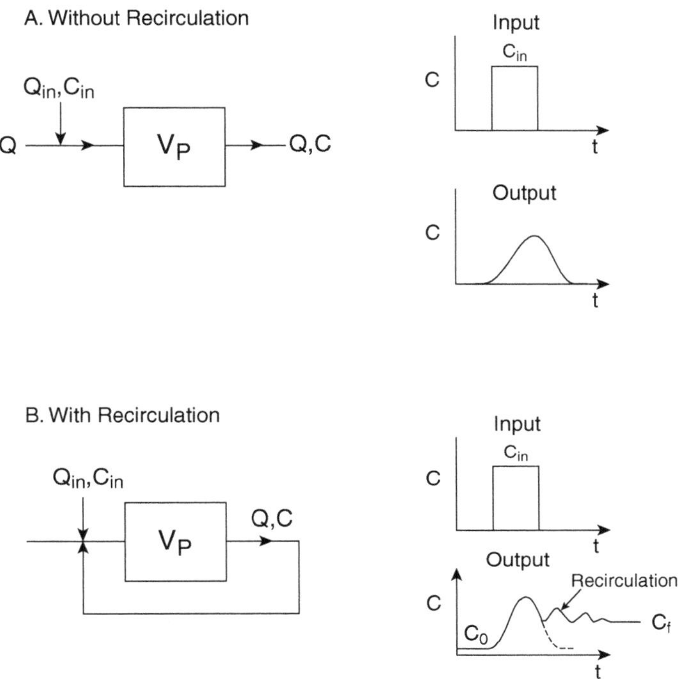

FIGURE 4.7 Plasma volume V and cardiac output Q by dye dilution: (A) Without recirculation, (B) With recirculation.

soluble in the fluid (i.e., is not absorbed by the surrounding tissue). Suppose we wish to find the total volume V of blood plasma in the body. This is done by injecting a measured amount of dye of known concentration into the circulatory system and monitoring its concentration with time at some convenient downstream location (Figure 4.7). Figure 4.7B shows a typical concentration vs. time record for both input and output. The input consists of a square wave of constant concentration, whereas output concentration exhibits both time lag and a drop in concentration due to the diluting effect of the plasma. The first and principal peak seen in the output is followed by several diminished secondary peaks as the circulating blood adds more dye to the sample region. Eventually, the dye concentration subsides to a constant steady state value C_f, which is recorded and used to compute the unknown plasma volume V_p. The calculation proceeds as follows.

Let us assume that the injection takes place at a flow rate Q_{in}, which may be constant or variable, but is generally not known or recorded. This does not pose a problem, however, because the expression containing it can be converted, as we shall shortly see, into the injection volume V_{in}, which is known and can be measured with a great degree of precision. We can then write the following mass balance for the period of injection:

$$\text{Rate of dye in} - \text{Rate of dye out} = \text{Rate of Change of dye content}$$

$$Q_{in}C_{in} - 0 = V_p \frac{dC}{dt} \tag{4.16a}$$

where C is the average concentration of the compartment, and V_p the volume of plasma.

Formal integration over the injection period from 0 to t and assuming zero initial concentration then yields the expression:

$$V_p \int_0^{C_f} dC = C_{in} \int_0^t Qdt \tag{4.16b}$$

where the integral on the right equals the volume of injected solution. We arrive at the relation:

$$V_p C_f = V_{in} C_{in} = mass\ m_d\ of\ injected\ dye$$

or alternatively

$$V_p = \frac{m_d}{C_f} \tag{4.16c}$$

Thus, the plasma volume is given simply by the ratio of total mass of dye injected to the final steady state concentration in the plasma, C_f. Several recirculations are usually necessary to attain that value. Because blood volume turns over approximately once every minute, a short time is sufficient to obtain reliable values of C_f.

Let us now consider the hypothetical case of no recirculation. In Figure 4.8A, the square input of dye results in an output consisting of a single Gaussian peak. That peak corresponds to the first and principal peak seen in an actual system with recirculation (Figure 4.8B), and can be reproduced there by extrapolation along the dotted line. It turns out that this simulated case of a system without recirculation can be used to deduce the cardiac output or flow rate produced by the heart muscle. Let us examine the model describing this case.

We consider injection of a dye solution of concentration C_a on the "upstream" side of the heart, which results in a corresponding output concentration C_v on the "downstream" side. A mass balance applied between the arterial and venous sides results in the following expression:

$$\text{Rate of dye in} - \text{Rate of dye out} = \text{Rate of change of the dye contents}$$

$$QC_a - QC_v = V_h \frac{dC_h}{dt} \tag{4.16d}$$

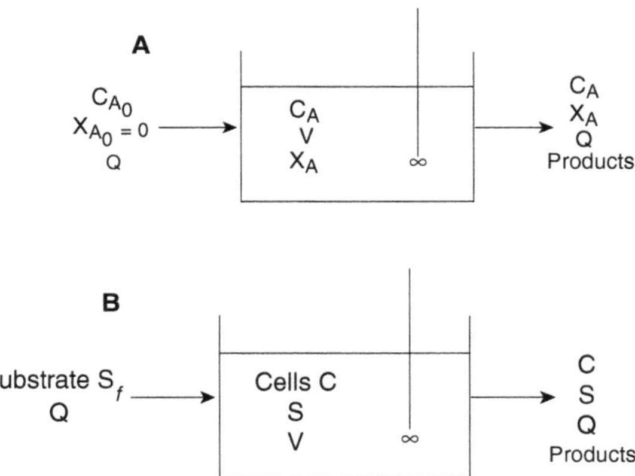

FIGURE 4.8 Two tank reactors: (A) The classical chemical CSTR. (B) The bioreactor.

where V_h and C_h are the liquid volume and average dye concentration in the heart, respectively. If this volume is assumed to be well mixed, C_h will equal C_v. This assumption, which is commonly used in all simple compartmental models, is not needed here. To show this, we proceed as follows.

Formally integrating Equation 4.16d, we obtain:

$$V_h C_h(t) = Q\int_0^t (C_a - C_v)dt \tag{4.16e}$$

To eliminate the unknowns V_h and $C_h(t)$, we consider the steady state $t \to \infty$ when all dye has left the system, that is, $C_h(t) = 0$. We obtain:

$$0 = Q\int_0^\infty (C_a - C_v)dt \tag{4.16f}$$

As C_a need not be monitored, it can be expressed in terms of the known total mass of the dye, m_d. Thus, we write:

$$m_d = Q\int_0^\infty C_a dt \tag{4.16g}$$

Combining 4.16f and Equation 4.16g, we arrive at the following expression:

$$Q = \frac{m_d}{\int_0^\infty C_v dt} \tag{4.16h}$$

The cardiac output Q is thus given by the simple ratio of the total mass of injected dye to the integral under the first extrapolated peak. That extrapolation is needed to cancel out the effects of recirculation, and can be effected by plotting the data on semilogarithmic paper, extrapolating the resulting straight line, and then replotting the data on arithmetic paper. Extrapolation is usually carried to a small finite value C_0, which is induced by preloading the system with a small amount of dye. This yields the dashed curve shown in Figure 4.8B.

Illustration 4.10 Two Tank Reactors: The Chemical Reactor and the Bioreactor

1. *The chemical reactor (CSTR):*
The classical chemical tank reactor, also known as a continuous (flow) stirred tank reactor (CSTR) is a widely used device for carrying out chemical reactions. A schematic diagram of the unit is shown in Figure 4.8A. Feed with a concentration C_{A0} enters continuously with a volumetric flow rate Q, undergoes a reaction in the well-stirred tank of volume V, and leaves with a residual concentration C_A. We assume a system of constant density and volume in which the reaction follows the scheme:

$$A \rightarrow \text{Products (B, C, etc.)}$$

with a reaction rate r_A in units of moles reacted per unit time and volume. Polymerization reactions fall in this category.

Irrespective of whether the reactor is heated or cooled, or contains a catalyst, the following mass balance holds at steady state:

$$\text{Rate of A In} - \text{Rate of A Out} = 0$$

$$QC_{Ao} - [QC_A + r_A V] = 0 \tag{4.17a}$$

and similarly for other species. When the reactor is in a transient state, for example, during start-up or because of changes in the feed, an unsteady term V (dC_A/dt) is added to the right-hand side.

We now introduce two variables designed to put Equation 4.17a into a terser and more practical form. The first is the so-called conversion X_A, which is defined as:

$$X_A = \frac{\textit{Initial moles} - \textit{Moles left}}{\textit{Initial moles}} = \frac{C_{Ao} - C_A}{C_{Ao}} \tag{4.17b}$$

X_A denotes the fraction of the incoming feed, which has been converted to product. It is zero when no reaction has taken place, and is equal to one when all the feed has been converted to product.

The second concept is that of *residence* or *holding time,* τ, in seconds, which is defined as the ratio of reactor volume to volumetric flow rate. Thus:

$$\tau = v/Q \tag{4.17c}$$

The longer the reactant stays in the tank, i.e., the higher the residence time, the greater will be the conversion to product. Hence large values of τ are good, and low values are undesirable.
With these definitions in hand, Equation 4.17a becomes:

$$\tau = V/Q = \frac{C_{Ao}}{r_A} X_A \tag{4.17d}$$

This equation can be used in reactor design to calculate reactor volume V for a specified conversion, predict conversion for prescribed values of τ, or to extract kinetic parameters lurking into r_A from performance data.

2. *The bioreactor*

The bioreactor, which is also a CSTR, shares many features with the classical chemical tank reactor but differs from it in some important and intriguing ways.

To begin with, the reaction itself is of a special type. It involves live cell culture (for example, yeast) contained in the tank, which are fed a nutrient termed the *substrate* (for example, sugar). Two events then take place: one is the conversion of the nutrient to some desired product (for example, alcohol); the other is the use of the nutrient to grow new cells, termed *biomass*. Both of these reactions are promoted by enzymes (i.e., catalysts) carried by the cells.

The most commonly used rate expression is that proposed by Monod in 1942, which has the form:

$$\mu = \frac{1}{m}\frac{dm}{dt} = \frac{\mu_{Max} S}{K_S + S} \tag{4.18a}$$

where μ is the rate of cell growth per unit mass of cells, with units of reciprocal time, and S is the substrate concentration (mass/volume). μ_{Max} is the maximum rate of production, which occurs when the enzyme catalyst is saturated with substrate.

For our purposes, it will be convenient to convert Equation 4.18a to the standard form of reaction rate in units of (mass/volume time), which we obtain by writing:

$$r = \mu C = \frac{\mu_{Max} SC}{K_S + S} \tag{4.18b}$$

where C is the cell concentration in mass/volume.

We are now in a position to make a standard mass balance on the cell population, which reads

Rate of cells in or produced – Rate of cells out = 0

$$\mu CV - QC = 0 \tag{4.18c}$$

which, after introduction of the rate expression (Equation 4.18b) yields:

$$\left[\frac{\mu_{Max} S}{K_S + S} - D\right] C = 0$$

or:

$$S = \frac{DK_S}{\mu_{Max} - D} \tag{4.18d}$$

where

$$D = Q/V = \tau^{-1} \tag{4.18e}$$

D, the so-called dilution rate, is the preferred variable used in biochemical engineering calculations, and is seen to be the reciprocal of the residence time τ we had previously encountered in the standard chemical CSTR (cf. Equation 4.17c). It has some rather surprising properties, which we shall address shortly.

A similar substrate mass balance leads to the expression:

$$S = S_f - \frac{C}{Y_{CS}} \tag{4.18f}$$

where Y_{CS} is defined as mass of cells produced per mass of substrate consumed, a quantity that is obtained by laboratory experiment.

Eliminating S between the two preceding expressions yields the cell concentration as a function of dilution rate:

$$C = Y_{CS}\left[S_f - \frac{DK_S}{\mu_{Max} - D}\right] \tag{4.18g}$$

One notes that the second term in the bracket increases with increasing dilution rate D until it becomes equal to the feed substrate concentration, S_f. At this point, the cell concentration C drops to zero. This happens at the critical dilution rate D_{crit} obtained from Equation 4.18d and given by:

$$D_{crit} = \frac{\mu_{Max} S_f}{K_s + S_f} \tag{4.18h}$$

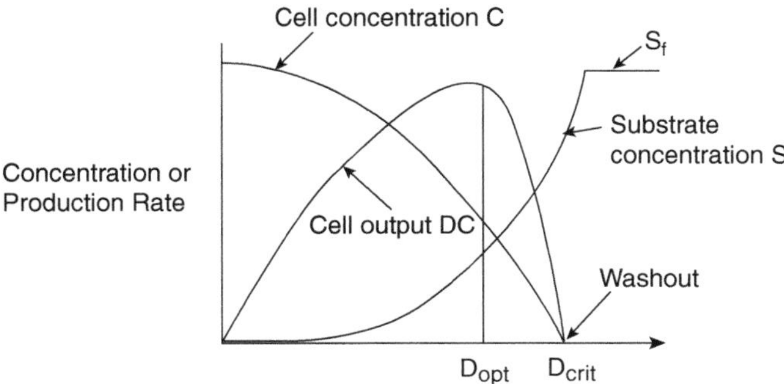

FIGURE 4.9 Concentration and production of cells in a continuous flow stirred bioreactor.

Physically, what happens here is that an increase in D causes a rise in the rate of removal of cells, which is only partly offset by the rate of cell growth due to the higher substrate concentration. Further increases in D accelerate the divergence of the two rates until a point is reached at D_{crit} where all cells are swept away. This condition, termed *washout*, is shown graphically in Figure 4.9.

Let us next examine what happens to a second quantity, the product of cell concentration and dilution rate DC. That quantity represents the cell output per unit reactor volume and time. As C declines only slowly at first, an increase in D will initially cause DC to increase. As the decrease in C accelerates toward ultimate washout, however, the rise in DC will slow down and, after passing through a maximum, will decrease at an accelerated pace, ultimately dropping to zero. The peak in DC defines an optimum dilution rate D_{opt}, which optimizes cell output and hence the reactor efficiency (see Figure 4.9).

Comments: The bioreactor reveals two features of reactor operation we have not yet encountered. The first is the existence of an optimum residence time (the inverse of the dilution rate D), which is not seen in the operation of conventional chemical reactors involving single reactions. The second novel feature is the occurrence of complete reactor shutdown at a certain feed rate. These factors, and the use of the unconventional operational parameter D by biotechnologists, require an adjustment in our thinking.

We had been accustomed to analyzing reactor behavior in direct proportion to the residence time $\tau = V/Q$. The situation here was simple: large values of τ were good, giving high yields; small values of τ were bad (see Equation 4.17d). No maximum occurred between the two extremes of zero and infinite residence time. In a bioreactor, the effect of the inverse quantity, the dilution rate $D = 1/\tau$, is more complex. Here high values of D (i.e., small residence times) can be very good, in fact leading to optimum conversion; but a relatively small increase in D beyond that point leads to complete shutdown of the reactor (i.e., washout). Thus, small residence times can be both very good and very bad, and these conditions are in close proximity to each other. These are all novel features not encountered in conventional reactor engineering and deserve the attention of the analyst.

The problem we address next is drawn from a broader class of processes that involve interphase mass transfer brought about in a stirred tank. Crystallization, which is the reverse of the dissolution process considered here, is one such operation. Others are liquid–liquid extraction, the leaching of solids, and the reverse process of adsorption from a liquid solution. The latter two operations often involve transient particle diffusion, which may lead to PDE models. We will address these in a subsequent chapter.

Illustration 4.11 Dissolution of Granular Solids in an Agitated Vessel

The assumption made at the outset is that the concentration at the surface of the particles equals the saturation concentration C_s of the solid material, and that the mass transfer is driven by the linear potential $(C_s - C)$, where C is the prevailing concentration in the liquid at any particular instant.

An initial unsteady mass balance over the solid particle leads to the following expression:

$$\text{Rate of solid in} - \text{Rate of solid out} = \text{Rate of change of solid mass}$$

$$0 - k_c A_s (C_s - C) = \frac{d}{dt} m \tag{4.19a}$$

Note that both the surface area A_s and concentration in the liquid C vary with time, or indirectly with the remaining mass m. For the area, which can be quite irregular, we stipulate that it varies with the two-thirds power of volume, so that:

$$A_s = \alpha V^{2/3} = \frac{\alpha}{\rho^{2/3}} m^{2/3} = \beta m^{2/3} \tag{4.19b}$$

where α is some shape factor and equals 4.83 for spherical particles.

To obtain an expression for the external concentration C, we apply a simple cumulative mass balance, which reads:

$$\text{Initial solid} = \text{Solid left over} + \text{Solid in Solution}$$

$$m_o = m + CV \tag{4.19c}$$

and consequently:

$$C = \frac{m_o - m}{V} \tag{4.19d}$$

Substituting Equation 4.19b and Equation 4.19d into the original mass balance (Equation 4.19a), we obtain:

$$-k_c \beta m^{2/3} \left(C_s - \frac{m_o - m}{V} \right) = \frac{dm}{dt} \tag{4.19e}$$

which yields, after integration by separation of variables:

$$At = \int_{m_o}^{m} \frac{dm}{m^{2/3}(m/V + C_s - m_o / V)} \tag{4.19f}$$

where $A = k_c\beta = k_c\alpha/\rho_s^{2/3}$.

Analytical evaluation of the integral can be achieved by Mathematica:

```
In[1]:= integral =
    Integrate[ -1 / (x^ (2/3)(-mo/V + x/V + Cs)), {x, m, mo},
        Assumptions → {mo > 0 && Re[Cs V] > Re[mo]}]
    // TraditionalForm
Out[1]//TraditionalForm=
```

$$\frac{1}{2(mo - Cs\,V)^{2/3}}\left((-1)^{2/3}V\left(-2\sqrt{3}\tan^{-1}\left(\frac{1-\frac{2\sqrt[3]{-1}\sqrt[3]{m}}{\sqrt[3]{mo-Cs\,V}}}{\sqrt{3}}\right) + 2\sqrt{3}\tan^{-1}\left(\frac{1-\frac{2\sqrt[3]{-mo}}{\sqrt[3]{mo-Cs\,V}}}{\sqrt{3}}\right) + 2\log\left(\sqrt[3]{-1}\sqrt[3]{m} + \sqrt[3]{mo - Cs\,V}\right) - 2\log\left(\sqrt[3]{-mo} + \sqrt[3]{mo - Cs\,V}\right) - \log\left((-1)^{2/3}m^{2/3} - \sqrt[3]{-1}\sqrt[3]{mo-Cs\,V}\sqrt[3]{m} + (mo-Cs\,V)^{2/3}\right) + \log\left((-mo)^{2/3} - \sqrt[3]{mo - Cs\,V}\sqrt[3]{-mo} + (mo - Cs\,V)^{2/3}\right)\right)\right)$$

Here, we consider, instead, the case in which vessel volume V and solubility C_s are sufficiently high, and the term $(m_o - m)/V$ in Equation 4.19f can be neglected compared to C_s. The result (Equation 4.19f) then becomes:

$$t = \frac{3\rho_s^{\,2/3}}{\alpha k_c C_s} m_o^{\,1/3} \tag{4.19g}$$

where t is now the total dissolution time, and m_o the initial mass of the particle.

Comments: There are several points of note in the final relation. First, time of dissolution varies inversely with the mass transfer coefficient k_c and the solubility C_s. This is in line with physical reasoning: High values of these coefficients imply a high mass transfer rate, which results in shorter dissolution times. A more startling result is the one-third power dependence on initial mass. This implies that an eightfold increase in particle mass by doubling the diameter will increase dissolution time only by a factor of two. This was certainly not anticipated on physical grounds, and is a direct consequence of the area–volume relation introduced in Equation 4.19b. We see here yet another example of the power of modeling to reveal the unexpected.

In the three illustrations just completed, uniformity within a compartment was in each instance achieved by rapid stirring or circulation. The next example we take up demonstrates that even if flow is slow, and the system inherently distributed, compartmental behavior may nevertheless be attained under certain limiting conditions.

Illustration 4.12 Percolation: An Unusual Compartment

Percolation processes involve the passage of a fluid through or over a bed of granular solids during which an exchange of mass takes place between the two phases. The purification of water by adsorption and ion-exchange, or of air and other gases in beds of adsorbents, are both common industrial applications practiced on a large scale. In the environment, percolation mass transfer is seen in the contamination and clearance of river beds, and in the transfer of toxins from contaminated groundwater to the surrounding soil. The analysis we present here is given in the context of adsorption purification of water.

Adsorption in a fixed bed of granular material is a complex system to model. Concentrations evidently vary with distance, and although they ultimately attain a steady form of distribution, they also vary with time. The model would consequently consist of two mass balances, one for the fluid phase and a second one for the solid phase, and both of these would be partial differential equations in time and distance, which generally have to be solved numerically. To avoid this complication, a procedure has come into use in which mass transfer resistance is neglected and the two phases are assumed to be in equilibrium everywhere. The concentration then propagates in the shape of a rectangular front, shown in Figure 4.10 and denoted as "Equilibrium". As both the fluid and solid phases are uniform in concentration, the contents of the bed up to the rectangular front may be viewed as a stirred tank or compartment, albeit one of variable volume. The movement of this front and its dependence on flow rate and feed concentration can be analyzed by means of a simple cumulative mass balance, which takes the following form:

$$\underset{\text{Amount introduced to time t}}{Y_F v \rho_f A_C t} = \underset{\text{Amount retained by adsorbent}}{X_F \rho_b A_C z} + \underset{\text{Amount leaving with fluid}}{Y_F \rho_f A_C z} \quad (4.20a)$$

Here v and A_C are the fluid velocity and cross-sectional area of the bed, respectively, and ρ_f and ρ_b are the fluid and bed densities. X and Y are the solute mass ratios in the solid and fluid phases, respectively.

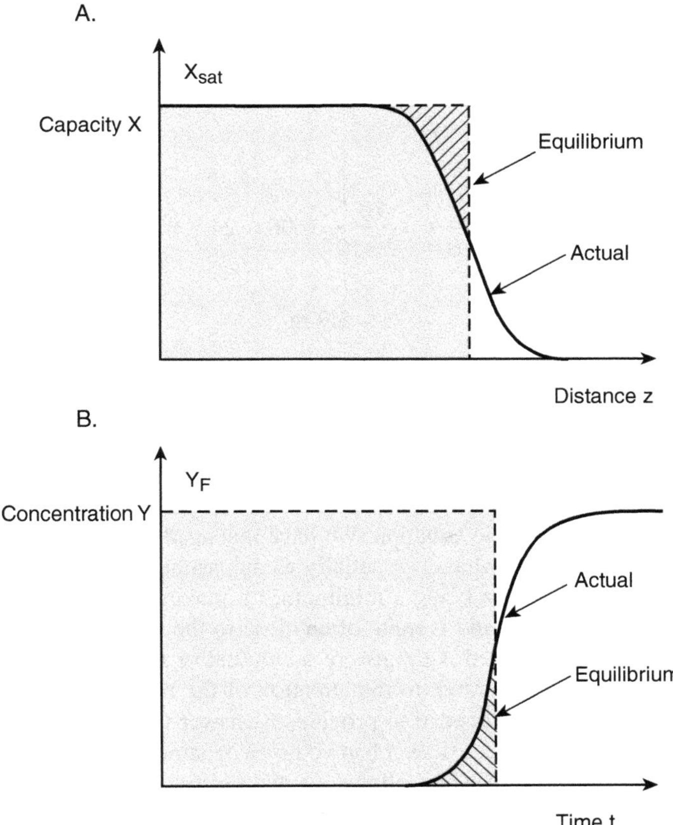

FIGURE 4.10 Adsorption in a fixed bed: (A) Adsorbent concentration profiles; (B) Fluid-phase concentration breakthrough curve.

The last term in this equation is generally negligible, as the bulk of the impurity will reside in the adsorbent. It would otherwise not be a very efficient adsorbent. Equation 4.20a can then be recast in the form:

$$z/t = \frac{\rho_f v}{\rho_b H} \tag{4.20b}$$

where we have assumed that X and Y are in a constant ratio H, i.e., the equilibrium is linear (Henry's law). This assumption is valid at low concentrations of the solute.

Equation 4.20b can be used to calculate the time t it takes the front to reach a particular position or, conversely, the position attained after a prescribed time interval. These quantities are, by necessity, limiting values as full saturation will, in fact, be retarded by the mass transfer resistance. However, in many instances the fluid flow is sufficiently slow that local equilibrium is attained, or will be nearly attained, during the interval of contact.

Suppose, for example, bed-to-fluid densities are in the ratio 2:1, v = 1 mm/sec, H = 10^3, and it is desired to keep the adsorber on stream for 90 d. Then the minimum bed height z required is given by:

$$z = \frac{\rho_f v}{\rho_b H} t = \frac{10^{-3}}{2 \times 10^3} 3600 \times 24 \times 90 \tag{4.20c}$$

$$z = 3.9 \text{ m} \tag{4.20d}$$

The actual bed height will have to be somewhat greater because the mass transfer resistance will cause the solute to break through earlier than stipulated (see Figure 4.10b).

Comments: There are several features of note in this example. One is the ingenious use of the equilibrium assumption, which has the effect of reducing the model from the PDE level to an algebraic equation. We have managed to convert a time dependent, distributed system to what is essentially a compartment. The calculated values are perforce limiting ones, z being a minimum, t a maximum, but this is still very useful information to have and is quite often close to the truth.

Another point to be noted is the use of a cumulative mass balance. This much neglected tool, which was drawn to the attention of the reader in Chapter 1, turns out to be exactly what was needed to provide an answer to this complex problem.

The illustration we have just seen had some environmental as well as industrial content and implications. In what follows we address two problems of exclusively environmental concern.

Illustration 4.13 Bioconcentration in Fish

In Figure 4.11, we display the various mechanisms by which toxic solutes can enter and leave fish. They are largely self-explanatory and are all expressed in terms of

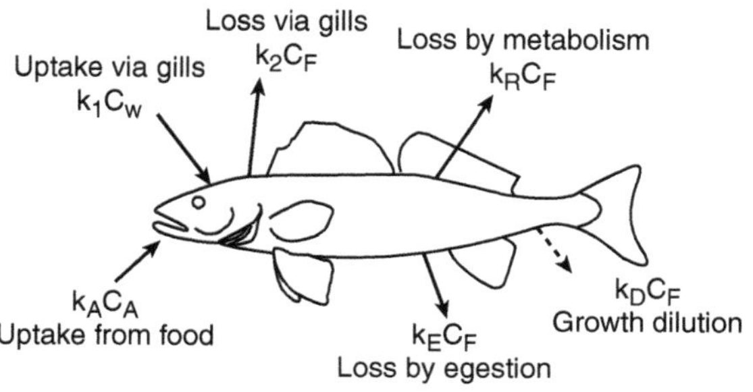

FIGURE 4.11 Uptake and loss of a toxic substance in fish.

first-order rate laws. The first of these was quantified in 1979 and used in measurements for parameter estimation. It defines the uptake and return of the solute through the gills as:

$$\frac{dC_F}{dt} = k_1 C_W - k_2 C_F \tag{4.21a}$$

where the subscript F denotes the fish.

If the concentration C_W in the water is assumed constant, integration by separation of variables yields the expressions:

For uptake:

$$C_F/C_W = k_1/k_2[1 - \exp(-k_2 t)] \tag{4.21b}$$

For clearance ($k_1 = 0$):

$$C_F/C_{Fo} = \exp(-k_2 t) \tag{4.21c}$$

We note that as $t \to \infty$, the uptake concentration ratio C_F/C_W approaches k_1/k_2. This ratio is termed the *bioconcentration factor*, BCF. Parameter estimation is usually done by running the uptake to saturation so that C_F/C_W = BCF, following this up by clearance to establish the value of k_2.

When all mechanisms are considered with the exception of growth dilution, a long-term effect, the rate expression becomes:

$$\frac{dC_F}{dt} = (k_1 + k_A)C_W - (k_2 + k_R + k_E)C_F \tag{4.21d}$$

where it has been assumed that $C_A \approx C_W$, which is often the case.

Integration by separation of variables then leads to the expression:

$$C_F/C_W = \frac{k_1 + k_A}{k_2 + k_R + k_E}\{1 - \exp[-(k_2 + k_R + k_E)t\} \tag{4.21e}$$

Bioconcentration factors can be of considerable magnitude. Studies of toxin uptake by bluegill fish has shown that the largest uptake occurs in the gall bladder, which has a BCF of 18,000 for the aromatic chemical benzopyrene. The solubility of this substance in water is of the order 10^{-4} mg/L, yielding a concentration in the gall bladder of

$$C_F = C_w(BCF) - 10^{-4} \times 18{,}000 \tag{4.21f}$$

$$C_F = 1.8\ mg/\mathrm{L} \tag{4.21g}$$

Illustration 4.14 The Rayleigh Equation in the Environment: Mercury Volatilization from Water

Although the present example deals with evaporation of a pollutant from a water basin, we provide the reader with a slight detour through the related topic of distillation to convey a sense of the underlying principles.

A special type of distillation arises when the vapor, which is at all times fully equilibrated with a well-stirred liquid, is slowly withdrawn in small amounts while the liquid undergoes a correspondingly slow change in composition. A full model of this process consists of the relevant mass balances, an energy balance, and an equilibrium relation, which relate vapor and liquid compositions. That relation is given by the expression:

$$p = \gamma x P^0 = y P_T \tag{4.22a}$$

where p is the partial pressure of the evaporating component, x its mole fraction in the liquid, P^0 the vapor pressure of the pure solvent, y the mole fraction in the vapor, and P_T total pressure. γ is the so-called activity coefficient which expresses deviations from ideality. Substances with low solubility, such as mercury in water, generally have high values of γ.

The complete model of the process usually requires a numerical solution. Rayleigh avoided this complication by considering the simpler problem of relating liquid composition to the amount of liquid remaining in the still W. This reduced problem does not require an energy balance and can be solved by analytical means. We use this approach to examine the following problem.

Consider a body of water containing dissolved mercury at or near the saturation level; we wish to calculate the reduction of mercury content that occurs when a mere 0.01% of the solution is evaporated. We shall view this as a Rayleigh distillation type of problem, with the evaporated liquid assumed to be in equilibrium with the well-stirred liquid of uniform concentration. The equilibrium relation, Equation 4.22a, contains the activity coefficient γ, which is not usually known or easily measured. We circumvent the difficulty by relating γ, to solubility, which is a well-established known quantity. This is done by considering a system consisting of pure mercury in equilibrium with its saturated aqueous solution. Chemical thermodynamics require that for such a system, the so-called chemical potentials μ of the two phases must be equal. This equality is expressed by the following relation:

$$\underset{\text{Mercury}}{\mu^0(T, p)} = \underset{\text{Mercury in water}}{\mu^0(T, p) + RT \ln \gamma x_{sat}} \tag{4.22b}$$

from which it follows that:

$$\gamma\, x_{sat} = 1 \tag{4.22c}$$

or

$$\gamma = 1/x_{sat}$$

This result agrees with the intuitive notion that the smaller the solubility x_{sat}, the higher its "escaping tendency" in terms of partial pressure or fugacity.

We now proceed to derive the usual mass balances:

Total mass balance:

Total moles in – Total moles out = Rate of change in total content

$$0 - D = \frac{d}{dt}W \tag{4.22d}$$

Component mass balance:

Moles mercury in – Moles mercury out = Rate of change in mercury content

$$0 - y_{Hg}\, D = \frac{d}{dt}(x_{Hg}\, W) \tag{4.22e}$$

where y and x are the vapor and liquid phase mole fractions. These two relations are supplemented by a statement of phase equilibrium. Combining (4.22a) and (4.22c) we obtain:

Equilibrium relation:

$$y_{Hg} = \frac{p_{Hg}}{p_{Hg} + P^{\circ}_{H_2O}} \cong \frac{\gamma_{Hg} x_{Hg} P^{\circ}_{Hg}}{P^{\circ}_{H_2O}} = \frac{x_{Hg} P^{\circ}_{Hg}}{x_{sat}\; P^{\circ}_{H_2O}} \tag{4.22f}$$

Note that the partial pressure of air is excluded from this expression, as we are concerned with equilibrium compositions of the binary mercury–water system.

We now resort to a neat trick, which was briefly alluded to in Chapter 1. Taking the ratio of the two mass balances, we eliminate the unknown D and time t. We obtain:

$$\frac{x_{Hg} P^{\circ}_{Hg}}{x_{sat} P^{\circ}_{H_2O}} = x_{Hg} + W\frac{dx_{Hg}}{dW} \tag{4.21g}$$

Integration by separation of variables yields the expression:

$$\left[\frac{P^{\circ}_{Hg}}{x_{sat} P^{\circ}_{H_2O}} - 1\right] \ln\frac{W}{W_0} = \ln\frac{x}{x_0} \tag{4.22h}$$

The data to be introduced at this stage are as follows:
Mercury solubility: 3×10^2 mg/l $\cong 2.7 \times 10^9$ mol fraction (25°C)
Mercury vapor pressure: 0.173 Pa (25°C)
Water vapor pressure: 3.17×10^3 Pa (25°C)
Substitution of these values into Equation 4.22h yields:

$$\left[\frac{0.173}{2.7\times10^{-9}(3170)}-1\right]\ln(1-10^{-4})=\ln\frac{x}{x_0}$$

where we have substituted 0.01% = 10^4 for the fraction evaporated.

Taylor series expansion of the logarithmic term on the left yields:

$$\ln\,(1-10^4)\cong 10^4$$

so that

$$\ln\left(\frac{x}{x_0}\right)_{Hg}=-2.02$$

and

$$(x/x_0)_{Hg}=0.132$$

Thus the fraction of mercury volatilized, $1 - x/x_0$, equals ≈ 0.87, a phenomenal amount considering that only 0.01% of the solution has been evaporated. This surprising result is confirmed by composing the ratio of vapor to liquid compositions y/x, which is a measure of the enrichment resulting from the evaporation process. We have from Equation 4.22f:

$$y/x=\frac{P^o_{Hg}}{x_{Sat}P^o_{H_2O}}=\frac{0.173}{2.7\times10^{-3}\times3.17\times10^3} \tag{4.22i}$$

or

$$y/x = 2.0 \times 10^4 \tag{4.22j}$$

Thus, the vapor is enriched by a factor of 20,000 over mercury concentration in the liquid.

Comments: A number of pitfalls were astutely avoided by drawing on background knowledge from appropriate subdisciplines. The fact that γ_{Hg} was unknown and could not be located in the literature on vapor–liquid equilibria could have brought the proceedings to a halt. Instead, we drew on intuitive reasoning that γ should be related to solubility, and knowing that the latter was tabulated, established the relation by means of elementary thermodynamics.

It would have been tempting to calculate γ_{Hg} from the conventional relation $p_i = y_i\, P_{Tot}$, with $P_{Tot} = 1$ atm. This would have led to the wrong result because the partial pressure of air would have been included. It required some thought to realize that γ_{Hg} refers to the mole fraction in the system H_2O-Hg, not H_2O-Hg-air.

Finally, even though the logarithm of $1 - 10^4 = 0.9999$ could have been easily evaluated by a pocket calculator, we prefer drawing on first year Calculus to obtain the same result in a more elegant way.

4.2.2 Distributed Systems

The one-dimensional pipe model we had introduced in Chapter 1 leads to distributions of concentrations in one principal direction and arises, as we saw, in a wide range of applications. In the three illustrations that follow, we start with the classical example of an industrial gas scrubber and follow this up with an example each drawn from the environmental and biomedical domains. The model equations are all of the algebraic or ODE-level and can be solved by graphical or analytical means.

Illustration 4.15 The Gas Scrubber Revisited

Previously, in Illustration 1.3, we drew the reader's attention to the multitude of mass balances that can be performed to describe the concentration changes in a gas scrubber. Both integral and differential balances were among the candidates, but it was only the latter which provided the tools necessary to design a column. We now return to the topic and examine these equations in greater detail. Figure 4.12A repeats some of the details of column operation and variables.

We start by composing a mass balance over the gas phase contained in the difference element (z, z + Δz) and obtain:

Rate of solute in – Rate of solute out = 0

$$G_sY|_z - [G_sY|_{z+\Delta z} + N_{avg}] = 0 \tag{4.23a}$$

where Y = kg solute/kg carrier and G_s = kg carrier/m^2sec.

Carrier units are chosen in order to obtain a constant gas flow rate and, thus, reduce the number of variables. We now introduce the auxiliary relation for N_{avg}, which takes the form of the product of a mass transfer coefficient K_Y, surface area in the element, aΔz, and a solute driving force $(Y - Y^*)_{avg}$:

$$N_{avg} = K_Y a\ \Delta z\ (Y - Y^*)_{avg} \tag{4.23b}$$

Here the surface area *a* is conveniently taken in volumetric units (m^2/m^3 packing), and Y* is the gas phase solute mole ratio in equilibrium with the liquid-phase concentration X at that point. Upon dividing by Δz and letting Δz → 0, we obtain:

$$G_s \frac{dY}{dz} + K_Y a(Y - Y^*) = 0 \tag{4.23c}$$

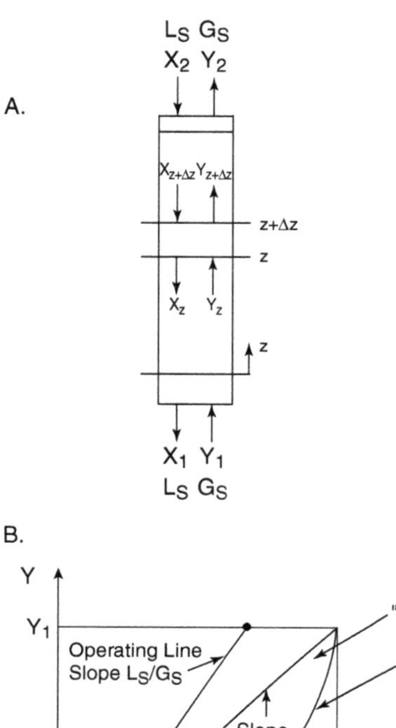

FIGURE 4.12 The countercurrent packed gas scrubber: (A) Column variables, (B) Operating diagram.

The liquid phase balance is obtained in similar fashion and yields:

$$L_s \frac{dX}{dz} + K_Y a(Y - Y^*) = 0 \tag{4.23d}$$

We note again that the driving force (Y–Y*) has now become a point quantity at the position z, i.e., it is not to be differentiated.

As we now have three state variables X, Y, and Y* to deal with, it follows that a third equation will be required, which is given by the equilibrium relation:

$$Y^* = f(X) \tag{4.23e}$$

Thus, the most general model consists of two differential and one algebraic equations and yields the profiles of X, Y and Y* in the longitudinal direction. It yields the most general solution which can be used for design, parameter estimation as well as performance analysis. For design purposes, it is often more convenient to replace the liquid-phase ODE by an integral algebraic balance (see Figure 1.3A).

Rate of solute in − Rate of solute out = 0

$$(G_sY + L_sX_2) - (G_sY_2 + L_sX) = 0 \tag{4.23f}$$

This expression is conveniently regrouped to yield the so-called operating line with slope L_s/G_s:

$$\frac{Y - Y_2}{X - X_2} = \frac{L_s}{G_s} \tag{4.23g}$$

To solve Equations 4.23c, 4.23e, and 4.23f in the variables X, Y, and Y*, we start by formally integrating Equation 4.23c. One obtains by separation of variables

$$H = \int_{z_1}^{z_2} dz = \frac{G_s}{K_Y a}\int_{Y_2}^{Y_1} \frac{dY}{Y - Y*} \tag{4.23h}$$

It has become the convention to separate the integral from its coefficient and express the right side as the product of the so-called height of a transfer unit HTU = $\frac{G_s}{K_Y a}$, and the number of transfer units NTU = $\int_{Y_1}^{Y_2} \frac{dY}{Y-Y*}$. Thus, the height H of packed column required to achieve the prescribed purification is given by the expression:

$$H = \frac{G_s}{K_y a}\int_{Y_1}^{Y_2} \frac{dY}{Y - Y*} = HTU \times NTU \tag{4.23i}$$

The relations Equation 4.23g and Equation 4.23i are conveniently plotted on an "operating diagram" shown in Figure 4.12B. In this diagram, prescribed solvent and carrier gas flow rates L_s and G_s are used to establish the slope of the operating line, which passes through the point (X_2, Y_2), i.e., solute concentration in the solvent entering the top of the column and the prescribed exit concentration of the gas, Y_2. The vertical distance between operating line and equilibrium curve establishes the driving force Y−Y*, which is used for the graphical or numerical evaluation of the NTU integral. Together with the value for HTU based on experimental values of

K_Ya, this yields the desired height of the column, H. The operating diagram can also be used for a quick assessment of the effect of a change in variables. A reduction in solvent flow rate L_s, for example, lowers the slope of the operating line, resulting in an increase in the NTU and, hence, height H of the column. Ultimately, the operating line intersects the equilibrium curve. The driving force there converges to zero, thus NTU and H become infinite. The value of L_s at which this occurs represents the minimum solvent flow rate, which will achieve the required separation and result in a column of infinite height.

We note, in addition, that the Equation 4.23i is only suited for design or for estimation of the parameter K_Ya from measured values of Y_2. Analysis of the performance of an existing column would require a trial and error procedure. This can be avoided by using the solutions to the two ODEs (Equation 4.23c and Equation 4.23d) which are quite general and suited for all tasks.

Illustration 4.16 The Streeter–Phelps River Pollution Model: The Oxygen Sag Curve

In this classical 1925 study, probably the first attempt to model the fate of a chemical in the environment, Streeter and Phelps derived an equation that described the oxygen profile in a river that undergoes a steady influx of pollutant at some point upstream. Initially, biodegradation of the pollutant causes a decline in dissolved oxygen C or, viewed slightly differently, an increase in the oxygen deficit D = C* − C, where C* is the equilibrium solubility of oxygen in water. As the pollutant concentration L decreases through biodegradation, the decline in oxygen concentration slows and ultimately passes through a minimum, the so-called *critical point*, as oxygen supply from the atmosphere replenishes the river. Further "reaeration" ultimately restores the oxygen concentration to full saturation levels (Figure 4.13).

In their model, Streeter and Phelps did not consider pollutant adsorption on river sediment, an important removal mechanism that will be addressed in the following

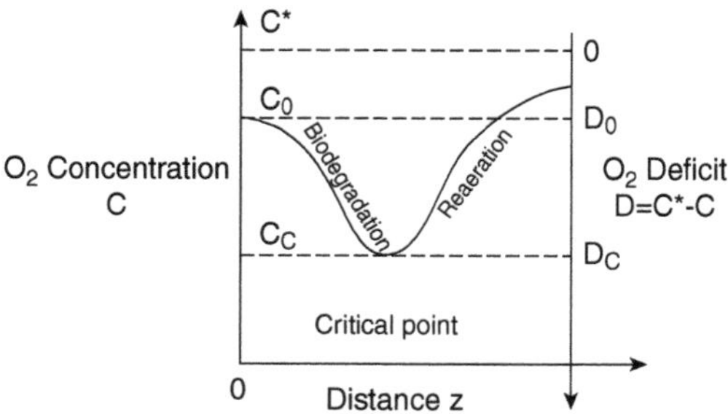

FIGURE 4.13 The Streeter–Phelps model. Dissolved oxygen profiles in a river with a steady influx of a pollutant.

illustration. Also neglected were the effects of runoff and respiration by algae. Thus, only biodegradation and reaeration rates needed to be considered, both of which were assumed to be first order in concentration. The model equations are then as follows:

Oxygen mass balance:

Rate of oxygen in – Rate of oxygen out = 0

$$\begin{bmatrix} Q\,C\big|_z \\ +k_L a(C^*-C)_{avg} A_C \Delta z \end{bmatrix} - \begin{bmatrix} Q\,C\big|_{z+\Delta z} \\ +k_r L\, A_C \Delta z \end{bmatrix} = 0 \tag{4.24a}$$

which becomes, in the limit $\Delta z \to 0$:

$$v\frac{dC}{dz} + k_r L - k_L a(C^*-C) = 0 \tag{4.24b}$$

where $v = Q/A_C$ which is the superficial river velocity, $k_L a$ is the volumetric mass transfer coefficient, k_r the reaction rate constant, and L the pollutant concentration.

Alternatively we can write in terms of the oxygen deficit $D = C^* - C$,

$$v\frac{dD}{dz} \quad - \quad k_r L + k_L a D \quad = \quad 0 \tag{6.2.31}$$

Flow Reaction Mass Transfer

Pollutant mass balance:

Rate of pollutant in – Rate of pollutant out = 0

$$Q\,L\big|_z - \begin{bmatrix} Q\,L\big|_{z+\Delta z} \\ +k_r L\, A_C \Delta z \end{bmatrix} = 0 \tag{4.24c}$$

which yields the ODE:

$$v\frac{dL}{dz} + k_r L = 0 \tag{4.24d}$$

The latter equation can be immediately integrated by separation of variables and we obtain:

$$L = L_0 \exp\left(-\frac{k_r}{v} z\right) \tag{4.24e}$$

This intermediate result gives the pollutant concentration profile in the river. Substitution of L into Equation 4.24c then leads to the ODE in the oxygen deficit:

$$v\frac{dD}{dz}+k_L a\, D=k_r L_0 \exp\left(-\frac{k}{v}z\right) \tag{4.24f}$$

This equation is of the form:

$$y'+f(x)y=g(x) \tag{4.24g}$$

which has the solution (see Item 6, Table 2.4):

$$y=\exp\int -f(x)dx\left[\int g(x)\exp\int f(x)dx+K\right] \tag{4.24h}$$

Upon evaluation of the integrals we obtain:

$$D=\exp\left(-\frac{k_L a}{v}z\right)\left[\frac{k_r L_0}{k_L a-k_r}\exp\left(\frac{k_L a-k_r}{v}z\right)+K\right] \tag{4.24i}$$

Using the boundary condition $D = D_0$ at $z = 0$ to evaluate the integration constant K finally yields:

$$D=\left(D_0-\frac{k_r L_0}{k_L a-k_r}\right)\exp\left(-\frac{k_L a}{v}z\right)+\frac{k_r L_0}{k_L a-k_r}\exp\left(-\frac{k_r}{v}z\right) \tag{4.24j}$$

This is the oxygen deficit profile in the river, shown in Figure 4.13. Let us now evaluate this result with the aid of Mathematica. The integration of equation (4.24f) and implementation of the boundary condition can be all integrated in one command using the `DSolve` function:

```
In[1]:= FullSimplify[
         d[z] /.
           First[DSolve[{v d'[z] + kLa d[z] ==
           kr Lo Exp[-k / v z],
                d[0] == do}, d[z], z]]
        ]
```

$$\text{Out[1]=}\ \frac{-(\text{kr Lo})+E^{((k-\text{kLa})\,z)/v}(\text{do}\,(k-\text{kLa})+\text{kr Lo})}{E^{(k\,z)/v}(k-\text{kLa})}$$

Comments: Equation 4.24j is often cast into the more convenient form:

$$D = \left(D_0 - \frac{L_0}{f-1} \right) \exp\left(-\frac{fk_r}{v} z \right) + \frac{L_0}{f-1} \exp\left(-\frac{k_r}{v} z \right) \tag{4.24k}$$

where $f = k_L a/k_r$ is the so-called self-purification rate (dimensionless). When f = 1, D is indeterminate ($\infty - \infty$) and has to be evaluated by L'Hopital's rule.

The critical point, or minimum shown in Figure 4.13, is evaluated by setting dD/dz = 0. There results:

Critical distance:

$$Z_C = \frac{V}{K_r(f-1)} \ln \left\{ f \left[1 - (f-1) \frac{D_0}{L_0} \right] \right\} \tag{4.24l}$$

and Critical oxygen deficit:

$$D_c = D_0 \{ f[1 - (f - 1)(D_0/L_0)] \}^{1/1-f} \tag{4.24m}$$

Two alternative terms used in environmental work are worth noting.

1. Pollutant concentrations are often expressed in terms of the so-called *biochemical oxygen demand* (BOD), which is the oxygen consumed by the pollutant in mg/L, established in a standard test. Thus, BOD is seen to be proportional to pollutant concentration L.
2. Distance z from the pollution source is frequently replaced by the quantity t = z/v, appropriately called *flow time*. As z always occurs in the combination z/v, all previously cited solutions can be expressed in terms of flow time t.

Illustration 4.17 Dialysis Revisited

It will be recalled that Chapter 1 made a brief reference to the medical procedure known as *dialysis*, which involves passing the blood of a patient through an assembly of tubular membranes where accumulated toxins are removed prior to returning the blood to the body (see Figure 1.9). The principal point made there was that the system was to be regarded as a combination of an internal blood compartment undergoing a slow change in concentration, linked to an external exchanger operating under quasi-steady-state conditions (Illustration 1.8). The model was consequently expected to consist of an ODE in time describing the internal events; whereas the external concentration distribution in the membrane bundle would come from a "quasi-steady-state" ODE in distance. The aim here is to derive and solve these equations, and to establish the time course of toxin concentration, as well as the duration of the dialysis procedure.

We start with a steady state toxin mass balance over a difference element Δx of the tubular membrane. The external concentration is low enough to be neglected. We have:

Rate of toxin in – Rate of toxin out = 0

$$QC\Big|_x - \Big[QC\Big|_{x+\Delta x} + k_C n\pi d\Delta x(C_{avg})\Big] = 0 \qquad (4.25a)$$

where k_C is the mass transfer coefficient in units of m/sec, and n = number of tubes in the bundle.

Dividing by Δx and going to the limit, we obtain:

$$Q\frac{dC}{dx} + k_C n\pi dC = 0 \qquad (4.25b)$$

which upon integration by separation of variables yields:

$$C/C_b = \exp(-k_C A/Q) \qquad (4.25c)$$

where A = total membrane area = nπdL.

Note that C_b is the inlet concentration received from the blood compartment, and C the outlet concentration, which returns to the same location. A toxin mass balance taken over that compartment consequently yields:

$$\text{Rate of toxin in} - \text{Rate of toxin out} = \frac{d}{dt}(\text{toxin contents})$$

$$QC - QC_b = V\frac{dC_b}{dt} \qquad (4.25d)$$

We thus have two equations in the two unknowns C and C_b. Eliminating C between them, we obtain:

$$-[1 - \exp(-k_C A/Q)]QC_b = V\frac{dC_b}{dt} \qquad (4.25e)$$

Integrating by separation of variables and solving for t, there results

$$t = \frac{V}{Q}\left[1 - \exp\left(-\frac{k_C A}{Q}\right)\right]^{-1} \ln\frac{C^o}{C} \qquad (4.25f)$$

where C^o is the initial toxin concentration in the patient.

A typical value for k_CA of the membrane dialyzer is 1 cm^3/sec, and that for the blood flow rate Q 5 cm^3/sec. Blood volume V of an adult is about 5 liters. We then have, for a tenfold reduction in toxin concentration:

$$t = \frac{5000}{5}\left[1-\exp\left(-\frac{1}{5}\right)\right]\ln\frac{10}{1} \tag{4.25g}$$

$$t = 12{,}690 \text{ sec} = 3.52 \text{ h} \tag{4.25h}$$

This is the duration of a typical dialysis procedure.

Comments: The quasi-steady-state assumption was crucial to obtaining a quick answer to the problem. Without it, we would have been dealing with a coupled PDE/ODE model, which would require elaborate analytical or numerical solution methods, and would not have improved the accuracy of the result by much. Deviations from experiment are much more likely to arise from inherent uncertainties in the values of V, Q (which fluctuates with blood pulse), and the mass transfer coefficient k_C. The approach used is therefore quite satisfactory for our purposes.

Equation 4.25f can be used in other important ways. It enables us to assess the effect of percent toxin to be removed for the duration of treatment, the advantages in increasing membrane, and the impact of flow rate on dialysis time. The model serves even better in this capacity than in predicting the length of the procedure.

4.3 APPLICATIONS OF ENERGY BALANCES

In formulating energy balances, we draw on the same two broad categories of models we had seen in the application of the law of conservation of mass: The compartmental or lumped-parameter model, which leads to ODEs or algebraic equations, or the one-dimensional distributed type (our 1-D pipe), which likewise led to ODEs under steady state conditions. If taken over finite parts of the 1-D pipe, the steady state balances revert to algebraic form.

Where the energy balances differ from their mass balance counterparts is in the wide variety of energy forms and transport mechanisms. Whereas mass transport takes place in only two major modes, convective and diffusive, the list becomes much longer when one seeks to apply the law of conservation of energy. In the first place, a distinction has to be made among at least four major categories of energy: thermal, mechanical, electrical, and nuclear. Thermal energy alone is host to three major transport mechanisms: conduction, convection, and radiation. Mechanical energy comprises kinetic and potential energies and work done by or on machines (pumps, turbines, lifts, etc.) Electrical energy is commonly transported through conductors, but can also be conveyed by electromagnetic radiation (e.g., microwaves). The various energy forms can also be converted one into another, giving rise to further extensions in the form and application of energy balances. Each of these categories can be accommodated within the framework of compartmental and distributed models.

The intent here is to present a limited but carefully chosen selection of examples, which convey the richness of the topic without overwhelming the reader with detail. This is done within the context of both classical and contemporary topics.

4.3.1 Compartmental Models

Illustration 4.18 Return to the Thermocouple

We did not wish to abandon this device, which was repeatedly mentioned in an abstract sense in Chapters 1 and 2 without presenting a concrete numerical example of its behavior. Rather than calculating the response of a particular device to a disturbance, we invert the problem and ask what its dimension should be in order to produce a prescribed response. The physical parameters we use are as follows:

Specific heat $C_p = 0.419$ kJ/kg K
Density $\rho = 8800$ kg/m^3
Heat transfer coefficient h = 0.455 kJ/m^2 sec K
The dimension (diameter d) should be such that a 15°C step increase in ambient temperature T_a will be registered with 0.5°C of the final value in no more than 4 sec. We have, for a balance around the thermocouple:

$$\text{Rate of energy in} - \text{Rate of energy out} = \frac{d}{dt}\text{energy content}$$

$$hA(T_a - T) - 0 = \frac{d}{dt}m\bar{H} = \rho V C_p \frac{dT}{dt} \tag{4.26a}$$

where m, A, V are the mass, surface area, and volume of the exposed wires, and $\bar{H}$ its specific enthalpy (kJ/kg).

Integration by separation of variables yields in the first instance:

$$t = \frac{\rho CV}{hA}\ln\frac{T_a - T_i}{T_a - T_f} = \frac{\rho C_p d}{4h}\ln\frac{T_a - T_i}{T_A - T_f} \tag{4.26b}$$

where i and f refer to the initial and final states of the thermocouple, respectively. Solving for d we obtain:

$$d = \frac{4ht}{\rho C_p \ln\dfrac{T_a - T_i}{T_a - T_f}} = \frac{4 \times 0.455 \times 4}{8800 \times 0.419 \ln\dfrac{15}{0.5}} \tag{4.26c}$$

$$d = 0.58\ mm \tag{4.26d}$$

Comments: Of note here is that we have in essence solved a design problem. The diameter d is just one of several pieces of information that can be extracted from the model, as we had indicated in Chapter 1. Response time and temperature history are others.

ILLUSTRATION 4.19 THE HEATED TANK AND THE BOILING POT

We use these two examples to illustrate two specific transport mechanisms: heat transfer by natural convection and heat transfer to a boiling liquid. The novelty here lies not in the models, which are simple and quickly composed, but rather in the nature of the heat transfer coefficients. These are not the usual constants one sees in normal applications, but instead vary with the temperature driving force in a significant way. This is because ΔT affects the onset and ultimate intensity of the detailed transfer mechanisms and, hence, influences the heat transfer coefficient itself. We assume water to be the working fluid and use its relevant physical parameters.

A. *Heat-up time of an unstirred tank:*

Natural convection heat transfer occurs as a result of the volumetric expansion of the heated fluid and its consequent movement in the upward direction. Upon cooling, the fluid contracts and reverses direction, thus setting in motion a circulatory transfer of thermal energy.

Heat transfer coefficients are measured experimentally and correlated in terms of the volumetric coefficient of expansion β(1/K), which is responsible for setting the convection process in motion, as well as other physical properties of the system. For a vertical cylinder, which is the appropriate geometry here, the correlation is given by:

$$h_{nc} = 0.13k\left(\frac{\rho^2 g\beta\Delta T}{\mu^2}\right)^{1/3}\left(\frac{Cp\mu}{k}\right)^{1/3} \tag{4.27a}$$

where k and μ are the thermal conductivity and viscosity of the fluid.

Of note here is the aforementioned dependence on ΔT, which, although a weak one, has nevertheless a marked effect on the heat-up time (see comments).

We consider a tank of diameter d = 1.5 m filled to a height H = 3 m, which is to be heated from 20°C to 100°C using external jacketed steam of 120°C.

Using the relevant physical parameters for water at an average temperature of 60°C, and an expansion coefficient of 4.8×10^{-4} K^{-1},we obtain from Equation 4.27a

$$h_{nc} = 321\ \Delta T^{1/3}\ J/m^2\ sec\ K \tag{4.27b}$$

From the dimensions of the tank, we obtain a surface area A = 14.2 m^2 and water volume V = 5.30 m^3. These values are used in the following energy balance about the tank contents:

$$\text{Rate of energy in} - \text{Rate of energy out} = \frac{d}{dt} energy\,contents$$

$$h_{nc}A(T_s - T) - 0 = \rho VC_p \frac{dT}{dt} \tag{4.27c}$$

Upon integration by separation of variables and solving for time t we obtain:

$$t = \frac{\rho VC_p}{A}\int_{T_i}^{T_f} \frac{dT}{h_{nc}(T_s - T)} \tag{4.27d}$$

Introducing numerical values into Equation (4.27d) and noting that

$$\int_{T_i}^{T_f} \frac{dT}{(T_s - T)^{4/3}} = -\int_{T_i}^{T_v} \frac{d(T_s - T)}{(T_s - T)^{4/3}} = 3\left[\frac{1}{(T_s - T)_f^{1/3}} - \frac{1}{(T_s - T)_i^{1/3}}\right]$$

we obtain:

$$t = \frac{(1000)(5.3)(4.2\times 10^3)}{(14.2)(321)} 3\left[\frac{1}{20^{1/3}} - \frac{1}{(100)^{1/3}}\right]$$

$$t = 1700 \text{ sec (i.e., nearly 40 min.)}$$

Comments: The effect of the dependence of the heat transfer coefficient on $(T_s - T)^{1/3}$, seemingly a weak one, is in fact not trivial. During the initial heat-up period, with $\Delta T \sim 100°C$, its inclusion increases the heat transfer rate by $100^{1/3}$, i.e., a factor of nearly 5. Even near the termination of the heat-up process the increase is by a factor of $20^{1/3} = 2.7$. This evidently brings about a fairly radical change in the results of the model.

On the other hand, the effect of the temperature dependence of the physical properties, is considerably milder if we choose values midway between the initial and final process temperatures. Choosing μ^2 as the most severe case, we have for the ratio $(\mu_{20}/\mu_{60})^{2/3} \sim (0.95/0.50)^{2/3} \cong 1.5$. This is, however, nearly compensated at the high-temperature end, where the relevant ratio is $(0.28/0.50)^{2/3} \cong 0.68$, hence $1.5 \times 0.68 = 1.01$. This compensating effect evidently does not exist in the case of the factor $(T_s - T)^{1/3}$, which causes an increase in heat transfer during the entire heat-up period.

B. *The boiling pot:*

We use this simple example of a pot boiling on a stove or hot plate to introduce the reader to what is termed *boiling heat transfer*, i.e. the heat-transfer rate, flux, and coefficients which are associated with boiling. Imagine, then, a pot brought to a boil on a hot element whose power input can be adjusted so that the pot bottom assumes various surface temperatures T_s. The boiling heat transfer is then driven by the temperature difference $\Delta T_b = T_s - T_b$, where T_b is the (constant) temperature of the boiling liquid. Let us examine what happens when this driving force is changed by adjusting the power input to the element. It turns out that the boiling mechanism undergoes various complex transitions as ΔT_b is increased, and that the heat transfer coefficient h_b not only depends on ΔT_b, as was the case in natural convection, but does so in a complex manner.

At low values of ΔT_b of less than ~ 5°C, the mechanism is one of natural convection with very few bubbles formed to disturb the normal natural convection. The dependence of h_b on ΔT_b is then roughly with the power 1/3, i.e., $h_b \propto (\Delta T_b)^{1/3}$, and we speak of convective boiling.

When ΔT_b is raised to the range of 5 to 25°C, bubble production increases, and with it there is an increase in the degree of turbulence and liquid circulation. This causes a dramatic increase in the temperature dependence of ΔT_b to $h_b \propto (\Delta T_b)^2 - (\Delta T_b)^3$. We speak of the process being in the nucleate boiling range.

This trend does not continue indefinitely, however. A stage is reached at which evaporation is so fast that there is insufficient time for the bubbles to detach themselves. The pot bottom is then blanketed with a layer of steam, which causes a sharp decline in the dependence of h_b on ΔT_b and the associated heat transfer because of the lower conductivity of the vapor.

This situation persists until a driving force of $\Delta T_b \sim 100°C$ is reached. The steam bubbles are now able to detach themselves as fast as they are formed, so that heat transfer takes place through a thin liquid film adjacent to the metal surface out to a region that is well-mixed and of uniform temperature because of the action of the bubbles. This stage is referred to as the region of film boiling. Further increases in ΔT bring an additional contribution due to radiation heat transfer.

The values of h for boiling are quite high. At the beginning of the nucleate boiling region they range from 6000 to approximately 11,000 J/m² sec K (for water). At the peak of nucleate boiling, before the decline due to blanketing sets in, it reaches a maximum value of approximately 60,000 J/m² sec K.

Let us next examine, using an actual numerical example, how this affects the evaporation rate of water. Extensive analysis of boiling rate data has led to the following two correlations for convective and nucleate boiling, respectively:

$$\text{Convective Boiling:} \quad h = 1043\,(\Delta T, K)^{1/3};\ q/A < 16\ kJ/m^2 \tag{4.28a}$$

$$\text{Nucleate Boiling:} \quad h = 5.56\,(\Delta T, K)^3;\ 16 < q/A < 240 \tag{4.28b}$$

We first choose a $\Delta T = 5$, characteristic of the convective boiling region. Thus,

$$h = 5.56\,(5)^{1/3} = 1.78\ kJ/m^2 \text{ sec } K;\ q/A = (1.78)(5) = 8.92\ kJ/m^2 \tag{4.28c}$$

which places it within the range of validity of the correlation, Equation 4.28a.

Using a latent heat of evaporation at the boiling point of $\Delta H = 2447$ kJ/kg, and assuming a pot surface A = 0.02 m², we obtain:

$$Q = \frac{q}{\Delta H} = \frac{hA\Delta T}{\Delta H} = \frac{(1.78)(0.02)(5)}{2447}3600 = 0.28\,kg\ steam\ produced/h \tag{4.28d}$$

Choosing next a $\Delta T = 12$, we obtain:

$$h = 5.56\,(12)^3 = 0.61\ kJ/m^2sK;\ q/A = (9.61)(12) = 115\ kJ/m^2 \tag{4.28e}$$

which is within the range of the correlation, Equation (4.27b). The corresponding evaporation rate becomes:

$$q = \frac{hA\Delta T}{\Delta H} = \frac{(9.61)(0.02)(12)}{2260}3600 = 3.7\,kg\ steam\ produced/h$$

These startling results indicate that by raising the pot bottom temperature by only 7°C, i.e., from a ΔT_b of 5 to a ΔT_b of 12, a nearly 18-fold increase in evaporation rate is achieved. This is an indication of the powerful influence of the cubic dependence of h on ΔT.

Comments: We use this example to demonstrate that mathematical complexity is not a necessary precursor of startling revelations. A simple relation, coupled with physical observation, can be equally effective in achieving the same result.

Illustration 4.20 Another Moving-Boundary Problem: Freeze-Drying of Meat

We had, on previous occasions, outlined the nature of so-called moving-boundary problems and provided a systematic way of modeling these processes (Illustration 1.8 and Illustration 2.3). It consists of starting with unsteady mass and energy balances around the receding or growing core, followed by quasi-steady-state balances applied to a surrounding stagnant film.

In the process to be considered here, we wish to derive a model, which would allow us to obtain relevant heat and mass transport coefficients from freeze-drying rate data. The food to be dried, e.g., a slab of frozen poultry meat, has an initial (frozen) water content of m_0 kg. It is heated with an electric heater, and in the experiment in question, provided with thermocouples to measure surface temperature T_g. Sublimation of the ice takes place in a vacuum chamber and water loss is monitored by means of a spring balance. A sketch of the configuration appears in Figure 4.14. As sublimation progresses, the core ice front, assumed to be at the constant temperature T_i, recedes into the interior, exposing an ice-free matrix that increases in thickness with time. Heat conduction through this matrix is assumed to be at a quasi-steady-state so that a linear temperature gradient prevails at any given instant.

We start with a mass balance around the core and add equations as the need arises.

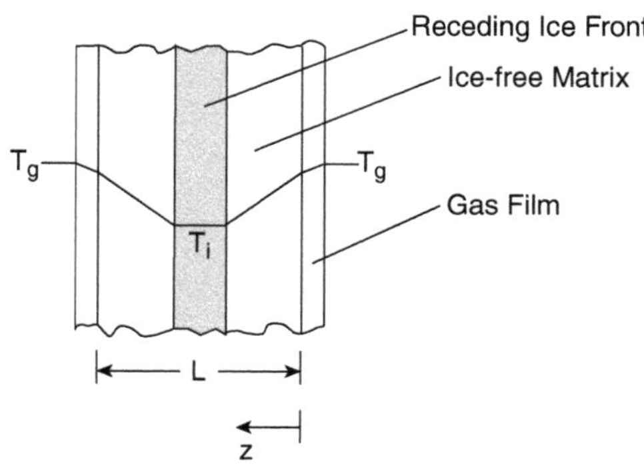

FIGURE 4.14 Temperature profiles in the freeze-drying of a slab of meat.

Core mass balance:

Rate of moisture in − Rate of moisture out = Rate of change of moisture content

$$0 - F = \frac{d}{dt}m \tag{4.29a}$$

where F equals the rate of sublimation of ice.

Core energy balance:

Rate of energy in − Rate of energy out = Rate of change of energy content

$$q - F\Delta H_v = \frac{d}{dt}(H_{ice} + H_{matrix}) \tag{4.29b}$$

We choose the reference temperature as that of the ice and its matrix, so that $H_{ice} = H_{matrix} = 0$ and the right side of Equation 4.29b becomes zero. We then obtain, after introduction of the pertinent auxiliary relation for q,

$$U2A\ (T_g - T_i) - \Delta H_s F = 0 \tag{4.29c}$$

where ΔH_s denotes the enthalpy of sublimation of ice. We note that the heat transfer coefficient U depends on the distance of conduction z and is given by the relation:

$$\frac{1}{U} = \frac{1}{h} + \frac{z}{k} \tag{4.29d}$$

with h and k denoting the external film coefficient and the matrix thermal conductivity, respectively. Combination of these three equations yields:

$$\frac{2A(T_g - T_i)}{\frac{1}{h} + \frac{z}{k}} = -\Delta H_s \frac{d}{dt}m \tag{4.29e}$$

We note that we still have, at this stage, two dependent variables z and m. An additional equation that relates z to m, therefore, will be required. Some reflection will show that this relation can be obtained from a cumulative mass balance on the core ice. We have:

Initial ice = Ice sublimated + Ice left over

$$m_0 = m_0 \frac{L - 2z}{L} + m \tag{4.29f}$$

from which there results:

$$z = \frac{m_0 - m}{m_0}\frac{L}{2} = f\frac{L}{2} \tag{4.29g}$$

where f is the fraction of ice removed.

The ODE (Equation 4.29e) now becomes:

$$\frac{2A(T_g - T_i)}{\frac{1}{h} + \frac{L}{2k}f} = -\frac{\Delta H_s}{m_0}\frac{df}{dt} \tag{4.29h}$$

which upon integration by separation of variables leads to the expression:

$$\frac{2A(T_g - T_i)}{\Delta H_s}t = \frac{1}{h}f + \frac{L}{4k}f^2 \tag{4.29i}$$

Symbolically we can write:

$$t/f = af + b \tag{4.29j}$$

Plots of experimental t/f values vs. f can then be used to evaluate matrix conductivity k from the slope a and the film coefficient h from the intercept b.

Experimental data have shown good agreement with this model up to values of f ≈ 0.85. Beyond that point, deviations from the straight line relation (Equation 4.29j) occur, which have been attributed to moisture adsorption on the matrix. This causes an increase in ΔH_s, which is now the enthalpy of desorption, and a resultant change in slope and intercept.

Comment: We note that this problem was not placed in the category of simultaneous mass and energy balances, dealt with in the next section. The reason for this is that the balances here are uncoupled and can be solved independently.

Illustration 4.21 Melting of a Silver Sample: Radiation

We use this example of the melting of a silver sample in a high-temperature furnace to introduce the reader to some simple concepts of radiation.

Thermal radiation is an important mode of heat transfer, especially so when large temperature differences occur as, for example, in furnaces, driers, metallurgical processes, and other high-temperature operations. It is a form of electromagnetic radiation, and its rate of transfer depends on the temperature of the emitting and receiving objects. Part of the radiation transmitted is absorbed by the receiving body, and part of it is reflected so that:

$$\alpha + \rho = 1 \tag{4.30a}$$

where α denotes absorptivity, the fraction absorbed, and ρ denotes reflectivity, the fraction reflected.

Bodies, whether they are the source or the receiver of radiation, emit radiation of their own, and this depends on the temperature of the body. To provide a measure of emission radiation, in general, we use that of a so-called black body as a reference, and define

$$Emissivity\ \varepsilon = \frac{Emission\ radiation\ of\ an\ arbitrary\ object}{Emission\ radiation\ of\ a\ black\ body} \tag{4.30b}$$

where the black body refers to an entity with an absorptivity $\alpha = 1$, i.e., one which reflects none of the incident radiation. It can be approximated by blackening the surface of an object with charcoal. All real bodies have an emissivity $\varepsilon < 1$ and are referred to as gray bodies.

Emissivity is high for dull surfaces, $\varepsilon \approx 0.6$ to 0.95 and low for highly reflective or polished surfaces, $\varepsilon \approx 0.01$ to 0.2. Oil paints of all colors, for example, have emissivities $\varepsilon = 0.92$ to 0.96. For highly polished iron, ε rises to 0.74, but, if dulled by oxidation, dramatically drops to a value of $\varepsilon = 0.052$.

The rate of radiation heat emission is given by the Stefan–Boltzmann law:

$$q_r = \varepsilon\sigma A T^4 \tag{4.30c}$$

where σ = Stefan–Boltzmann constant = 5.767 x 10^{-8} J/m^{2}sK, and T is in K. Note the strong fourth power dependence on temperature, compared to the linear dependence on driving force in the case of convective and conductive heat transfer.

To derive the radiative heat transmission rate between two bodies, one argues that the radiative emission from the hot body, T_2, reaching the colder body, T_1 and area A is given by:

$$q_{21} = \varepsilon\sigma A T_2^4 \tag{4.30d}$$

This amount has to be diminished by the amount emitted by the cold body, again with area A, and absorbed by the hot body:

$$q_{12} = \alpha\sigma A T_1^4 \tag{4.30e}$$

As absorptivity is usually very nearly equal to emissivity, $\alpha \approx \varepsilon$, the net heat flow received by the cold body is the difference of the two, i.e.,

$$q_r = \varepsilon\sigma A(T_2^4 - T_1^4) \tag{4.30f}$$

To calculate the time required to melt the silver sample under consideration, we use the following data:

Latent heat of fusion of silver ΔH_s = 89 J/g
Specific heat of silver C_p = 0.24 J/g K
Melting point of silver T_1 = 1230 K
Furnace temperature T_2 = 1500 K
Crucible charge m = 1000 g silver
Crucible surface area A = 10^{-2} m^2
Furnace emissivity ε = 0.6

The total requirement time will be made up of the heat-up time to bring the sample to its melting point plus the time needed to melt the charge. We have:

For the heat-up time:

$$\text{Rate of energy in} - \text{Rate of energy out} = \frac{d}{dt}(\text{energy contents})$$

$$\varepsilon\sigma A(T_2^4 - T_1^4) - 0 = mC_p \frac{dT_1}{dt} \tag{4.30g}$$

We use separation of variables to solve the heat-up ODE:

$$dt = \frac{mC_p}{\varepsilon\sigma A} \frac{dT_1}{T_2^4 - T_1^4} \tag{4.30i}$$

The solution to this equation can be found by direct integration.

The heat-up time is:

```
In[1]:= heatuptime = Integrate[m Cp/(∈ σ A) 1/(T2^4 - T1^4), T1]

          Cp m (2 ArcTan[T1/T2] - Log[-T1 + T2] + Log[T1 + T2])
Out[1]= ------------------------------------------------------
                            4 A T2^3 ∈ σ
```

Assuming an initial temperature of 298 K, the heat-up time for the silver sample can be calculated as $t_{\text{heat up}} = 148.6$ sec.

For the melting time (which takes place at constant temperature):

$$t = \frac{\Delta H_s m}{q_r} = \frac{\Delta H_s m}{A\varepsilon\sigma(T_2^4 - T_1^4)} \tag{4.30h}$$

Based on the data, the melting time is 92.7 sec. Therefore, the total time for melting the silver sample is 241.3 sec.

4.3.2 Distributed Models

We start this section by presenting a general energy equation that is distributed in the sense that it interrelates the variables at two different locations in space rather than as a continuous function of distance. In this it resembles the Bernoulli equation

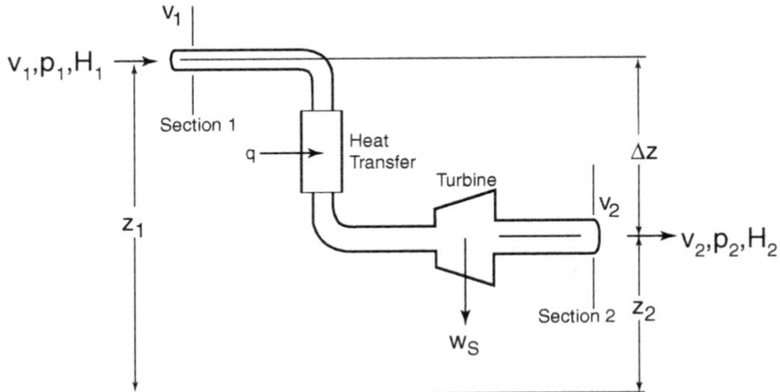

FIGURE 4.15 Diagram depicting the various terms of the energy balance.

referred to in Practice Problem 4.6 and, in fact, contains some terms that are common to both. Its major feature is that it establishes a relation between mechanical and thermal energy — other forms, such as electrical and surface energies, are excluded, and we exploit this feature to convey a sense of the magnitude of these two energy forms.

Illustration 4.22 The Steady State Integral Balance for Mechanical and Thermal Energy Forms

The energy forms involved in this balance are indicated in Figure 4.15. Fluid enters and leaves the system with a kinetic energy $v^2/2$ per unit mass and enthalpy H per unit mass. There is also a difference in potential energy $g\Delta z$ between the two locations. In addition, discrete amounts of thermal energy q or "shaft" work w_s are added or withdrawn between entrance and exit. The energy balance then reads:

$$\Delta H + \Delta \frac{v^2}{2} + g\Delta z = \pm q \pm w_s \tag{4.31a}$$

Both kinetic and potential energy terms appear in the Bernoulli equation as well. The pressure term seen there lurks in the enthalpy difference of Equation 4.29a, but the Bernoulli equation does not include either exchange of thermal energy or shaft work with the surroundings. Note that each term, both here, and in the Bernoulli equation, carries units of J/kg fluid.

Let us now examine the relation between the various energy forms:

a. Kinetic energy and enthalpy:
   Consider a hurricane force wind moving at 75 m/sec which is suddenly brought to an abrupt halt. If the conversion is entirely into thermal energy, the corresponding temperature rise would be given by:

$$\Delta H + C_p \Delta T = \frac{v^2}{2} \tag{4.31b}$$

or, with the specific heat C_p for air being 1000 J/kg K:

$$\Delta T = \frac{75^2}{2 \times 1000} = 2.8^\circ C \tag{4.31c}$$

b. Potential energy and enthalpy:
Consider next a waterfall of 100-m height. If, again, conversion is entirely into thermal energy, we have:

$$\Delta H = C_p \Delta T = g \Delta z \tag{4.31d}$$

or with $C_p = 4180$ J/kg water:

$$\Delta T = \frac{9.81 \times 100}{4180} = 0.24^\circ C \qquad (4,31e)$$

It is seen from these two examples that mechanical energy transforms into relatively modest thermal equivalents. Thus, in most heat-transfer processes, at least of the industrial variety, kinetic and potential energy changes can be neglected. If, in addition, no shaft work is involved, we obtain from Equation 4.31a:

$$\Delta H = \pm\, q \tag{4.31f}$$

This expression will be immediately recognized as the equation that, in differential or integral form, is used in the analysis of heat exchange processes (see Illustration 1.2 and Illustration 4.18).

c. Kinetic energy and work:
We provide two examples of the interconversion of these two energy forms. In Illustration 4.23, which follows, we examine the conversion of the kinetic energy of wind into shaft work (wind turbine). Illustration 4.26 addresses the conversion of kinetic energy — that of a moving car — into work done to overcome friction with the pavement. The analysis in both cases is simple, yet surprisingly revealing.

Illustration 4.23 Kinetic Energy Effects: The Idealized Wind Turbine

Wind-driven machines, whose history dates back many centuries, have in recent years emerged as an attractive, emission-free source of power (Figure 4.16A). Although the interaction between wind and machine is a highly complex one, some important limiting parameters can be established through the use of simple, idealized models. The oldest such model which describes wind turbine dynamics is due to Rankine and Froude (1889). The Rankine–Froude actuator-disk model, as it is known, considers the turbine to be made up of an infinite number of infinitely thin blades, in effect a porous disk of zero thickness. The flow field around such a device is shown in Figure 4.16B.

A.

B.

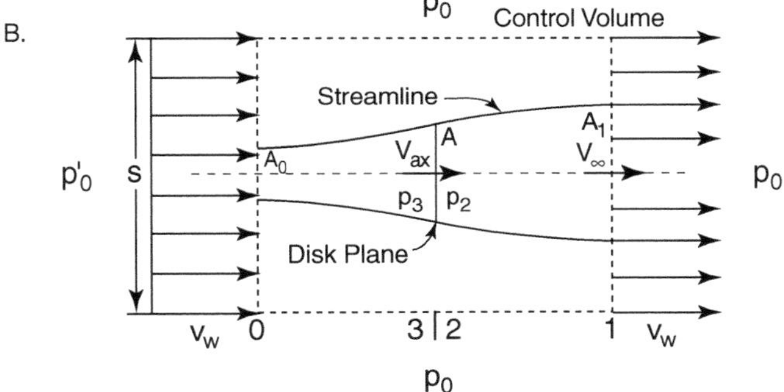

FIGURE 4.16 (A) A wind farm in California. (B) Flow field around the actuator-disk.

Wind approaching with a velocity V_w expands as it nears the disk and slows to a velocity V_{ax} when it attains its target. This is followed by further expansion until the wind resumes uniform flow with a downstream velocity V_∞. V_{ax} is taken to be halfway between the upstream and downstream values.

If we define a fractional decrease f in velocity at the disk over that of the approaching wind velocity V_w, we obtain:

At the disk: $$V_{ax} = V_w\,(1 - f) \tag{4.32a}$$

Far downstream: $$V_\infty = V_w\,(1 - 2f) \tag{4.32b}$$

For f = 0 in Equation 4.32a, the wind is not decelerated and no power is extracted. For f = 0.5 in Equation 4.32b, we have a similar situation. The far-wake velocity V_∞ vanishes, and without the presence of flow behind the turbine, the extracted power is again zero. Between these two vanishing extremes, the power extracted must perforce pass through a maximum. This occurs at some value f_{opt} such that:

$$0 < f_{opt} < 0.5 \tag{4.32c}$$

It is this maximum power we wish to calculate here.

Power P is defined as energy per unit time. In terms of the variables being dealt with here, we can write:

$$P = \textit{Mass Flow Rate}\ (kg/sec) \times \textit{Kinetic Energy}\ (J/kg) \tag{4.30d}$$

and for the power recovered by the actuator disk:

$$\begin{pmatrix} \textit{Power} \\ \textit{recovered} \end{pmatrix} = \begin{pmatrix} \textit{Mass flow} \\ \textit{through disk} \end{pmatrix} \times (\textit{Kinetic energy change})$$

$$P = (\rho V_{ax} A) \times \left(\frac{1}{2} V_n^{\,2} - \frac{1}{2} V_\infty^{\,2} \right) \tag{4.32e}$$

Introducing V_{ax} and V_∞ from Equations 4.30a and 4.30b, we obtain the following expression for the extracted power:

$$P = \frac{1}{2} \rho A V_w^{\,3} 4f(1-f)^2 \tag{4.32f}$$

As $\frac{1}{2}\rho A V_w^{\,3}$ represents the total power available in the approaching wind, we can also write:

$$E = \frac{P}{\frac{1}{2}\rho A V_w^{\,3}} = 4f(1-f)^2 \tag{4.32g}$$

where E is the fractional "extraction" efficiency of the turbine.

E will be a maximum for:

$$\frac{dE}{df} = 0 = 4(10f)(1-3f) \tag{4.32h}$$

which yields:

$$f_{opt} = \frac{1}{3} \tag{4.32i}$$

and hence, by back substitution into Equation 4.32g:

$$E_{Max} = 0.593 \tag{4.32j}$$

Comments: What this result shows is that the maximum power one can hope to recover amounts to slightly less than 60% of the kinetic energy conveyed by the wind per unit time. This is an extremely valuable upper limit of what can be produced by wind power. It also sets an upper limit to what designers should be aiming at. The efficiency value given by Equation 4.32j is referred to as the Lanchester–Betz limit.

Illustration 4.24 Distributions due to Conduction: Maximum Temperature in a Nuclear Reactor Fuel Rod

The core of a nuclear reactor consists of bundles of cylindrical fuel rods, typically 1 cm in diameter and several meters long, surrounded by a "coolant" which is circulated to the power generating system (e.g., steam turbine plus electrical generator). The coolant may consist of pressurized water (PW), boiling water (BW), or, in the case of the Canadian CANDU reactor, heavy water (D_2O). Both PW and BW reactors use uranium oxide fuel enriched to approximately 3% U^{235} content. The CANDU reactor avoids this step by using a combination of natural uranium (0.7% U^{235}) and heavy water.

A critical parameter in the operation of a nuclear reactor is the internal temperature of the fuel rods. If excessive, it may result in deformation of the elements and ultimately ends in meltdown.

The temperature distribution in a cylindrical rod is the result of a balance between heat generated and heat conducted to the coolant. We had addressed this problem in Illustration 2.10 and arrived at the following expression:

$$T - T_s = \frac{SR^2}{4k}\left[1 - \left(\frac{r}{R}\right)^2\right] \tag{2.21g}$$

where S = heat produced (J/sec m^3 or W/m^3) and T_s = surface temperature.

We consider here a boiling-water reactor (BWR) which has a typical value for S of 400 MW/m^3 (the entire reactor produces 3000 to 4000 MW). Thermal conductivity of the fuel is 1.8 W/m K, and radius R = 0.5 cm. The maximum temperature will be at the central axis, i.e., at r = 0. We immediately obtain from Equation 2.21g.

$$T - T_s = \frac{4 \times 10^8 \times (0.5 \times 10^{-2})^2}{4 \times 1.8}$$

$$T_{r=0} = T_s + 1389$$

With a typical coolant temperature T_s of 500°C, this becomes T = 1889°C. This is the maximum temperature in the fuel element.

Comments: A number of assumptions were slipped in, perhaps unnoticed, which need to be addressed:

- Our model assumed that no heat was lost form the cylinder ends. This is completely justified here since rod length is 300 times that of the diameter.
- External thermal resistances due to the cylinder wall containing the fuel and the coolant were tacitly neglected. We had seen in Illustration 4.19

that typical boiling heat transfer coefficients for water are in the range 6000-11,000 W/m²K. The equivalent heat transfer coefficient for the fuel rod h_f is obtained approximately by dividing thermal conductivity k by rod radius R:

$$h_f = \frac{k}{R} = \frac{1.8}{0.5 \times 10^{-2}} = 360\, W/m^2K$$

which is a small fraction (3 to 5%) of the boiling heat transfer coefficient. Taking account of this resistance would raise the core temperature by some 40 to 70°C. The resistance due to the cylinder wall, which is typically less than 1 mm thick and has seven times the thermal conductivity of the fuel, is even less of a contributing factor. The tacit assumptions we made can therefore be considered reasonable.

- Of much greater consequence to the internal temperature is the quadratic dependence on cylinder radius (cf. Equation 2.21g). We see here that a mere increase of 1 mm in R results in a dramatic temperature increase of (36/25 – 1)1389 = 611°C. It is precisely this type of parameter sensitivity that the model is designed to reveal. It also acts as a crucial guide for the initial choice of fuel rod dimensions. A cylinder of 5-cm diameter, a seemingly reasonable option, would have been immediately ruled out owing to unacceptably high temperature levels.

Illustration 4.25 An Industrial Problem: Coating of a Pipe

It often comes about that an informal opinion is sought on a particular industrial problem without going through the formalities of a secrecy agreement. This happens when the problem is not a particularly serious one, and it is merely desired to obtain a second opinion to confirm the validity of solutions arrived at in-house. In such cases, it is customary not to reveal details of the process involved, and the problem is outlined only in general terms and often circumlocutory language. We reproduce the language used in a particular situation verbatim:

> A length of pipe L with possible ID ranging from d_1 to d_2 and possible thickness from t_{p1} to t_{p2} is to be coated with two layers of dissimilar material A and B [see Figure 4.17, which is the original drawing provided]. The thickness of the two layers is the same (t_c), while the thermal conductivities may be expressed as k_1 and k_2. The original temperature of the pipe is T_1, and it needs to be cooled down to T_2 with water spray. Now the complications are as follows: material A is applied first, followed by material B in a continuous process (see diagram). Due to the temperature of the pipe, the coatings will be molten and need to be solidified by the spray of water before the pipe can be handled. The question is what line speed (v) can be used if this process is to be carried out on a continuous [basis], i.e., the pipe enters one end at T_1 [and] must be at T_2 before it can be handled again.

A rigorous model for the process would require three coupled PDEs in the axial and radial directions subject to boundary conditions that are far from simple. As one

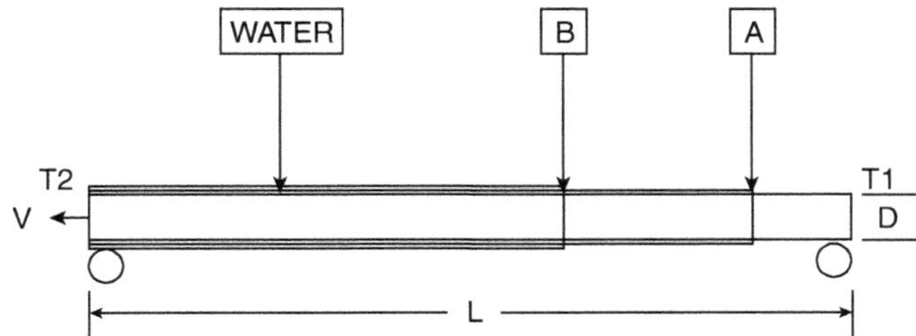

FIGURE 4.17 Coating of a steel pipe with a double layer of molten plastics A and B.

usually has no more than an afternoon's contemplation to provide advice, we resorted to the following simplifying assumptions.

Heat transfer to the water spray is controlled by the resistances of the two coatings, i.e., the steel pipe is taken to have a uniform temperature at any particular position, and the surface temperature of the coatings approaches that of the water impinging on the pipe and evaporating. These are reasonable assumptions considering the high values of boiling heat-transfer coefficients, and the high conductivity of steel compared to that of the coatings.

Latent heats of solidification are either not involved or are small compared to the sensible heat changes of the steel pipe. Axial conduction in the steel wall and heat given off to the interior of the pipe are negligible.

With these simplifications in hand, the model reduced to the following energy balance:

$$\text{Rate of heat in} - \text{Rate of heat out} = 0$$

$$H\big|_z - \begin{bmatrix} H\big|_{z+\Delta z} \\ +q_{avg} \end{bmatrix} = 0$$

$$\rho_s v_s\, \pi D t_s\, C_{ps}\,(T_s - T_{ref})\big|_z \begin{bmatrix} \rho_s v_s\, \pi D t_s\, C_{ps}\,(T_s - T_{ref})\big|_{z+\Delta z} \\ +\dfrac{(T_s - T_w)_{avg}\, \pi D \Delta z}{t_c(1/k_1 + 1/k_2)} \end{bmatrix} = 0 \qquad (4.33a)$$

where t = thickness and the subscripts s and c refer to steel and coating, respectively. Note that $\rho_s v_s\ \pi D t_s$ is the pipe "flow rate" in kg steel/suc and the heat transfer coefficient $U = [t_c(1/k_1 + 1/k_2)]^{-1}$.

Dividing by $\pi D \Delta z$ and going to the limit yields the ODE:

$$(\rho v t C_p)_s \frac{dT_s}{dz} + \frac{T_s - T_w}{t_c(1/k_1 + 1/k_2)} = 0 \qquad (4.33b)$$

which can be immediately integrated by separation of variables to yield:

$$z = (\rho v t C_p)_s [t_c(1/k_1 + 1/k_2)] \ln \frac{T_{s1} - T_w}{T_{s2} - T_w} \tag{4.33c}$$

where z denotes the length of cooling section.

Comments: The line velocity v can be calculated for a specified length of cooling section z, or conversely z determined for a prescribed value of v. z is seen to be proportional to both the line velocity and wall thickness t. This is as expected. Somewhat less expected is the fact that z does not depend on pipe diameter d, which cancels out in Equation 4.33a. This is at variance with the results obtained in-house.

The value to be used for T_w is somewhat of a question mark. The principal heat transfer mechanism will be one of evaporation of water, as latent heat effects outweigh sensible heating of the water by a factor of at least seven. A first approximation would be to set T_w equal to the water temperature at the source. This will provide a *minimum* value of the length of cooling section required, or the maximum permissible line velocity for a given z.

A first estimate of the required water flow rate F_w can be obtained from an integral energy balance by setting the sensible heat change of the steel wall equal to the cumulative latent heat of evaporation. Thus,

$$\rho_s v \pi D t_s z C_{ps}(T_{s1} - T_{s2}) = F_w \Delta H_w \tag{4.33d}$$

where ΔHw is the latent heat of evaporation per unit mass of water. Fw does not include that portion of water that fails to make contact with the pipe or evaporative losses in transit, and is thus to be considered a minimum value.

Taking latent heat of solidification of the plastic coating into account brings about a considerable escalation in the complexity of the problem. Let us see whether we were justified in neglecting these heat effects by examining the ratio of sensible heat changes in the pipe wall q_s to possible latent heat effects in the coating q_c. We have:

$$\frac{q_s}{q_c} = \frac{\rho_s v \pi D t_s z C_{ps}(T_{s1} - T_{s2})}{\rho_c v \pi D t_c z \Delta H_c} \tag{4.34}$$

$$\frac{q_s}{q_c} = \frac{\rho_s t_s}{\rho_c t_c} \frac{C_{ps}(T_{s1} - T_{s2})}{\Delta H_c}$$

Conservative estimates of the density and thickness ratios of steel and coating lead to the value:

$$\left(\frac{\rho_s}{\rho_c}\right)\left(\frac{t_s}{t_c}\right) \cong (10)(5) = 50$$

For the thermal ratio $C_{ps}\Delta T/\Delta H_c$, we assume ΔT = 100°C and ΔH_c = 100 J/g (approximately one third that of the corresponding value for water). We obtain:

$$\frac{C_{ps}\Delta T}{\Delta H_c} = \frac{(0.473)(100)}{100} = 0.473$$

with a total value for the ratio of q_s/q_c of:

$$\frac{\textit{Sensible heat of steel } q_s}{\textit{Latent heat of coating } q_c} = (50)(0.4730) = 24$$

Thus, according to these conservative estimates, latent heat effects account for about 4% of the total heat load. We feel justified, therefore, in focusing our attention on the sensible heat changes undergone by the pipe.

It will have been noted that our model only yields minimum values for the length of cooling section z and water flow rate F_w, and an upper limit for the line velocity v. These are nevertheless useful boundaries to have and can be easily moved by imposing more severe (though in our opinion artificial) conditions.

4.4 SIMULTANEOUS APPLICATIONS OF THE CONSERVATION LAWS

When more than one conservation law has to be applied to describe a particular system or process, the result is a combination of balances that have to be solved simultaneously. Such combinations arise, in the first instance, in the form of pairs of balances and ultimately lead, in the case of more complex systems, to the simultaneous use of three or more balances.

Perhaps the most commonly used pair of balances is the mass–energy balance combination. Any substantial heat effects that arise in the course of mass transport or of a chemical reaction will almost invariably entail the use of twinned mass and energy balances. The two moving-boundary problems we presented in Illustration 2.4 and Illustration 4.20 did, in fact, make use of such combined balances, but were easily handled because the equations were not coupled and could be solved in isolation. The present section will deal with both coupled and uncoupled systems.

Mass balances coupled to force balances arise most commonly in flowing systems. The mass balance, known here as the continuity equation, has the following steady state form:

$$W(kg/s) = v_1\rho_1A_1 = v_2\rho_2A_2 = const. \tag{4.35}$$

where v and ρ refer to the average velocity and density of the fluid, and A is the cross-sectional area normal to the flow. Subscripts 1 and 2 refer to two different locations in the flow field.

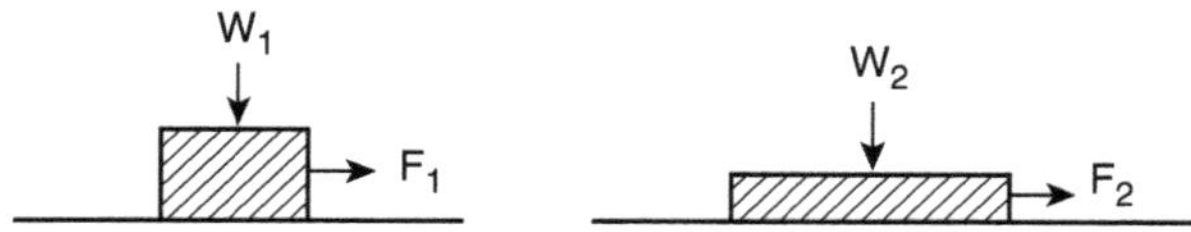

FIGURE 4.18 Amonton's law.

Extensions to three balances (mass, energy, and force or momentum balances) again arise in flowing systems and are usually, at the ODE level, distributed in one direction. Several examples of this case will also be presented.

Illustration 4.26 Two Simple Uncoupled Balances: Analysis of Skid Marks and the Hot-Air Balloon

We use these examples to demonstrate to the reader that combined balances can often be of a most elementary form. They can nevertheless yield results of considerable practical significance and provide insight into seemingly complex problems.

A. *Analysis of skid marks:*

The proposition that the length of the skid marks left by a vehicle brought to a sudden braking stop is related to the velocity at which it was traveling makes good intuitive sense. The longer the skid marks, the higher the initial velocity and the greater the likelihood the vehicle was exceeding speed limits. Police investigations of traffic accidents often make use of this fact.

To provide a theoretical basis for the relation, one must reach back to the notion of a *coefficient of friction*, μ, introduced by the French engineer Amonton in 1699. μ is defined as the ratio of the frictional force F between two surfaces and the load W that rests on them. Thus:

$$\mu = F/W \tag{4.36a}$$

Amonton further asserted that μ is a constant for two given surfaces, independent of the area of contact and the velocity at which the load is displaced. Thus, if $W_1 = W_2$ in Figure 4.18, then $F_1 = F_2$, i.e., the frictional force one must overcome to set and maintain the loads in motion is the same, in spite of the difference in contact areas. Although it was first met with skepticism, the law has stood up remarkably well over the course of time. Amonton proposed a constant value of $\mu = 1/3$, and this is, in fact, the observed value for a number of materials and surfaces. Evidently there are some exceptions: For wet surfaces and some plastics it may be as low as 0.2 and even lower, 0.03, on ice. Very clean surfaces, on the other hand, lead to higher values, such as 0.

To arrive at a quantitative relation between skidding distance d and velocity v just prior to braking, we combine a force balances with an energy balance. We have: Force balance (Amonton's law):

$$F = \mu W = \mu\, mg \tag{4.36b}$$

and

Mechanical energy balance:

$$\underset{\text{Kinetic energy change}}{\frac{mv^2}{2} - 0} = \underset{\text{Work done against friction}}{Fd} \tag{4.36c}$$

where d is the braking distance.

Eliminating F between the two equations and solving for v we obtain:

$$v = (2\mu g d)^{1/2} \tag{4.36d}$$

Comments: The remarkable feature of this result is that the mass of the vehicle is not involved, neither are the size (width) of the tire treads, nor how hard the brakes were applied. Evidently, the condition of the road surface will have some effect on the values of μ, and this is taken into account in accident investigation. It has now become a routine matter to obtain skid mark data at the scene of serious accidents.

B. *The hot-air balloon:*

The hot-air balloon relies on the fact that the density of the hot combustion gases enclosed in the balloon is less than that of the surrounding colder air. We have, in other words, a situation where the weight of the displaced fluid is greater than that of the displacing agent. It follows from Archimides' principle (see Illustration 4.2), that the combustion gases, will experience a net upward force. To the mass of these gases one must, of course, add the balloon load (basket, passengers, and equipment) to obtain a sense of the lift capability. We express this by means of the following force balance:

$$\begin{array}{c}\text{Buoyancy force} \\ \text{(weight of displaced air)}\end{array} = \begin{array}{c}\text{Weight balloon} \\ \text{gases and load}\end{array}$$

$$\rho_{air} V_{balloon} g = \rho_{gases} V_{balloon} g + m_{load} g \tag{4.37a}$$

and hence:

$$m_{load} = (\rho_{air} - \rho_{gases}) V_{balloon} \tag{4.37b}$$

It is clear that in order to determine the density of the combustion gases, it is first necessary to estimate their temperature. This is done using the following energy balance:

$$\begin{array}{c}\textit{Heat given off by} \\ \textit{combustion at } 25^\circ C\end{array} = \begin{array}{c}\textit{Heat required} \\ \textit{to raise product gases} \\ \textit{from } 25^\circ C \textit{ to } T_a\end{array}$$

$$\Delta H_C^\circ = \Sigma n_i Cp_i \, (T_a - 25) \tag{4.37c}$$

where $\Delta H_C°$ is the so-called standard enthalpy of combustion (J/mol) tabulated in the literature, and n_i and Cp_i are the moles and molar specific heats of the combustion gases.

We have, in other words, assumed that there are no heat losses to the surroundings, i.e., that the process is "adiabatic", and that the entire energy released by the combustion of the fuel (usually propane C_3H_8) goes toward heating the product gases.

The temperature attained in this manner is termed the adiabatic flame temperature.

The combustion process is described by the chemical reaction equation:

$$C_3H_8 + 5\ O_2 = 3\ CO_2 + 4\ H_2O$$

and the relevant thermal data are as follows:

Compund	CO_2	H_2O (g)	N_2	O_2
Cp (J/mol K)	40	35	30	30

$$\Delta H_C° = 2146 \text{ kJ/mol } C_3H_8$$

Let us consider the following specific example: We set balloon volume at 1000 m³, external air density at 1.17 kg/m³, and external pressure at 100 kPa. A 10% excess of oxygen is to be used in the combustion process. We obtain, using Equation 4.37c:

$$(T_a - 25)[\underset{CO_2}{3 \times 40} + \underset{H_2O}{4 \times 35} + \underset{O_2}{0.5 \times 30} + \underset{N_2}{20.7 \times 30}] = 2146 \times 10^3 \quad (4.37d)$$

and hence:

$$T_a = 2421°C = 2694\ K \quad (4.37e)$$

The gas density at this temperature, derived from the ideal gas law, works out to:

$$\rho_{gas} = \frac{\rho M_{avg}}{RT} = \frac{10^5(Pa) \times 28.1}{8.314 \times 2694} = 0.125\ kg/m^3 \quad (4.37f)$$

where the average molar mass M of the combustion gases, consisting mostly of nitrogen, was taken to be the arithmetic average of the components.

Using this value, we finally obtain from Equation 4.23b:

$$m_{load} = (1.17 - 0.125)103 \quad (4.37g)$$

$$m_{load} = 1045 \text{ kg} \quad (4.37h)$$

Comments: The balloon, which has an approximate diameter of 10 m, is shown to be capable of lifting three passengers and an additional load of about 800 kg. This is a reasonable result.

A subsidiary question, which is of some interest, is whether much would be gained by using pure oxygen in the combustion. Note that this would eliminate nitrogen as a "product gas," resulting in a higher adiabatic flame temperature T_a. Without resorting to detailed calculations, it is immediately seen from Equation 4.37g that the gain would be no more than marginal. Even if T_a were doubled by this procedure and gas density dropped to 0.063 kg/m^3 as a consequence, the payload would increase only by some 60 kg, less than would be required to carry the extra oxygen on board. The alternative is thus not a viable one.

Illustration 4.27 Coupled Mass and Energy Balances: Heat Effects in a Stirred Tank Reactor–The van Heerden Diagram

The case we consider here, a classic one drawn from chemical reactor theory, involves an exothermic reaction carried out in a stirred tank reactor with constant inflow of reactants A and B, and continuous removal of products P and excess reactant. We show this situation in Figure 4.19.

To analyze the events that take place in this process, we resort to some creative doodling. We start by composing a plot of the rate of heat production (J/sec) against the temperature in the tank, which equals that of the product stream. The result, which is shown in Figure 4.20A, exhibits an initial rise, followed by an inflection, and ultimate leveling off of the heat production curve. The reason for this lies in the existence of two opposing factors. The first is the strong exponential dependence on temperature of the reaction rate constant k_r, expressed through the well-known Arrhenius equation:

$$k_r(T) = A \exp(-E_A/RT) \tag{4.38a}$$

which is linked to the overall reaction rate r through the expression:

$$r = k_r(T)\, C_A^m\, C_B^n \tag{4.38b}$$

The second and opposing factor lurks in the term $C_A^m C_B^n$ of Equation 4.38b, which indicates that the reaction rate r will diminish as reactants are consumed to

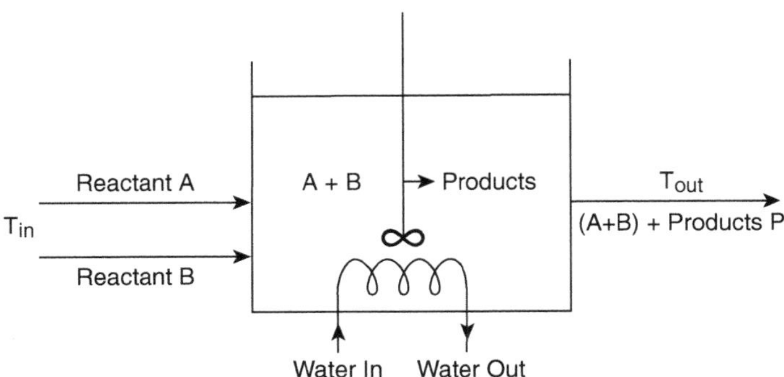

FIGURE 4.19 Exothermic reaction in a stirred tank.

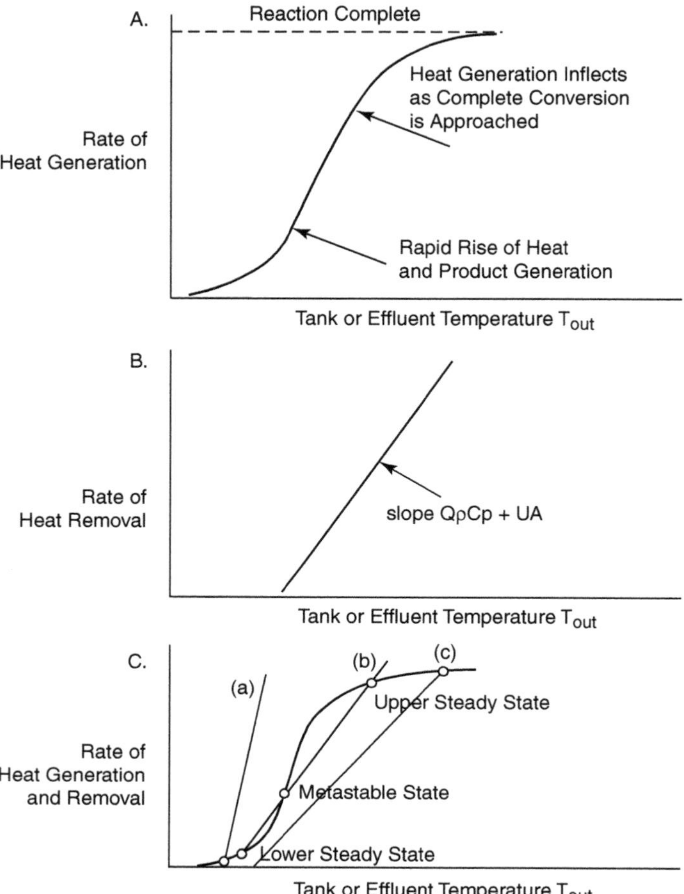

FIGURE 4.20 Genesis of a van Heerden diagram: (A) Rate of heat production for an exothermic irreversible reaction, (B) Rate of heat removal, and (C) Superposition of A and B leading to the van Heerden diagram.

form product. This causes the production rate curve to inflect and to approach a constant asymptotic value, which corresponds to the rate of heat produced by complete conversion to products.

It is customary in such exothermic reactions to provide external cooling (Figure 4.19) to prevent undesirable side effects due to excessive temperature levels (evaporation, decomposition, degradation, and extraneous reactions). One can then argue that the heat generated by the reaction will, for a steady state operation, be exactly balanced by the heat removed by the cooling medium and the heat content (or enthalpy) difference between entering and exiting streams. We can express this formally by writing the following energy balance:

Rate of energy in – Rate of energy out = 0

$$\begin{bmatrix} -\Delta H_r V \\ +Q\rho H_{in} \end{bmatrix} - \begin{bmatrix} Q\rho H_{out} \\ +UA(T_{out} - T_{ext}) \end{bmatrix} = 0 \tag{4.38c}$$

Replacing $H_{out} - H_{in}$ by $C_p(T_{out} - T_{in})$ and rearranging Equation 4.38c then leads to the following convenient form:

Rate of Heat Generation = Rate of Heat Removal

$$-\Delta H_r rV = -(Q\rho C_p T_{in} + UAT_{ext}) + (Q\rho C_p + UA)T_{out} \tag{4.38d}$$

The left-hand side represents the heat generation curve previously shown in Figure 4.20A. The right side can be plotted as a straight line of rate vs. T_{out} and is shown in Figure 4.20B. We now proceed to superpose the two plots and arrive at the composite shown in Figure 4.20C, often referred to as a van Heerden diagram. Several features of this diagram are noted in the following text.

We start with the observation that depending on the location of the rate-of-removal line, one can have one or three intersections with the rate of production curve; that is, one or three steady state tanks or effluent temperatures. At very low or very high cooling rates (low or high values of the product UA), or at very low or high flow rates Q, only one steady state results. It is at intermediate values of these variables that one encounters several steady states. Of these, only the upper and lower states are stable, whereas the middle state is metastable, i.e., does not materialize spontaneously. This can be shown as follows:

Suppose the middle state in Figure 4.20C is slightly disturbed upward by a small temperature rise. Heat produced will then exceed the heat removed, leading to a further increase in temperature. The process repeats itself until the upper steady state is reached. That point is stable, for a further increase in temperature would result in heat removal exceeding heat production, and a consequent return to the point C. By similar reasoning, if a slight disturbance lowers the temperature of the middle state, such a process will continue until the lower (and stable) steady state is reached.

The existence of three steady states, two stable and one metastable, is fairly common for highly exothermic reactions in stirred tanks. Perhaps even more common is the existence of only one steady state. In the production of polystyrene from styrene, a highly exothermic reaction carried out extensively in the plastics industry, all three steady states exist only for a limited range of the feed temperature T_{in}. If T_{in} is sufficiently high, only the upper "runaway" condition can be realized. For intermediate values of T_{in}, all three steady states are possible.

Let us consider the example of styrene polymerization in more detail, using a feed temperature T_{in} = 300°C and no external cooling (UA = 0). Cooling is instead provided by the incoming feed and can still result in three steady states. In addition the following information is provided:

Residence time τ = Reactor volume/Flow rate = 2 h

Reaction rate $r = k_r C_{out} = 10^{10} \exp(10^4/T_{out}) C_{out}$ moles/h l and $C_{in} \Delta H_r/\rho C_p = 400$ K

We note that the energy balance (Equation 4.38d) alone does not suffice to solve the problem, as it contains two variables, temperature T_{out}, and rate r or concentration C_{out}. We must supplement it with a second equation, for which we draw on a styrene mass balance about the reactor. We obtain:

$$\text{Rate of styrene in} - \text{Rate of styrene out} = 0$$

$$Q\,C_{in} - \begin{bmatrix} Q\,C_{out} \\ +k\,C_{out}V \end{bmatrix} = 0 \tag{4.38e}$$

The two equations (Equation 4.38d and Equation 4.38e) comprise the model for the required solution. We rearrange them by dividing them by flow rate Q and solving for C_{out}/C_{in}. This yields the following set:

$$\frac{C_{out}}{C_{in}} = \frac{1}{1+\tau k} = \frac{1}{1+2\times 10^{10}\exp(-10^4/T_{out})} \tag{4.38f}$$

and

$$T_{out} - T_{in} = T_{out} - 300 = -\frac{C_{out}\Delta H_r}{\rho C_p}\tau k = 400\times 2\times 10^{10}\exp(-10^4/T_{out})\frac{C_{out}}{C_{in}} \tag{4.38g}$$

These two nonlinear equations can be solved for C_{out} and T_{out} by the Newton–Raphson method using the Mathematica package, which yields the results given in Table 4.3, so that there are three steady state solutions. In the low-temperature solution, the reactor acts merely as a storage vessel, and no significant reaction occurs. The high-temperature solution represents an upper runaway condition where the reaction goes to near completion. The middle steady state is metastable. Surprisingly, industrial practice is to operate at this metastable state rather than at the stable upper state, because the latter produces a lower grade product brought about by the high operating temperature and excessive reaction rate. The control necessary

TABLE 4.3
Values of C_{out} and T_{out} for Equation 4.38f and Equation 4.38g

T_{out}	Conversion $X = C_{out}/C_{in}$
300.03	0.00007
404	0.262
699.97	0.99992

to maintain the system in this inherently unstable condition is achieved through autorefrigeration (i.e., cooling by boiling). The reactor pressure is set so that the styrene, which is a liquid, boils when the desired operating temperature is exceeded. The latent heat of vaporization rapidly reduces the temperature of the reactor below the boiling point. Boiling then stops until the heat of reaction returns the temperature to the boiling point. Steady state is thus never attained. Instead, the temperature cycles about the metastable state, causing a cyclic variation in the outlet product concentration. Although this is not generally considered to be a desirable condition, the need to ensure the required quality makes this type of operation necessary.

A priori, the number of roots for Equation 4.38e and Equation 4.38g may not be obvious. For the numerical solution, we start by plotting these equations to determine the number of roots. Let us combine Equation 4.38f and Equation 4.38g:

$$T_{out} = 300 + \frac{8\times 10^{12}\exp(-10^4/T_{out})}{1+2\times 10^{10}\exp(-10^4/T_{out})} \tag{4.38h}$$

Plotting the right-hand and left-hand sides of this equation on the same graph will provide valuable information regarding both the number and approximate values of the roots of this equation:

```
In[1]:= rhs = 8 10^12 Exp[-10^4/t] / (1 + 2 10^10Exp[-10^4 / t]);
In[2]:= << Graphics'Graphics'
        p1 = LogPlot[rhs + 300, {t, 0.1, 1000}, PlotRange → All,
           DisplayFunction → Identity];
        p2 = LogPlot[t, {t, 0, 1000}, DisplayFunction → Identity];
        Show[p1, p2, DisplayFunction → $DisplayFunction, Frame → True,
          TextStyle → {FontFamily → "Times", FontSize → 14},
          FrameLabel → {"T", "f(T)", "", ""}]
```

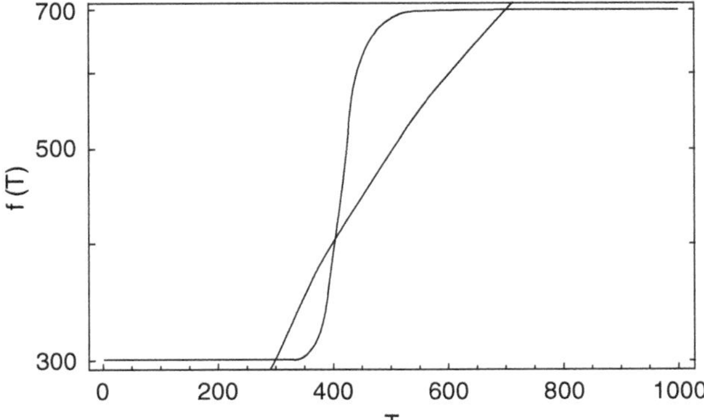

According to this plot, there are three solutions at approximately 300, 400, and 700°C. We use these values as an initial guess to solve the set of Equation 4.38f and Equation 4.38g. Here, c represents the C_{out}/C_{in} , and t replaces T_{out}:

```
In[6]:= FindRoot[{c == 1 / (1 + 210^10Exp[-10^4/t]), t - 300 ==
        400 2 10^10Exp[-10^4/t]c}, {{c, 0}, {t, 100}}]
Out[6]= {c -> 0.999933, t -> 300.027}
In[7]:= FindRoot[{c == 1 / (1 + 210^10Exp[-10^4/t]), t - 300 ==
        400 2 10^10Exp[-10^4/t]c}, {{c, 0}, {t, 400}}]
Out[7]= {c -> 0.740647, t -> 403.741}
In[8]:= FindRoot[{c == 1 / (1 + 210^10Exp[-10^4/t]), t - 300 ==
        400 2 10^10Exp[-10^4/t]c}, {{c, 0}, {t, 700}}]
Out[8]= {c -> 0.0000800619, t -> 699.968}
```

Illustration 4.28 Adiabatic Compressible Flow in a Pipe

It will be recalled from previous examples (see, e.g., Illustration 1.6 and Illustration 4.7) that in steady liquid flow through a duct of constant cross-section A, the velocity is constant and remains unchanged over the entire length of conduit. This also follows from the continuity Equation 4.24, and is a requirement for the maintenance of a constant mass flow. The same equation shows that in the flow of gases (i.e., compressible flow), the decrease in density ρ occasioned by the frictional pressure drop will cause the velocity to rise in order to maintain the required constant mass flow rate. Simultaneously, the gas will undergo a change in temperature because of expansion in the direction of flow. We assume here that this change remains confined to the pipe and that no heat is exchanged with the surroundings, i.e., the process is adiabatic.

As we are dealing with four variables, v, ρ, p, and T, four equations will be required to model the process. Three of these relations are provided by the conservation laws, and a fourth one, the ideal gas law, interrelates density, pressure, and temperature of the medium. They are presented below, followed by some brief comments on their form.

Mass balance:

$$W\ (kg/sec) = \rho v A = constant \tag{4.39a}$$

or

$$G\ (kg/m^2\ sec) = \rho v = constant \tag{4.39b}$$

or

$$\frac{d\rho}{\rho} = -\frac{dv}{v} \tag{4.39c}$$

Energy balance (q = 0):

$$\frac{v_2^{\,2} - v_1^{\,2}}{2} = \frac{\gamma}{\gamma - 1}\left(\frac{p_1}{\rho_1} - \frac{p_2}{\rho_2}\right) \tag{4.39d}$$

or $$\frac{p}{\rho}\left(1+\frac{\gamma-1}{2\gamma}\frac{v^2\rho}{p}\right)=constant \tag{4.39e}$$

Extended Bernoulli equation (force balance):

$$\frac{dp}{\rho}+vdv+2f\frac{v^2}{d}dz=0 \tag{4.39f}$$

Ideal gas law:

$$\rho=\frac{pM}{RT} \tag{4.39g}$$

Both the energy and force balances require some elaboration. The right-hand side of the former looks unfamiliar, but does in fact represent the enthalpy change ΔH which also appears in the energy Equation 4.39a. This comes about as a result of the dual expressions for ΔH given in Table 1.2, which we repeat:

$$\Delta H = C_p\Delta T = C_v\Delta T - (p/\rho) \tag{4.39h}$$

Here C_p and C_v represent specific heats at constant pressure and volume, which are nearly identical in the case of liquids but differ markedly for gases. Their ratio $\gamma = C_p/C_v$ equals 1.4 for air. Using this ratio in Equation 4.39h and eliminating ΔT leads to the following expression:

$$\Delta H = \frac{\gamma}{\gamma-1}\Delta(p/\rho) \tag{4.39i}$$

which is identical to the right-hand side of Equation 4.39d.

The force balance (Equation 4.39f), which we term the *extended Bernoulli equation*, has the same pressure and velocity terms we will see in Practice Problem 4.6, but these are written here in differential form to allow for a continuous change in these variables. The third expression is an empirical friction term, with f being the so-called friction factor, which is of the order 10^{-3} for flow in standard steel pipes.

Frequently one adds to this model a fifth variable, the so-called Mach number, Ma, which represents the ratio of gas velocity to the velocity of sound, c, which can be recalled as the speed at which a pressure disturbance propagates through a medium. For an ideal gas, the Mach number is represented by the relation:

$$Ma \equiv v/c = v\Big/\left(\gamma\frac{RT}{M}\right)^{1/2} \tag{4.39j}$$

Full solutions of various forms of this formidable set exist, of which a particularly useful one has been cast into graphical form, to be discussed below. One relation, that of the pressure drop as a function of Mach number, is of special interest and can be obtained by differentiating Equation 4.39e and using the other two balances to eliminate dv and dρ. We obtain:

$$\frac{dp}{dz} = \frac{-2f\rho v^2}{D}\left[\frac{1+(\gamma-1)Ma^2}{1-Ma^2}\right] \tag{4.39k}$$

Several features of pipe flow are revealed by this relation. For small Mach numbers, Ma → 0, or ∂p/∂ρ → 0, it reduces to the equation for incompressible flow in a horizontal pipe. For more substantial Mach numbers, 0.3 < Ma < 1, the pressure drop is negative in the direction of flow and compressible flow can proceed. As Ma is increased toward unity, pressure drop and, hence, flow rate increase rapidly until, at a value of Ma = 1, the pressure gradient becomes infinite, pressure itself is at a minimum, and the fluid velocity equals the velocity of sound. Beyond that point, at Ma > 1, the pressure drop becomes positive and flow ceases. We conclude from this that the maximum velocity attainable in a duct of constant cross section is the velocity of sound. Since velocity increases with distance along the pipe, this maximum will be reached at the pipe outlet. To obtain supersonic flows in a duct, the cross-sectional area must increase in the direction of flow. This will be demonstrated in Practice Problem 4.2.

To provide a physical underpinning to this phenomenon of limiting sonic flow, we display in Figure 4.21 the pressure profiles which result from a progressive increase in inlet pressure. Figure 4.21A shows a modest pressure drop obtained at low flow rates. An increase in inlet pressure propagates downstream, adjusting pressure to a steeper gradient and, hence, higher flow rates (Figure 4.21B). As pressure is increased further, a point is ultimately reached where the flow at the outlet reaches sonic velocity. An additional increase in inlet pressure at this point will no longer be able to propagate downstream to adjust the pressure gradient, because to do so it would have to exceed the speed of sound. The pressure increase beyond the sonic level will be held back as a standing discontinuity or pressure shock wave at or near the pipe inlet (Figure 4.21C).

A consequence of this phenomenon is that a pipe of given length and diameter cannot accommodate arbitrarily high gas flow rates. Beyond a certain value, $G^* = G_{Max}$, an increase in inlet pressure will not result in increased flow. To accommodate flow rates higher than G_{Max}, one has to either increase the diameter, or reduce the effective length of the pipe by recompressing the gas at intervals. The latter method is standard practice in the operation of long-distance gas transmission lines.

If, instead of increasing inlet pressure, we progressively reduce outlet pressure, a similar sequence of events will take place. Flow will increase until we attain sonic velocity at the outlet. Further reductions in pressure will result in a standing shock wave that is now at or near the outlet, and flow will remain constant at G_{Max}. Another way of putting this is to say that the flow rate anticipated from an increase in pressure is "choked" down to the value G_{Max}.

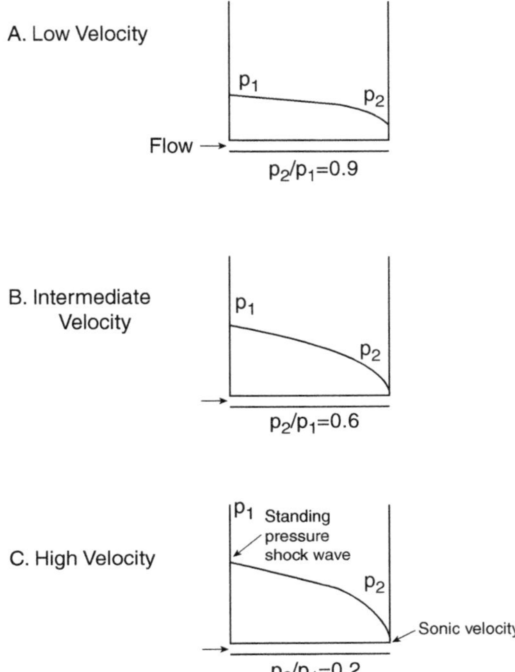

FIGURE 4.21 Pressure profiles in compressible duct flow with increasing inlet pressure: (A) Low flow rates, (B) Intermediate flow rates, and (C) Sonic velocity at the exit-standing shock at inlet.

The question arises as to whether such limiting velocities materialize in liquid flows. Liquid velocities in pipe flow, i.e., those used in conventional operations with standard pumps and construction materials, are of the order 1 m/sec. Sonic velocities, in liquids, on the other hand, are three orders of magnitude higher. The pressures required to generate sonic flow in liquids would therefore be enormous and beyond the ordinary strength of the containing pipes. Hence, limiting flow considerations do not, in general, arise in the case of liquids.

Illustration 4.29 Compressible Flow Charts

The complexity of the full model for compressible flow has led to the solution by numerical methods, and the compilation of the results in the form of convenient flow charts. A typical result is shown in Figure 4.22, which represents a plot of the downstream to upstream pressure ratio p_2/p_1 against the dimensionless mass velocity $G/G^* = G/G_{Max}$, where G_{Max} is the flow corresponding to the sonic velocity in frictionless flow, given by:

$$G^* = G_{Max} = p_1\left[\frac{M\gamma}{RT_1}\left(\frac{2}{\gamma+1}\right)^{(\gamma+1)/(\gamma-1)}\right]^{1/2} \tag{4.40a}$$

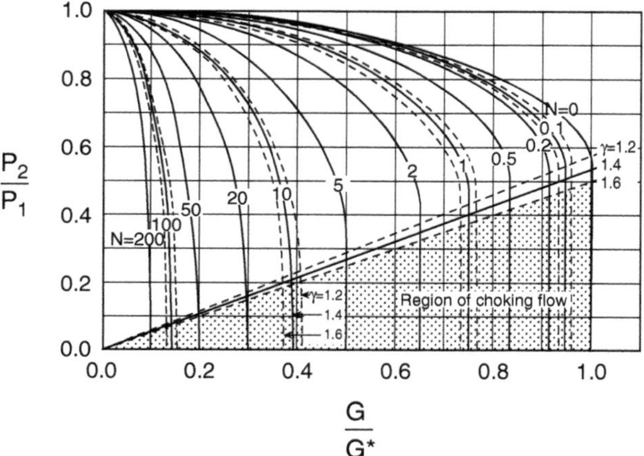

FIGURE 4.22 Chart for the computation of compressible flow in ducts. P_2/P_1 = downstream-to-upstream pressure ratio, G/G^* = dimensionless mass velocity, and N = dimensionless pipe number fL/d.

The parameter $N = f\frac{L}{d_h}$, also called the *pipe number*, represents the piping configuration with d_h is the hydraulic diameter, equal to d for a cylindrical pipe, L represents length, and f is the friction factor. The plots show that for a given pipe configuration N, the mass velocity G through a pipe of length L and diameter d initially increases as inlet pressure p_1 is increased or p_2 is dropped. Flow is subsonic throughout the pipe under these conditions. Further reductions in the pressure ratio lead to an additional increase in G until a critical ratio $(p_2/p_1)_c$ is reached. At this point G has reached a maximum and remains constant at that value with further decrease in the pressure ratio. We are now in the domain denoted as "region of choking flow." Note that the sonic velocity at which this occurs decreases with increasing friction. For frictionless flow, N = 0, the critical pressure ratio reaches a maximum and is given by:

$$(p_2/p_1)_c = \left(\frac{2}{\gamma+1}\right)^{\gamma/\gamma-1} \tag{4.40b}$$

with

$$(T_2/T_1)_c = 2(\gamma+1) \tag{4.40c}$$

For air, $\gamma = 1.4$, so that for frictionless flow of air the critical ratios are given by $(p_2/p_1)_c = 0.53$ and $(T_2/T_1)_c = 0.83$. These are useful limiting numbers to remember for all gas flows, as γ has only a marginal effect on their values.

Suppose now that we wish to establish whether a pipe of given length L and diameter d will accommodate a prescribed air flow. Let us set L = 10 m, d = 0.0525 m,

$G = 7500$ kg/m^2 sec, $T_o = 20°C$, and $P_1 = 1.0$ MPa. f is found to be 0.003 for the flow rate and pipe in question. We obtain from Equation 4.39a:

$$G^* = G_{Max} = p_1 \left[\frac{M\gamma}{RT_o} \left(\frac{2}{\gamma+1} \right)^{(\gamma+1)/(\gamma-1)} \right]^{1/2}$$

$$= 1 \times 10^6 \left[\frac{(29)(1.4)}{(8314)(293)} \left(\frac{2}{1.4+1} \right)^{(1.4+1)/(1.4-1)} \right]^{1/2}$$

$$G^* = 1970 \; kg/m^2 s$$

where 8314 is the value for R in units of kg m^2/kmol sec^{2}K.

Thus $G/G^* > 1$, and the prescribed flow does not materialize. Instead, it is reduced or choked to a value set by the pipe number N. We have:

$$N = fL/d = (0.003)(10/0.0525) = 0.57$$

The corresponding value of G/G^* read from the chart is 0.825, so that the actual flow rate which materializes is

$$G = 0.825\; G^* = (0.825)(2360) = 1947 \; kg/m^2 s$$

i.e., 26% of the prescribed value.

Illustration 4.30 Film Condensation on a Vertical Plate

Although condensation is, like evaporation, a two-phase phenomenon, it has considerably simpler heat transfer characteristics than those seen in boiling (cf. Illustration 4.19B). A distinction is made between dropwise condensation, in which the liquid does not wet the cooled surface, and film condensation, in which a smooth continuous film of liquid is formed. The former process has much higher heat transfer rates than film condensation, because of the direct exposure of the vapor to the cold surface, and would thus be the preferred mode of operation. In practice, however, dropwise condensation is difficult to maintain because the droplets tend to coalesce into a more or less coherent film before dropping off the surface. Film condensation is thus the principal mode to be dealt with, and has the advantage of being amenable to a relatively simple treatment.

An elegant analysis of film condensation on a vertical surface was first given by Nusselt in 1916. The physical process considered is shown in Figure 4.23, and the aim is to derive an expression for the local heat-transfer coefficient, which is a function of film thickness $h = k/\delta(z)$.

A preliminary analysis is needed here to establish the variables and the balances which make up the model. We argue as follows:

- A primary variable will be the flow rate L of the condensate, as it will determine the thickness of the falling film. That flow rate is derived by a

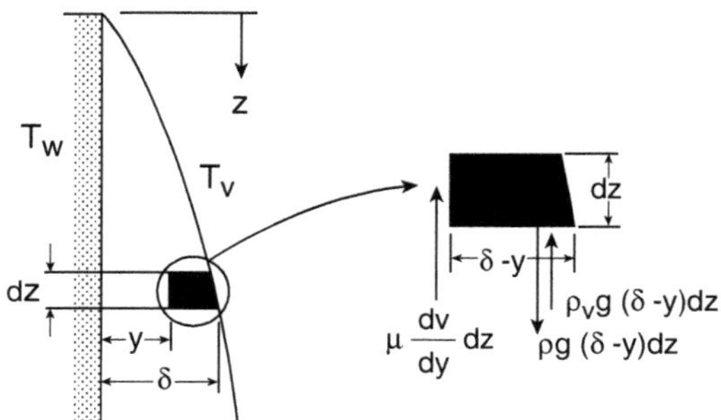

FIGURE 4.23 Variables and forces in filmwise condensation on a vertical plane.

balance of the shear and gravity forces acting on the descending condensate and yields, in the first instance, the velocity distribution which can be integrated to obtain the local flow rate L(z). The expression will contain δ(z) as a second state variable, and a second equation, at the very least, will be required.

- We turn to a mass balance, taken over an increment Δz, as the logical source for a second equation. That balance will contain the rate of condensation, ΔN, which is a new state variable, as it depends on conduction through the film thickness δ, which is, in turn, a function of vertical distance. A third relation is therefore required.
- We draw on the last available balance, that of energy, which equates the liberated heat of condensation of ΔN to the rate of conduction through the film thickness δ(z). No new variables are brought in, and the model is consequently complete.

The detailed sequence is as follows.

Force balance:

Here the diagram of Figure 4.23 reveals a subtlety which is not immediately apparent. In addition to the obvious forces of gravity and the opposing viscous shear, a third force, that of the buoyancy of the displaced vapor, needs to be taken into account. We obtain:

$$\text{Gravity} - (\text{viscous force} + \text{buoyancy}) = 0$$

$$\rho_L g(\delta - y)dx - \mu \frac{dv}{dy}dx - \rho_v g(\delta - y)dx = 0 \tag{4.41a}$$

which, upon integration between y = 0 (v = 0) and y, leads to the distribution:

$$v(y) = \frac{(\rho_L - \rho_v)g}{\mu}\left(\delta y - \frac{1}{2}y^2\right) \tag{4.41b}$$

A second integration of this expression over the cross-sectional area of the film, assuming unit depth, yields the local mass flow rate L(δ):

$$L(kg/\text{sec}) = \rho_L \int v dA = \rho_L \int_0^\delta (\rho_L - \rho_v) g \left(\delta y - \frac{1}{2} y^2 \right) dy \qquad (4.41c)$$

$$L = \rho_L(\rho_L - \rho_v)g\delta^3/3\mu \qquad (4.41d)$$

Mass balance:

Rate of liquid in − Rate of liquid out = 0

$$(L_z + \Delta N) - L_{z+dz} = 0$$

or

$$dL = d[\rho_L(\rho_L - \rho_v)g\delta^3/3\mu] = dN \qquad (4.41e)$$

We refrain from integrating this expression because it is dN that will be required in the energy balance, and instead evaluate the differential. We obtain:

$$d[\rho_L(\rho_L - \rho_v)g\delta^3/3\mu] = [\rho_L(\rho_L - \rho_v)g\delta^2/\mu]d\delta = dN \qquad (4.41f)$$

Energy balance:

Rate of energy in − Rate of energy out = 0

$$dN\,\Delta H_v - kdz\,(T_v - T_w)/\delta = 0$$

Latent heat − Conductive cooling = 0

Upon introducing dN from Equation 4.41e and assuming a linear temperature distribution in the film, there results:

$$[\rho_L(\rho_L - \rho_v)g\delta^2/\mu]d\delta\Delta H_v = kdz\,(T_v - T_w)/\delta \qquad (4.41h)$$

When this expression is integrated over z with the film thickness at the inlet δ(0) set equal to zero, we obtain the desired film thickness distribution and, hence, the local heat-transfer coefficient h(z):

$$\delta(z) = \left[\frac{4\mu kz(T_V - T_w)}{g\Delta H_v \rho_L(\rho_L - \rho_v)} \right]^{1/4} \qquad (4.41i)$$

$$h(z) = \frac{k}{\delta(z)} = \left[\frac{\rho_L(\rho_L - \rho_v)g\Delta H_v k^3}{4\mu z(T_v - T_w)} \right]^{1/4} \qquad (4.41j)$$

Nusselt also considered horizontal tubes and reported both results in terms of mean integral film coefficients $\bar{h} = \frac{1}{z}\int_0^L h(z)d(z)$:

Vertical plate:

$$\bar{h} = 0.943\left[\frac{\rho_L(\rho_L - \rho_v)g\,\Delta H_v k^3}{\mu L(T_v - T_w)}\right]^{1/4} \tag{4.41k}$$

Horizontal tube:

$$\bar{h} = 0.725\left[\frac{\rho_L(\rho_L - \rho_v)g\,\Delta H_v k^3}{\mu D(T_v - T_w)}\right]^{1/4} \tag{4.41l}$$

Of note here is the dependence in both cases of $\bar{h}$ on $(T_v - T_w)^{1/4}$. This is an inverse dependence, in contrast to the free convection coefficient (see Illustration 4.19A) which varied directly with $\Delta T^{1/4}$.

PRACTICE PROBLEMS:

4.1 Forces Acting on a Storage Bin

Three identical solid cylinders, each weighing 400 N, are stacked within a bin as shown in Figure 4.24A. It is required to calculate the forces at the points where the cylinders touch one another and the walls of the bin.

Hint: Start by calculating the contact forces shown in Figure 4.24B.

4.2 Force on a Submerged Hinged Gate

A hinged submerged rectangular steel gate 1 m × 0.5 m separates a water reservoir from an underground cavern connected to the atmosphere. The gate is inclined by 30° to the horizontal water surface, located 10 m above the upper edge of the gate. Calculate the force required to lift the gate in addition to its weight.

Answer: 5.02×10^4 N

4.3 Minimum Wall Thickness of a Gas Cylinder

Calculate the minimum wall thickness of a commercial nitrogen cylinder (d = 0.15 m) made of carbon steel (τ = 450 MPa) filled to a pressure of 15 MPa.

4.4 Solids Removal in a Settling Tank

A dilute aqueous suspension of spherical particles of density ρ = 2000 kg/m^3 and ranging in size from 0.05 to 5 mm is fed into a settling tank which is filled to a height of 10 m. Calculate the minimum time required for all particles to settle. (Note: We speak of minimum time since, in the later stages, the particles encounter additional resistance from nearby neighbors. This is termed hindered settling.)

Hint: Consult Table 4.1 ($\mu_{H_2O} = 10^{-3}\ Pa\,s$).

Answer: 7350 sec

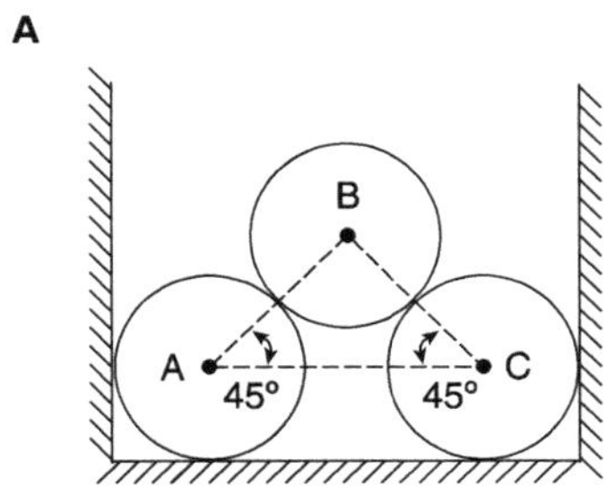

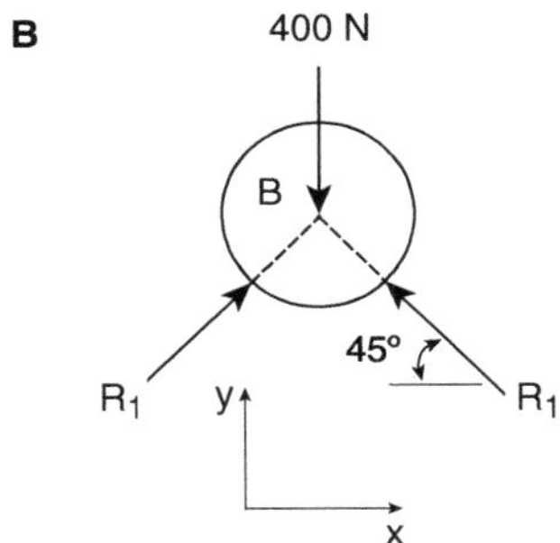

FIGURE 4.24 Force acting on a storage bin. (A) The bin. (B) Forces diagram for the upper cylinder.

4.5 Marker Particles for Scuba Diving

An entrepreneur with an interest in scuba diving had the idea of marketing marker particles that, upon release under water, would rise at a velocity of 60 ft/min = 0.305 m/s, the maximum value recommended for the safe ascent of divers. The particles are meant to be released by the diver who would then adjust his rate of ascent to that of the particles. Spheres in the range 2 to 4 cm in diameter are the preferred geometry, and the particle to fluid density ratio ρ_p/ρ_f is to be kept below 0.9 so as not to be unduly influenced by small water density variations.

Can these requirements be met?

Hint: Table 4.1 shows that the conditions fall in the turbulent regime, for which $C_D = 0.44$.

4.6 The Bernoulli Equation

The Bernoulli equation, given below, describes the changes in pressure, velocity, and elevation z, which occur in frictionless flow of a liquid between two points

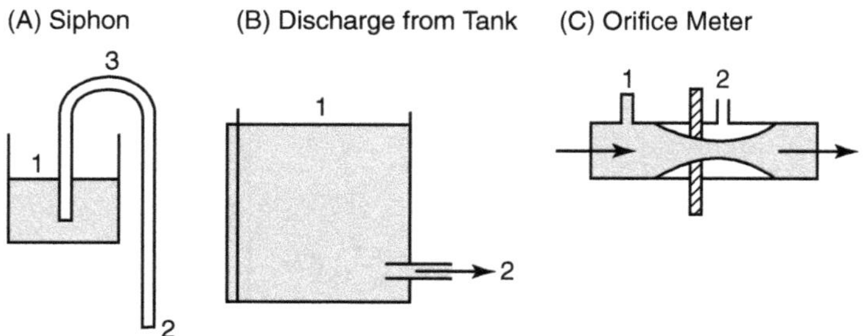

FIGURE 4.25 Applications of the Bernoulli equation.

located on the same stream line or stream tube. It applies when the locations are in fairly close proximity to each other:

$$\frac{p_2 - p_1}{\rho} = \frac{v_2^{\,2} - v_1^{\,2}}{2} + g(z_2 - z_1) = 0$$

Three devices to which the Bernoulli equation can be applied are shown in Figure 4.25.

(a) Show why a siphon works.
(b) Derive the expression known as Toricelli's equation for the discharge velocity from the tank.
(c) Show that the orifice meter relates flow rate Q to a measured pressure drop $p_1 - p_2$ via the relation:

$$p_1 - p_2 = \frac{\rho Q^2}{2A_1^{\,2}}\left(\frac{d_1^{\,4}}{d_2^{\,4}} - 1\right)$$

(d) The height to which a suction pump can deliver a liquid is limited by its vapor pressure. Why? What is the maximum height to which a solvent with a vapor pressure of 400 mmHg, $\rho = 730$ kg/m^3 can be pumped?

4.7 A Nanodevice: The Atomic Force Microscope

The atomic force microscope (AFM), inaugurated in 1986, has since become a widely used tool for the study of protein and cell structures. The device, shown in Figure 4.26, consists of a microscopic cantilever beam, typically 100 to 400 μm in length carrying a tip of 10- to 50-nm radius. Fabrication is by silicon-based nanotechnology. In a typical application, the tip is used to indent the surface of the sample, which causes the cantilever beam to be deflected upward. The deflection is

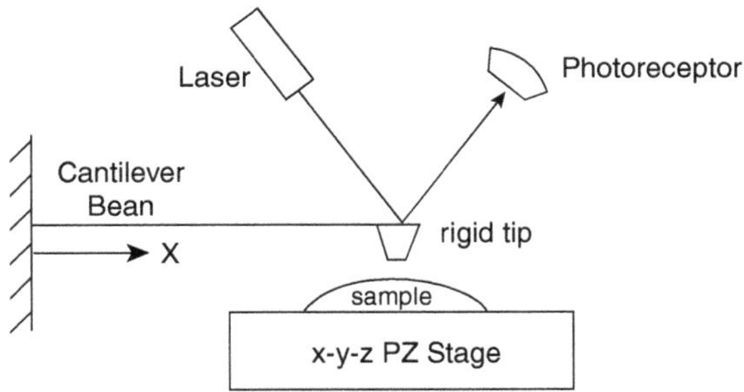

FIGURE 4.26 The Atomic Force Microscope (AFM).

measured by the laser photoreceptor assembly shown in Figure 4.26. The sample itself is mounted on a piezoelectric stage that can be displaced in x-y-z. In the application described here, measured force-indentation depth data provide information on the local mechanical properties of the sample. The behavior of the device can be analyzed using the classical theory of the deformation of beams. Show that for a cantilever beam of length L, the deflecting force F is given by:

$$F = \frac{3EI}{L^3}\delta$$

where δ = deflection.

Hint: Consult Equation 2.9.

4.8 Sizing of a Pump

A viscous oil with a viscosity μ = 7.36 Pa sec is to be pumped over a distance of 15 km at the rate of 10^{-3} m^3/sec. It is desired to calculate the pressure drop and the required horsepower of the pump (1 hp = 0.746 kW).

Hint: Power equals the product of pressure drop and flow rate.

Answer: 6.13 hp

4.9 Deflection of an Electron in Thomson's Experiment

The deflection of an electron in a purely electrical field, used by Thomson in his famous experiment, is shown in Figure 4.27.

Derive Equation 4.15d, which describes the parabolic trajectory of the particle.

Hint: Consult Illustration 4.6.

4.10 Production of a Fine Water Spray

A fine mist of droplets 0.1 mm in diameter is to be produced at the rate of 0.10 kg/sec by forcing water through a perforated nozzle. Calculate the power requirement and the upstream pressure needed.

Hint: $\gamma_{H_2O} = 7.3 \times 10^{-2}\ N/m$

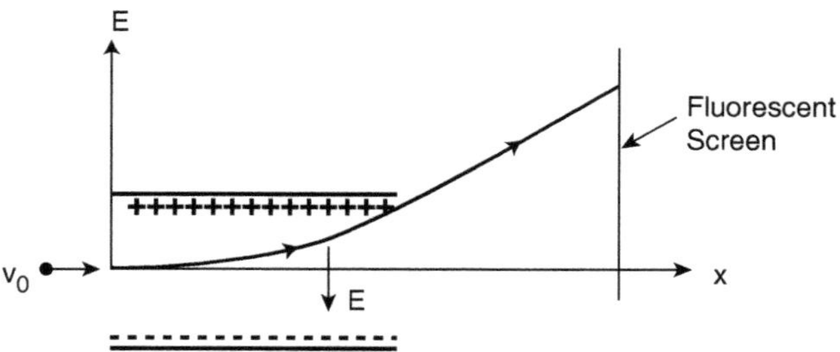

FIGURE 4.27 Deflection of an electron in an electrical field.

4.11 An Example from Industry: Decontamination of a Nuclear Reactor Coolant

Organic coolants of nuclear reactors have to be purified periodically by passage through an adsorbent bed to remove degradation products and other impurities that reduce the efficiency of the coolant. As it is desirable to maintain continuous operation of the reactor, the coolant cannot be removed to a separate facility for treatment. It is customary to circulate it continuously through a purifier and return the decontaminated coolant to the reactor (Figure 4.28). Twenty four hours are to be allowed for completion of the operation, which requires reduction in impurity concentration from 2500 ppb to 75 ppb. What is the required flow rate to achieve this? use $V = 141\ m^3$ and $\rho = 1200\ kg/m^3$.

4.12 Batch Filtration: The Ruth Equation

The accompanying diagram, Figure 4.29, sketches the equipment and procedures used in batch filtration. The slurry enters at a flow Q and with a solid content C_o. The solids are deposited on the filter to depth z which varies with time, while the solvent

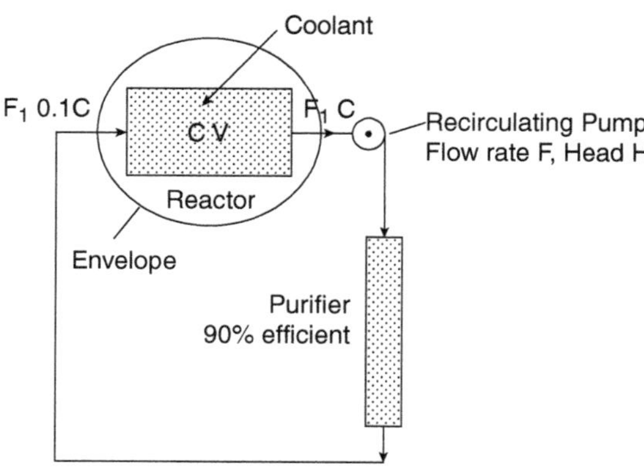

FIGURE 4.28 Flow diagram for the purification of an organic coolant for a nuclear reactor.

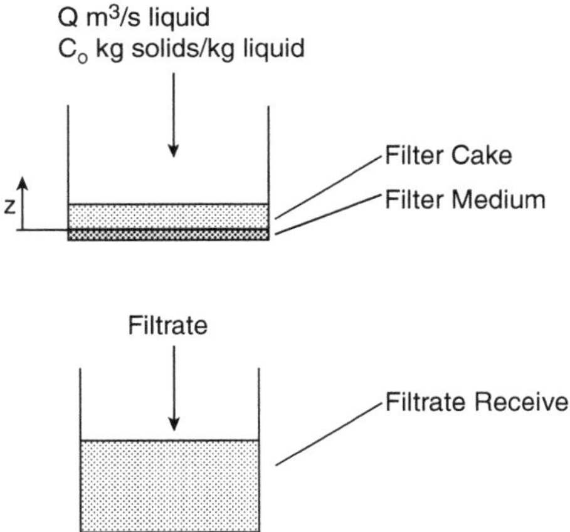

FIGURE 4.29 Diagram of batch filter and filtrate receiver.

(usually water) percolates through and accumulates in the filtrate receiver with a time-varying volume V.

Derive the following equation due to Ruth, which gives the filtration time t as a function of filtrate volume V:

$$t = \frac{1}{2}\frac{\alpha\mu C_o}{A^2 \Delta p}V^2$$

where Δp is the constant pressure drop across the filter cake, A is the filter area, and α is related to the properties of the cake. The equation is used for design purposes (t and V are specified and A is to be determined), or for predicting filter performance (A is given and t is to be determined). Note that, contrary to expectations, filtration time varies nonlinearly with filtrate volume.

Hint: Use D'Arcy's law given in Illustration 1.4 and then eliminate Q and z by means of a cumulative balance.

4.13 Simple Analysis of a Drying Process

Dry warm air at the rate of 1 kg/sec is to be used to evaporate a total of 100 kg water by passing it through a bed of wet nonporous granular solids. After an initial adjustment period, the surface moisture stabilizes at the so-called wet-bulb temperature for which the vapor pressure of water is 50 mm Hg. Estimate the minimum time required to evaporate the entire mass of water.

Hint: Convert vapor pressure to humidity Y (kg water/kg dry air).

4.14 One-Compartment Pharmacokinetics

Pharmacokinetics concerns itself with the fate of a drug in the animal or human body. It follows the time course of the administered drug as it is taken up by the

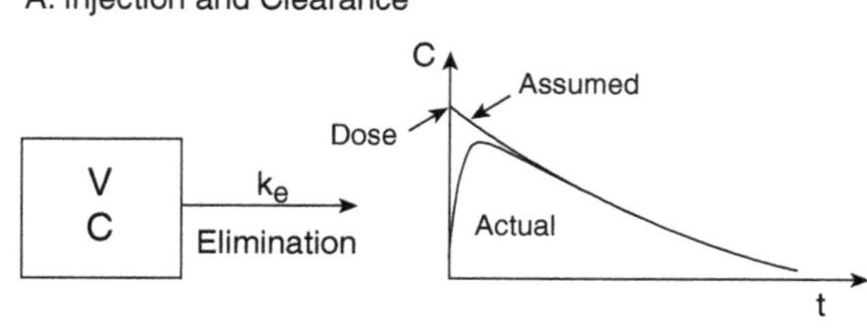

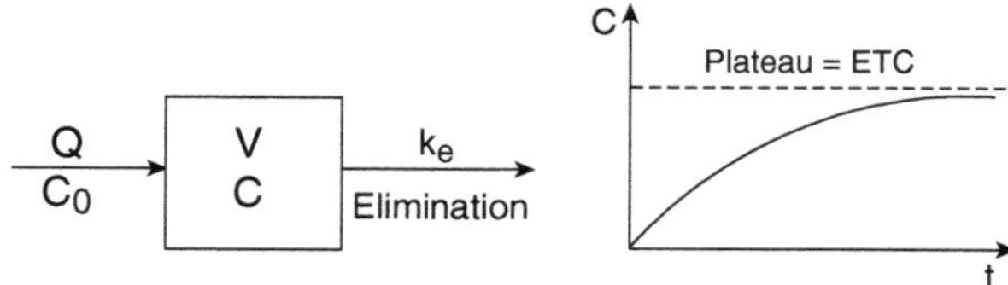

FIGURE 4.30 One-compartment models: Time course of drug concentration during (A) Injection and Clearance and (B) Infusion.

blood stream, conveyed to various parts of the body, and ultimately metabolized and excreted.

One distinguishes between two modes of drug administration, shown in Figure 4.30. Part A shows the time course of drug concentration following the injection of the drug as a one-shot dose. As blood volume and the cardiac output in the adult human are approximately 5 L and 5 L/min, the drug is "turned over" or distributed within the blood about once very minute. This allows us to assume that the distribution of drugs within the plasma approaches "stirred-tank" configuration shortly after injection. Following injection, the drug is slowly "cleared" from the body. In the simplest one-compartment model, that clearance process is assumed to follow a first-order rate law, $r_e = k_e C$.

Part B shows the time history of drug concentration when the drug is continually infused into the patient at a steady flow rate and concentration. The aim of infusion is to reach a plateau concentration which will be therapeutically beneficial. When that plateau, termed the *effective therapeutic concentration*, or ETC, is reached, influx and elimination are exactly in balance, and the system is at steady state.

Pharmacokinetic data are often obtained by single dose injection followed by continuous monitoring of drug concentration in the plasma as it is gradually cleared from the body.

Suppose that a clinical trial has yielded the following clearance data:

t(h)	1	2	5	6	8	10
C(mg/L)	8.53	7.28	4.52	3.85	2.81	2.04

a. Use these results to estimate the elimination rate constant k_e.
b. If clinical trials had shown the ETC to be 50 mg/l, determine the required infusion rate in mg/h and the time necessary to reach 95% of that value.
Answer: $k_e = 0.159\ h^{-1}$

4.15 Maximum Conversion of an Intermediate in a CSTR

Consider the series of consecutive irreversible first-order reactions:

$$A \xrightarrow{k_1} R \xrightarrow{k_2} S$$

It will be recognized that the concentration of the intermediate R will at first rise, pass through a maximum, and ultimately decline to zero as it is converted to the final product S.

Show that the optimum residence time in the reactor is given by

$$\tau_{opt} = (k_1 k_2)^{-1/2}$$

4.16 Release of a Solute into Tubular Flow

In certain applications, a soluble substance is released from the wall of a channel into a flowing liquid. Examples are the release of anticoagulants into blood from coated vascular grafts and the underground recovery of potash by "solution mining".

Derive expressions for the concentration profiles which arise in the following two cases:

a. Solute is released at the constant rate N (kg/sec m^2) into turbulent flow in a rectangular channel.
b. As in (a), but the concentration at the wall corresponds to the solubility C* of the solute.

4.17 Optimum Solvent Flow Rate in a Gas Scrubber

Consider qualitatively the various capital and operating costs as a function of solvent flow rate to a scrubber. Graph these items and show that their sum passes through a minimum, resulting in an optimum solvent flow rate.

4.18 Clearance of a River Bed

A river bed has been contaminated over a 10 km stretch with a toxin having a Henry constant $H = 10^3$ kg/kg. Calculate the minimum time of recovery using the following data:

$$v = 1\ m/sec,\ \rho_b/\rho_f = 2,\ A_b/A_f = 10^{-2}$$

Identify and justify A_b/A_f.
Hint: Consult Illustration 4.12.
Answer: 2×10^5 sec

4.19 Evaporation of a Chemical from a River Carrying Suspended Solids

Rivers have the ability to carry considerable amounts of suspended matter. The Yellow River in China and the Rio Grande in the U.S. may contain as much as 10 g solids/l, whereas a clear mountain stream will have as little as 1 mg/l.

Consider a stream carrying a volatile chemical, both dissolved in the water as well as adsorbed on the suspended solids. Show that under steady state conditions, the fraction f of the chemical desorbed by release to the atmosphere is given by the expression:

$$f = 1 - \exp\left[-\frac{Ka\tau}{1+HC_s}\right]$$

4.20 Another Moving Boundary Problem: Melting of a Charge of Glycerol

Industrial grade glycerol often gels into a quasi-solid at temperatures below 18°C. A steam-jacketed tank is to be used to render a charge of mass m fluid. Show that the time required to achieve this is given by:

$$t = \frac{1}{4\pi}\frac{m\Delta H}{kL\Delta T}$$

$$\text{Hint: } \int x \ln x = \frac{x^2}{2}\ln x - \frac{x^2}{4}$$

4.21 Freezing of a Liquid

A liquid at its freezing point T_f is exposed to colder air at $T < T_f$. Show that the relation between dimensionless time τ and ice thickness z* is given by

$$\tau = \frac{1}{2}(z^*)^2 + z$$

where

$$z^* = \text{dimensionless solid thickness} = \frac{h_1}{k}z$$

$$\tau = \text{dimensionless time} = th_1^2 \frac{T_f - T_\infty}{\rho(-\Delta H_f)k}$$

ΔH_f = latent heat of freezing (negative)
k = thermal conductivity of solid

Hint: Consult Illustration 4.20.

4.22 Potential Freezing of a Water Pipeline

In this example, again from an industrial source, it was proposed to pump water from a plant to a supply of exhaust steam 500 m away where the water would be heated and subsequently returned to the plant. A "double jeopardy" situation was to be considered in which flow is interrupted due to pump failure, and the stationary water (20°C) simultaneously exposed to an ambient temperature of −5°C for as long as 10 h. The question posed was whether this would result in freezing of the water and ultimate failure of the pipe. The following data were assembled:

Pipe:
Internal diameter $d_i = 15.7$ cm
Wall thickness $w = 0.71$ cm
Density $\rho_s = 7800$ kg/m^3
Heat capacity $C_{ps} = 0.473$ kJ/kg K
Water–ice:
Heat capacity of water $C_{pw} = 4.2$ kJ/kg K
Compressibility of water $\beta = -\frac{1}{V}\frac{dV}{dP} \cong 4.4 \times 10^{-6}\ atm^{-1}$
Density of ice $\rho_i = 920$ kg/m^3
Latent heat of freezing $\Delta H_f = 335$ kJ/kg
Heat-transfer coefficients:
External h_0 = Average for still air = 10 W/m^{2}K
Internal: assume natural convection
$h_i = 4.69$ W/m^{2}K
$U = [(1/h_i + 1/h_0]^1 = 3.20$ W/m^{2}K

Hint: Start by neglecting the cooling down time for the pipe wall. This will yield a conservative answer, i.e., t less than actual time required.

4.23 Pumping Power Requirements

Calculate the minimum power required to pump water to a height of 500 m at the rate of $Q = 1$ m^3/sec.

4.24 Conduction in Systems with Heat Sources

a. Derive the equations that apply to conduction in (1) a slab, and (2) a sphere with uniformly distributed heat sources generating heat at the constant rate of S_s (J/m^{3}sec).
b. For a slab, consider the faces at x = 0 and x = L to be held at the temperature T_o by appropriate cooling. Integrate the model equations to obtain the temperature distribution.

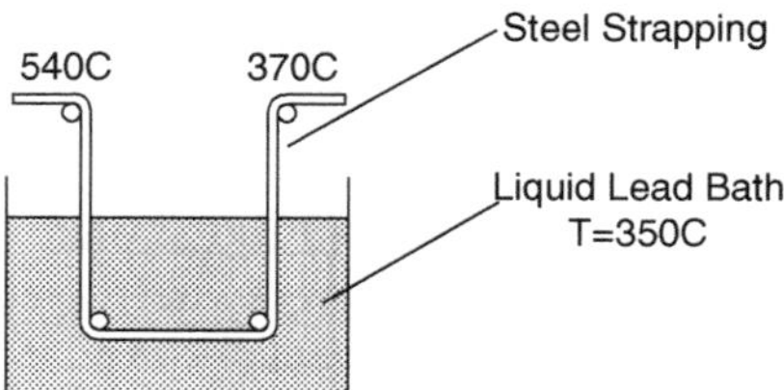

FIGURE 4.31 Annealing of steel strapping in a model lead bath.

c. Show that if the slab is immersed in a bath of temperature T_b, the difference between the surface and bath temperatures is given by $T_s - T_b = S_s L/h$, where h denotes the heat transfer coefficient to the bath.

Answer: (b) $T - T_s = \frac{S_s L^2}{2k}\left[\left(\frac{x}{L}\right) - \left(\frac{x}{L}\right)^2\right]$

4.25 Thermal Treatment of Steel Strapping

In this example, also drawn from industry, one is asked to establish the dimensions of a liquid lead bath used in the annealing of steel strapping (Figure 4.31). Such thermal treatments of newly cast or drawn metal forms are designed to relieve stresses created during the forming process and, in general, to improve the physical properties of the material. They often involve a slow, controlled cooling over a prescribed temperature interval, or prolonged exposure to a fixed temperature. Such procedures are common in the metallurgical industries.

The strapping of width 6.4×10^{-3} m (1/4 in.) enters at a flow rate of 0.189 kg/sec and temperature of 540°C and is required to be cooled to a temperature of 370°C during its passage through the bath. The dimensions of that bath will be determined by the submerged length, or what one might term the residence length of the strapping, and it is this length which will have to be established.

Data:
h (free convection) = 5.56 kW/m^2 C
C_p = 0.15 J/g K

4.26 Potential Thermal Stress Cracking of a Weld

During erection of a catalytic cracking unit in an oil refinery, concern arose that thermal gradients in the steel supports might cause stress cracking of the connecting welds (see Figure 4.32). The reactor, itself insulated, was known to reach temperatures of about 500°C, and ambient temperatures as low as −30°C could be anticipated in the winter months. Structural engineers specified a maximum permissible temperature gradient at the weld of 15°C/cm. A suggested remedy was to cover the supports with insulation, but it was not known how thick the insulation should be, or indeed whether this would solve the problem.

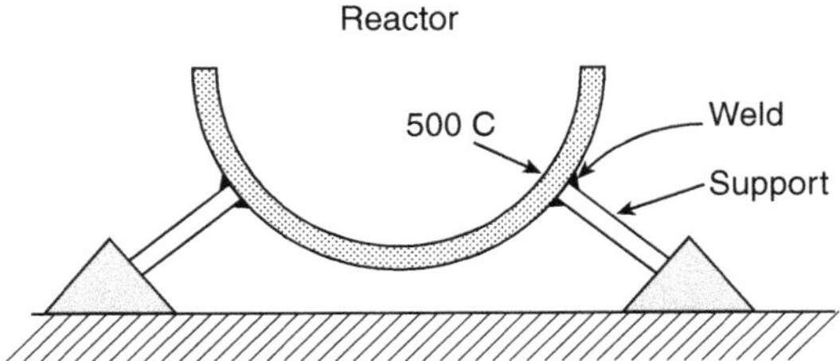

FIGURE 4.32 Welded support of a catalytic cracking reactor.

Data:
Thermal conductivity of steel: 43 W/mK
Thermal conductivity of insulation: 0.12 W/mK
Heat transfer coefficient to air: 50 W/m^2 K
Diameter of support: 0.1 m
Length of Support: 0.5 m

(Hint: Attempt an "asymptotic" solution, with the support bare, a temperature at ground level of −30°C, and a gradient at the reactor end of 15°C/cm. Extract a heat transfer coefficient from the solution. Its value is proportional to the maximum permissible rate of heat transfer. Calculate the equivalent thickness of insulation that will match this value.)

4.27 Transients in a Solar Collector: A Numerical Analysis

Solar energy collectors have come into increasingly wide-spread use as one of several naturally driven clean energy sources. Figure 4.33 shows a diagram of a flat-plate collector used for heating circulating water or some other liquid.

Collectors operate almost entirely under transient conditions, but particularly so during the early morning start-up period. The question to be answered here is the time required for the collector to reach a working temperature of 60°C after an inactive night-time period.

The conventional model for this process consists of an unsteady integral energy balance around the collector, which is assumed to have a uniform temperature. The fluid is initially taken to be at rest.

Taking account of incident and emitted heat, the energy balance reads:

$$\text{Rate of heat in} - \text{Rate of heat out} = \frac{\mathrm{d}}{\mathrm{dt}}\text{heat contents}$$

$$A_C S(t) - U(t)A_C\left[T - T_A(t)\right] = mC_p\frac{dT}{dt}$$

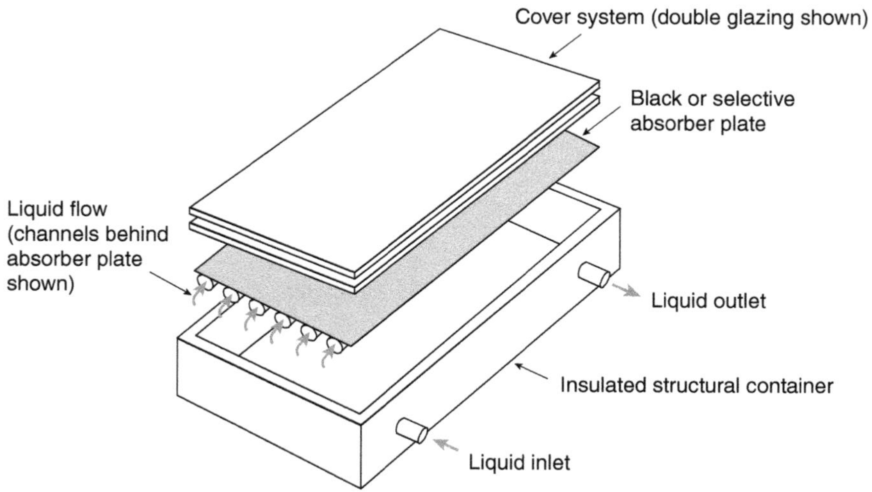

FIGURE 4.33 Diagram of a flat-plate solar collector.

where S is the incident solar energy (W/m^2), A_C the collector area, T_a the ambient temperature, and U the effective heat transfer coefficient to the surroundings. The complicating factor here is that three of the parameters (S, U, T_a), are time dependent.

Solar incident energy depends in a complex way on the position of the sun, cloud cover, inclination of the collector plate and, of course, time of day and year. It is calculated in a semiempirical way from the relevant geometry, and reported data from national weather services which are found compiled for various geographic locations in monographs on solar energy.

Reradiation and other heat losses from the collector are equally complex. The Stefan–Boltzmann law, Equation 4.29c, is rarely used and is replaced instead by a linear driving force $T - T_a$ and an entirely empirical heat transfer coefficient U(t). U depends in a complex way on wind velocity, collector type and emissivity, and ambient temperature, but more importantly also on collector temperature T. The model will therefore require an iterative procedure for its solution. The following limited data are to be used to compute the heat-up time, corresponding to a January day in Denver, Colorado.

t	T_a°C	S(kJ/m^{2}S)
7:30 AM	6	471
8:00 AM	0	747
9:00 AM	5	1714
1:00 AM	8	2332

Collector temperature T°C	−10	0	20	40	60	80
Heat transfer coefficient U (W/m^2 °C)	6	5	6.3	6.8	7.2	8.0

We note that the model ODE is of the form:

$$y' + f(x)y = g(x)$$

which has the following analytical solution (see item 6 of Table 2.4):

$$T = T_o \exp(-t/\tau) + \exp(-t/\tau)\int_0^t \left(\frac{A_C S}{mC_p} - \frac{T_a}{\tau}\right)\exp(t'/\tau)dt'$$

where τ is the time constant mC_p/UA_C, and its initial temperature T_o equals that of the ambient air, T_a, at sunrise (7:30 AM).

4.28 A Biological Switch and the van Heerden Diagram

The enzyme thrombin, which exists in minute quantities in the blood stream, is the principal agent responsible for triggering the coagulation process. When blood comes in contact with an injured site or a foreign surface, thrombin levels are greatly increased by a series of enzymatic reactions, tipping the scales toward coagulation. The rate at which thrombin is produced plots into an inflecting curve of the type shown in Figure 4.20A. Countering this process is the tendency for the thrombin to be swept away by mass transfer into the flowing blood. It has been proposed that the combination of the two processes acts as a "biological switch" that causes coagulation to proceed or subside depending on circumstances. Using the van Heerden diagram as a guide, provide an argument in support of this notion.

4.29 Sonic Velocity in Steam: Discharge of Superheated Steam

(a) Calculate the velocity of sound in saturated steam at 150°C, 475 kPa, assuming ideal gas behavior. $\gamma = 1.31$ under these conditions.
(b) Using the compressible flow chart, Figure 4.22, and assuming frictionless flow, calculate the maximum initial discharge mass velocity of steam from a tank through a valve.
Answer: (a) 506 m/sec

4.30 Compressor Stations in Long Distance Natural Gas Transmission Lines

Natural gas is transmitted through a 1m diameter pipe at the rate of $G = 800\ kg/m^2sec$. Pressure at the inlet is 50 atm. Calculate the maximum possible length of pipe which will accommodate this flow before it has to be recompressed. Friction factor is estimated at $f = 0.002$, and $\gamma \approx 1.4$ for methane. What is the pressure at the outlet?

Answer: $L_{Max} = 100$ km; $p_2 \cong 2.5$ atm

4.31 Annular Condensation in a Vertical Tube

Use the Nusselt film condensation model to derive the local heat transfer coefficient for condensation inside a small tube where the film builds up as an annulus.

4.32 The Converging-Diverging Duct

For adiabatic frictionless gas flow in a duct of variable cross-section A, the relation between A and the Mach number Ma is given by:

$$dA = -\frac{A(1 - Ma^2)}{Ma^2[2 + (\gamma^{-1})Ma^2]} d(Ma^2)$$

Show that flow is always subsonic in the converging section of the duct. What is it on the divergent part and in the transition section (throat) between the two?

5 Partial Differential Equations: Classification, Types, and Properties — Some Simple Transformations

The reader will have noted our avoidance, up to this point, of any use of partial differential equations (PDEs). We limited ourselves to ODEs that arose from unsteady state "stirred tank" and from steady state "one-dimensional pipe" models, and from algebraic integral or cumulative balances. These equations served us well in a good many cases, yielding close approximations of the exact solutions or, at the very least, upper or lower bounds to them. Cases do arise, however, in which PDEs can no longer be avoided or circumvented by valid simplifications and assumptions. In particular, PDEs will have to be addressed in the following situations:

- Thermal conduction or diffusive processes in which temperature or concentration vary with time and distance or, if at steady state, vary in more than one dimension. Exceptions occur when transport coefficients are large or system dimensions small, so that the system may be approximated as a stirred tank at the ODE level. This was done in the case of the thermocouple response given in Illustration 4.18, and was valid there by virtue of the high thermal conductivity and small dimension of the device. When size is more substantial, as in the quenched steel billet shown in Figure 1.3, internal temperature gradients can no longer be ignored, and the full PDE model has to be applied. To distinguish between these two cases, we have provided a criterion, the so-called Biot number, which allows us to determine *a priori* whether it is appropriate to use the reduced ODE model (Illustration 1.7). When this is no longer possible, solutions to the full PDE model must be sought, and this task will be addressed in the subsequent chapters. Time-dependent temperature and concentration distributions may also arise in systems containing instantaneous or continuous heat or mass sources. This important subcase, which also calls for the use of the full PDE model, is taken up in Chapter 6.
- Transport of mass, energy, or momentum in external viscous flow around submerged objects, or in internal duct flow under steady or transient

conditions. In duct flow, this leads to the so-called Graetz problem for heat and mass transfer, or to the Navier–Stokes equation for momentum transfer. The classical example for the use of PDEs in external flows is the derivation of the drag coefficient for flow around a sphere (Stokes' law). The reader will note that we specified viscous or laminar flow for these systems. For turbulent flow, PDE models are usually replaced by ODEs or algebraic equations containing lumped transport parameters, such as heat and mass transfer, or drag coefficients that are determined experimentally. Numerous examples of this treatment appear in the preceding chapters.

- All one-dimensional-pipe processes that operate under unsteady or transient conditions. These include time-varying or fluctuating feed conditions to heat exchangers, packed columns, or tubular reactors. A river exposed to a transient influx of pollutants falls in this category. So does the unsteady pressure distribution that arises in pulsatile blood flow. An important class in its own right is the so-called percolation process, in which concentrations or temperature are distributed in both time and distance within the one-dimensional pipe. We have addressed this problem at a simple algebraic level in Illustration 4.12 and will return to the topic to analyze the impact of the full PDE model.
- Viscous compressible flow in which velocity and pressure vary with distance and time or, if at steady state, vary in more than one dimension. Both internal and external flows, as well as sonic and supersonic conditions are included. This is a vast and complex field usually treated under the heading "Aerodynamics." We note again that under steady turbulent flow conditions, one reverts to ODE or algebraic models with empirical friction factors or drag coefficients. Examples of these cases appeared in Chapter 3, Section 3.5.
- Inviscid, i.e., frictionless flow, in more than one dimension. The slightly more restrictive case of so-called irrotational or potential flow is included here. The one-dimensional case of inviscid flow was taken up at the ODE level and led to the formulation of Bernoulli's equation (Practice Problem 4.6).
- Unsteady, transverse displacements of fluids and solids. This includes wave propagation and the vibrations of strings or membranes, and leads to the so-called wave equation.

The preceding summary includes some of the more important sources of PDEs. PDEs arise in many other contexts as well that were not touched upon. Some of these will become apparent throughout the rest of this chapter and those that follow.

We now turn to the task of describing various properties, classes, and boundary conditions of PDEs, and follow this up with some simple transformations and with a first glimpse of certain elementary solutions and solution methods. Properties and classes are largely those we have seen in an ODE context, but some new subcategories make their appearance, which we present and justify. The solution methods are at this stage kept at a simple level and make no heavy demands in terms of prior knowledge or the use of exotic techniques.

5.1 PROPERTIES AND CLASSES OF PDEs

5.1.1 Order of a PDE

The order of a PDE is defined by the order of its highest derivative. This is the same definition as that used for ODEs. Thus,

Fourier's equation:

$$\alpha \frac{\partial^2 T}{\partial x^2} = \frac{\partial T}{\partial t} \tag{5.1a}$$

is a second-order PDE.

Biharmonic equation:

$$\frac{\partial^4 u}{\partial x^4} + 2\frac{\partial^2}{\partial x^2}\left(\frac{\partial^2 u}{\partial y^2}\right) + \frac{\partial^4 u}{\partial y^4} = 0 \tag{5.1b}$$

is a fourth-order PDE.

Continuity equation:

$$\frac{\partial v_x}{\partial x} + \frac{\partial v_y}{\partial y} + \frac{\partial v_z}{\partial z} = 0 \tag{5.1c}$$

is a first-order PDE.

We describe briefly the occurrence of PDEs of various orders, which again parallels that seen in ODEs.

5.1.1.1 First-Order PDEs

First-order PDEs are found in the description of convective transport, i.e., in the absence of diffusive (second-order) transport terms. The continuity Equation 5.1c, which represents a convective flow mass balance in three-dimensional Cartesian space, is an example of a first-order PDE. Such equations also arise in a host of convective one-dimensional-pipe processes operating under unsteady conditions. A heat exchanger with fluctuating inlet temperature or flow rate is the classical example in this category. First-order PDEs are also seen in the description of convective mass and heat transfer between a fluid and a stationary solid medium (percolation processes).

5.1.1.2 Second-Order PDEs

The classical examples of this type are Fourier's and Fick's equations. In one-dimensional rectangular coordinates, they take the form:

Fourier:

$$\alpha \frac{\partial^2 T}{\partial x^2} = \frac{\partial T}{\partial t} \tag{5.2a}$$

Fick:

$$D\frac{\partial^2 C}{\partial x^2} = \frac{\partial C}{\partial t} \tag{5.2b}$$

The celebrated Navier–Stokes equation, to be described in more detail later, consists of three second-order PDE momentum balances.

5.1.1.3 Higher-Order PDEs

PDEs of order greater than two occur with a much lesser frequency. They are seen in solid mechanics and elasticity theory (cf. the biharmonic Equation 5.1b). They also make their appearance in the course of introducing the so-called stream function ψ, which is defined in terms of fluid velocities in the x and y directions, v_x and v_y.

$$v_x = -\frac{\partial \psi}{\partial y} \qquad v_y = \frac{\partial \psi}{\partial x} \tag{5.2c}$$

Introduction of ψ serves the purpose of coalescing velocity components in a mass or momentum balance into a single term involving ψ, thus simplifying the underlying PDE. The penalty to be paid is an increase in the order of the equation by one.

5.1.2 Homogeneous PDEs and BCs

This definition again parallels that given for ODEs, i.e., it refers to equations that do not contain separate terms in the independent variable or constant. Thus, Fourier's equation (Equation 5.2a) is homogeneous, but if one adds to it a heat source S that is constant, or dependent on time and distance, the equation becomes nonhomogeneous. Hence:

$$\alpha\frac{\partial^2 T}{\partial x^2} + S(x,t) = \frac{\partial T}{\partial t} \tag{5.3a}$$

is a nonhomogeneous PDE. Similarly, the following initial and boundary conditions are both nonhomogeneous:

$$T(x,0) = f(x)$$

$$T(0,t) = g(t) \tag{5.3b}$$

5.1.3 PDEs with Variable Coefficients

As in the case of ODEs, the term "variable" implies changes in the coefficients of the derivative as a function of the independent, not the dependent variable. Such models arose as we had seen, whenever diffusive transport took place through an area that varied with distance. Thus, for radial diffusion in a long cylinder, the unsteady Fick's diffusion Equation 5.2b becomes:

$$D\left[\frac{\partial^2 C}{\partial r^2}+\frac{1}{r}\frac{\partial C}{\partial r}\right]=\frac{\partial C}{\partial t} \tag{5.4}$$

where the quotient 1/r is the variable coefficient. The steady state version of this equation was used, together with a reaction term, to derive concentration profiles in a cylindrical catalyst pellet (Illustration 2.12). The solution there was given in terms of Bessel functions, and we can expect a similar appearance of these functions in the solution of the PDE (Equation 5.4). This is in fact the case, as will be seen in Chapter 8.

5.1.4 Linear and Nonlinear PDEs: A New Category — Quasilinear PDEs

The classification here is again very much akin to what we have seen with ODEs — a PDE is linear if the dependent variable and its derivatives appear in linear combination. Thus, the most general version of a second-order linear PDE in two independent variables has the form:

$$A(x,y)u_{xx}+B(x,y)u_{xy}+C(x,y)u_{yy}+D(x,y)u_x+E(x,y)u_y+F(x,y)u+G(x,y)=0 \tag{5.5a}$$

where the subscripts on the dependent variable u denote differentiation with respect to x and y. Note that here again the variable coefficients A through G can be arbitrarily nonlinear without violating the linearity of the PDE itself. According to this definition, Fourier's equation (Equation 5.1a), the biharmonic equation (Equation 5.1b), the continuity equation (Equation 5.2a), and Fick's equation (Equation 5.2b) are all linear.

When the PDE is not linear, a distinction is made between the so-called quasilinear PDEs and fully nonlinear PDEs. The former is defined as an equation in which the highest derivative is still linear, but not necessarily the lower derivatives or the dependent variable itself. The following examples illustrate these categories:

Linear: $x^2u_{xx}+\exp(y)u_{yy}=0$

Quasilinear: $u_{xx}=u_{yy}+u_t^2$ (5.5b)

Fully nonlinear $u_{tt}=(u_{xx}+u_{yy}{}^2)$

The motivation for introducing this new category of quasilinear PDEs lies in their behavior, which differs from that of fully nonlinear PDEs, and in the fact that a fairly complete theory for them has been developed (see Chapter 8: Method of Characteristics). No such comprehensive treatment exists as yet for fully nonlinear PDEs.

What sets linear PDEs apart is that here one can again apply the superposition principle, i.e., if a set of independent solutions is known, their sum will also be a solution. The famous (and much-dreaded) Fourier series solutions of the linear Fourier and Fick's equations are the result of precisely such a superposition procedure. Details will appear in Chapter 8 under the Separation of Variables Method.

Superposition cannot, in general, be applied to either quasilinear or fully nonlinear PDEs. It is, however, an immensely useful tool for addressing linear problems. We emphasize this fact by setting aside the next chapter for a detailed description of the method and its application in solving PDEs.

5.1.5 Another New Category: Elliptic, Parabolic, and Hyperbolic PDEs

These categories draw their nomenclature from a similar classification for algebraic equations in two variables x and y. The reader may recall that for the general equation:

$$ax^2 + 2bxy + cy^2 + d = 0 \tag{5.6a}$$

one obtains:

An ellipse if:	$b^2 - ac < 0$
A parabola if:	$b^2 - ac = 0$
A hyperbola if:	$b^2 - ac > 0$

To conform to this classification, the PDE, which here is a second-order one with variable coefficients, is arranged in the form:

$$A(x,y)u_{xx} + 2B(x,y)u_{xy} + C(x,y)u_{yy} = F(x,y,u_x,u_y) \tag{5.6b}$$

The resulting categories, with properties and examples, are presented in Table 5.1. We note that the classification is not a trivial one, because it reflects the nature of the problem (boundary value BVP or initial value IVP, singly or in combination), and through it the solution methods to be used. The Laplace transformation, for example, is usually reserved for IVPs only. A similar motivation for classifying equations arose at the ODE level, where a distinction was made between second-order equations with constant and variable coefficients. The former were solved by the D-operator method, whereas the latter required the use of series solutions.

Quasilinear first-order PDEs, or rather sets of them, have the same three categories, but they are arrived at in a slightly different fashion. We write the set in vector-matrix form:

$$\underset{\sim}{A}(\underset{\sim}{u},x,y)\underset{\sim}{u}_x + \underset{\sim}{B}(\underset{\sim}{u},x,y)\underset{\sim}{u}_y + \underset{\sim}{C}(\underset{\sim}{u},x,y) = 0 \tag{5.6c}$$

and set the criteria as follows:

TABLE 5.1
Elliptic, Parabolic, and Hyperbolic Second-Order PDEs

Criterion	Type of PDE	Example	Properties
$B^2 - AC < 0$	Elliptic	Laplace's equation $\frac{\partial^2 u}{\partial x^2} + \frac{\partial^2 u}{\partial y^2} = 0$	Boundary value problem
$B^2 - AC = 0$	Parabolic	Fourier's equation $\alpha \frac{\partial^2 u}{\partial x^2} = \frac{\partial u}{\partial t}$	Mixed BV and IV problem
$B^2 - AC > 0$	Hyperbolic	Wave equation $c^2 \frac{\partial^2 u}{\partial x^2} = \frac{\partial^2 u}{\partial t^2}$	Mixed BV and IV problem or IV problem

The set is:	if the eigenvalues of det $\lvert \underset{\sim}{A} - \lambda \underset{\sim}{B} \rvert = 0$ are
Elliptic	Imaginary
Parabolic	Real and identical
Hyperbolic	Real and distinct

Application of this criterion is best studied in the context of the example given in Illustration 5.1.

5.1.6 Boundary and Initial Conditions

Classification of BCs and ICs follows that established for ODEs, but now have the names of mathematicians attached to them. Dimensionality is also increased so that a more general formulation is called for. We summarize the main features for convenience.

There are three major types of boundary and initial conditions.

1. Type I (Dirichlet) BCs contains the dependent variable only. The initial conditions usually fall into this category, and in rectangular coordinates have the form:

$$u(0, S) = u_0$$

or more generally,

$$u(0, S) = f(x, y, z) \tag{5.7a}$$

where S denotes a bounding surface of the system.

The novelty here, compared to the ODE case, is that the initial condition need not be a constant, but can have an initial distribution in space f(x, y, z). Thus, the quenched steel billet shown in Chapter 1 (Figure 1.3) could have, at the start of the operation, a temperature distribution f(0, x), rather than a constant and uniform value T_0.

2. Type II (Neumann) BCs contain the derivative only, usually taken normal to a surface S and denoted by ∂u/∂n. The general condition is of the form:

$$\frac{\partial u}{\partial n}(t,S) = f(t) \tag{5.7b}$$

and includes the special case:

$$\frac{\partial u}{\partial n}(t,S) = 0 \tag{5.7c}$$

The latter condition applies when u is a maximum or minimum in a particular location, e.g., the center of symmetry, or when the boundary is impermeable to mass, energy, or momentum flux.

3. Type III (Robin) or Mixed BC. This condition contains both the derivative ∂u/∂n and the dependent variable u and has the general form:

$$\frac{\partial u}{\partial n}(t,S) = f[u(S),t] \tag{5.7d}$$

It frequently arises at phase boundaries where the rate of convective transport in one phase (moving fluid) must equal diffusive transport in the other phase (stationary fluid or solid). For this particular case, Equation 5.7d becomes:

$$-K_1 \frac{\partial u}{\partial n}(t,S) = K_2[u(t,S) - u_0(t)] \tag{5.7e}$$

Specific forms of Equation 5.7e will be presented in Illustration 5.2, which follows.

Two additional points need to be noted. The number of BCs required usually equals the sum of the highest order of the derivatives with respect to a particular independent variable. Thus, for the one-dimensional Fourier's Equation (Equation 5.1), we require two BCs for the second-order derivative, and one BC (or rather IC) for the first-order time derivative. A total of three boundary conditions are therefore required to solve this equation. It often happens that additional BCs can be specified but are not used. They must nevertheless be satisfied by the solution.

Most analytical and numerical methods can easily handle complex initial conditions but have difficulty with complex boundary conditions, particularly those occasioned by unusual geometries. There is thus an incentive to simplify or transform difficult boundary conditions, even if this results in a more complicated initial condition. An example of this type of transformation is taken up in Illustration 5.6a.

Illustration 5.1 Classification of PDEs

Properties and classifications of the following PDEs are to be established.

1. $$\frac{\partial^2 u}{\partial x^2} + \frac{\partial^2 u}{\partial y^2} + S(x,y) = 0 \tag{5.8a}$$

   This PDE, known as Poisson's equation, is a second-order, nonhomogeneous equation in the two independent variables x and y. It is linear because the derivatives appear in linear combination. The nonhomogeneous term S(x,y), although arbitrary in form, is a function of the independent variables only and therefore does not affect linearity.
   Comparison with Equation 5.6b yields the criterion:

   $$B^2 - AC = 0 - (1)(1) = -1 < 0$$

   Hence, according to Table 5.1, Poisson's equation is elliptic.

2. $$-G_S \frac{\partial Y}{\partial z} = [\varepsilon \rho_g + \rho_b f'(Y)\rho_b]\frac{\partial Y}{\partial t} \tag{5.8b}$$

   This PDE is a special case of the so-called chromatographic equation, which expresses variations in solute concentration Y of a fluid flowing through a stationary sorptive medium under equilibrium conditions (see Section 5.2). The term f′(Y) is the derivative of the equilibrium relation:

   $$q = f(Y) \tag{5.8c}$$

   and is generally nonlinear in form.

   The equation is a first-order, homogeneous PDE in the two independent variables, distance z and time t. It is not linear because of the product f′(Y)(∂Y/∂t), but neither is it fully nonlinear. Because the highest derivatives, ∂Y/∂z and ∂Y/∂t, appear in linear form, it is a quasilinear PDE.
3. Two-dimensional compressible irrotational (i.e., frictionless) flow is described by:

$$(v_x^{\,2} - c^2)\frac{\partial v_x}{\partial x} + (v_x v_y)\frac{\partial v_y}{\partial x} + v_x v_y \frac{\partial v_x}{\partial y}(v_y^{\,2} - c^2)\frac{\partial v_y}{\partial y} = 0$$

$$-\frac{\partial v_x}{\partial y} + \frac{\partial v_y}{\partial x} = 0 \tag{5.8d}$$

These are two first-order homogeneous PDEs in the two velocity components v_x and v_y, with c = velocity of sound. The second equation is linear and the first one quasilinear by virtue of the fact that its highest derivatives appear in linear form.

To determine whether the set is elliptic, parabolic, or hyperbolic, we identify the coefficient matrices of the partial derivatives as follows:

$$\underset{\sim}{A} = \begin{pmatrix} v_x^2 - c^2 & v_x v_y \\ 0 & 1 \end{pmatrix} \tag{5.9a}$$

and

$$\underset{\sim}{B} = \begin{pmatrix} v_x v_y & v_y^2 - c^2 \\ -1 & 0 \end{pmatrix}$$

Hence,

$$\det|\underset{\sim}{A} - \lambda \underset{\sim}{B}| = \begin{vmatrix} v_x^2 - c^2 - \lambda v_x v_y & v_x v_y - \lambda(v_y^2 - c^2) \\ \lambda & 1 \end{vmatrix} = 0 \tag{5.9b}$$

or

$$\lambda_{1,2} = \frac{v_x v_y \pm c\left[v_x^2 + v_y^2 - c^2\right]^{1/2}}{(v_y^2 - c^2)} \tag{5.9c}$$

This results in the following classification:

Eigenvalues $\lambda_{1,2}$	**Velocities**	**Flow**	**PDE**
Imaginary	$v_x^2 + v_y^2 < c^2$	Subsonic	Elliptic
Identical and real	$v_x^2 + v_y^2 = c^2$	Sonic	Parabolic
Distinct and real	$v_x^2 + v_y^2 > c^2$	Supersonic	Hyperbolic

Thus, the PDEs can fall into any one of the three categories (elliptic, parabolic, and hyperbolic), depending on whether the flow regime is subsonic, sonic, or supersonic. Consequently, one can expect changes in both solution methods and solution behavior as the flow regimes change.

Illustration 5.2 Derivation of Boundary and Initial Conditions

Leaching of a slurry in a stirred tank — A slurry of porous solid particles assumed to be thin flakes and containing a soluble component is to be leached in a stirred tank of solvent (Figure 5.1A). We wish to derive BCs and ICs for the full PDE model in terms of solid and external fluid concentrations C_s and C_f.

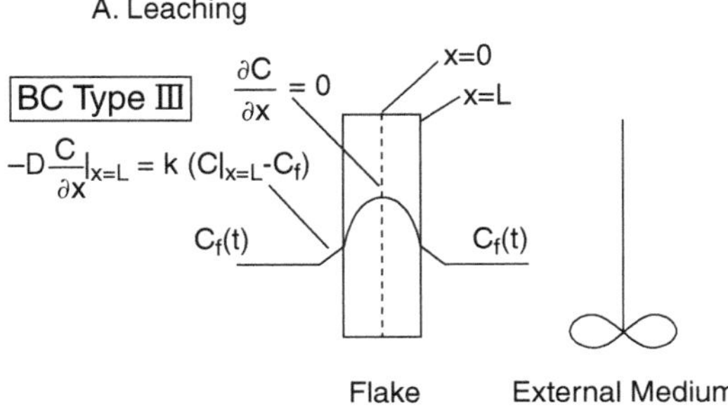

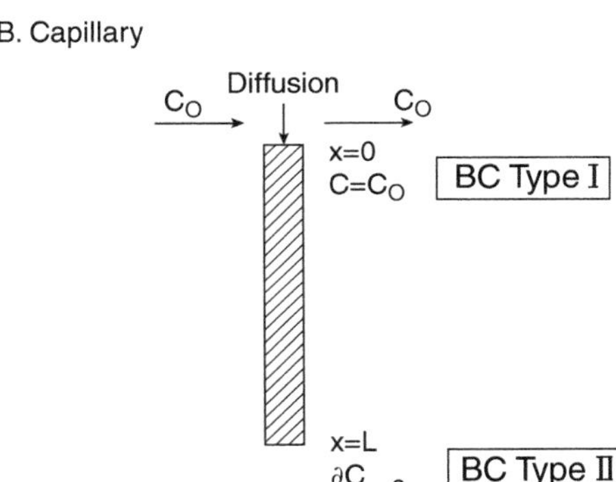

FIGURE 5.1 Examples of physical systems with BCs of Type I, Type II, and Type III (Dirichlet, Neumann, and Robin).

The internal mass transfer is assumed to be Fickian. Equation 5.2b applies and requires three BCs/ICs, two for the second-order derivative and one for the first-order time derivative. We obtain:

IC $$C_s(x, 0) = C_s^0 \text{ (assumed uniform)} \tag{5.10a}$$

BC I $$\left.\frac{\partial C_s}{dx}\right|_{x=0} = 0 \text{ (symmetry)} \tag{5.10b}$$

or $$C_s(0, t) = \text{bounded} \tag{5.10c}$$

BC II $$-D\left.\frac{\partial C_s}{\partial x}\right|_{x=L} = k\left[C_s(L) - C_f(t)\right]$$

where k is a mass transfer coefficient.

BC I is of Type II, and BC II of Type III or mixed type. The latter expresses equality of diffusive flux at the surface, and convective removal through a film resistance. Fick's equation is coupled to an external integral unsteady mass balance, which leads to a first-order ODE in the fluid concentration C_f, requiring an initial condition: IC – $C_f(0) = 0$, assuming clean solvent at the start of the operation.

Systems of PDEs coupled to ODEs arise frequently in diffusive processes taking place in a confined, well-stirred medium. In these cases, four boundary and initial conditions are required for each of the participating species. Solutions of such systems are taken up in Chapter 7, Illustration 7.12.

Diffusion into a narrow capillary — A liquid containing a solute at a concentration C^0 flows over the opening of a capillary containing pure solvent (Figure 5.1B). The solute diffuses into the capillary, which is sealed at the bottom. This type of arrangement has been used to determine diffusivities from average solute concentrations in the capillary measured at various time intervals. Substitution of the values into the solution of Fick's equation (Equation 5.2b) allows the relevant diffusion coefficient D to be extracted (see Practice Problem 8.14).

The required boundary and initial conditions, which are three in number, are as follows:

$$\text{IC} \qquad C(0, x) = 0 \text{ (pure solvent at } t = 0) \tag{5.10d}$$

$$\text{BC I} \qquad C(t,0) = C^0 \text{ (concentration at open end)}$$

Here, it is assumed that flow of the liquid is fast enough that there are no significant changes in external solute concentration. The same effect may be obtained by contacting the capillary with a large volume of a well-stirred fluid.

$$\text{BC II} \qquad \left.\frac{\partial C}{\partial x}\right|_{x=L} = 0 \text{ (no flux at sealed end)} \tag{5.10e}$$

Alternatively, we have:

$$C(t, L) - \text{bounded}$$

5.2 PDEs OF MAJOR IMPORTANCE

The previous section gave the reader a first glimpse of PDEs and their properties. The reassuring fact emerged that, except for one or two new features, these properties were very much like those we had seen at the ODE level. We now open the door slightly wider and expose the reader to a listing of PDEs of major importance. These are presented in scalar Cartesian form to make them easier to decipher. The more general vectorial representation is deferred to Chapter 7.

Although still somewhat intimidating in appearance, the reader will take comfort from the fact that these equations share some familiar features with ordinary differential equations. Thus, convective transport is still described by first-order derivatives, diffusive transport by second-order derivatives, and interphase transport is, as before, proportional to a linear driving force. These PDEs are thus, for the most part, similar in appearance to ODEs we have seen in previous chapters, and may be viewed as multidimensional extensions of the balances we have encountered at the ODE level. In fact, we shall see that if we reverse the process by reducing dimensionality, we can, in many instances, recover an ODE balance seen before.

5.2.1 First-Order Partial Differential Equations

Unsteady tubular operations (turbulent flow)— We noted in the introduction to the chapter that all systems of the one-dimensional pipe category, which led to first-order ODEs under steady state conditions, become PDEs when time-dependent inlet conditions are imposed. They will be first-order PDEs because the rate-of-change term will be a first-order derivative. Thus, for the steam-heated pipe described in Illustration 1.2, the relevant energy balance now reads:

$$F(t)C_p \frac{\partial T}{\partial z} - U(t)\pi D(T_s - T) = \rho C_p \frac{\partial T}{\partial t} \tag{5.11}$$

with inlet temperature in general given by:

$$T(t,0) = f(t)$$

Flow rate f(t) and inlet temperature can vary individually or in combination with time t and affect the heat transfer coefficient U(t) as well.

Similar first-order PDEs arise in tubular reactors, and in cocurrent and countercurrent continuous mass and heat transfer operations. Turbulent flow conditions have been specified because laminar flow leads to radial diffusive processes and results in second-order PDEs. These are described in Section 5.2.2, the Graetz problem.

One notes that for the steady state, Equation 5.11 reduces to the ODE (Equation 1.4e) for the steam-heated pipe and we are on familiar ground again. Thus, the two equations differ only by the single unsteady term $\rho C_p \partial T/\partial t$. If, on the other hand, we drop the convective term F(t)Cp(∂T/∂z), we recover an unsteady stirred tank balance that leads again to familiar territory.

The chromatographic equations — These equations describe concentration changes that occur when a fluid containing a solute, or devoid of it, flows through or over a stationary matrix of solid sorptive material. The latter may be "clean," or loaded with solute. We have considered problems of this type at an algebraic level in Illustration 4.12 without going into the details of the underlying PDEs. We now consider these both under equilibrium and nonequilibrium conditions where the fluid phase is a gas.

Equilibrium (no mass transfer resistance): We have already seen this case in Illustration 5.1, where the PDE was of the form:

$$-G_s \frac{\partial Y}{\partial z} = [\varepsilon \rho_g + \rho_b f'(Y)] \frac{\partial Y}{\partial z} \tag{5.12a}$$

where f'(Y) is the derivative of the equilibrium relation:

$$q = f(Y) \tag{5.12b}$$

and $\varepsilon\rho_g$ can usually be neglected.

A more general formulation for several solutes is given by the two-phase mass balance:

$$-G_s \frac{\partial Y_i}{\partial z} = \rho_b \frac{\partial q_i}{\partial t} + \varepsilon\rho_g \frac{\partial Y_i}{\partial t} \tag{5.12c}$$

with the equilibrium relation:

$$q_i = f(Y_1, Y_2 \ldots Y_n) \tag{5.12d}$$

and I = 1, 2, … , n. Here, q is the solid-phase concentration (kg solute/kg solid) and Y denotes the gas-phase concentration (kg solute/kg carrier). G_s is the carrier mass velocity (kg/sm²), ρ_b and ρ_g the bed and carrier densities (kg/m³), and ε the bed void fraction.

Thus, we obtain, after substitution of Equation 5.12d into Equation 5.12c, n coupled first-order PDEs in the fluid concentrations Y_1, Y_2, … , Y_n.

Nonequilibrium (with mass transfer resistance) — Here, the relevant equations are given by:

Gas-phase mass balance:

$$-G_s \frac{\partial Y}{\partial z} - K_{0Y}a(Y - Y^*) = \varepsilon\rho_g \frac{\partial Y}{\partial t} \tag{5.12e}$$

Flow Mass transfer Unsteady term

Solid-phase mass balance:

$$K_{0Y}a(Y - Y^*) = \rho_b \frac{\partial q}{\partial t} \tag{5.12f}$$

Equilibrium (inverted):

$$Y^* = f(q) \tag{5.12g}$$

Here, again, it is of some comfort to see that omission of the unsteady term in Equation 5.12e yields the familiar steady state gas-phase mass balance, Equation 4.23c, which we had encountered while modeling a countercurrent gas absorber. The reader should attempt to overcome the initial mistrust of PDEs by reducing them, whenever possible, to ODEs that are familiar from more conventional steady state and one-dimensional situations.

Movement of traffic — It seems at first sight surprising that movement of traffic, which consists of discrete vehicular entities, can be described by a differential equation. This is achieved by defining a vehicle density C (number per unit distance x) and a vehicle flux q (number crossing x per unit time), corresponding to concentration C and flux N_A in ordinary mass transport. A simple "mass balance" then yields the expression:

$$-\frac{\partial q'}{\partial x} = \frac{\partial C}{\partial t} \tag{5.12a}$$

where flux q′ is given by the product of density C and velocity v:

$$q' = Cv \tag{5.12b}$$

v in turn depends on density, because high densities, for example, lead to low velocities. A simple relation that is often used to express this fact has the form:

$$v = v_{Max}\left(1 - \frac{C}{C_{Max}}\right) \tag{5.12c}$$

Thus, when $C \cong 0$, traffic moves at its maximum speed. Conversely, when $C = C_{Max}$, traffic comes to a halt, i.e., we have a bumper-to-bumper traffic jam.

Substitution of Equation 5.12b and Equation 5.12c into the mass balance Equation 5.12a yields a single PDE in the vehicle density C, as a function of time and position:

$$v_{Max}\left(\frac{C}{C_{Max}} - 1\right)\frac{\partial C}{\partial x} = \frac{\partial C}{\partial t} \tag{5.12d}$$

The equation is first order, quasilinear, and homogeneous. We shall encounter it again in Chapter 8 (Method of Characteristics).

Sedimentation of particles — The treatment of this case is analogous to that of traffic problems, i.e., one performs in the first instance a mass balance that leads to Equation 5.12a, where C is now the number of particles per unit volume, and flux q is in terms of number of particles settling per unit area and time. An auxiliary relation is then introduced, much like that given by Equation 5.12b and Equation 5.12c to relate flux to concentration:

$$v = v_{Max}\left(1 - \frac{C}{C_{Max}}\right) \tag{5.12e}$$

The final PDE is then identical to that given by Equation 5.12d.

The auxiliary relation (Equation 5.12c) holds in simple situations, but more complex relations, both for traffic and sedimentation problems, are needed in other cases. These are described in specialized monographs.

5.2.2 Second-Order PDEs

Laplace's equation — This equation, which is among the most important relations of mathematical physics, is given in Cartesian coordinates by the expression:

$$\frac{\partial^2 u}{\partial x^2}+\frac{\partial^2 u}{\partial y^2}+\frac{\partial^2 u}{\partial z^2}=0 \tag{5.13a}$$

The solution of this equation yields the steady state distribution of the state variable u. That variable is often referred to as a *potential* or *potential function*, and the study of its behavior as *potential theory.* The reason for this designation is that in many applications of Laplace's equation, u represents a driving potential for a particular physical process leading to the movement of mass and heat. In other cases, it serves to generalize force fields that are related to the movement of mass or electrical charges. We summarize these applications in the following.

Conduction — Here, Laplace's equation has the specific form in rectangular coordinates:

$$\frac{\partial^2 T}{\partial x^2}+\frac{\partial^2 T}{\partial y^2}+\frac{\partial^2 T}{\partial z^2}=0 \tag{5.13b}$$

The solution represents the steady state temperature distribution in three-dimensional Cartesian space. It can be displayed as a "flow net" (Figure 5.2A) in which the lines T = constant are the equipotential lines or isotherms that carry no heat, and the orthogonal set of flux lines q = constant are the pathways along which conduction takes place with the components of q given by Fourier's law, i.e.,

$$q_x=-k_x\frac{\partial T}{\partial x},\quad q_y=-k_y\frac{\partial T}{\partial y},\quad q_z=-k_z\frac{\partial T}{\partial z} \tag{5.13c}$$

It is comforting to note that for the one-dimensional case, Equation 5.13b reduces to

$$\frac{d^2T}{dx^2}=0 \tag{5.13d}$$

with the solution:

$$T = Ax + B \tag{5.13e}$$

This is the well-known linear temperature profile that arises in unidirectional heat conduction between two isothermal plates.

Diffusion — The counterpart of Equation 5.13b for mass diffusion is given by:

$$\frac{\partial^2 C}{\partial x^2}+\frac{\partial^2 C}{\partial y^2}+\frac{\partial^2 C}{\partial z^2}=0 \tag{5.13f}$$

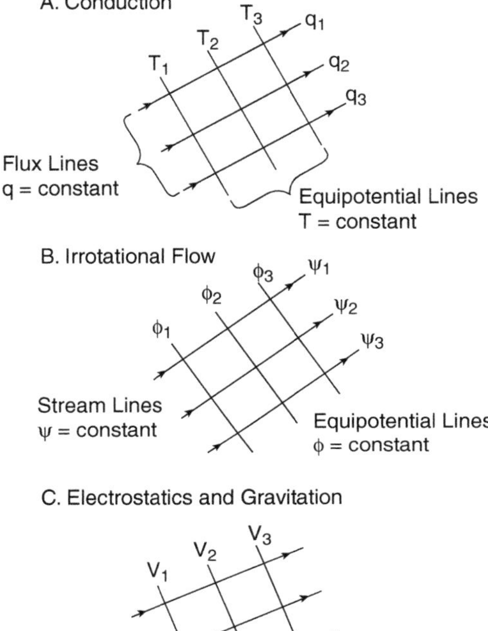

FIGURE 5.2 Physical systems described by Laplace's equation.

where concentration C is the potential corresponding to temperature T, and the mass flux $N_{x,y,z} = -D\,\partial C/\partial(x,y,z)$ takes the place of conductive flux q.

Flow through porous media — Here, the driving potential is the hydrostatic pressure p, and Laplace's equation takes the form:

$$\frac{\partial^2 p}{\partial x^2}+\frac{\partial^2 p}{\partial y^2}+\frac{\partial^2 p}{\partial z^2}=0 \tag{5.13g}$$

p is related to the velocity components v_x, v_y, and v_z via the auxiliary D'Arcy's law, which is the counterpart to Fourier's and Fick's law:

$$v_x=\frac{K_x}{\mu}\frac{\partial p}{\partial x},\quad v_y=\frac{K_y}{\mu}\frac{\partial p}{\partial y},\quad v_z=\frac{K_z}{\mu}\frac{\partial p}{\partial z} \tag{5.13h}$$

The Ks in this expression represent permeabilities of the porous medium in the three directions.

The solution of Equation 5.13g represents the pressure distribution in the porous medium that can in turn be used to derive the velocity distribution by back-substitution into Equation 5.13h, giving us a complete description of the flow field in a porous medium. Note that no flow takes place along the equipotential lines p = constant.

Irrotational or potential flow — Although a precise definition of irrotational flow requires the use of vector calculus (see Chapter 7), it suffices for our present purposes to identify it with frictionless flow. Some exceptions to this definition do occur, but they are quite rare.

The flow in question has a potential function associated with it, the so-called velocity potential ϕ, which is related to the two-dimensional velocity components v_x and v_y via the expressions:

$$v_x = \frac{\partial \phi}{\partial x} \quad v_y = \frac{\partial \phi}{\partial y} \tag{5.13i}$$

Alternative formulations using negative gradients may also be used.

The best way to obtain a physical sense of this quantity is to note that for ϕ = constant, the velocity components vanish, i.e., there is no flow along equipotential lines ϕ = constant. In this respect, ϕ has the same properties as the potential functions for conduction (T), diffusion (C), and flow in porous media (p). Orthogonal to this network of lines is the network of flow that is the streamlines of the flow field. These lines have the properties that a second function, the so-called stream function ψ, is invariant along them, ψ = constant (see Figure 5.2B). That function can also be related to the two-dimensional velocity components, as follows:

$$v_x = \frac{\partial \psi}{\partial y} \qquad v_y = -\frac{\partial \psi}{\partial x} \tag{5.13j}$$

Both ψ and ϕ satisfy Laplace's equation as can be seen by differentiating Equation 5.13i and Equation 5.13j. Thus:

$$\frac{\partial^2 \phi}{\partial x^2} + \frac{\partial^2 \phi}{\partial y^2} = 0 \tag{5.13k}$$

$$\frac{\partial^2 \psi}{\partial x^2} + \frac{\partial^2 \psi}{\partial y^2} = 0$$

One can solve for either ϕ or ψ and obtain, by direct substitution into Equation 5.13i and Equation 5.13j, the velocity distribution in the flow field. To obtain a complete description of the flow, we require, in addition, the pressure distribution p(x,y). This is arrived at from Bernoulli's equation, which for the present purposes takes the form (incompressible flow):

$$\frac{p_1}{\rho} + \frac{v_x^2 + v_y^2}{2} = \frac{p_2}{\rho} + \frac{v_x^2 + v_y^2}{2} \tag{5.13}$$

Gravitational and electrostatic fields — The potential function for gravitational and electrostatic fields derives from the empirical laws due to Newton (gravity) and Coulomb (electrostatics) that the attractive forces between two masses or two charges of opposite sign varies directly with the product of the mass m or charges q, and inversely with the square of the distance between them:

$$F_G = C_1 \frac{m_1 m_2}{r^2} \qquad F_e = C_2 \frac{q^+ q^-}{r^2} \tag{5.13m}$$

It can be shown that these empiricisms lead to the definition of a potential function V, which is related to the Cartesian force components as follows:

$$F_x = C\frac{\partial V}{\partial x} \qquad F_y = C\frac{\partial V}{\partial y} \qquad F_z = C\frac{\partial V}{\partial z} \tag{5.13n}$$

where we have generalized both the potential function and the proportionality constant into single entities (Figure 5.2C). One notes that these "auxiliary relations" are nearly identical in form to the relations applicable to conduction, diffusion, and flow through porous media, with the exception that the derivatives of the potential function V take a positive sign, as did the velocity potential. Differentiation along constant force lines converts Equation 5.13n to Laplace's equation:

$$\frac{\partial^2 V}{\partial x^2} + \frac{\partial^2 V}{\partial y^2} + \frac{\partial^2 V}{\partial z^2} = 0 \tag{5.13o}$$

For convenience, we have summarized in Table 5.2 the relevant potential functions for the various cases discussed.

TABLE 5.2
Potential Functions of Laplace's Equation

Process	Potential Function	Relation to Process Variables		
Conduction	Temperature T	$q_x = -k_x \frac{\partial T}{\partial x}$	$q_y = -k_y \frac{\partial T}{\partial y}$	$q_z = -k_z \frac{\partial T}{\partial z}$
Diffusion	Concentration C	$N_x = -D_x \frac{\partial C}{\partial x}$	$N_y = -D_y \frac{\partial C}{\partial y}$	$N_z = -D_z \frac{\partial C}{\partial z}$
Flow through porous media	Pressure p	$v_x = -\frac{K_x}{\mu}\frac{\partial p}{\partial x}$	$v_y = -\frac{K_y}{\mu}\frac{\partial p}{\partial y}$	$v_z = -\frac{K_z}{\mu}\frac{\partial p}{\partial z}$
Irrotational flow	Velocity potential ϕ	$v_x = \frac{\partial \phi}{\partial x}$	$v_y = \frac{\partial \phi}{\partial y}$	
Gravitation electrostatics	Potential V	$F_x = C\frac{\partial V}{\partial x}$	$F_y = C\frac{\partial V}{\partial y}$	$F_z = C\frac{\partial V}{\partial z}$

TABLE 5.3
Source Terms in Poisson's Equation

Process	A(x, y, z)	Source Term
Conduction	S(x, y, z)/k	$S(J/sm^3)$
Gravitation	$4\pi\rho(x, y, z)$	$\rho(kg/m^3)$
Electrostatics	$4\pi\rho(x, y, z)$	$\rho(C/m^3)$

Poisson's equation — Poisson's equation is obtained by adding what may be called a Source Term A to Laplace's equation. This yields the general form:

$$\frac{\partial^2 u}{\partial x^2}+\frac{\partial^2 u}{\partial y^2}+\frac{\partial^2 u}{\partial z^2}+A(x,y,z)=0 \tag{5.14}$$

Such nonhomogeneous terms A(x, y, z) arise in the steady conduction of heat through a medium with distributed or uniform heat sources, in gravitational fields with distributed masses, and in electrostatic fields with distributed charges. The exact form of A(x,y, z) is summarized for the convenience of the reader in Table 5.3.

Helmholz equation — Here, a linear term in the state variable u is added to Laplace's equation to yield:

$$\frac{\partial^2 u}{\partial x^2}+\frac{\partial^2 u}{\partial y^2}+\frac{\partial^2 u}{\partial z^2}+Au=0 \tag{5.15a}$$

Occurrence of this equation is more limited. It arises principally in diffusional processes accompanied by a first-order reaction. For example, diffusion in three dimensions in a solid matrix, in which the reactants undergo a first-order irreversible reaction is given by the expression:

$$\frac{\partial^2 C}{\partial x^2}+\frac{\partial^2 C}{\partial y^2}+\frac{\partial^2 C}{\partial z^2}-\frac{k_r}{D}C=0 \tag{5.15b}$$

An identical form arises in nuclear processes in which neutron density ϕ takes the place of concentration C, and the rate term is given by the expression:

$$\frac{v\Sigma_f-\Sigma_a}{D}\phi$$

where $\Sigma_{f,a}$ are the so-called fission and absorption cross-sections, and ν represents the number of neutrons produced per fission event. The reader will note that upon dropping the derivatives in y and z, we recover the ODE for diffusion and reaction in a catalyst slab.

Biharmonic equation — This equation, used as an example of a higher-order PDE, has the two-dimensional rectangular form:

$$\frac{\partial^4 u}{\partial x^4} + 2\frac{\partial^2}{\partial x^2}\left(\frac{\partial^2 u}{\partial y^2}\right) + \frac{\partial^4 u}{\partial y^4} = 0 \tag{5.16}$$

It arises in certain areas of solid mechanics, such as elasticity theory, in which u describes the displacement in the solid body. In fluid mechanics, the equation is encountered in the description of creeping flow, i.e., slow viscous flow, in which u becomes the stream function ψ (see Chapter 7, Equation 7.30c).

Fourier's equation — We use the term "Fourier's equation" to denote the unsteady version of the expression for steady state conduction, Equation (5.13b). For constant physical properties, it takes the rectangular form:

$$\alpha\left[\frac{\partial^2 T}{\partial x^2} + \frac{\partial^2 T}{\partial y^2} + \frac{\partial^2 T}{\partial z^2}\right] = \frac{\partial T}{\partial t} \tag{5.17}$$

where $\alpha = k/\rho Cp$ = thermal diffusivity. A derivation of the one-dimensional version of this equation is given in Illustration 5.3.

Fick's equation — Fick's equation is the diffusional counterpart to Fourier's equation and is identical to it in form, with concentration C replacing temperature, and mass diffusivity D taking the place of thermal diffusivity α. We have in Cartesian coordinates:

$$D\left[\frac{\partial^2 C}{\partial x^2} + \frac{\partial^2 C}{\partial y^2} + \frac{\partial^2 C}{\partial z^2}\right] = \frac{\partial C}{\partial t} \tag{5.18}$$

Although many solutions to this equation are identical in form to those of the conduction equation, there are sufficient differences in the two processes that a separate treatment of the topic becomes desirable. This is reflected in the existence of separate monographs devoted to the subject.

The wave equation — The wave equation describes not only, as the name implies, wave propagation phenomena, but applies quite generally to physical processes that result in the disturbance or displacement of fluid or solid elements. It has the Cartesian form:

$$\frac{\partial^2 u}{\partial x^2} + \frac{\partial^2 u}{\partial y^2} + \frac{\partial^2 u}{\partial z^2} = \frac{1}{c^2}\frac{\partial^2 u}{\partial t^2} \tag{5.19a}$$

where c is a constant and a function of the physical process under consideration. We summarize specific applications in the following.

Transverse vibrations in a string — The one-dimensional form of Equation 5.19a applies here, with u(x,t) describing the displacement of an element of the string in the transverse direction. The constant c^2 equals H/ρ, where H is the horizontal component of the tension along the string. The derivation of the PDE appears in Illustration 5.3.

One-dimensional sound propagation — Here, the constant c represents the velocity of sound:

$$c = \left(\frac{\partial p}{\partial \rho} \right)^{1/2} \tag{5.19b}$$

and the displacement u is related to the local pressure p by the relation:

$$p - p_0 = -c^2 \rho_0 \frac{\partial u}{\partial x} \tag{5.19c}$$

where the subscript 0 denotes the undisturbed state.

Transverse vibrations in a membrane — Vibrations of a thin membrane of width L are represented by the two-dimensional wave equation, with c^2 = TL and u(x,y,t) the displacement in the transverse direction. T is the uniform tension of the membrane. Other applications include the propagation of tidal and electromagnetic waves, as well as elastic waves in a solid.

The Navier–Stokes equations — The Navier–Stokes equation is a vectorial force or momentum balance for viscous fluid flow. It has three component balances, of which we reproduce one in the Cartesian z direction:

$$\underbrace{-\rho\left[v_x \frac{\partial v_z}{\partial x} + v_y \frac{\partial v_z}{\partial y} + v_z \frac{\partial v_z}{\partial z}\right]}_{\text{Momentum owing to flow}} + \underbrace{\mu\left[\frac{\partial^2 v_z}{\partial x^2} + \frac{\partial^2 v_z}{\partial y^2} + \frac{\partial^2 v_z}{\partial z^2}\right]}_{\text{Viscous forces}}$$

$$\underbrace{-\frac{\partial p}{\partial z}}_{\text{Pressure force}} \quad \underbrace{- \rho g_z}_{\text{Gravity}} \quad = \quad \underbrace{\rho \frac{\partial v_x}{\partial t}}_{\text{Unsteady term}} \tag{5.20a}$$

Together with the continuity equation, the Navier–Stokes equations comprise four PDEs in the state variables v_x, v_y, v_z, and p. Their solution represents a complete description of the flow field.

As usual, one can gain a better physical understanding of these formidable equations by reducing them to simpler and more familiar cases. Thus, for steady inviscid (μ = 0), one-dimensional steady flow, Equation 5.20a becomes:

$$v \frac{dv}{dz} + \frac{1}{\rho} \frac{dp}{dz} + g = 0 \tag{5.20b}$$

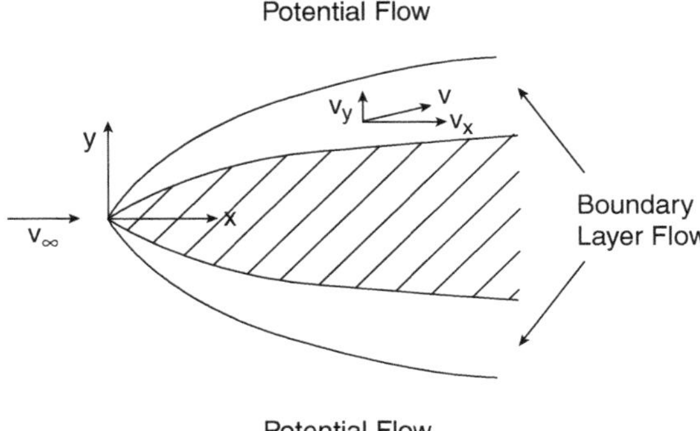

FIGURE 5.3 Flow about a submerged body. Division of the flow field into boundary layer flow and potential flow.

which is the differential form of the Bernoulli equation.

We shall return to the Navier–Stokes equations in Chapter 7, where they will be discussed in greater detail and refinement.

The Prandtl boundary layer equations — In boundary layer theory, it is assumed that viscous or frictional effects in exterior flows are confined to a thin layer adjacent to the surface of the submerged body. Outside this region, flow is essentially frictionless, or to be more precise, we are dealing with potential or irrotational flow (Figure 5.3). An analysis due to Prandtl established that viscous forces and those due to bulk flow in the y-direction are negligible compared to those operating in the x-direction. Under these conditions, the three Navier–Stokes momentum balances will reduce to a single two-dimensional balance that is supplemented by the corresponding continuity equation. We obtain:

Momentum balance:

$$-\rho\left[v_x\frac{\partial v_x}{\partial x}+v_y\frac{\partial v_x}{\partial y}\right]+\mu\frac{\partial^2 v_x}{\partial x^2}-\frac{dp}{dx}=0 \tag{5.21a}$$

Continuity equation:

$$\frac{\partial v_x}{\partial x}+\frac{\partial v_y}{\partial y}=0 \tag{5.21b}$$

The pressure gradient, which is significant only in the x-direction, is established from the solution of the potential flow equations in conjunction with Bernoulli's equation (Equation 5.20b). Here again we shall retrace the various steps in greater detail in Chapter 7, where we shall also provide a detailed derivation of these important equations.

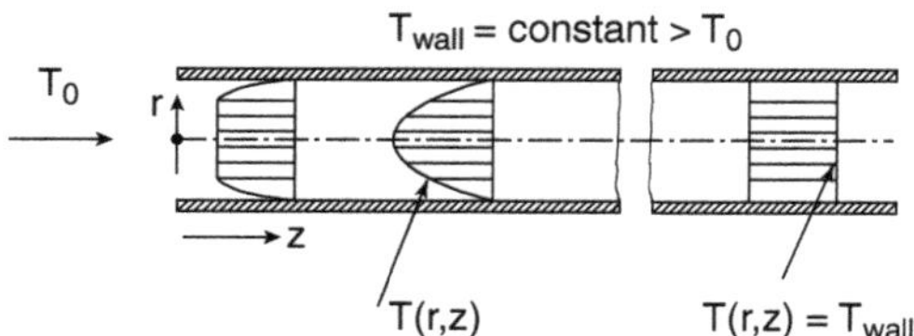

FIGURE 5.4 The Graetz problem in heat transfer. Development of steady axial and radial temperature distributions.

The Graetz problem — The nature of this problem may be deduced from Figure 5.4. It consists essentially of the derivation of the axial and radial temperature profiles that result when a fluid in viscous flow and with inlet temperature T_0 flows through a duct whose wall temperature is kept constant at some value T_w. T_w is assumed to be higher than T_0 in the figure, but it can equally well be set lower. A steady state energy balance in cylindrical coordinates leads to the PDE (see Illustration 5.3):

$$v(r)\frac{\partial T}{\partial z} = \alpha\left[\frac{\partial^2 T}{\partial r^2} + \frac{1}{r}\frac{\partial T}{\partial r}\right] \tag{5.22}$$

whose solution provides the desired profile T(r, z).

This classical problem has been extended to encompass a variety of wall boundary conditions, duct geometries, axial conduction, as well as non-Newtonian fluids. It will be evident to the reader that an expression identical in form to Equation 5.22 applies to mass transport that may arise in membrane processes, or when a substance is either released or reacted at the tubular wall.

Because of the scope and importance of the topic, entire monographs listing solutions and solution methods have been devoted to it. Further details and illustrations dealing with this problem will appear in Chapter 7.

Illustration 5.3 Derivation of Some Simple PDEs

The derivation of PDEs can be accomplished in several ways.

- At the simplest level, one uses an extension of the methodology established for algebraic and ordinary differential equations. This consists of invoking various conservation laws and applying the scheme: rate in – rate out = rate of change of contents.

  For force balances, use is also made of Newton's law. This method can be applied easily to simple PDEs, but becomes increasingly cumbersome as the geometry and the balances grow in complexity.

- One can use the preceding approach to derive the PDE for a simple geometry and generalize the results by means of vector calculus. This is the approach used in most texts on transport phenomena.
- One can start immediately with a vector/tensor representation of the basic conservation laws and apply these to various geometries of interest. This requires considerable background in the basic laws and manipulations of vector and tensor calculus. Some examples of this approach will be given in Chapter 7.

For the present, we limit ourselves to the simplest methodology and use it to derive some elementary PDEs.

Fourier's equation in one dimension — We apply an energy balance to the differential element shown in Figure 5.5A and obtain:

Rate of energy in – Rate of energy out = Rate of change of energy contents

$$q\Big|_x - q\Big|_{x+\Delta x} = \left(\frac{\partial H}{\partial t}\right)_{avg} \tag{5.23a}$$

A. Fourier's Equation

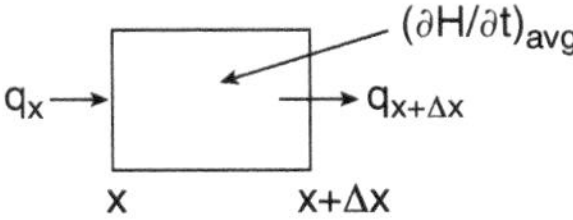

B. Graetz Problem

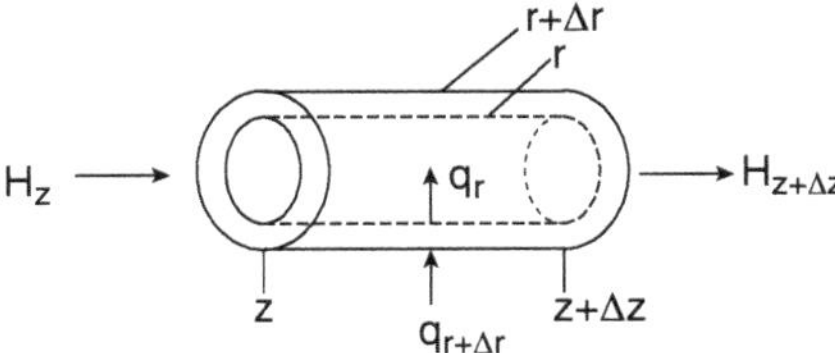

C. Vibrating String

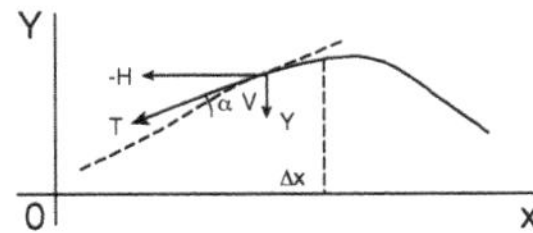

FIGURE 5.5 Difference elements for the derivation of PDEs.

and in expanded form:

$$-kA\left.\frac{\partial T}{\partial x}\right|_x - \left(-kA\left.\frac{\partial T}{\partial x}\right|_{x+\Delta x}\right) = \rho C_p A\Delta x\left(\frac{\partial T}{\partial t}\right)_{avg}$$

where it is to be noted that the time derivative is an average over the difference element.

Dividing by AΔx and letting $\Delta x \to 0$, there results:

$$\frac{\partial}{\partial x}k\frac{\partial T}{\partial x} = \rho Cp\frac{\partial T}{\partial t} \tag{5.23b}$$

where $(\partial T/\partial t)_{avg}$ has now become a point quantity.

For constant thermal conductivity k, this reduces to the form given previously:

$$\alpha\frac{\partial^2 T}{\partial x^2} = \frac{\partial T}{\partial t} \tag{5.23c}$$

where $\alpha = k/\rho Cp$ = thermal diffusivity.

The Graetz problem — In this instance, the energy balance is taken over the annular cylindrical element shown in Figure 5.5B. Energy in convective flow enters and leaves the element axially at z and z + Δz, and conductive flux enters and leaves radially at positions r + Δr and r. The process is at steady state, and we obtain:

Rate of energy in – Rate of energy out = 0

$$\begin{bmatrix} H|_z \\ +q|_{r+\Delta r} \end{bmatrix} - \begin{bmatrix} H|_{z+\Delta z} \\ +q|_{r+\Delta r} \end{bmatrix} = 0 \tag{5.24a}$$

or in expanded form:

$$\begin{bmatrix} v_{avg}(r)\rho Cp\, 2\pi r\Delta r(T-T_o)_z \\ +k\Delta x\left(2\pi r\frac{\partial T}{\partial r}\right)_{r+\Delta r} \end{bmatrix} - \begin{bmatrix} v_{avg}(r)\rho Cp\, 2\pi r\Delta r(T-T_o)_{z+\Delta z} \\ +k\Delta x\left(2\pi r\frac{\partial T}{\partial r}\right)_r \end{bmatrix} \tag{5.24b}$$

Several features are of note here:

- Velocity v varies in general with radial distance r, and an average value $v_{avg}(r)$ has to be specified over the radial increment Δr. That term becomes a point quantity v(r) upon going to the limit $\Delta r \to 0$.
- Fourier's law of heat conduction takes a positive sign because the process takes place in the negative radial direction.
- The radial perimeter 2πr, as well as the temperature gradient ∂T/∂r, have to be differentiated upon going to the limit $\Delta r \to 0$. Note that the perimeter

in the conduction term has been left as $2\pi r$, not $[2\pi(r + \Delta r), 2\pi r]$. The latter formulation, often used by novices, leads to messy or incorrect results. Similarly, the area of the radial increment is given as $2\pi r\,\Delta r$, rather than $\pi(r + \Delta r)^2 - \pi r^2$, to avoid unnecessary complications.

- The axial enthalpy flow F (kg/sec) H (J/kg) becomes, in expanded form, $v\rho AC_p(T - T_0)$.

We now proceed to divide Equation 5.24b by $2\pi r\Delta r\Delta x$ and obtain

$$-\left[\frac{v_{avg}(r)\rho\, Cp(T-T_0)_{z+\Delta z} - v_{avg}(r)\rho\, Cp(T-T_0)_z}{\Delta z}\right] + \left[\frac{1}{r}\,\frac{k\left(r\dfrac{\partial T}{\partial r}\right)_{r+\Delta r} - k\left(r\dfrac{\partial T}{\partial r}\right)_r}{\Delta r}\right] = 0 \tag{5.24c}$$

Letting Δz and Δr go to zero, and taking care to differentiate the entire product $(r\, \partial T/\partial r)$, we obtain after rearrangement:

$$v(r)\frac{\partial T}{\partial z} = \alpha\left(\frac{\partial^2 T}{\partial r^2} + \frac{1}{r}\frac{\partial T}{\partial r}\right) \tag{5.24d}$$

where $\alpha = k/\rho Cp$ = thermal diffusivity, and v(r) has now become a point quantity. Graetz provided an initial treatment to the problem, but it was not until 1956 that a full solution to the variable case v(r) became available.

The vibrating string — This system requires a force balance of which the principal components and the attendant deflection of the string are depicted in Figure 5.5C. V(x,t) is the restoring vertical force that varies with time and distance x, whereas the horizontal component H of the tension T is constant over the length of the string and time invariant. It is further assumed that the angle α remains small, so that the incremental length of string is given by its projection Δx. In other words, we are limiting ourselves to vibrations of small amplitude, such as those experienced by the strings of musical instruments.

With these considerations in place, we obtain the following relation between horizontal and vertical force components H and V(x,t):

$$V(x,t) = -H\tan\alpha = -H\frac{\partial Y}{\partial x} \tag{5.25a}$$

where Y is the vertical displacement of the string. Applying this relation to a difference element Δx of the string and invoking Newton's law to describe its motion, we obtain:

$$\rho\,\Delta x\frac{\partial^2 Y}{\partial t^2} = -H\left(\left.\frac{\partial Y}{\partial x}\right|_x - \left.\frac{\partial Y}{\partial x}\right|_{x+\Delta x}\right) \tag{5.25b}$$

where ρ denotes the, mass per unit length of string, and $\partial^2Y/\partial t^2$ is the acceleration of the element.
Dividing by ρ Δx and letting Δx → 0, we finally obtain:

$$\frac{\partial^2 Y}{\partial x^2} = \frac{1}{c^2}\frac{\partial^2 Y}{\partial t^2} \tag{5.25c}$$

where $c^2 = H/\rho$. Equation 5.25c is seen to be the one-dimensional version of the general wave equation (Equation 5.19a).

5.3 USEFUL SIMPLIFICATIONS AND TRANSFORMATIONS

Because the task of solving PDEs is, in many cases, a considerable undertaking, it is worthwhile and indeed often mandatory to simplify the PDEs prior to embarking on a solution. These simplifications may take various forms, which we summarize in the following.

- Elimination of independent variables or reduction to an ODE
- Elimination of dependent variables or reduction in number of PDEs
- Reduction to homogeneous form
- Change of independent variables and a reduction to canonical form
- Simplification of geometry
- Nondimensionalization

Some of these simplifications had already been practiced at the ODE level. Thus, our favorite trick of dividing simultaneous first-order ODEs had led to the elimination of the independent variable, and sometimes of a dependent variable as well (see Section 2.4.1 and Illustration 4.14). Dependent variables could also, on occasion, be eliminated algebraically to reduce the number of ODEs in a set at the cost of raising the order of the equations (Practice Problem 2.5). Both the D-operator method and the Laplace transform were, in a sense, simplifying devices that reduced the ODEs to algebraic forms. These techniques have their counterparts at the PDE level. Let us now consider these methods in some detail:

5.3.1 Elimination of Independent Variables: Reduction to ODEs

The device of eliminating independent variables or reducing the PDE to lower dimensionality is used in virtually all major analytical solution methods. This includes the method of separation of variables, integral transforms of various types, similarity transformation, and the method of characteristics. We sketch the principal steps of some of these procedures in the following. More detailed coverage is provided in Chapter 8.

Separation of variables — The basic line of attack here is to assume that the solution is made up of the product of functions of the independent variables:

$$u(x, y, z, t) = f(x)g(y)h(z)k(t) \tag{5.26a}$$

This assumption has the effect of reducing the PDE to an equivalent set of ODEs. We demonstrate this for the simple case of the one-dimensional Fourier's equation. Introducing the assumed solution form T = f(x)g(t) into the PDE $\alpha\ \partial^2T/\partial x^2 = \partial T/\partial t$, we obtain the result:

$$\frac{\partial T}{\partial t} = f(x)g'(t), \ \frac{\partial^2 T}{\partial x^2} = f''(x)g(t) \tag{5.26b}$$

Substitution into Fourier's equation and some rearrangement leads to the relation:

$$\frac{f''(x)}{f(x)} = \frac{1}{\alpha}\frac{g'(t)}{g(t)} \tag{5.26c}$$

where we have cleverly grouped functions of the same variable together.
Now functions of two different independent variables cannot be equal unless they are constant. We therefore obtain:

$$\frac{f''(x)}{f(x)} = \frac{1}{\alpha}\frac{g'(t)}{g(t)} = -\lambda^2 \text{ (a constant)} \tag{5.26d}$$

and hence the equivalent set of ODEs:

$$\frac{d^2 f}{dx^2} + \lambda^2 f = 0 \tag{5.26e}$$

$$\frac{dg}{dt} + \alpha\lambda^2 g = 0$$

These equations can be solved independently and subsequently developed into a general solution of the PDE. This will be demonstrated in Chapter 8.

Laplace transform — We retain the example of the one-dimensional Fourier's equation and formally apply the transformation. One obtains:

$$\int_0^\infty \frac{\partial T}{\partial t} e^{-st} dt = \alpha \int_0^\infty \frac{\partial^2 T}{\partial x^2} e^{-st} dt \tag{5.27a}$$

These two integrations provide the mechanism for reducing the PDE to an ODE. The left side yields the customary expression for the Laplace transform of a first-order derivative. Thus,

$$\int_0^\infty \frac{\partial T}{\partial t} e^{-st} dt = s\overline{T}(s) - T(0) \tag{5.27b}$$

where $\overline{T}(s)$ is the transformed temperature, and T(0) the initial condition.

The right side is evaluated by reversing the order of differentiation and integration. This procedure is justified under some mild conditions of continuity and existence. We obtain:

$$\int_0^\infty \frac{\partial^2 T}{\partial x^2} e^{-st} dt = \int_0^\infty \frac{d^2}{dx^2} T e^{-st} dt = \frac{d^2}{dx^2} \int_0^\infty T e^{-st} dt = \frac{d^2 \overline{T}}{dx^2} \tag{5.27c}$$

Thus, the transform of the derivative has become the derivative of the transform. Combining the results of Equation 5.27b and Equation 5.27c, we obtain the following ODE in the transformed variable $\overline{T}$:

$$\alpha \frac{d^2 \overline{T}}{dx^2} = s\overline{T} - T(0) \tag{5.27d}$$

This is a linear second-order nonhomogeneous ODE in $\overline{T}(s)$, which can be solved by standard techniques given in Chapter 2. Subsequent inversion of the result $\overline{T} = f(x,s)$ then yields the solution of the PDE T(x,t). A more detailed exposition of the method will be given in Chapter 8. Other integral transforms that are described in the same chapter act in a similar fashion to reduce PDEs to ODEs.

Similarity or Boltzmann transformation (combination of variables) — In this method, two independent variables of the PDE, for example, x and y or x and t, are combined into a single new independent variable η, termed the *similarity variable*. This is achieved by means of the following transformation:

$$\eta = y/x^n \qquad \text{or} \qquad \eta = x/t^n \tag{5.28a}$$

Although other functional forms exist and can be applied, this form (Equation 5.28a) is by far the most extensively used. The first such transformation, proposed by Boltzmann in 1894 to solve a nonlinear conduction problem, had the form:

$$\eta = x\, t^{-1/2} \tag{5.28b}$$

It has been used in a number of other contexts since.

The reason for choosing the particular form (Equation 5.28a) resides in the requirement that in combining the two independent variables one must, at the same time, bring about a coalescence of two boundary conditions into a single BC. Such coalescence is difficult to achieve in finite geometries, but if one limits oneself to a semi-infinite or infinite medium, the transformation (Equation 5.28a) will give us the desired result. Thus, in the conduction problem considered by Boltzmann, an initial condition at t = 0 would be available, as well as a boundary condition at x → ∞, and the temperatures for both conditions would be the same. If the form (Equation 5.28b) is chosen, these conditions will lead to the same value of the similarity variable, i.e., η = ∞, thus bringing about the desired merger of initial and boundary conditions.

The view we have presented here is a very limited one, designed to give the reader a first introduction to the technique of similarity transformation. The underlying theory is much broader and more profound, and is aimed at deducing general transformations that will lead to a simplification and contraction of PDEs. It is based on concepts of group transformation and invariant groups, and is the subject of specialized monographs.

In the course of performing a similarity transformation, one is required to express partial derivatives in the old independent variables x and y (x and t) in terms of the new similarity variable η. This is done by what is known as the *chain rule of partial differentiation*, which is described in standard calculus texts. The derivation of the pertinent formulas is somewhat cumbersome, and we have therefore summarized the more important results for convenience in the accompanying Table 5.4. These can be used directly to transform "old" derivatives into "new" ones.

TABLE 5.4
Transformation of Partial Derivatives

Change $u = f(x, y)$ to $u = g(X,Y)$

A. First Derivatives

$$\frac{\partial u}{\partial x} = \frac{\partial u}{\partial X}\frac{\partial X}{\partial x} + \frac{\partial u}{\partial Y}\frac{\partial Y}{\partial x}$$

$$\frac{\partial u}{\partial y} = \frac{\partial u}{\partial X}\frac{\partial X}{\partial y} + \frac{\partial u}{\partial Y}\frac{\partial Y}{\partial y}$$

$$\frac{\partial u}{\partial z} = \frac{\partial u}{\partial X}\frac{\partial X}{\partial z} + \frac{\partial u}{\partial Y}\frac{\partial Y}{\partial z}$$

or, in general,

$$\begin{matrix} \textit{Old derivative of u with} \\ \textit{respect to old independent} \\ \textit{variable i} \end{matrix} = \sum_j \left\{ \begin{matrix} \textit{Partial derivative of u} \\ \textit{with respect to new} \\ \textit{independent variable j} \end{matrix} \quad x \ \frac{\partial j}{\partial i} \right\}$$

B. Second Derivatives

$$\frac{\partial^2 u}{\partial x^2} = \frac{\partial u}{\partial X}\frac{\partial^2 X}{\partial x^2} + \frac{\partial^2 u}{\partial X^2}\left(\frac{\partial X}{\partial x}\right)^2 + 2\frac{\partial^2 u}{\partial X \partial Y}\frac{\partial X}{\partial x}\frac{\partial Y}{\partial x} + \frac{\partial u}{\partial Y}\frac{\partial^2 Y}{\partial x^2} + \frac{\partial^2 u}{\partial Y^2}\left(\frac{\partial Y}{\partial x}\right)^2$$

$$\frac{\partial^2 u}{\partial y^2} = \frac{\partial u}{\partial X}\frac{\partial^2 X}{\partial y^2} + \frac{\partial^2 u}{\partial X^2}\left(\frac{\partial X}{\partial y}\right)^2 + 2\frac{\partial^2 u}{\partial X \partial Y}\frac{\partial X}{\partial y}\frac{\partial Y}{\partial y} + \frac{\partial u}{\partial Y}\frac{\partial^2 Y}{\partial y^2} + \frac{\partial^2 u}{\partial Y^2}\left(\frac{\partial Y}{\partial y}\right)^2$$

$$\frac{\partial^2 u}{\partial x \partial y} = \frac{\partial u}{\partial X}\frac{\partial^2 X}{\partial x \partial y} + \frac{\partial^2 u}{\partial X^2}\frac{\partial X}{\partial x}\frac{\partial X}{\partial y} + \frac{\partial^2 u}{\partial X \partial Y}\frac{\partial Y}{\partial x}\frac{\partial X}{\partial x} + \frac{\partial u}{\partial Y}\frac{\partial^2 Y}{\partial x \partial y}$$

$$+ \frac{\partial^2 u}{\partial X \partial Y}\frac{\partial Y}{\partial y}\frac{\partial X}{\partial x} + \frac{\partial^2 u}{\partial Y^2}\frac{\partial Y}{\partial x}\frac{\partial Y}{\partial y}$$

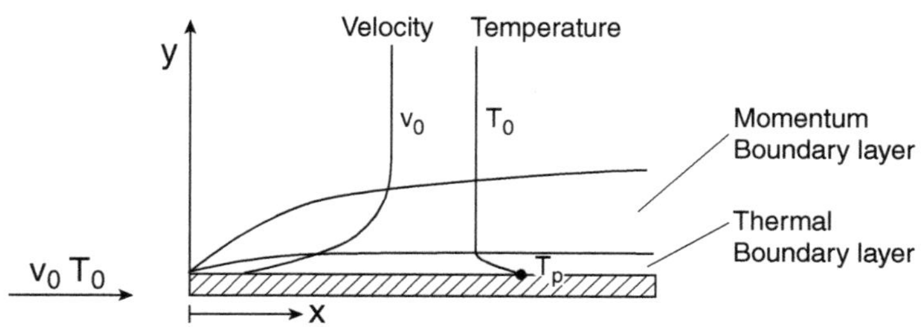

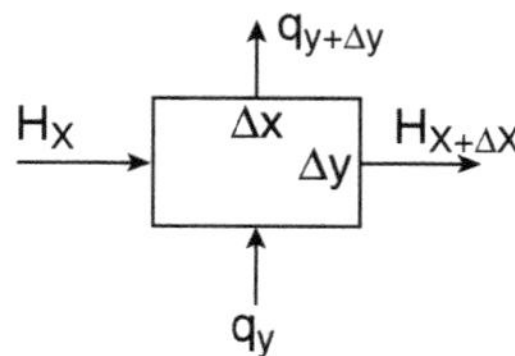

FIGURE 5.6 Heat transfer in boundary layer flow around a flat plate.

ILLUSTRATION 5.4 HEAT TRANSFER IN BOUNDARY-LAYER FLOW OVER A FLAT PLATE: SIMILARITY TRANSFORMATION

To illustrate the use of the similarity transformation and the attendant simplifications that result, we consider the situation depicted in Figure 5.6. A flat plate maintained at a constant temperature T_p is exposed to a flow of colder air with an approach temperature of T_0. Heat transfer takes place through the so-called thermal boundary layer and into the semi-infinite medium of flowing air. That thin layer adjacent to the surface of the exposed plate contains the temperature variations that result from the heat transfer process, and which range from T_p at the plate to T_0 at the outer edge of the layer. The thermal boundary layer has a counterpart, the so-called momentum boundary layer, which expresses momentum transfer from the air to the stationary plate and contains the velocity variations in the flowing air. They range from a value of $v = 0$ at the surface of the plate, to the free-stream velocity $v = v_0$ at the edge of the layer and beyond.

Theoretical analysis of these two boundary layers has revealed that the thermal layer is thinner than the momentum boundary layer, and lies within the linear portion of the velocity profile. This brings about considerable simplifications of the energy balance required to derive the temperature profile in the boundary layer. The balance is taken over the element shown in Figure 5.6B, leading in the first instance to the following expression:

$$\text{Rate of energy in} - \text{Rate of energy out} = 0$$

$$[H_x + q_y] - [H_{x+\Delta x} + q_{y+\Delta y}] = 0 \tag{5.29a}$$

In expanded form, we obtain:

$$\begin{bmatrix} v\rho Cp W\Delta y(T - T_o) \\ -kW\Delta x \left.\dfrac{\partial T}{\partial y}\right|_y \end{bmatrix} - \begin{bmatrix} v\rho Cp W\Delta y(T - T_o)_{x+\Delta x} \\ -kW\Delta x \left.\dfrac{\partial T}{\partial y}\right|_{y+\Delta y} \end{bmatrix} = 0 \tag{5.29b}$$

where W represents the width of the plate and $v = \dot{\gamma}y$. $\dot{\gamma}$ is the shear rate of the air at the plate and equals the slope $\partial v/\partial y$ of the linear portion of the velocity profile.

Upon dividing by $W\Delta x\Delta y$ and letting Δx, $\Delta y \to 0$, there results:

$$-\dot{\gamma}y\rho Cp\frac{\partial T}{\partial x} + k\frac{\partial^2 T}{\partial y}^2 = 0 \tag{5.29c}$$

We now introduce the dimensionless temperature $\theta = (T_p - T)/(T_p - T_0)$ and obtain, after slight rearrangement, the compact form:

$$\frac{\partial\theta}{\partial x} = \frac{A}{y}\frac{\partial^2\theta}{\partial y^2} \tag{5.29d}$$

with $A = \alpha/\dot{\gamma}$ and $\alpha = k/\rho Cp$ = thermal diffusivity.

The associated boundary conditions are as follows:

At the plate edge: $\theta(0, y) = 1$ (5.29e)

At the plate surface: $\theta(x, 0) = 0$ (5.29f)

In the free stream: $\theta(x, \infty) = 1$ (5.29g)

This represents the system to be solved.

The conditions (5.29e) and (5.29g), given as they are at values of x = 0 and $y = \infty$, indicate that a similarity transformation may be possible. We therefore specify a similarity variable of the form:

$$\eta = y/x^n \tag{5.29h}$$

and proceed to evaluate n. That parameter has to be such that the transformed PDE becomes an ODE in η, only and is free of any terms containing the old independent variables x and y. We use Table 5.4 to transform old derivatives in x and y to new derivatives with respect to η. This yields:

$$\frac{\partial\theta}{\partial x} = \frac{\partial\theta}{\partial\eta}\frac{\partial\eta}{\partial x} = -n\frac{\partial\theta}{\partial\eta}yx^{-(n+1)} = -n\frac{\partial\theta}{\partial\eta}\eta x^{-1} \tag{5.29i}$$

$$\frac{\partial^2\theta}{\partial y^2} = \frac{\partial\theta}{\partial\eta}\frac{\partial^2\eta}{\partial y^2} + \frac{\partial^2\theta}{\partial\eta^2}\left(\frac{\partial\eta}{\partial y}\right)^2 = \frac{\partial^2\theta}{\partial\eta^2}x^{-2n}$$

Substitution of these expressions into the PDE (Equation 5.29a) gives the result:

$$-n\frac{\partial\theta}{\partial\eta}\eta x^{-1} = \frac{Ax^{-2n}}{\eta x^n}\frac{\partial^2\theta}{\partial\eta^2} \tag{5.29j}$$

One notes that the terms in the old independent variable x will cancel if we impose the condition n = 1/3. The resulting ODE is then given by:

$$\frac{d^2\theta}{d^2} + \frac{\eta^2}{3A}\frac{d\theta}{d\eta} = 0 \tag{5.29k}$$

with the similarity variable η given by:

$$\eta = y/x^{1/3} \tag{5.29l}$$

The boundary conditions are now only two in number, with the first one resulting from the merger of Equation 5.29e and Equation 5.29g and having the value:

$$\theta = 1 \quad \text{at} \quad \eta = \infty \tag{5.29m}$$

Added to this is the old BC equation (Equation 5.29f), which now becomes:

$$\theta = 0 \quad \text{at} \quad \eta = 0 \tag{5.29n}$$

Equation 5.29k is seen to be a second-order linear ODE, with variable coefficients and devoid of terms in θ other than the derivatives. This suggests a p-substitution (item 9 of Table 2.4) and results in the reduced ODE:

$$\frac{d\theta}{d\eta} + \frac{\eta^2}{3A}\theta = 0 \tag{5.29o}$$

This equation is immediately integrable by separation of variables to yield:

$$p = \frac{d\theta}{d\eta} = C_1 \exp(-\eta^3/9A) \tag{5.29p}$$

A second integration, also by separation of variables, leads to the result:

$$\theta = C_1 \int_0^\eta \exp(-z^3/9A)dz + C_2 \tag{5.29q}$$

where z is the integration variable.

From the boundary conditions (Equation 5.29n) we obtain:

$$C_2 = 0 \tag{5.29r}$$

and
$$C_1 = \left(\int_0^\infty \exp(-z^3/9A)dz \right)^{-1}$$

giving as the final result:

$$\theta(\eta) = \frac{\int_0^\eta \exp(-z^3/9A)\,dz}{\int_0^\infty \exp(-z^3/9A)\,dz} \tag{5.29s}$$

The integrals in this expression are both convergent. The result can be expressed in terms of the old coordinates x and y by making use of Equation 5.29. An identical solution arises in the so-called Lévêque solution of the Graetz problem, which is taken up in Chapter 7.

Comments: The temperature profiles in the boundary layer change in the direction of flow (x), but only in scale, not in shape. They are for this reason termed *self-similar,* and the associated solution method is, as we have seen, referred to as a *similarity transformation.* Boundary layer profiles of velocity and concentration are likewise self-similar, as are certain profiles that arise in unsteady conduction and diffusion. One consequence of this behavior is that the wall shear rate $\dot{\gamma} = (\partial v_x/\partial y)_{y=0}$, which is needed in the result of Equation 5.29s, is a constant, which can be extracted from the drag coefficient C_D of the relevant drag force. Thus,

$$F_D = C_D A\rho \frac{v_0^2}{2} \tag{5.29t}$$

and
$$\tau_w = \mu\dot{\gamma} = F_D/A = C_D\rho \frac{v_0^2}{2} \tag{5.29u}$$

where τ_w represents the shear stress at the plate.

Hence,

$$\dot{\gamma} = C_D(\rho/\mu)\frac{v_0^{\ 2}}{2} \tag{5.29v}$$

Values of C_D can be found in tables of drug coefficients.

ILLUSTRATION 5.5 BOLTZMANN TRANSFORMATION WITH MATHEMATICA

Using the Boltzmann transformation ($\eta = \frac{y}{x^n}$), we solve the following PDE:

$$\frac{\partial^2\theta}{\partial y^2} + \frac{1}{2y}\left(\frac{\partial\theta}{\partial x}\right)^2 = 0 \tag{5.30a}$$

with the following boundary conditions:

$$\theta(x,0) = 0 \tag{5.30b}$$

$$\theta(\infty, y) = 0 \tag{5.30c}$$

$$\theta(x,\infty) = 1 \tag{5.30d}$$

Let us first define the PDE in Mathematica:

$$\text{In[1]:=}\ \mathbf{eqn} = \partial_{y,y}\theta[x, y] + \frac{1}{2y}(\partial_x\ \theta[x, y])^2$$

$$\text{Out[1]=}\ \theta^{(0,2)}[x, y] + \frac{\theta^{(1,0)}[x, y]^2}{2\,y}$$

Next, use the chain rule to rewrite this equation in terms of η :

$$\text{In [3]:=}\ \mathbf{Dx\theta} = \mathbf{D[\theta[\eta[x, y]], x]\ /.\ \frac{y}{x^n} \to \eta}$$

$$\text{Out [3]} = -(n\ x^{-1-n} y\ \theta'[\eta])$$

$$\text{In [4]:=}\ \mathbf{Dyy\theta} = \mathbf{D[\theta[\eta[x, y]], \{y, 2\}]\ /.\ \frac{y}{x^h} \to \eta}$$

$$\text{Out [4]} = \frac{\theta''[\eta]}{x^{2n}}$$

$$\text{In [5]:=}\ \mathbf{eqn} = \mathbf{eqn}\ /.\{\partial_{y,y}\theta[x, y] \to \mathbf{Dyy\theta},\ \partial_x\ \theta[x, y] \to \mathbf{Dx\theta}\}$$

$$\text{Out [5]} = \frac{n^2\ x^{-2-2n}\ y\ \theta'[\eta]^2}{2} + \frac{\theta''[\eta]}{x^{2n}}$$

We now multiply this equation by x^{2n} :

```
In[6]:= eqn = Expand[x^(2 n) eqn]
```

$$\text{Out[6]} = \frac{n^2\, y\, \theta'[\eta]^2}{2\, x^2} + \theta''[\eta]$$

To transform the remaining x and y terms into η, n has to be equal to 2:

```
In[7]:= eqn = (%/. n → 2)/. y/x^2 → η
```

$$\text{Out[7]} = 2\,\eta\,\theta'[\eta]^2 + \theta''[\eta]$$

Now we solve this ODE with new boundary conditions $\theta(0) = 0$ and $\theta(\infty) = 1$:

```
In[8]:= sol = θ[η]/. First
[DSolve[{eqn == 0, θ[0] == 0, θ[∞] == 1}, θ[η], η]
        ]
```

$$\text{Out[8]} = \frac{2\,\text{ArcTan}\left[\frac{2\,\eta}{\text{Pi}}\right]}{\text{Pi}}$$

Let us now plot the result:

```
In[9]:= Plot[sol, {η, 0, 10}, Frame → True,
        FrameLabel → {"y/x2", "θ(x,y)", " ", " "}]
```

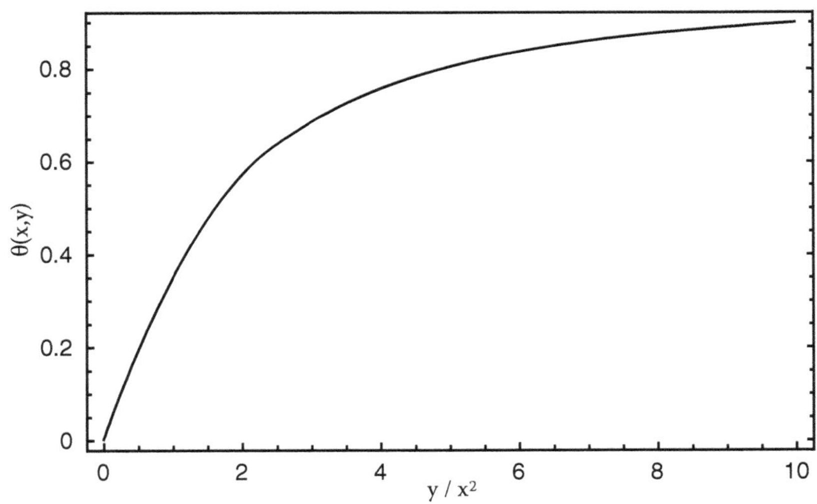

Out[9]=-Graphics-

5.3.2 Elimination of Dependent Variables: Reduction of Number of Equations

Dependent variables can, on occasion, be reduced in number, particularly when they represent the components of a vector quantity. The classical example is that of the stream function ψ, which serves to combine the velocity components v_x and v_y of two-dimensional fluid flow into a single entity via the previously given relation

$$v_x = -\frac{\partial\psi}{\partial y} \qquad v_y = \frac{\partial\psi}{\partial x} \tag{5.31}$$

Evidently, some skill and imagination are required to formulate a new dependent variable that will bring about the desired reduction in both the number of equations and dependent variables. We illustrate an example of this in the following.

Illustration 5.6 Use of the Stream Function in Boundary Layer Theory: Velocity Profiles Along a Flat Plate

External boundary layer flow in two dimensions was previously described by the Prandtl boundary layer equations (Equation 5.21a and Equation 5.21b). For flow over a flat plate, the pressure term dp/dx vanishes and we obtain:

Momentum balance:

$$-\left[v_x\frac{\partial v_x}{\partial x} + v_y\frac{\partial v_x}{\partial y}\right] + \nu\frac{\partial^2 v_x}{\partial x^2} = 0 \tag{5.32a}$$

Continuity equation:

$$\frac{\partial v_x}{\partial x} + \frac{\partial v_y}{\partial y} = 0 \tag{5.32b}$$

where $\nu = \mu/\rho$ = kinematic viscosity.

Introduction of the stream function into the momentum balance leads to the relation

$$-\left[\frac{\partial\psi}{\partial y}\frac{\partial^2\psi}{\partial x\partial y} - \frac{\partial\psi}{\partial x}\frac{\partial^2\psi}{\partial y^2}\right] + \nu\frac{\partial^3\psi}{\partial y^3} = 0 \tag{5.32c}$$

A similar substitution into the continuity equation brings about the following interesting result:

$$\frac{\partial^2\psi}{\partial x\partial y} - \frac{\partial^2\psi}{\partial y\partial x} \equiv 0 \tag{5.32d}$$

That is, the difference in the derivatives is identically zero. The continuity equation is eliminated, and the desired reduction in the number of equations is thus accomplished. The price to be paid is that the new Equation 5.32c is one order higher than the previous momentum balance. Such equations, however, are easier to handle than the previous coupled pair of PDEs.

We note that Equation 5.32c is homogeneous and quasilinear because the highest derivative appears in the linear form. We further note that the geometry and boundary conditions are similar to those we saw in Illustration 5.4. Similarity transformation therefore suggests itself as a possible solution route, and this does, in fact, turn out to be the case. It leads to a third-order ODE, known as the Blasius equation, which is of the form:

$$f'''(\eta) + \frac{1}{2} f(\eta) f''(\eta) = 0 \tag{5.32e}$$

where f is a modified nondimensionalized stream function.

The original system of two PDEs, Equation 5.32a and Equation 5.32b, one of which was quasilinear, has thus been reduced to a single nonlinear ODE of boundary value type. This represents a considerable simplification, even though the order has increased by one.

5.3.3 Elimination of Nonhomogeneous Terms

Nonhomogeneous terms can complicate the solution of differential equations, and in some instances render a solution method inapplicable. At the ODE level, the presence of such terms required the evaluation of a particular integral for use in the D-operator method. The Laplace transform, on the other hand, did not encounter such difficulties, and was capable of handling both homogeneous and nonhomogeneous ODEs with equal ease, although the solution became more complex in form. At the PDE level, integral transforms are again well-suited for handling nonhomogeneous equations, but the powerful method of separation of variables, previously sketched in Section 5.3.1.a, requires both the PDE and the boundary conditions to be homogeneous for its successful application. The presence of nonhomogeneous terms in the PDE usually renders the assumption of separable variables, expressed by Equation 5.26a, invalid, whereas nonhomogeneous boundary conditions complicate the evaluation of undefined constants that arise in the course of the solution. Thus, there is considerable incentive to eliminate nonhomogeneous terms. Another prominent example of the benefits of this procedure is the reduction of the nonhomogeneous Poisson equation to Laplace's equation. The latter has been studied more extensively and has a host of known solutions.

A promising line of attack when dealing with such equations is to introduce a new dependent variable v of the form:

$$v = u + g(x, y, z, t) \tag{5.33}$$

where u is the original dependent variable, and g(x,y,z,t) is of the same functional form as the nonhomogeneous term in the original PDE or the boundary condition.

This function g should contain a sufficient number of unknown constants to satisfy the condition of homogeneity for both the PDE and the BCs. There is no requirement for the initial condition to be homogeneous, which can therefore serve as a convenient repository for functional forms that take the place of the eliminated nonhomogeneous terms.

The method suggested here is particularly well suited for nonhomogeneous terms of polynomial form. We illustrate its use in the example that follows:

Illustration 5.7 Conversion of a PDE to Homogeneous Form

We consider the system:

$$\frac{\partial^2 u}{\partial t^2} = c^2 \frac{\partial^2 u}{\partial x^2} + Ax \tag{5.34a}$$

with BCs and ICs

$$u(0,t) = 0 \tag{5.34b}$$

$$u(1,t) = 0 \tag{5.34c}$$

$$\frac{\partial u}{\partial t}(x,0) = 0 \tag{5.34d}$$

$$u(x,0) = 0 \tag{5.34e}$$

where x has been normalized to the interval (0,1).

Physically, this set of equations can be seen as representing a string that is clamped at both ends and is initially horizontal, with zero deflection. A force that is linear in the horizontal distance x is applied to it and results in time-dependent deflections u(x,t). The task is to eliminate the nonhomogeneous term Ax, while maintaining the boundary conditions (Equation 5.34b to Equation 5.34e) in homogeneous form. Because four homogeneity conditions are to be satisfied, a third-order polynomial with four arbitrary constants suggests itself as a trial function g(x). Therefore, we specify:

$$v = u + a' + b'x + c'x^2 + d'x^3 \tag{5.34f}$$

Trials with this function, which we omit for brevity, indicate that the quadratic term has to be dropped for the constants to be independent of x. We consequently make the new specification:

$$v = u + a + bx + dx^3 \tag{5.34g}$$

and obtain the following results:

For the PDE:

$$\frac{\partial^2 v}{\partial t^2} = c^2 \frac{\partial^2 v}{\partial x^2} - 6dx + Ax \rightarrow d = \frac{A}{6} \tag{5.34h}$$

For BC (Equation 5.34b):

$$u(0, t) = v(0, t) - a = 0 \rightarrow a = 0 \tag{5.34i}$$

For BC (Equation 5.34c):

$$u(1, t) = v(1, t) - (a + b + d) = 0 \rightarrow b = -\frac{A}{6} \tag{5.34j}$$

For BC (Equation 5.34d):

$$\frac{\partial u}{\partial t}(x,0) = \frac{\partial v}{\partial t}(x,0) = 0 \tag{5.34k}$$

i.e., the boundary condition remains homogeneous without the appearance of any of the coefficients. Three such coefficients will therefore suffice to render the system homogeneous, with the exception of the IC (Equation 5.34e). The final transformed model is then as follows:

$$\frac{\partial^2 v}{\partial t^2} = c^2 \frac{\partial^2 v}{\partial x^2} \tag{5.34l}$$

with BCs and ICs

$$v(0, t) = 0 \tag{5.34m}$$

$$v(1, t) = 0 \tag{5.34n}$$

$$\frac{\partial v}{\partial t}(x,0) = 0 \tag{5.34o}$$

$$v(x, 0) = -\frac{A}{6}(x - x^3) \tag{5.34p}$$

and the relation between the old and new dependent variables given by:

$$v = u - \frac{A}{6}(x - x^3) \tag{5.34q}$$

Comments: The nonhomogeneous term in Equation 5.34p is more complex than the term that was eliminated, Ax. This is nevertheless no disadvantage because a high degree of nonhomogeneity can be tolerated in the initial condition without seriously affecting the ease of solution.

Application of the method was not totally automatic. Some trial-and-error was required, which is not unusual for the successful application of transformations involving unknown functions.

5.3.4 Change in Independent Variables: Reduction to Canonical Form

A change in independent variables can occasionally result in a simplification of the PDE through a reduction in the number of terms that appear in it, or by bringing it into a form more suitable for analysis. A systematic way of achieving possible simplifications is by the procedure known as reduction to canonical, or "regular" form. The original purpose of this method was to cast PDEs into the same form as the classical Laplace, Fourier, and wave equations, and to extend the results obtained from the study of these relatively simple equations to more complex forms. In the process of this change in variables, one encounters the so-called characteristic equations that also find a place in the method of characteristics, a solution technique to be discussed in greater detail in Chapter 8.

Justification for the steps used in the reduction of PDEs to canonical form can be found in various sources. For our present purposes, we limit ourselves to an outline of these steps, without providing either justification or proof, and address only two classes of equations: (1) single second-order PDEs in two independent variables, and (2) sets of two quasilinear first-order PDEs.

The general form considered here is

$$A(x,y)u_{xx} + 2B(x,y)u_{xy} + C(x,y)u_{yy} = F(x,y,u_x,u_y) \tag{5.35a}$$

We note that the equation is identical to that used to arrive at the categories of elliptic, parabolic, and hyperbolic PDEs. There is, as we shall see, a direct link between these categories and the number of characteristics that arise in the course of reduction to canonical form.

The steps in the application of the procedure are as follows:

Step one — Formulate an ODE in the independent variables x and y, which has a quadratic form with coefficients identical to those in the PDE (Equation 5.35a). Thus, we form:

$$A\left(\frac{dy}{dx}\right)^2 + 2B\left(\frac{dy}{dx}\right) + C = 0 \tag{5.35b}$$

Step two — Solve the quadratic equation to obtain the two roots in dy/dx:

$$\left(\frac{dy}{dx}\right)_{1,2} = \frac{-B \pm (B^2 - AC)^{1/2}}{A} \tag{5.35c}$$

These two ODEs are the so-called *characteristic equations* or *characteristic directions*.

Step three — Integrate the two equations (Equation 5.35c) to obtain:

$$f_1(x,y) = C_1 \quad \text{and} \quad f_2(x,y) = C_2 \tag{5.35d}$$

where C_1 and C_2 are the integration constants. These two algebraic equations form two families of curves that are referred to as characteristics of the PDE (Equation 5.35a). Physically, they may be regarded as propagation pathways in (y,x) or (x,t) space of a particular physical entity, such as the deflection u of a certain magnitude in a vibrating string, or of a certain wave height v in wave propagation problems. The reader will note that for hyperbolic PDEs with $(B^2 - AC) > 0$, there are two real and distinct characteristics equations and characteristics; for parabolic PDEs ($B^2 = AC$), there is only one of each; and for elliptic PDEs ($B^2 - AC < 0$), none. We summarize these features for convenience in Table 5.5.

Step four — Set the new independent variables $\bar{x}$ *and* $\bar{y}$ equal to the characteristic function $f_{1,2}$. Thus,

$$\bar{x} = f_1(x,y) \qquad \bar{y} = f_2(x,y) \tag{5.35e}$$

When the roots of the quadratic are real and equal, there is only one characteristic function and, hence, only one new independent variable. The second one is then established by making it as close as possible in form to, but independent from, the first variable. An example of this procedure appears in Illustration 5.8.

Difficulties also arise when the roots are complex conjugates. A second change in variables is then performed, which takes the form

$$\alpha = \frac{1}{2}(\bar{x} + \bar{y}) \quad \text{and} \quad \beta = \frac{1}{2}i(\bar{x} - \bar{y}) \tag{5.35f}$$

This procedure serves to eliminate the imaginary part of $\bar{x}$ *and* $\bar{y}$, and is taken up in Practice Problem 5.7.

TABLE 5.5
Characteristics of Second-Order PDEs

Type of PDE	$B^2 - AC$	Roots of Equation 5.35b	Number of Characteristics
Elliptic	< 0	Complex conjugate	0
Parabolic	$= 0$	Real and equal	1
Hyperbolic	> 0	Real and distinct	2

Step five — Reformulate the old PDE in terms of the new independent variables $\bar{x}$ *and* $\bar{y}$, or α and β, so that u(x,y) now becomes $v(\bar{x},\bar{y})$ or $v(\alpha, \beta)$. This completes the reduction to canonical form.

Step six — As a final step, the new PDE is solved and the old independent variables x,y reintroduced into the solution to obtain the solution in the form u = F(x, y).

The second reduction to canonical form we consider is that of a set of two first-order quasilinear PDEs. We have encountered this system in a general form as Equation 5.6c, which is repeated here for convenience:

$$\underset{\sim}{A}(\underset{\sim}{u},x,y)\underset{\sim}{u}_x + \underset{\sim}{B}(\underset{\sim}{u},x,y)\underset{\sim}{u}_y + \underset{\sim}{C}(\underset{\sim}{u},x,y) = 0 \tag{5.35g}$$

where u_x and u_y were first-order partial derivatives. A characteristic equation, $\det|\underset{\sim}{A} - \lambda\underset{\sim}{B}| = 0$, was then introduced whose eigenvalues λ determined whether the set was elliptic, parabolic, or hyperbolic. It turns out that the same eigenvalues also represent solutions of a quadratic equation in dy/dx that is completely analogous to Equation 5.35b used for second-order PDEs. We therefore can obtain the characteristic equation by writing:

$$\left(\frac{dy}{dx}\right)_{1,2} = \lambda_{1,2} \tag{5.35h}$$

and these equations can in turn be integrated to obtain the characteristics

$$F_1(x, y) = C_1 \qquad F_2(x,y) = C_2 \tag{5.35i}$$

The new independent variables $\bar{x}$ *and* $\bar{y}$ are then established, again in completely analogous fashion to the case of second-order PDEs, by writing:

$$\bar{x} = F_1(x,y) \quad and \quad \bar{y} = F_2(x,y) \tag{5.35j}$$

We note that the characteristic Equation 5.35h can only be integrated if the coefficients of the PDE, $\underset{\sim}{A}$ *and* $\underset{\sim}{B}$, and hence $\lambda_{1,2}$, are independent of $\underset{\sim}{u}$. When this is not the case, other techniques have to be applied.

Illustration 5.8 Reduction of PDEs to Canonical Form

As a first example, we consider the reduction of the following second-order PDE to canonical form:

$$\frac{\partial^2 u}{\partial x^2} + 2\frac{\partial^2 u}{\partial x \partial y} + \frac{\partial^2 u}{\partial y^2} = 0 \tag{5.36a}$$

Step one — We form the quadratic equation in the characteristic direction dy/dx and obtain:

$$\frac{d^2y}{dx^2} + 2\left(\frac{dy}{dx}\right) + 1 = 0 \tag{5.36b}$$

Step two — Equation 5.36a is solved to yield the characteristic directions:

$$\left(\frac{dy}{dx}\right)_{1,2} = -1 \tag{5.36c}$$

These roots are real and identical, and the PDE is therefore parabolic.

Step three — The Equation 5.36a is next integrated, resulting in a single characteristic:

$$f_1(x, y) = y + x = C_1 \tag{5.36d}$$

Step four — The new independent variable $\overline{x}$ is set equal to $f_1(x,y)$. Thus,

$$\overline{x} = y + x \tag{5.36e}$$

A second new variable, $\overline{y}$, distinct and independent from $\overline{x}$ and close to it in form, is established by writing:

$$\overline{y} = y - x \tag{5.36f}$$

Step five — The old PDE is reformulated in terms of the new independent variables $\overline{x}$ *and* $\overline{y}$. To aid in this procedure, use is made of the conversion formulas given in Table 5.4. This is fairly routine work, which is omitted here for brevity. The final result is given by:

$$\frac{\partial^2 v}{\partial \overline{x}^2} = 0 \tag{5.36g}$$

Step six — Equation 5.36g can be immediately integrated to yield the result:

$$v = \overline{x}\, f_1(\overline{y}) + f_2(\overline{y}) \tag{5.36h}$$

Note that because $v = F(\overline{x}, \overline{y})$, the integration "constants" f_1 and f_2 must in fact be functions of the second new independent variable $\overline{y}$. Reintroduction of the old variables then leads to the solution:

$$u = (x + y)f_1(y - x) + f_2(y - x) \tag{5.36i}$$

This expression, referred to as the *fundamental solution of the PDE*, is, of course, not a final result, because $f_{1,2}$ still need to be evaluated from the boundary conditions, which may not be a simple task. Knowing the functional form of the solution, however, is in itself a very useful result that provides a starting point for analyzing the physical process underlying the original PDEs. Note that Equation 5.37i was arrived at by a simple transformation of independent variables, carried out in a rather mechanical fashion. This is no reason to disdain it.

Let us now use Mathematica to tackle the previous problem. We apply the DSolve function on Equation 5.36a:

```
In[1]:= u[x, y] = u[x, y] /. First
          [DSolve [∂x,x u[x, y] + 2 ∂x,y u[x, y] + ∂y,y u[x, y] ==
               0, u[x, y], {x, y}, GeneratedParameters → f]
     ]

Out[1] = f[1] [-x + y] + x f[2] [-x + y]
```

Let us check if it satisfies the PDE:

```
In[2]:= Simplify [
          D[ u[x, y], {x, 2}] + 2D[D[u[x, y], x], y] +
          D[ u[x, y], {y, 2}]
                  ]

Out[2] = 0
```

This result is different from what we found earlier, but it is an acceptable answer.

As a second example, we consider the reduction to canonical form of the chromatographic equations consisting of fluid- and solid-phase mass balances (Equation 5.12e and Equation 5.12f) and the equilibrium relation (Equation 5.12g). For our present purposes we assume the latter to be linear, so that:

$$Y^* = mq \tag{5.37a}$$

We introduce this relation into the two mass balances and rewrite them to conform to the general formulation of (5.35g). There results

$$\frac{G_s}{\rho_g}\frac{\partial Y}{\partial z} + 0 + \frac{\partial Y}{\partial t} + 0 + (K_{0Y}a/\varepsilon\rho_g)(Y - Y^*) = \tag{5.37b}$$

$$0 + 0 + 0 + \frac{\partial Y^*}{\partial t} - \frac{mK_{0Y}a}{\rho_b}(Y - Y^*) = 0 \tag{5.37c}$$

These are two linear PDEs in the two dependent variables Y and Y*, which we now proceed to reduce to canonical form using the procedure outlined in the previous section.

Step one — We start by establishing the coefficient matrices $\underset{\sim}{A}$ *and* $\underset{\sim}{B}$, which appear in the general formulation of the PDEs given in Equation 5.6c. They are relatively simple, as the coefficients here are constant, and take the form:

$$\underset{\sim}{A} = \begin{pmatrix} G_s/\rho_g & 0 \\ 0 & 0 \end{pmatrix} \tag{5.37d}$$

$$\underset{\sim}{B} = \begin{pmatrix} 1 & 0 \\ 0 & 1 \end{pmatrix}$$

so that

$$\det \left| \underset{\sim}{A} - \lambda \underset{\sim}{B} \right| = \begin{vmatrix} G_s/\rho_g - \lambda & 0 \\ 0 & 0 - \lambda \end{vmatrix} = 0 \tag{5.37e}$$

Solution of the quadratic in λ resulting from this expression leads to the two distinct and real roots:

$$\lambda_1 = 0 \qquad \lambda_2 = G_s/\rho_g \tag{5.37f}$$

The set of PDEs is therefore hyperbolic.

Step two — We now establish the characteristic equations by setting the characteristic directions dy/dx equal to the eigenvalues λ. Thus,

$$\left(\frac{dy}{dx}\right)_1 = 0 \qquad \left(\frac{dy}{dx}\right)_2 = G_s/\rho_g \tag{5.37g}$$

Step three — This step involves the integration of the ODEs (Equation 5.37g) and yields the following characteristics:

$$f_1(x,t) = z = C_1 \tag{5.37h}$$

or

$$f_1(x,t) = t = C_1$$

and

$$f_2(x,t) = z - \frac{G_s}{\rho_g} = C_2 \tag{5.37i}$$

or

$$f_2(x,t) = t - \frac{\rho_g}{G_s} z = C_2$$

Step four — We can now proceed to the formulation of the new independent variables $\bar{x}$ *and* $\bar{y}$, which are equal to the characteristic function $f_{1,2}$. Thus,

$$\bar{x} = z \qquad or \qquad \bar{x} = t \tag{5.37j}$$

and

$$\bar{y} = z - \frac{G_s}{\rho_B} t \qquad or \qquad \bar{y} = t - \frac{\rho_g}{G_s} z \tag{5.37k}$$

Step five — The new independent variables are introduced into the PDE, and the new derivatives evaluated by means of the formula of Table 5.4. We choose the combination:

$$\bar{x} = z \qquad \bar{y} = t - \frac{\rho_g}{G_s} z$$

and obtain:

$$G_s \frac{\partial Y}{\partial \bar{x}} + K_{0Y} a(Y - Y^*) = 0 \tag{5.37l}$$

and

$$\frac{\partial Y}{\partial \bar{y}} - \frac{K_{0Y}}{\rho_b} a(Y - Y^*) = 0$$

Step six — The constant coefficients of this set can be taken into the independent variables, leading to the following compact form:

$$\frac{\partial Y}{\partial \alpha} + (Y - Y^*) = 0 \tag{5.37m}$$

$$\frac{\partial Y^*}{\partial \beta} - (Y - Y^*) = 0$$

where

$$\alpha = \frac{K_{0Y} a}{G_s} z$$

$$\beta = \frac{m\, K_{0Y} a}{\rho_b} \left(t - \frac{\rho_g}{G_s} z \right)$$

In most practical applications, the second term in β can be neglected. We shall encounter these parameters again in Illustration 7.15.

We have thus achieved a considerable simplification of the original set of PDEs, Equation 5.37j and Equation 5.37k, with relatively little effort. The solution of these equations is addressed in Chapter 8, Practice Problem 8.15.

One notes from Equation 5.37j and Equation 5.37k that, although one of the old independent variables is retained unchanged, the linear combination (Equation 5.37j) of z and t is merged into a single new variable $\overline{y}$. This type of merger can, in principle, be applied to any model PDEs that contain the following combination of first-order derivatives:

$$A\ \partial u/\partial x + B\ \partial u/\partial y \tag{5.38}$$

This quantity is sometimes referred to as a *convective derivative* and arises naturally in all unsteady convective processes. We shall encounter it again in the context of atmospheric dispersion in Illustration 5.7.

5.3.5 Simplification of Geometry

The geometry of a system and the boundary conditions associated with it, have a major impact on both the ease of solution and its form. Simplifications often appear to be trivial or self-evident, whereas on other occasions the changes effected are clearly of major importance.

Suppose, for example, that one wishes to monitor the penetration into the ground of the daily temperature variations that occur at the surface of a particular location of the Earth. Because the depth of penetration is, at most, of the order of meters, it is evident that the geometry can be considered planar and semi-infinite. The curvature of the Earth or its spherical shape do not enter into the picture. The effect of these obvious simplifications, which may at first sight appear to be minor, is in fact quite considerable. The solution procedure itself is simplified, and the result can be expressed in the compact form of a single function, rather than an infinite series of functions.

Consider next the case of a buried steam line, assumed to be isothermal, which loses heat to the nearby surface of the ground. Here there is an immediate difficulty that resides in the mixed nature of the geometry, planar for the surface, and circular or cylindrical for the pipe. The underlying model itself is simple and adequately described by Laplace's equation, but the boundary conditions have to be specified at $z = 0$ and $r = R$, and hence belong to different coordinate systems. There are ways of interrelating the two, which lead to rather messy solutions. A more elegant and powerful method is that of *conformal mapping*, in which functions of a complex variable are used to introduce new independent variables that transform the system into a simpler geometry. Laplace's equation remains invariant under the transformation, and is solved by standard techniques once the geometry has been sufficiently simplified. The method requires some background in complex variable theory and will not be taken up in detail here. We, do however, present a short list of solutions, shown in Figure 5.7, which may be deemed of interest to the reader. Additional

1. Half-Plane

$$u=\frac{1}{\pi}\tan^{-1}\left(\frac{2ay}{x^2+y^2-a^2}\right)$$
$$0\leq \tan^{-1} t\leq\pi$$

2. Quadrant

$$u=\frac{2}{\pi}\tan^{-1}(y/x)$$
$$0\leq\tan^{-1} t\leq\pi$$

3. Insulated Corner

$$u=\frac{2}{\pi}\sin^{-1}\frac{1}{2}\left[\sqrt{(x+1)^2+y^2}-\sqrt{(x-1)^2+y^2}\right]$$
$$0\leq\sin^{-1} t\leq\pi$$

4. Interior of Half-Circle

$$u=\frac{2}{\pi}\tan^{-1}\left(\frac{1-x^2-y^2}{2y}\right)$$

5. Exterior of Half-Circle

$$u=\frac{1}{\pi}\tan^{-1}\left(\frac{\sin y}{\sinh x}\right)$$
$$0 \tan^{-1} t \ \pi$$

6. Half-Ring

$$u=\frac{4}{\pi}\sum_{n=1}^{\infty}\frac{r^m-r^{-m}}{r_0^m-r_0^{-m}}\,\frac{\sin m\theta}{m}$$
$$m=2n-1$$

7. Rectangle

$$u=\frac{4}{\pi}\sum_{m=1}^{\infty}\frac{\sinh(m\pi y/a)}{m\sinh(m\pi b/a)}\sin\frac{m\pi x}{a}$$

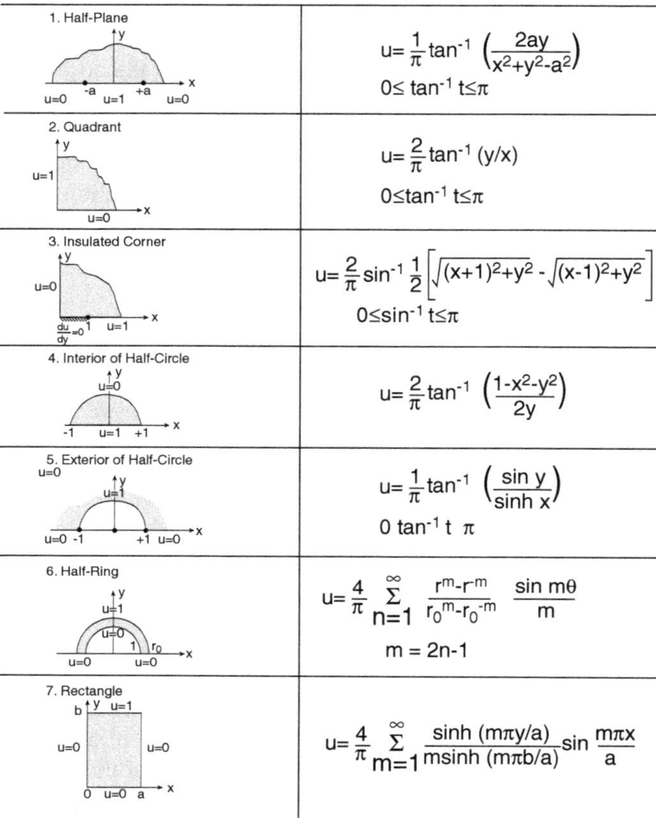

FIGURE 5.7 Solutions of Laplace's equation for various geometries and boundary conditions.

solutions in the form of "shape factors" for more complex geometries will be given in Chapter 7. Note that in Figure 5.7, the dependent variable has been nondimensionalized and represents, for example, dimensionless temperature or electrostatic potential. Thus, any one solution can be applied to several different physical situations described by Laplace's equation.

Illustration 5.9 Use of Laplace's Equation: Cooling Requirements for a Hemicylinder

To obtain some practice in the use of Figure 5.7, consider the following problem. A hemicylindrical structure of radius 1.5 m, has its flat base surface exposed to a temperature of 200°C. Its curved surface is to be maintained at 15°C, and the task we set ourselves is to calculate the heat flux at the center of the flat base. The value of k for the structure is 50 J/ms°C.

Inspection of Figure 5.7 reveals that the solution to the problem has to be derived from the nondimensionalized distribution listed under item 4:

$$u = \frac{2}{\pi}\tan^{-1}\left(\frac{1-x^2-y^2}{2y}\right) \tag{5.39a}$$

The first step in the solution is to redimensionalize the variables u, x, and y to accommodate the stated conditions.

We have:

$$u = \frac{T-15}{200-15} \tag{5.39b}$$

and hence:

$$T = 185\ u + 15 \tag{5.39c}$$

Similarly:

$$x = \frac{x'}{1.5} \quad and \quad y = \frac{y'}{1.5} \tag{5.39d}$$

which yields the dimensional coordinates:

$$x' = 1.5\ x \text{ and } y' = 1.5\ x \tag{5.39e}$$

To obtain the steady state heat flux at the center of the flat base, we first derive an expression for the local base heat flux, $k(\partial T/\partial y')_{y'=0}$:

$$Flux\ q/A = -185\,k\left(\frac{\partial u}{\partial y'}\right)_{y'=0} \tag{5.39f}$$

or

$$q/A = -185\,k\frac{2}{\pi}\frac{\partial}{\partial y'}\tan^{-1}\left.\left(\frac{1-\left(\frac{x'}{1.5}\right)^2-\left(\frac{y'}{1.5}\right)^2}{2(y'/1.5)}\right)\right|_{y'=0} \tag{5.39g}$$

For unit length of the structure, the flux becomes:

$$q_{L=1} = -\frac{185\times 50\times 2}{\pi}\left.\frac{12\,(4x'^2-9)}{16x'^4-72x'^2+81}\right|_{x'=0} = 7852\ \ \text{W/m} \tag{5.39h}$$

For the remainder of this section, we confine ourselves to the description of some simple transformations applicable to second-order PDEs. Most of these are of an obvious type but are often overlooked in the quest for an easier life.

We start by noting that the difficulties caused by geometrical complexity usually increase, for various domains of interest, in the following sequence: [doubly infinite] < semi-infinite < one-dimensional planar < circular or radial cylindrical < radial spherical < two- and three-dimensional cartesian < angular cylindrical and spherical < other. The doubly infinite domain sometimes gives rise to difficult boundary conditions and is, therefore, placed in brackets. We present in the following a number of transformations designed to move the system toward lesser complexity.

5.3.5.1 Reduction of a Radial Spherical Configuration into a Planar One

This simplification can be accomplished by introducing a new dependent variable v that is related to the old variable by the product ur, where r = radial variable:

$$v = ur \tag{5.40a}$$

The substitution has the effect of bringing about the following transformation:

$$\frac{\partial^2 u}{\partial r^2} + \frac{2}{r}\frac{\partial u}{\partial r} \Rightarrow \frac{\partial^2 v}{\partial r^2} \tag{5.40b}$$

A double simplification is thereby effected. The number of terms is reduced, and the variable coefficient 2/r is eliminated. Note that boundary conditions have to be altered accordingly. Thus, u(R) becomes v(R) = u(R)R, etc.

An example of this procedure applied to the ODE level is addressed in Practice Problem 5.9.

5.3.5.2 Reduction of a Radial Circular or Cylindrical Configuration into a Planar One

When a process is confined to a narrow region near the circumference of a circle or cylinder, and boundary conditions are specified on both sides of the strip, the geometry can be reduced to a planar one by cutting it open and unfolding it. An example of the application of this simplification occurs in connection with the freezing of water in a pipe, which is considered in Practice Problem 4.22. Because the thickness of ice required to cause rupture is small compared to the diameter of the pipe, we can unravel it into a planar configuration with boundary conditions set externally by the subzero ambient temperature and internally by the freezing point of water. Although we still have to deal with the complexities of a moving boundary, this task is considerably eased by the fact that the geometry is now planar rather than cylindrical.

5.3.5.3 Reduction of a Radial Circular or Cylindrical Configuration to a Semi-Infinite One

The same confinement of the principal events to a narrow surface strip seen in the previous case occurs here, and is exploited in the same way, by cutting open and unfolding the surface region. It differs, however, in its internal boundary condition, which is specified at infinity. This results in the reduction of the cylindrical geometry to a semi-infinite configuration. An example of its application appears in Illustration 7.10, dealing with the Lévêque solution of the Graetz problem. In this example, concentration changes in the flowing fluid are confined to a thin boundary layer at the duct wall, while far away at $r \to \infty$, its value remains constant. The solution is consequently the same as that given for boundary layer flow over a flat plate (Illustration 5.4).

5.3.5.4 Reduction of a Planar Configuration to a Semi-Infinite One

This simplification finds its principal application in time-dependent diffusional processes in which the planar surface is exposed to a brief change of the state variable, and when the resulting variation of near-surface values are to be monitored only for a brief interval. The changes will thus not have had the time to penetrate into the interior of the planar geometry, which can therefore be treated as a semi-infinite domain. An example of this type of behavior is seen in Illustration 6.8, which deals with the response of a coated sheet of paper to a brief burst of heat, and the simultaneous changes in surface temperature in response to that burst. The surprising fact is that the sheet, which is only a fraction of a millimeter thick, can be treated as a semi-infinite medium for the interval in question. This results in a considerable reduction in complexity of both the method of solution and the form of the result.

5.3.6 Nondimensionalization

Nondimensionalization can be applied to both independent and dependent variables and results in a considerable simplification of the model. It is generally achieved by combining the variables with their boundary or initial values in the form of dimensionless ratios. Let us illustrate the procedure and the resulting benefits by means of the following example.

Illustration 5.10 Nondimensionalization of Fourier's Equation

Suppose we wish to nondimensionalize the one-dimensional Fourier's equation:

$$\frac{\partial T}{\partial t} = \alpha \frac{\partial^2 T}{\partial x^2} \tag{5.41a}$$

with associated boundary and initial conditions:

$$T(0, t) = T_s \tag{5.41b}$$

$$T(L,t) = T_s \tag{5.41c}$$

$$T(x,0) = T_0 \tag{5.41d}$$

The model describes the response of a slab of thickness L and initial temperature T_0 to a step change in surface temperature T_s. The reader may recall it as the quenched steel billet of Chapter 1, Figure 1.3.

We start by nondimensionalizing the distance variable x by dividing it by thickness L, forming the new dimensionless variable X = x/L. X is said to be normalized, i.e., it has values ranging from zero to one $0 \le X \le 1$. Fourier's equation now reads:

$$\frac{\partial T}{\partial t} = \frac{\alpha}{L^2}\frac{\partial^2 T}{\partial X^2} \tag{5.42e}$$

Next, we proceed to nondimensionalize time t by dividing it by L^2/α, which also has the dimension of time. The resulting dimensionless group is known as the Fourier number $Fo = \alpha t/L^2$. Note that Fo is not normalized, as it ranges from zero to infinity.

Finally, we turn to the nondimensionalization of the dependent variable T. After some thought, we choose $\theta = (T - T_0)/(T_0 - T_s)$, as this combination normalizes, as well as nondimensionalizes, the temperature variable. The model now takes the following form:

$$\frac{\partial \theta}{\partial (Fo)} = \frac{\partial^2 \theta}{\partial X^2} \tag{5.41f}$$

with boundary conditions:

$$\theta(0, Fo) = 0 \tag{5.41g}$$

$$\theta(1, Fo) = 0 \tag{5.41h}$$

$$\theta(X,0) = 1 \tag{5.41i}$$

The benefits of this procedure, which may not be immediately obvious, are as follows:

- Both temperature T and distance x have been normalized, in addition to being nondimensionalized. This makes for much greater convenience in presenting the solution in both graphical and numerical form (see Figure 6.3 in this connection).
- Time t has been nondimensionalized and merged with the parameters α and L, making for a more compact solution.
- The boundary conditions (a) and (b) have been rendered homogeneous, which now makes it possible to solve the problem by the separation of variables technique. We demonstrate this in Illustration 8.3 of Chapter 8.

We note in closing that the various transformations that we have presented can all be applied with relative ease. The benefits to be derived by far exceed the effort required to implement them. Two among them, the similarity transformation and the reduction to canonical form, have more profound roots because they are linked to the broad topics of group transformations and characteristics. We do not enter into a discussion of the former, but will take up the topic of characteristics again in Chapter 8, in which they are used to arrive at solutions of first-order quasilinear PDEs.

5.4 PDEs PDQ: LOCATING SOLUTIONS IN THE LITERATURE

The topic to be taken up here diverges somewhat from the principal themes of this chapter. Its purpose is to encourage the reader to seek solutions to a particular problem in a related discipline and exploit any unusual features or unexpected behavior shown in the reported solution. We call this PDE PDQ (pretty damn quick), and consider it an important part of an analyst's skills.

We have, on a number of occasions, drawn the reader's attention to the fact that systems that are totally dissimilar physically may, in fact, have models that are identical in form, and consequently have identical solutions. The two dissimilar systems of a vibrating mechanical assembly and an oscillating electrical RLC circuit (presented in Illustration 2.17) are a striking and classical example of this type of behavior. The present chapter contains several additional examples. Laplace's equation, for instance, describes and can be applied to at least four different physical phenomena: conduction, diffusion, electrostatics, and gravitation, as can Poisson's equations and other classical PDEs. Thus, to solve a problem in diffusion, a useful approach is to look into the related topic of conduction and vice versa; or, if one is of an adventurous bent, into electrostatics or even celestial mechanics. In other cases, the source of a potential answer is not so self-evident, and it takes a good nose and a broad knowledge of the literature and underlying sciences, as well as persistence, to achieve one's goal.

Illustration 5.11 Pressure Transients in a Semi-Infinite Porous Medium

In the field of gas reservoir engineering, an important question to be answered is the lifetime of a gas field. To estimate this quantity, one has to derive the variations with time and distance within the porous medium of the gas pressure p(x,y,z,t). In this example, we consider the simple case of a semi-infinite medium, initially at a pressure p_0. The well pressure is to be maintained at a constant value of p_w during the lifetime of the reservoir. We start by composing a mass balance for compressible, one-dimensional flow. Taken over a difference element Δz, this becomes:

Mass balance:

$$\text{Rate of gas in} - \text{Rate of gas out} = \text{Rate of change of contents}$$

$$(\rho vA)_x - (\rho vA)_{x+\Delta x} = \varepsilon A\Delta x\left(\frac{\partial \rho}{\partial t}\right)_{avg} \tag{5.42a}$$

where $\rho vA = F$ = mass flow rate and ε is the porosity of the medium.

Upon dividing by AΔz and going to the limit, we obtain the continuity equation for flow through a porous medium:

$$-\frac{\partial(\rho v)}{\partial x}=\varepsilon\frac{\partial\rho}{\partial t} \tag{5.42b}$$

Velocity of flow is related to pressure via the empirical D'Arcy's law:

$$v=-\frac{K}{\mu}\frac{dp}{dx} \tag{5.42c}$$

To complete the model, which contains the three dependent variables ρ, v, and p, the density–pressure relation has to be expressed in terms of an appropriate gas law. We assume isothermal flow and write:

Ideal gas law:

$$\rho=\frac{pM}{RT} \tag{5.42d}$$

Upon substituting Equation 5.42c and Equation 5.42d into Equation 5.42b, the following quasilinear PDE is obtained:

$$\frac{\partial p}{\partial t}=B\frac{\partial}{\partial x}\left(p\frac{\partial p}{\partial x}\right)=B\left[\left(\frac{\partial p}{\partial x}\right)^2+p\frac{\partial^2 p}{\partial x^2}\right] \tag{5.42e}$$

where B = K/εμ.

The boundary conditions to this PDE are as follows:

$$p(0,t)=p_w$$

$$p(\infty,t)=p_0 \text{ (bounded)} \tag{5.42f}$$

$$p(x,t)=p_0$$

Success in locating a literature solution hinges on recognizing the equivalence of the term p(∂p/∂x) and the diffusional flux D(C)(∂C/∂x), where D(C) is a variable, concentration-dependent diffusivity. It further requires the knowledge that the latter systems exist and have been successfully solved. At any rate, there is no harm in quickly consulting the pertinent literature, and if need be, locating solutions to the corresponding conduction problem containing the term k(T)(∂T/∂x).

The case of a temperature-dependent thermal conductivity k(T), although rare, has been addressed, but no convenient solutions are immediately available in standard reference texts. Concentration-dependent diffusivities D(C) arise with much greater frequency, for example, in the diffusion through polymers, and a number of solutions to Fick's equation with variable diffusivity are available. One that comes closest to our requirements has the form:

$$\frac{\partial(C/C_0)}{\partial t} = \frac{\partial}{\partial x}\left[D_0 + \alpha D_o(C/C_0)\right]\frac{\partial C/C_0}{\partial x} \tag{5.42g}$$

with BCs for desorption:

$$C/C_0(0, t) = 1$$

$$C/C_0(x, 0) = 0 \tag{5.42h}$$

$$C/C_0(\infty, t) \text{ bounded}$$

To bring Equation 5.42e in line with this formulation, we introduce the dimensionless variable $\bar{p} = (p_0 - p)/(p_0 - p_w)$ and obtain the revised model:

$$\frac{\partial \bar{p}}{\partial t} = \frac{\partial}{\partial x}\left[Bp_0 + B(p_0 - p_w)\bar{p}\right]\frac{\partial \bar{p}}{\partial x} \tag{5.42i}$$

with BCs:

$$\bar{p}(0,t) = 1$$

$$\bar{p}(x,0) = 0 \tag{5.42j}$$

$$\bar{p}(\infty,t) \textit{ bounded}$$

Correspondence between the two cases is established via the expressions:

$$D_0 = Bp_0 \qquad \alpha = B(p_0 - p_w) \qquad C/C_0 = \bar{p} \tag{5.42k}$$

The literature solutions were obtained numerically, and are given as plots of C/C_0 vs. $x/(4D_0t)^{1/2}$, with α as a parameter (see Crank, 1978).

Comments: The reader will have noted the skills that were required to arrive at a solution. They included an awareness of the equivalence of $\partial/\partial x[p\partial p/\partial x]$ and $\partial/\partial x[D(C)\partial p/\partial x]$, the knowledge (or at least an intuitive feeling) that such solutions existed, and the patience to look for them.

The solutions can be used to calculate the time it takes for the reservoir pressure to drop to a prescribed value. This is often taken to be 25% of the initial pressure p_0, because a further drop makes the recompression costs required at the surface unattractive. Secondary methods are often used to recover the residue.

One notes the appearance of the group $x/t^{1/2}$ in the parameters of the solution, which suggests that similarity transformation was used to arrive at a solution. It is left to the exercises to show that this results in the ODE:

$$B\left[\bar{p}\frac{d^2\bar{p}}{d\eta^2}+\left(\frac{d\bar{p}}{d\eta}\right)^2\right]+\frac{1}{2}\eta\frac{d\bar{p}}{d\eta}=0 \tag{5.43}$$

PRACTICE PROBLEMS

5.1 Classification of PDEs: Oscillations of a Hanging Chain

a. Consider a heavy chain of uniform density suspended vertically from one end. If one takes the origin at the position of equilibrium of the lower, free end, and the x-axis along the equilibrium position of the chain, then small oscillations of the chain in the horizontal direction y are described by the following PDE:

$$\frac{\partial^2 y}{\partial t^2}=g\frac{\partial}{\partial x}\left(x\frac{\partial y}{\partial x}\right)$$

where g = gravitational constant.
Give a complete classification of this equation as to linearity, homogeneity, type of coefficients, and whether it is elliptic, parabolic, or hyperbolic.

b. Classify the following PDE:

$$(ku_x)_x = u_y + ku$$

Consider the cases where (a) k = constant, (b) k = f(x), and (c) k = g(u).

5.2 Boundary and Initial Conditions: Physical Interpretations

a. Consider the three-dimensional Laplace, Fourier, and wave equations. How many boundary and initial conditions does each require?

b. Classify the boundary and initial conditions, and give a physical interpretation of the following Fourier system:

$$\frac{\partial T}{\partial t}=k\frac{\partial^2 T}{\partial x^2}+K_1e^{-t} \qquad 0<x<L \qquad t>0$$

$$\frac{\partial T}{\partial x}(0,\,t)=-K_2$$

$$T(L, t) = K_3$$

$$T(x,0) = f(x)$$

5.3 Derivation of Simple PDEs

a. Using the procedures given in Illustration 5.3 for deriving the PDE for a vibrating string, derive the equation for the oscillation of a suspended chain (Practice Problem 5.1).
b. Show that the temperature distribution in an electrical conductor of specific resistivity S and carrying an electrical current i is given by:

$$k\frac{\partial^2 T}{\partial x^2} + \frac{Si^2(t)}{(\pi R^2)^2} - \frac{2h}{R}(T - T_s) = \rho Cp\frac{\partial T}{\partial t}$$

where h is the convective heat loss to the surroundings of temperature T and R represents the wire radius.

Hint: $q(J/s) = R_e i^2$, where R_e = electrical resistance = $SL/\pi R^2$.

5.4 Transformation of Independent Variables

a. Starting with Laplace's equation in Cartesian coordinates (Equation 5.13b), show that introduction of the cylindrical coordinates r, φ, z, defined by the relations:

$$x = r\cos\varphi,\ y = r\sin\varphi,\ z = z$$

leads to the form:

$$\frac{\partial^2 u}{\partial r^2} + \frac{1}{r}\frac{\partial u}{\partial r} + \frac{1}{r^2}\frac{\partial^2 u}{\partial y^2} + \frac{\partial^2 u}{\partial z^2} = 0$$

b. Show that the three-dimensional polar coordinates r, θ, φ are related to the Cartesian coordinates by the relations:

$$x = r\sin\theta\cos\varphi,\ y = r\sin\theta\sin\varphi,\ z = r\cos\theta$$

Using these relations, derive Laplace's equation for a sphere in three dimensions:

Answer:

$$\frac{1}{r^2}\left(\frac{\partial}{\partial r}r^2\frac{\partial u}{\partial r}\right) + \frac{1}{r^2\sin\theta}\frac{\partial}{\partial\theta}\left(\sin\theta\frac{\partial u}{\partial\theta}\right) + \frac{1}{r^2\sin^2\theta}\frac{\partial^2 u}{\partial y^2} = 0$$

5.5 Similarity Transformation for Non-Fickian Diffusion

Non-Fickian diffusion is the term used to describe diffusional transport in which the diffusion coefficient depends on concentration, D = f(C). This case arises in the diffusion of solutes present in high concentrations, particularly in high-density gases and in liquids, and in the gas- and liquid-phase diffusion of solutes through polymeric substances.

Consider non-Fickian diffusion in a one-dimensional semi-infinite medium with an initial concentration $c(x,0) = C_0$, and an imposed surface concentration of $C_s(0,t) = C_s$ at $t > 0$. Derive a similarity variable η for the system, and show that the similarity transformation reduces the non-Fickian diffusion equation:

$$\frac{\partial}{\partial x} D(C) \frac{\partial C}{\partial x} = \frac{\partial C}{\partial t}$$

to the form:

$$\frac{d}{d\eta} D(C) \frac{dC}{d\eta} + 2\eta \frac{dC}{d\eta} = 0$$

Describe a physical problem in heat conduction that can be solved by the same method.

5.6 Transformation of Nonhomogeneous Boundary Conditions to Homogeneous Form

The constant-rate drying of a porous slab $0 < x < L$ can be described by the following system of equations:

$$D_{eff} \frac{\partial^2 C}{\partial x^2} = \frac{\partial C}{\partial t}$$

$$\left(\frac{\partial C}{\partial x}\right)_0 = -\left(\frac{\partial C}{\partial x}\right)_L = \frac{W}{D_{eff}}; \quad \left(\frac{\partial C}{\partial x}\right)_{L/2} = 0; \quad C(x,0) = C_0$$

where W denotes the rate of drying.

Convert these expressions to a system with homogeneous boundary conditions without introducing nonhomogeneous terms into the PDE.

Hint: Because four expressions have to be made or kept homogeneous, a third-order polynomial with four undetermined coefficients suggests itself as a trial function for a new dependent variable C′. Show that if the polynomial is defined in x only, incompatible conditions arise. Remedy this by including time t in the trial function C′ = C + f(x, t).

5.7 Reduction to Canonical Form

a. Show that the two-dimensional wave equation is hyperbolic. Identify the characteristics and reduce the PDE to canonical form.
   Answer: $\frac{\partial^2 u'}{\partial \bar{x} \partial \bar{y}} = 0$
b. Reduce the following PDE to canonical form:

$$\frac{\partial^2 u}{\partial x^2} = x_2 \frac{\partial^2 u}{\partial y^2}$$

Answer: $$\frac{\partial^2 u'}{\partial \bar{x} \partial \bar{y}} = \frac{1}{4(\bar{x} - \bar{y})}\left(\frac{\partial u'}{\partial \bar{x}} - \frac{\partial u'}{\partial \bar{y}}\right)$$

Comment: Although the new PDE is more complex in appearance, it has the virtue of having the same structure as the canonical form of a generalized wave equation.

5.8 Steady State Temperature in a Quadrant: Solution of Laplace's Equation

Using item 3 of the solutions to Laplace's equation shown in Figure 5.7, calculate the dimensionless temperature along the insulated portion at a position x = 0.5.

Answer: u = $\frac{1}{3}$

5.9 Laplace's Equation and the Flow of Electricity

Consider the hemicylinder of Illustration 5.9. A potential of 100 V is applied to the base, while the circumferential area is grounded. Calculate the current flow per unit length between the two surfaces. The specific resistance is taken to be 10^{-2} Ωm.

5.10 Simplifying the Geometry: The Catalyst Pellet Again

a. Derive an expression for the effectiveness factor E of a spherical catalyst pellet undergoing a first-order reaction. Hint: Invoke Equation 5.40b.
b. Catalyst pellets are sometimes cast in the form of hollow cylinders (Raschig rings) to reduce the pressure in the reactor and to increase the effectiveness factor E. If the ratio of internal to external radius is 0.8, and the length five times the internal radius, how would you go about estimating E?

5.11 PDEs PDQ: Translating Conduction Problems into Meaningful Diffusion Problems

The following problem statement appears in a standard work on heat conduction (Carslaw and Jaeger, 1959):

Heat is emitted at the origin for times $t > 0$ at the rate q heat units per unit time, and an infinite medium moves uniformly past the origin with a velocity U parallel to the x-axis.

For the limiting steady state case, Carslaw and Jaeger report the solution (their symbols):

$$v = \frac{q}{4\pi KR} \exp[-U(R - x)/2K]$$

where v represents temperature, K is the thermal conductivity, κ thermal diffusivity, and $R^2 = x^2 + y^2 + z^2$.

Translate this solution into a meaningful real-world mass diffusion problem. Hint: Consider emission from a pollutant source.

6 Solution of Linear Systems by Superposition Methods

We have previously (in Section 2.1.2) drawn the reader's attention to the important superposition principle, which is one of the principal tools listed in dealing with linear systems. It states, in essence, that the general solution of a linear differential equation can be composed of the sum of all independent particular solutions. Its application in that section was restricted to ordinary differential equations (ODEs), and in Section 2.3.2, we used it successfully to compose solutions of linear ODEs with constant coefficients from the sum of exponential functions of the so-called eigenvalues (D-operator method). Later, in Section 2.3.4, we extended the method to linear systems with variable coefficients and saw the emergence of infinite sums of particular solutions, a feature we shall encounter again at the PDE level.

We now extend and amplify this principle and apply it to the solution of partial differential equations (PDEs). We make the following distinctions.

1. *Superposition by addition:* This is the standard definition of the term as it was originally applied at the ODE level. Superposition of simple flows, which is the first item to be taken up, is the classical example of this method.
2. *Superposition by integration:* This procedure views the integration as one infinite set of superpositions by addition. Integration is with respect to time or distance, and its most fruitful application is found in the treatment of time-varying sources and boundary conditions. The use of Green's functions, to be taken up in Chapter 7, also falls within this category.
3. *Superposition by multiplication:* This is a somewhat unconventional category. It is not generally viewed as a superposition method, but has all the necessary features of one and is, therefore, included here. It consists of composing solutions of multidimensional problems as the product, rather than the sum, of solutions to one-dimensional linear PDEs and is also refered to as the Neumann Product Solution. Its justification is to be found in a particular PDE solution method known as *separation of variables*. We shall amplify on this aspect at the appropriate time.

We start our deliberations with an exposition of superposition of simple flows. This classical method has the unusual feature of not addressing a particular problem, but rather seeking out situations that might fit a particular solution. The reader will find this a new and intriguing experience.

6.1 SUPERPOSITION BY ADDITION OF SIMPLE FLOWS: SOLUTIONS IN SEARCH OF A PROBLEM

The technique we apply here starts by defining some simple stream functions ψ, which can be shown to satisfy Laplace's equation in two Cartesian coordinates (x,y) or in polar coordinates (r, θ). These stream functions represent certain simple, albeit somewhat artificial, flow patterns and are summarized in Figure 6.1.

The stream function $\psi = v_\infty y$ = constant represents rectilinear flow parallel to the x-axis, with an orthogonal network of potential functions $\phi = v_\infty x$ = constant (Figure 6.1A). Both ϕ and ψ satisfy Laplace's equation.

The stream function $\psi = \frac{Q}{2\pi}\theta$ = constant represents flow emanating from the origin and proceeding radially outward (Figure 6.1B). The corresponding potential functions are represented by a set of orthogonal concentric circles. Q is the so-called strength of the source and physically equals the flow rate per unit depth into the paper. For a sink, the flow is taken to be radially inward, and both the stream function

System	Stream Function	Velocities
A. Uniform Flow 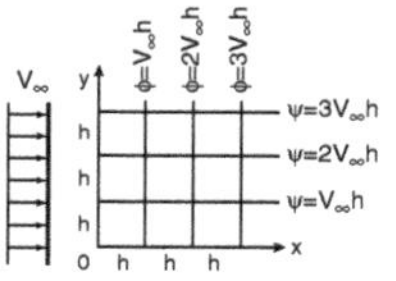	$\psi = V_\infty y$ $\psi = V_\infty r.\sin\theta$	V_x = constant $V_y = 0$
B. Line Source 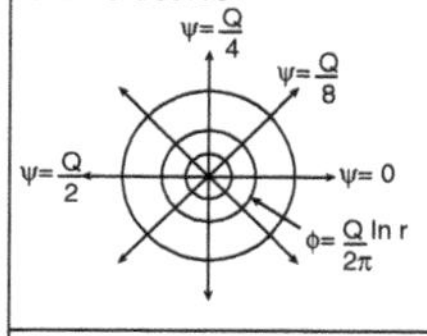	Source: $\psi = \frac{Q\theta}{2\pi}$ Sink: $\psi = -\frac{Q\theta}{2\pi}$	Source: $V_\theta = 0 \quad V_r = Q/2\pi$ Sink: $V_\theta = 0 \quad V_r = -Q/2\pi$
C. Line Vortex 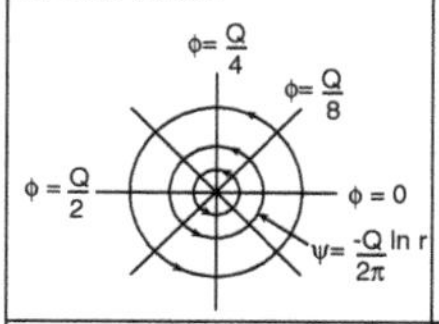	Clockwise $\psi = \frac{Q \ln r}{2\pi}$ Counter-Clockwise $\psi = -\frac{Q \ln r}{2\pi}$	Clockwise $V_\theta = Q/2\pi r \quad V_r = 0$ Counter-Clockwise $V_\theta = -Q/2\pi r \quad V_r = 0$
D. Doublet 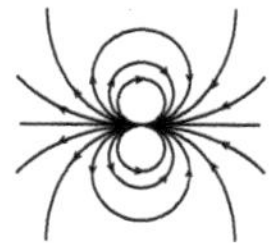	$\psi = -K\frac{\sin\theta}{r}$	$V_\theta = \frac{K}{r^2}\sin\theta$ $V_r = -\frac{K}{r^2}\cos\theta$

FIGURE 6.1 Compilation of some simple flows used in the solution of Laplace's equation for irrotational flow by superposition.

TABLE 6.1
Stream Functions, Velocity Potentials Velocities, and Laplace's Equation

Cartesian Coordinates	Polar Coordinates
(1) ψ and ϕ	
$\frac{\partial\psi}{\partial x} = -\frac{\partial\phi}{\partial y}$	$\frac{\partial\psi}{\partial\theta} = r\frac{\partial\phi}{\partial r}$
$\frac{\partial\psi}{\partial y} = \frac{\partial\phi}{\partial x}$	$\frac{\partial\psi}{\partial r} = -\frac{1}{r}\frac{\partial\phi}{\partial\theta}$
(2) Velocity Components	
$v_x = \frac{\partial\psi}{\partial y} = \frac{\partial\phi}{\partial x}$	$v_\theta = \frac{\partial\psi}{\partial r} = -\frac{1}{r}\frac{\partial\phi}{\partial\theta}$
$v_y = -\frac{\partial\psi}{\partial x} = \frac{\partial\phi}{\partial y}$	$v_r = \frac{1}{r}\frac{\partial\psi}{\partial\theta} = \frac{\partial\phi}{\partial r}$
(3) Laplace's Equation	
$\frac{\partial^2\psi}{\partial x^2} + \frac{\partial^2\psi}{\partial y^2} = 0$	$\frac{\partial^2\psi}{\partial\theta^2} + r\frac{\partial}{\partial r}\left(r\frac{\partial\psi}{\partial r}\right) = 0$
$\frac{\partial^2\phi}{\partial y^2} + \frac{\partial^2\phi}{\partial y^2} = 0$	$\frac{\partial^2\phi}{\partial\theta^2} + r\frac{\partial}{\partial r}\left(r\frac{\partial\phi}{\partial r}\right) = 0$

and the velocity component change sign. In both cases, ψ satisfies Laplace's equation in polar coordinates (Table 6.1).

The stream function $\psi = (Q/2\pi)\ln r$ = constant represents a clockwise circular flow pattern around the origin, with ϕ = constant making up the radial rays that emanate from the origin and are orthogonal to the circles ψ = constant (Figure 6.1C). This pattern is merely a reversal of that shown by the source, with the roles of ϕ and ψ having been interchanged, and is appropriately termed a *vortex*. Q represents the strength of the vortex and is again given in units of flow rate per unit depth into the paper. For counterclockwise vortices, the signs of both the stream function and the velocities are reversed.

The doublet, shown in Figure 6.1D, is more of an artifact and is not meant to represent any real flow pattern. The streamlines are made up of a set of circles tangent to the x-axis and satisfy Laplace's equation as required.

Other simple flows, such as point sources and point doublets, have been used in the literature but are not taken up here.

We now proceed to superpose various simple flows by adding the corresponding stream functions. The linearity of Laplace's equation guarantees that it is satisfied for any arbitrary sum of such stream functions. The question that arises, however, is the following: What, if any, real potential flow problem has been solved by superposing the simple flows? If no real flow pattern emerges from a particular superposition, new combinations are tried until something worthwhile emerges. Note that both addition and subtraction can be used resulting in a host of possible combinations.

We term this procedure *a solution in search of a problem.* This is evidently a reversal of the conventional process of first specifying a problem and then seeking its solution, but the procedure has a legitimate place in mathematical analysis and has led to many fruitful results. A summary of several classical superpositions appears in Figure 6.2.

System	Stream Function	Velocities
A. Tornado 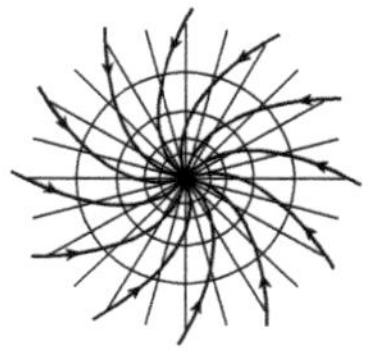	Superposition of Counter-clockwise Vortex and Sink $\psi = -\frac{Q\theta}{2\pi} - \frac{Q}{2\pi}\ln r$	$v_\theta = -\frac{Q}{2\pi r}$ $v_r = -\frac{Q}{2\pi r}$
B. Flow Around Cylinder 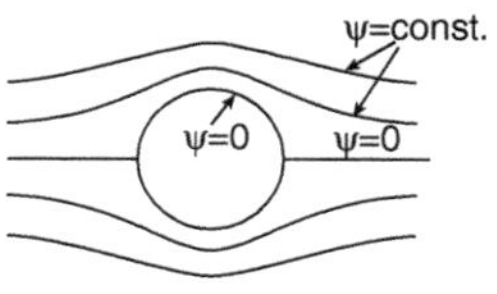	Superposition of Uniform Flow and Doublet $\psi = v_\infty r \sin\theta - \frac{K\sin\theta}{r}$	$v_\theta = v_\infty \sin\theta + \frac{K\sin\theta}{r^2}$ $v_r = v_\infty \cos\theta - \frac{K\cos\theta}{r^2}$
C. Rankine Half-Body 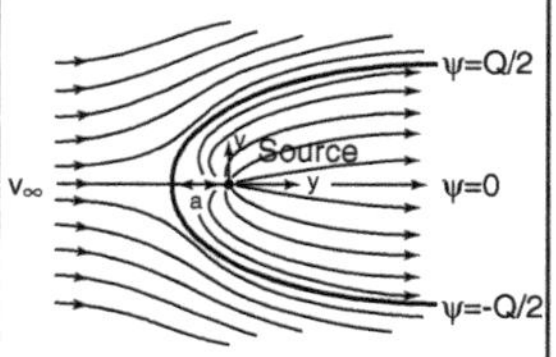	Superposition of Uniform Flow and Source $\psi = v_\infty r \sin\theta + \frac{Q\,\theta}{2\pi}$	$v_\theta = v_\infty \sin\theta$ $v_r = +v_\infty \cos\theta + \frac{Q}{2\pi r}$
D. Flow in a Corner 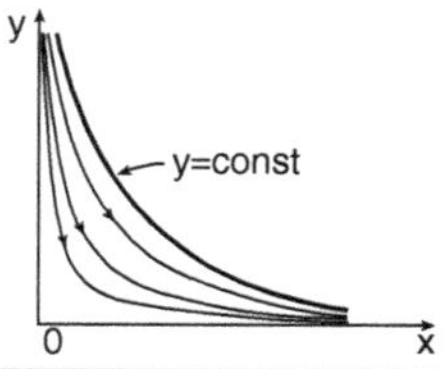	No Superposition $\psi = 2Axy$ $\psi = Ar^2 \sin 2\theta$	$v_\theta = 4Ar \sin 2\theta$ $v_r = 4Ar \cos 2\theta$

FIGURE 6.2 Flow fields resulting from the superposition of some simple flows.

The results of the superpositions displayed in this figure show several recognizable flow patterns.

In Figure 6.2A, a sink and a clockwise vortex have been superposed resulting in spiral flow into the origin. This resembles the flow pattern in a draining tank or in a tornado and agrees with the intuitive notion that the combination of circular and radial flow should result in streamlines of a spiral form converging to the origin.

In Figure 6.2B, uniform flow has been superposed onto a doublet. Here it is difficult to anticipate the resulting flow pattern, and a detailed analysis of the stream functions has to be undertaken. This is done in Illustration 6.1. The result indicates that the pattern represents potential flow around a circle or cylinder, or around humps, which diminish in height as one moves outward. The relations given in Table 6.1 enable us to calculate velocity distributions around the cylinder, which can then be substituted into the Bernoulli equation to arrive at a pressure distribution. These are highly useful results to have been obtained by simple manipulations of the stream function and Bernoulli's equation. A similar useful result arrived at by superposition appears in Illustration 6.2.

Figure 6.2C shows the results of the superposition of uniform flow and a source. Plots of various value of ψ = constant lead to the appearance of a near-elliptical shape known as the *Rankine half-body*, with upper and lower stream lines equal to 1/2 Q, where Q is the strength of the source. An increase in Q will thus result in an increase of the body thickness. The maximum velocity occurs at $\theta = 63°$ and is given by $|v|_{Max} = 1.26\ v_{\infty}$.

The streamlines shown in Figure 6.2D cannot be obtained by superposition of simple flows. They represent rectangular hyperbolas that were deduced to represent flow in a right-angled corner, also termed *stagnation flow*. This example can be truly termed a solution in search of a problem. A similar approach can be used to show that flow around corners of various angles can be represented by the general streamline:

$$\psi = Ar^n \sin n\,\theta \tag{6.1}$$

The parameter n is related to the angle α of the corner as follows:

n	1/2	2/3	3/2	2	3	4
α	360°	270°	135°	90°	60°	45°

The detailed deduction of the flow pattern from an assumed form of the stream function ψ is illustrated in the following example.

Illustration 6.1 Superposition of Uniform Flow and a Doublet: Flow Around an Infinite Cylinder or a Circle. Drag In Turbulent Flow

A convenient method of analysis is to start with the simplest possible stream function $\psi = 0$, i.e., we set the sum of ψ for uniform flow and a doublet equal to zero:

$$\psi = \sin\theta\left(v_{\infty} r - \frac{K}{r}\right) = 0 \tag{6.2a}$$

This relation is satisfied for:

$$\theta = 0, -\pi, -2\pi, ..., \text{any } r \tag{6.2b}$$

and

$$r = (K/v_\infty)^{1/2}, \text{ any } \theta \tag{6.2c}$$

These conditions represent the x-axis with an intervening circle (or an infinitely long cylinder) of radius $R = (K/v_\infty)^{1/2}$. The size of the cylinder or circle can thus be manipulated by adjusting the strength K of the doublet. Note also that the stream functions with values other than zero, i.e., $\psi = C_1$, C_2, etc., represent flow around humps whose height diminishes with an increase in the constant C (see Practice Problem 6.3).

Let us now deduce velocities and pressures at various points of the flow field, using Table 6.1 as an aid.

Radial velocity at the cylinder surface

$$v_r = \frac{1}{r}\frac{\partial\psi}{\partial\theta} = \frac{1}{r}\cos\theta\left(v_\infty r - \frac{K}{r}\right) \tag{6.2d}$$

At the surface, $r = R = (K/v_\infty)^{1/2}$, so that:

$$v_r = \frac{\cos\theta}{(K/v_\infty)^{1/2}}[(v_\infty K)^{1/2} - (v_\infty K)^{1/2}] = 0 \tag{6.2e}$$

i.e., the radial velocity component vanishes at the surface everywhere, as required.

Tangential velocity at the cylinder surface — From Table 6.1 and the relation $R = (K/v_\infty)^{1/2}$, we obtain:

$$v_\theta = \frac{\partial\psi}{\partial r} = \sin\theta\left(v_\infty + \frac{K}{r^2}\right) = 2v_\infty \sin\theta \tag{6.2f}$$

Thus, at the stagnation points $\theta = 0, -\pi$, the tangential velocity vanishes, as required. The maximum velocity is attained at $\theta = -\pi/2$ and has the value $|v|_{Max} = v_\theta = 2v_\infty$, i.e., twice the velocity of approach v_∞. This is an important result and applies as well to flow just outside a viscous boundary layer.

Pressure distribution at the cylinder surface — To calculate the pressure distribution in the flow field, we draw on Bernoulli's equation, which for our present purposes assumes the form:

$$\frac{p}{\rho} + \frac{v_\theta^2}{2} = \frac{p_\infty}{\rho} + \frac{v_\infty^2}{2} \tag{6.2g}$$

Substitution of the Equation 6.2e into Equation 6.2f yields the pressure distribution:

$$p = \frac{\rho v_\infty^2}{2}(1 - 4\sin^2\theta) + p_\infty \tag{6.2h}$$

For $\theta = 0, -\pi$, one obtains the stagnation pressure p_s:

$$p_s = \frac{\rho v_\infty^2}{2} + p_\infty \tag{6.2i}$$

Thus, p_s equals the pressure far from the cylinder, p_∞, augmented by the kinetic energy of the approaching fluid, which is completely converted into what might be termed *volumetric pressure energy.*

The minimum pressure p_{min} occurs at $\theta = \pm\pi/2$ and assumes the value:

$$p_{min} = p_\infty - 3\frac{\rho v_\infty^2}{2} \tag{6.2j}$$

i.e., the pressure here is diminished over the free stream pressure p_∞ by three times the kinetic energy of approach. The value approximates the actual value found in real flow around a cylinder. In fact, Equation 6.2g is an accurate description of the pressure distribution in the upstream half of the cylinder, and can be used to obtain an upper bound to the drag force in the absence of viscous effects (i.e., in turbulent flow). The latter is given by the relation (cf. Equation 4.66):

$$F_D = C_D A_p\, \rho v_\infty^2/2 = 1.2 \times 2\text{RL}\, \rho v_\infty^2/2 = 2.4\text{ RL}\, \rho v_\infty^2/2 \tag{6.2k}$$

where C_D is the empirical drag coefficient and A_p the projected cross section of a cylinder of radius R and length L.

A similar result is obtained by integrating the x-component of the pressure force distribution (Equation 6.2h) over the upstream half of the cylinder. We support this with the argument that, in turbulent flow, the rear half of the cylinder contributes very little to the total drag force because of flow separation. We obtain:

$$F_D = \int_{-\pi/2}^{\pi/2} p dA = \frac{\rho v_\infty^2}{2}\int_{-\pi/2}^{\pi/2} -4\sin^2\theta\cos\theta RLd\theta \tag{6.2l}$$

$$F_D = \frac{8}{3}RL\rho v_\infty^2\ /2 = 2.67RL\ \rho v_\infty^2/2 \tag{6.2m}$$

in close agreement with the empirical Equation 6.2k. Considering the simple tools used in arriving at these results, this is a remarkable accomplishment.

Illustration 6.2 Superposition of Stagnation Flow and a Source at the Origin

As in most simple flow superpositions, the pattern resulting from the addition of a source to flow in a corner is not immediately evident or easily predicted. One can however argue as follows: The flow emanating from the source will tend to counteract the flow approaching the corner, bringing it to a halt. One can therefore expect, at least qualitatively, that the pattern will be representative of stagnation flow near a blunt obstacle at the origin.

The flow field can be established from the combined stream functions:

$$\psi = Kx\,y + m\theta \tag{6.2a}$$

where K = 2A (Figure 6.2D and Figure 6.1B). If we define nondimensionalized distances:

$$x^* = x(K/m)^{1/2} \tag{6.3b}$$

$$y^* = y(K/m)^{1/2} \tag{6.3c}$$

we obtain the modified stream function

$$\frac{\psi}{m} = x^*\,y^* + \theta \tag{6.3d}$$

where

$$\theta = \tan^{-1}\left(\frac{y^*}{x^*}\right) \tag{6.3e}$$

The flow field can then be established by plotting $\psi^*(x^*,y^*)$ for various parameter values of the source strength. The results, obtained by means of the Mathematica package, are shown in Figure 6.3. The Mathematica code to generate this graph is:

```
<< Graphics'MultipleListPlot'
MultipleListPlot[
 Table[
  Table[
  {x,
   y/.FindRoot[xy + ArcTan[y/x] - c == 0,{y,0},
    MaxIterations → 1000][[1]]},
   {x,-2.95,2.95,0.1}]
   ,{c,-5,5,0.5}],PlotJoined → True,
SymbolShape → None,
TextStyle → {FontFamily → "Arial - Bold",FontSize → 14},
Frame → True,
PlotStyle → {{Thickness[0.001],GrayLevel[0]}}
]
```

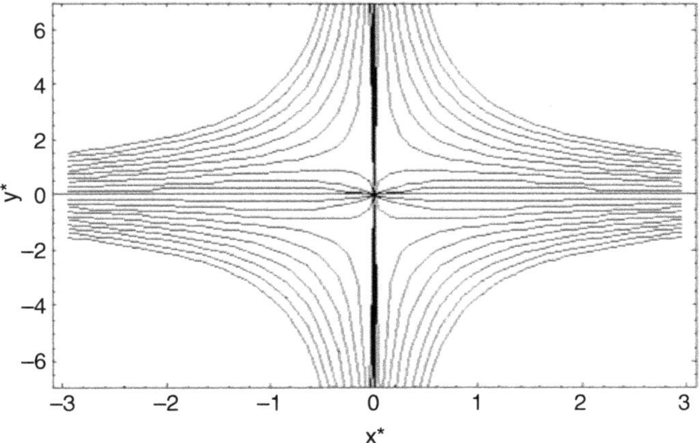

FIGURE 6.3 Streamlines for Illustration 6.3 as a function of dimensionless distances.

We note from Figure 6.1 and Figure 6.2 that the source strength Q has units of m^2/sec, and the stagnation parameter K has units of sec^{-1}, which assures the nondimensionality of x* and y*. To recover the dimensional quantities x and y, we use the floating parameter values for m, and evaluate K by specifying an upstream velocity v_0 on a particular streamline. We have, from Table 6.1, Equation 6.3d, and Equation 6.3e:

$$v_x = v_0 = \frac{\partial \psi}{\partial y} = \frac{\partial}{\partial y}\left[Kxy + m \tan^{-1}\left(\frac{y}{x}\right)\right] \tag{6.3f}$$

which allows us access to the parameter value K. That value is subsequently used on all other streamlines.

Comments: One notes immediately from Figure 6.3 that the streamlines for large values of y* are not truly representative of the actual flow field. The streamlines in the rear of the bump located at the origin are likewise unrealistic, given that the complex pattern that results there involves flow separation and the formation of vortices. The frontal streamlines, however, give a correct representation of the prevailing conditions. They allow us, furthermore, to calculate the all-important frontal stagnation pressure from Equation 6.2i:

$$\Delta p = (\rho\, v_1^2)/2 \tag{6.3g}$$

Given that the pressure in the rear of the bump is insignificant because of flow separation, integration of Δp over the frontal area will again, as in the case of the cylinder, provide a close approximation of the drag force on the bump or, for that matter, a tall building. Practice Problem 6.3 addresses a similar flow field, arrived at by different means, in the approach to a gently sloped (rather than abrupt) obstruction.

6.2 SUPERPOSITION BY MULTIPLICATION: THE NEUMANN PRODUCT SOLUTIONS

The term superposition by multiplication is not commonly used in the literature, which prefers to regard superposition as an additive process and uses the formal designation "Neumann Product Solutions". We adopted it nevertheless, because it contains the same features as classical superposition, i.e., the solution of a complex problem by the process of patching together solutions to simpler problems. In the application we consider here, the simpler problems are the solutions of the one-dimensional Fourier's or Fick's equations for unsteady conduction and diffusion, the complex counterparts being the two- and three-dimensional cases. The method is only applicable to linear PDEs, which can be solved by the method of separation of variables. We provide below two examples that are based on Fourier's equation.

First consider the case of rectangular one-dimensional unsteady conduction in a finite domain $x_1 < x < x_2$. The system is described by:

$$\frac{\partial T_x}{\partial t} = \alpha \frac{\partial^2 T_x}{\partial x^2} \tag{6.4a}$$

with general Type III boundary conditions:

$$A_x \frac{\partial T_x}{\partial x} - B_x T_x = 0 \quad at\ x = x_1$$

$$C_x \frac{\partial T_x}{\partial x} - D_x T_x = 0 \quad at\ x = x_2 \tag{6.4b}$$

and initial condition:

$$T_x(x,0) = f_x(x) \tag{6.4c}$$

Its solution will be denoted by:

$$T_x(x,t) = g_x(x,t) \tag{6.4d}$$

Extend this case now to three dimensions in the rectangular domain $x_1 < x < x_2$, $y_1 < y < y_2$, and $z_1 < z < z_2$ with boundary conditions in these dimensions identical in form to those described by Equation 6.4b. The solution for these three-dimensional temperature distributions T(x,y,z,t) can then be expressed as the product of three one-dimensional solutions, i.e.:

$$T(x, y, z, t) = g_x(x,t)g_y(y,t)g_z(z,t) \tag{6.4e}$$

with initial conditions:

$$T(x, y, z, 0) = f_x(x)f_y(y)f_z(z) \tag{6.4f}$$

This is a useful and quite a general result, except that the three-dimensional initial condition has to be of the separable form as in Equation 6.4f. In most practical applications, including that given in the next illustration, the temperature is uniform and constant over the entire three-dimensional body, so that Equation 6.4f becomes:

$$T(x, y, z, 0) = T_s \tag{6.4g}$$

Illustration 6.3 A Simple Application of Product Solutions: Heat Conduction in a Finite Cylinder or Disk

To obtain some practice in the mechanism of applying product solutions, we consider the following problem: three cylindrical objects, all of radius R = 5 cm and respective lengths of L = 5 cm, 10 cm, and 20 cm, are to be quenched from their initial temperature of 1000°C by immersion in a medium at 25°C. Solutions to the intermediate case (length equals diameter) are easily available in the literature, but any other length to diameter ratio usually requires a full analytical or numerical solution. This is where product solutions provide a quick and elegant alternative.

To obtain a first sense of the speed of the process, we propose to calculate the time required for the center of each object to cool from 1000°C to 500°C. Thermal diffusivity α is set at 10^{-5} m^2/sec, a typical value for steel, and the external resistance to heat transfer is taken to be negligible.

To ease the task of applying the product solutions, we propose to use the temperature response curves shown in Figure 6.4, which relate the temperature T_C at the center of various geometries to the dimensionless Fourier number Fo = $\alpha t/\delta^2$. The Fourier number, which we previously encountered in Illustration 5.9, is the result of the nondimensionalization of Fourier's equation and contains the characteristic dimension δ of the object in question.

We start by composing a product solution for the intermediate cylinder, having R = 5 cm and L = 10 cm (length equals diameter), and comparing the result with the analytical solution shown in Figure 6.4 (item 6, short cylinder). This provides a quick validation of the product solution.

We have, from the given data, a required dimensionless temperature response of

$$\theta_{SC} = (T_C - T_S)/(T_0 - T_S) = \frac{500 - 25}{1000 - 25} = 0.487 \tag{6.5a}$$

This value must be matched by the product of the responses for two parallel plates to account for conduction in the axial direction (item 1 in Figure 6.4), and an infinitely long cylinder to describe conduction in the radial direction (item 4). We must, in other words, have:

$$\underset{\substack{\text{Short cylinder}\\ \delta = R = 5\,cm,\ L = 10\,cm}}{(T_C - T_S)/(T_0 - T_S)_{SC}} = \underset{\substack{\text{Parallel planes}\\ \delta = 5\,cm}}{(T_C - T_S)/(T_0 - T_S)_{pp}} \times \underset{\substack{\text{Infinite cylinder}\\ \delta = R = 5\,cm}}{(T_C - T_S)/(T_0 - T_S)_{IC}} \tag{6.5b}$$

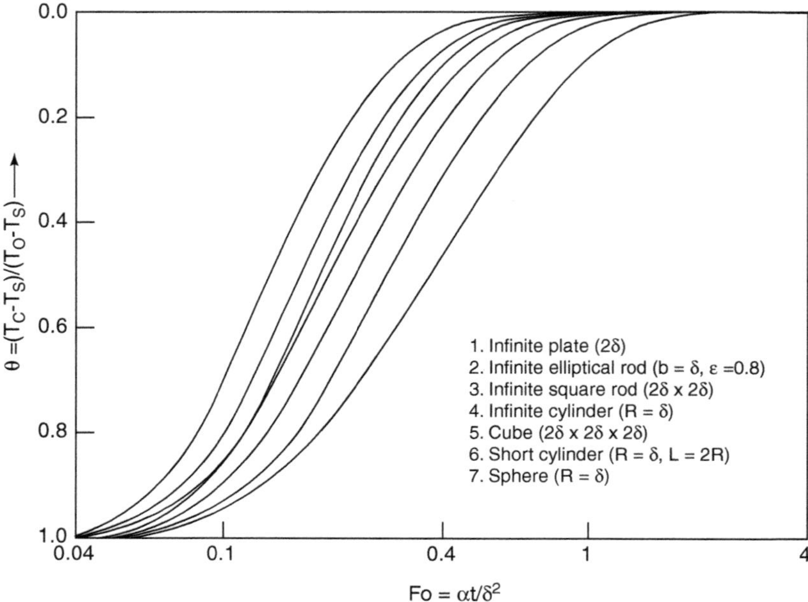

FIGURE 6.4 Temperature response at center of various body shapes after sudden change in external surface temperature from T_0 to T_S.

For the prescribed ordinate value $\theta_{SC} = 0.487$, Figure 6.4 yields a Fourier number of 0.17 for the cylinder and hence a response time of

$$t = (\delta^2/\alpha)Fo = [(0.05)^2/10^{-5}]0.17 = 42.5\ s \tag{6.5c}$$

This is the time required for the temperature at the center to drop from 1000 to 500°C. As δ and t remain unchanged for the simpler geometries, their Fourier numbers will likewise remain unchanged. The corresponding dimensionless temperatures θ_{pp} and θ_{IC}, however, will increase in magnitude, and so will the temperature at the center. This is a reflection of the fact that heat conduction in these simpler geometries is in one direction only, whereas the short cylinder loses heat much faster in both axial and radial directions.

The values of θ, read from the ordinate of Figure 6.4 at a value of Fo = 0.17 are:

$$\theta_{pp} = 0.825 \qquad \theta_{IC} = 0.59 \tag{6.5d}$$

and hence:

$$\theta_{SC} = \theta_{pp} \times \theta_{IC} = 0.487 \tag{6.5e}$$

in agreement with the value for the short cylinder. This confirms the validity of the product solution.

The ease of solution seen here was due in large measure to the invariance of the Fourier numbers for the three geometries. When length of the cylinder no longer equals its diameter, this invariance is lost, and the appropriate Fourier numbers must then be sought out by a trial-and-error procedure until the product solution matches the prescribed value of θ_{SC}. This lengthens the procedure somewhat, but not by much.

Let us consider the case of the longest cylinder of length L = 20, double that of the previous case. The rate of heat loss here will be somewhat slower as will be the response time. The Fourier number for the infinite cylinder, previously at 0.17, will rise somewhat, but that for the parallel planes will almost certainly diminish because δ. A brief trial-and-error procedure, lasting perhaps 15 min., produces the following final results.

Infinite cylinder:	Fo = 0.195, t = 48.8 sec, $\theta_{IC} = 0.515$
Parallel planes:	Fo = 0.053, t = 48.8 sec, $\theta_{pp} = 0.950$

Hence:

$$\theta_{SC} = \theta_{IC} \times \theta_{pp} = 0.515 \times 0.950 = 0.489 \tag{6.5f}$$

in close agreement with the prescribed value of 0.487 of Equation 6.5a. Response time is 48.8 sec, compared to 42.5 sec for the short cylinder.

Similar calculations can be made for the shortest cylinder of L = 5 cm. Here, the response time will be considerably faster and a first reasonable guess for Fo_{IC} is 0.1. The reader can verify the following final numbers:

Infinite cylinder:	Fo = 0.113, t = 28.3 s, $\theta_{IC} = 0.800$
Parallel planes:	Fo = 0.30, t = 28.3 s, $\theta_{pp} = 0.610$

and hence:

$$\theta_{SC} = \theta_{IC} \times \theta_{pp} = 0.8 \times 0.61 = 0.488 \tag{6.5g}$$

Response is now only a little over one-half that of the intermediate cylinder.

Comments: It is worth using these results to obtain a more general sense of the behavior seen in heat conduction processes. The first and most striking impression is the relative rapidity of the process. In cylinders, 50% of the final steady state is attained within a few seconds, the 90% level requires about twice as long (see Figure 6.4).

The effect of the cylinder radius is more pronounced and can be directly assessed from the Fourier number. Thus, doubling the radius to 10 cm will quadruple t, but this still leaves us with a response time of only a few minutes.

For nonconducting materials, such as plastics, wood, glass, or concrete, the prolongation of response time is more severe. Thermal diffusivities α are now of the order of 10^{-7} m^2/sec, and the corresponding response times of the order of hours, rather than minutes or seconds. This is, of course, no less than what one would wish, because these materials are precisely those used for insulating purposes.

We add one final note of caution: individual analytical solutions used in the construction of the product solution need to be carefully checked for their validity. They are usually expressed in terms of dimensionless variables, as shown in

Figure 6.4. The user should be aware, however, that these variables are frequently processed and transformed further for various reasons of convenience. Dimensionless temperature, for example, is often transformed into a slightly different entity, $1 - \theta = (T_0 - T_C)/(T_0 - T_S)$, which causes the ordinate values to invert. Although this new variable still provides a valid representation of the analytical solution, it is no longer capable of producing a product solution. What one requires for its construction is the primary result, and not some transformed version, no matter how innocuous the change. Overlooking this fact leads to endless frustrations and ultimate dismissal of the method.

6.3 SOLUTION OF SOURCE PROBLEMS: SUPERPOSITION BY INTEGRATION

Problems involving sources of mass or energy, and the distributions in concentration and temperature associated with them, arise in a wide variety of ways. A host of environmental problems involve sources of pollutants of one type or another. We have already encountered heat sources at the ODE, i.e., steady state level (Illustration 2.10 and Illustration 4.24) and in the formulations of the Poisson equation (Equation 5.14) for distributed steady sources. The types we wish to consider here initially consist of instantaneous point sources that give rise to unsteady temperature and concentration distribution in infinite or semi-infinite space. The point source results can be integrated over space or time to arrive at distributions due to instantaneous line or area sources, or those due to sources that are continuous over certain time intervals. This is the procedure referred to as *superposition by integration*.

In many of the solutions to source problems by integration, an important new entity called the *error function* makes its appearance and, therefore, we shall devote a brief preamble to its description. The error function, denoted as erf(x), is defined as the integral:

$$erf(x) = \frac{2}{\sqrt{\pi}} \int_0^x e^{-\lambda^2} d\lambda \tag{6.6}$$

where the integrand will be recognized as the Gaussian distribution $\frac{2}{\sqrt{\pi}} e^{-\lambda^2}$. Its properties are summarized in Table 6.2.

erf(x) cannot be determined analytically, and numerical integration is required for its evaluation. A brief table of values appears in Table 6.3.

The source problems we wish to address here are of two basic types: Instantaneous sources involve the release of a fixed amount of mass M_p (kg) or heat q_P (J) over an infinitesimally small time interval. Thereafter, the released substance or heat diffuses into the surrounding space, giving rise to time-dependent concentration and temperature distributions. These distributions can generally be described by negative exponential functions. Continuous sources involve a continuing release of mass M_c (kg/sec) or heat q_C (J/sec) over a certain time period. When release proceeds indefinitely, integration over time of the instantaneous source solutions lead to the aforementioned error functions.

TABLE 6.2
Properties of the Error Function

1. erf(0) = 0
2. erf(∞) = 1
3. erf(−x) = − erf(x)
4. 1 − erf(x) = erfc(x) (complementary error function)
5. $\dfrac{d}{dx} erf\ x = \dfrac{2}{\sqrt{\pi}} e^{-x^2}$
6. $\displaystyle\int_x^\infty erfc\, x\, dx = \frac{1}{\sqrt{\pi}} e^{-x^2} - x\ erfc\, x$
7. Approximation for small x:

   $erf\ x \cong 2\sqrt{x}$
8. Approximation for large x:

   $erfc\ x \cong \pi^{-1/2} e^{-x^2} \left[\dfrac{1}{x} - \dfrac{2}{x^3} + \ldots \right]$

TABLE 6.3
Values of the Error Function

x	erf(x)
0	0
0.05	0.05637
0.1	0.11246
0.15	0.16700
0.20	0.22270
0.25	0.27632
0.30	0.32863
0.35	0.37963
0.40	0.42839
0.50	0.52050
0.60	0.60386
0.70	0.67780
0.80	0.74210
0.90	0.79691
1.0	0.84270
1.2	0.91037
1.5	0.96611
2.0	0.99532
∞	1.00000

A. Instantaneous Source

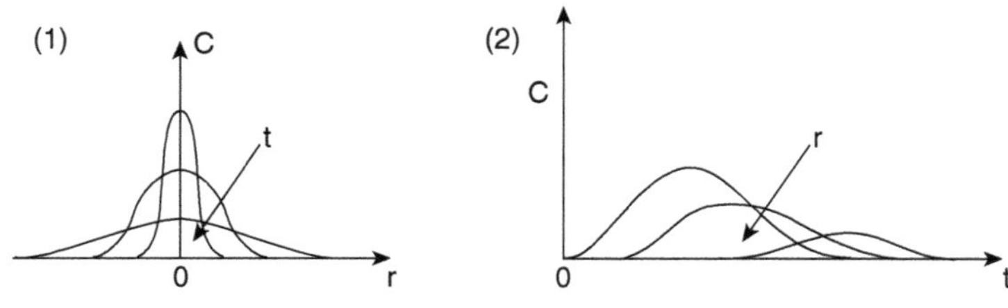

B. Continuous Source

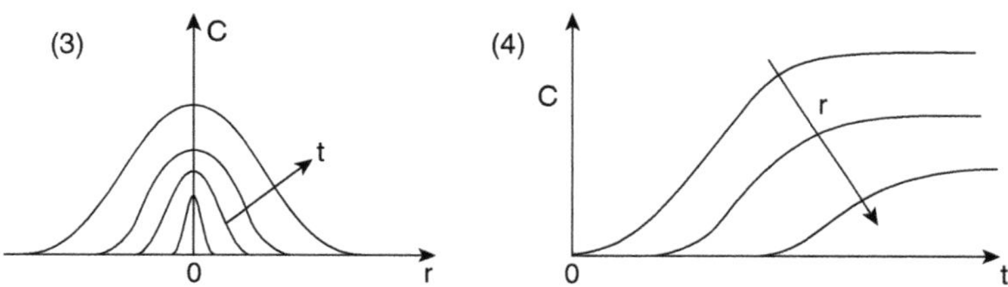

FIGURE 6.5 Concentration profiles and histories in source problems (infinite domain).

To obtain a sense of the spreading waves that result from the emissions of sources, we have sketched the concentration profiles and histories associated with instantaneous and continuous point mass sources emitting radially into infinite space (Figure 6.5). In the case of instantaneous point sources, the pulse emitted at $t = 0$ gradually spreads about the origin and diminishes in strength, i.e., concentration diminishes with the passage of time and distance (part 1 of Figure 6.5). Continuous sources result in a similar spreading of concentrations about the origin, but in this case there is a continuous increase in the concentration level, which ultimately reaches a plateau (part 4 of Figure 6.5). Of interest is the appearance of maxima at a particular location of the domain. In the case of the instantaneous source, the maximum occurs at finite times, whereas the concentrations from continuous sources attain their maximum only in the limit $t \to \infty$. We address these maxima in Illustration 6.5. Completely analogous behavior is shown by the temperature distributions that result from heat sources.

Illustration 6.4 Solutions to Some Simple Heat Source Problems

The instantaneous infinite plane source: Occasionally, simple solutions may be arrived at by inspired guesswork. The solution to the source problem in one-dimensional infinite space for Fourier's equation is a case in point. We consider the source to be on the infinite y, z plane, with temperature variations taking place in the x-direction only. The solution then takes the form:

$$T = \frac{A}{t^{1/2}} \exp(-x^2/4\alpha t) \tag{6.7a}$$

where A is an arbitrary constant. Qualitatively, this expression reflects the required symmetry of the temperature about the plane x = 0, and its decay with both time and distance. It can be shown by direct substitution that it satisfies Fourier's equation:

$$\frac{\partial T}{\partial t} = \alpha \frac{\partial^2 T}{\partial x^2} \tag{6.7b}$$

More importantly, it also satisfies the following initial conditions:

$$T(x,0) = 0 \tag{6.7c}$$

and

$$T(0,0) = \infty \tag{6.7d}$$

Here, Equation 6.7c and Equation 6.7d express the fact that at time zero the entire domain, except for the origin, is at zero temperature, while at the origin it momentarily rises to infinity because of the instantaneous and concentrated nature of the source. The integration constant A is evaluated from the total heat released per unit area of an infinite plane, q_{Tot}, i.e.,

$$q_{Tot} = \rho Cp \int_{-\infty}^{\infty} T dx = A\rho Cp \int_{-\infty}^{\infty} \frac{\exp(-x^2/4\alpha t)}{t^{1/2}} dx = 2A \rho Cp(\pi\alpha)^{1/2} \tag{6.7f}$$

where the integral is obtained from tables of integrals. Thus, the temperature distribution in the infinite x-domain is given by:

$$T = \frac{q_A}{2\rho Cp(\pi\alpha t)^{1/2}} \exp(-x^2/4\alpha t) \tag{6.7g}$$

where q_A is in units of J/m^2.

An expression identical in form applies to mass diffusion, i.e., the solution of Fick's equation for this case is given by:

$$C = \frac{M_A}{2(\pi D t)^{1/2}} \exp(-x^2/4Dt) \tag{6.7h}$$

where M_A has units of kg/m^2.

The instantaneous point heat source in three-dimensional infinite space — This situation involves a point source of strength q_P placed at some arbitrary coordinate point x_0, y_0, z_0. To solve this three-dimensional case, we invoke the principle of superposition by multiplication, described in Section 6.2. This involves extending the solution for the one-dimensional case, given by Equation 6.7g, to the three-dimensional case, by multiplication in the three coordinate directions x, y, and z. We obtain:

$$T(x,y,z,t) = \frac{q_P}{8\rho Cp(\pi\alpha t)^{3/2}} \exp\left[\frac{-(x-x_o)^2-(y-y_0)^2-(z-z_o)^2}{4\alpha t}\right] \tag{6.8a}$$

If we shift the source to the origin and introduce the radial variable:

$$r = (x^2 + y^2 + z^2)^{1/2} \tag{6.8b}$$

we obtain the simple alternative formulation:

$$T(r,t) = \frac{q_P}{8\rho Cp(\pi\alpha t)^{3/2}} \exp(-r^2/4\alpha t) \tag{6.8c}$$

where q_p = total heat emitted by the source, (J).

The finite plane source in three-dimensional infinite space — As we already have in the solution of the point source in three-dimensional space, Equation 6.8a, all we need to do is to extend that solution over the area of the plane source using superposition by integration. We obtain:

$$T(x,y,z,t) = \frac{q_{PL}}{8\rho Cp(\pi\alpha t)^{3/2}} \int_{y_0=c}^{y_0=d}\int_{z_0=a}^{z_0=b} \exp\left[\frac{-(x-x_0)^2-(y-y_0)^2-(z-z_0)^2}{4\alpha t}\right] dy_0 dz_0 \tag{6.9}$$

where q_{PL} represents the total heat emitted, (J).

Note that the integration variables are the coordinates of the point source, x_0, y_0, and z_0, rather than the general space coordinates x, y, and z. The plane source is thus seen to be made up of an infinite number of point sources, which are summed or superposed by integration over the area of the plane.

The continuous source — Here, our starting point is again the instantaneous point source, Equation 6.8c, which now has to be integrated over the lifetime t of the source. To do this, we introduce the integration variable t′ and write the following integral of Equation 6.8c:

$$T(r,t) = \frac{q_{CP}}{8\rho Cp(\pi\alpha)^{3/2}} \int_0^t \frac{e^{-r^2/4\alpha(t-t')}}{(t-t')^{3/2}} dt' \tag{6.10a}$$

Note that the argument of integration is (t – t′) rather than t′. Validation of this expression can be obtained by taking finite increments Δ(t – t′) for a heat release of Δq_{Tot}, summing the increments, and letting Δ(t – t′) → 0.

The form of the integral is similar to that of the error function, Equation 6.6a, and the simple transformation $\tau = (t - t')^{1/2}$ reduces Equation 6.10a to the compact form:

$$T(r,t) = \frac{q_{CP}}{8\rho Cp\pi\alpha r} erfc \frac{r}{(4\alpha t)^{1/2}} \tag{6.10b}$$

where, as noted before:

$$erfc(x) = 1 - erf(x) = 1 - \int_0^x e^{-\lambda^2} d\lambda$$

Equation 6.10b describes the transient radial temperature distribution in three-dimensional infinite space caused by a continuous point heat source q_{CP} (J/sec) at the origin.

The infinite plane source in the semi-infinite domain (*the method of images*) — With this example, we wish to introduce the reader to a graphical technique termed the *method of images*. It consists of placing physical entities on either side of a boundary, here set at x = 0 of a semi-infinite domain, $x > 0$ and proportioning, them in such a way that a prescribed boundary condition, for example, T = 0, is satisfied. In heat conduction, this is done with the help of heat sources and heat sinks. The latter can be viewed as cryogenic sources leading to temperatures below the datum of T = 0. In potential flow, a similar procedure applies in which images of the submerged object are placed on either side of a boundary that is required to have a prescribed value of the stream function, say $\psi = 0$ (see Practice Problem 6.2). The method is most easily applied to a semi-infinite medium, and becomes more complex in bounded geometries such as slabs, cylinders, or spheres.

In the problem considered here, a plane heat source is placed at a distance of $x = x_0$ from the boundary, which is maintained at T = 0 by appropriate cooling. To solve for the resulting temperature distribution by the method of images, a heat sink of equal but opposite strength is placed at a distance $x = x_0$ from the boundary. This sink gives rise to the temperature profile T(x, t) shown in Figure 6.6A. Superposition of source and sink, which is permitted because of the linearity of Fourier's equation, then leads to the profile depicted in Figure 6.6B. As the two entities are of equal but opposite strengths, temperatures at the boundary exactly cancel each other to give the required value of T = 0. We note that the resulting distribution for x > 0 satisfies the posed problem; whereas the entire profile over the range $-\infty < x < \infty$ is to be regarded as a solution to the two-source problem. The relevant equations are as follows:

For the cryogenic sink alone:

$$T = -\frac{q_P}{2\rho Cp(\pi\alpha t)^{1/2}} \exp[-(x + x_0)^2/4\alpha t] \tag{6.11a}$$

A. Cryogenic Sink

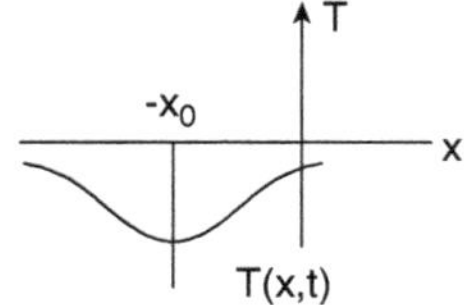

B. Superposition of Source and Sink

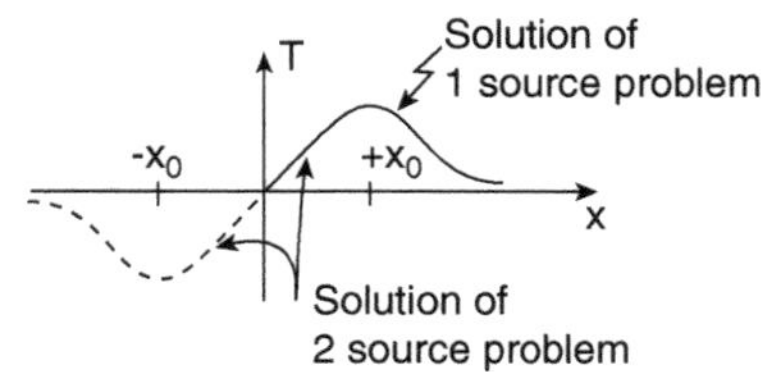

C. Superposition of Source and Source

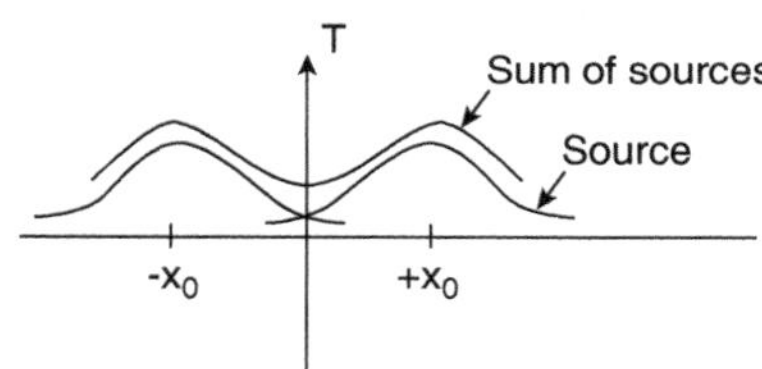

FIGURE 6.6 Superposition of sources and sinks by the method of images.

For the heat source and cryogenic sink:

$$T(x,t) = \frac{q_P}{2\rho Cp(\pi\alpha t)^{1/2}}\{\exp[-(x-x_0)^2/4\alpha t] - \exp[-(x+x_0)^2/4\alpha t]\} \qquad (6.11b)$$

Where g_p is in units of J/m^2.

Equation 6.11a represents the temperature distribution for the following two cases:

1. Heat source q_P at $x = x_0$ of a one-dimensional semi-infinite medium, with its surface $x = 0$ maintained at $T = 0$.
2. Heat source of strength q_P at $x = x_0$ of a one-dimensional infinite domain and a source of strength q_P at the position $x = x_0$.

The latter is unlikely to arise in practice but is added for completeness. Case C of Figure 6.6, on the other hand, involving the addition of two symmetrical sources, can be exploited in a practical way. As the two sources are equal in strength and equidistant from the dividing plane, their outputs will exactly cancel each other at the position x = 0. No heat will cross that position, and the temperature gradient at the plane will consequently vanish. We have, in other words, solved the problem of a source located at a distance $+x_0$ from an impermeable infinite plane. For a heat source, the temperature distribution for this case is given by:

$$T(x,t) = \frac{q_P}{2\rho Cp(\pi\alpha t)^{1/2}}[\exp\{-(x-x_0)^2/4\alpha t\} + \exp\{-(x+x_0)^2/4\alpha t\}] \quad (6.12)$$

We have tabulated these results, and some additional ones of interest in Table 6.4 and Table 6.5. The first is a compilation of solutions of mass source problems, whereas the second addresses the analogous case of heat emissions. Although the variables involved in the two cases differ, the form of the solution is, as it should be, identical.

The following illustration provides an interesting example of the application of various superposition methods:

Illustration 6.5 Concentration Distributions from a Finite Plane Pollutant Source in Three-Dimensional Semi-Infinite Space

Pollution from point, line, and area sources is a frequent occurrence of environmental concern. The resulting concentration profiles are usually modeled by means of PDEs that combine convective transport due to air movement caused by wind, and a diffusive mode of transport. A general formulation of the problem then leads to an extended form of Fick's equation:

$$D_x\frac{\partial^2 C}{\partial x} + D_y\frac{\partial^2 C}{\partial y^2} + D_z\frac{\partial^2 C}{\partial z^2} - v\frac{\partial C}{\partial x} = \frac{\partial C}{\partial t} \quad (6.13a)$$

where $D_{x,y,z}$ are empirical diffusion coefficients, and v the wind velocity. Variations with time and in the direction of v are not taken into account in this first model. Furthermore, we shall assume the dispersion coefficients in x and y to be equal D.

In the example considered here, an instantaneous plane source of strength $S(kg/m^2)$ and dimensions a, b is assumed to be located at ground level. This approximates conditions which arise because of brief periods of pollution, from an industrial area. Material diffuses into the semi-infinite region depicted in Figure 6.7, and is further dispersed by air movement.

We present a solution that makes use of several of the devices and methods that we described in previous sections.

First, we recognize the combination $v(\partial C/\partial x) + \partial C/\partial t$ as a convective derivative, Equation 5.39, which we were able to transform into a single derivative by identifying new independent variables, and carrying out a reduction to canonical form. We choose the combination:

$$u = x - vt \text{ and } \tau = t \quad (6.13b)$$

TABLE 6.4

1. Instantaneous Point Source Emitting into Infinite Space

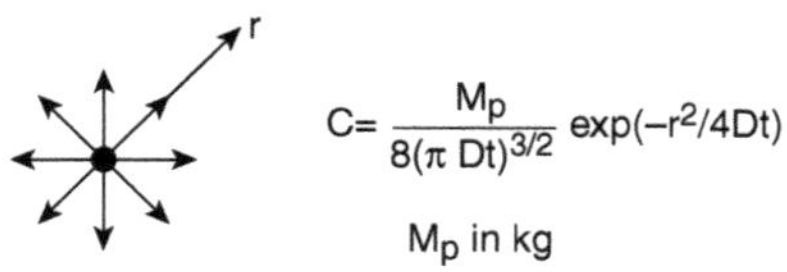

$$C = \frac{M_p}{8(\pi\, Dt)^{3/2}} \exp(-r^2/4Dt)$$

M_p in kg

2. Instantaneous Point Source on an Infinite Plane Emitting into Half-Space

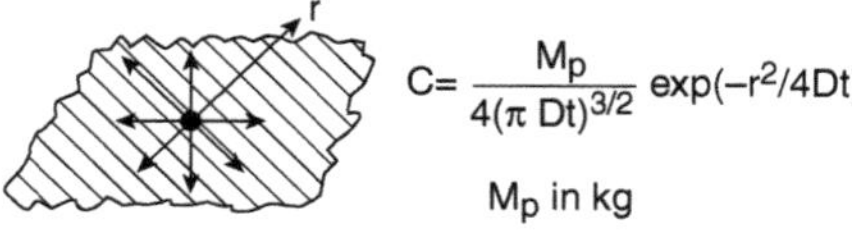

$$C = \frac{M_p}{4(\pi\, Dt)^{3/2}} \exp(-r^2/4Dt)$$

M_p in kg

3. Instantaneous Infinite Plane Source Emitting into Infinite Space

$$C = \frac{M_A}{2(\pi\, Dt)^{1/2}} \exp(-x^2/4Dt)$$

M_A in kg/m^2

4. Instantaneous Infinite Plane Source Emitting into Half Space

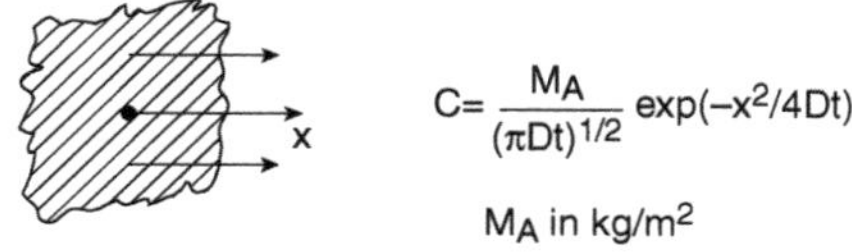

$$C = \frac{M_A}{(\pi Dt)^{1/2}} \exp(-x^2/4Dt)$$

M_A in kg/m^2

5. Continuous Point Source Emitting into Infinite Space

$$C = \frac{M_{CP}}{4\pi Dr} \operatorname{erfc} \frac{r}{2\sqrt{Dt}}$$

M_{CP} in kg/s

6. Continuous Point Source on a Semi-Infinite Plane Emitting into Half Space

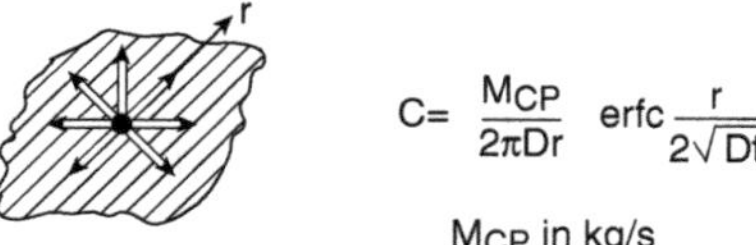

$$C = \frac{M_{CP}}{2\pi Dr} \operatorname{erfc} \frac{r}{2\sqrt{Dt}}$$

M_{CP} in kg/s

7. Continuous Infinite Plane Source Emitting into Half Space

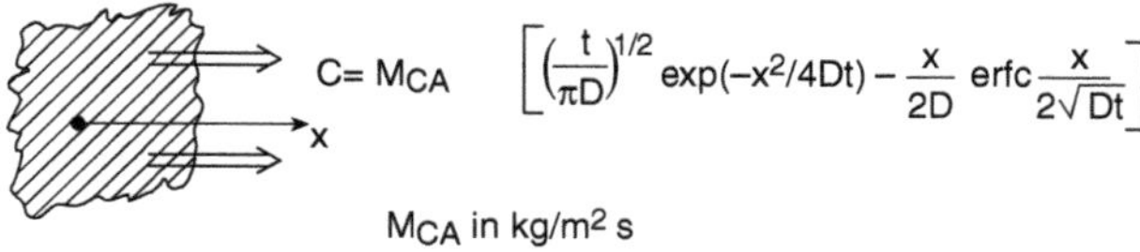

$$C = M_{CA} \left[\left(\frac{t}{\pi D}\right)^{1/2} \exp(-x^2/4Dt) - \frac{x}{2D} \operatorname{erfc} \frac{x}{2\sqrt{Dt}} \right]$$

M_{CA} in kg/m^2 s

TABLE 6.5

1. Instantaneous Point Source Emitting into Infinite Space

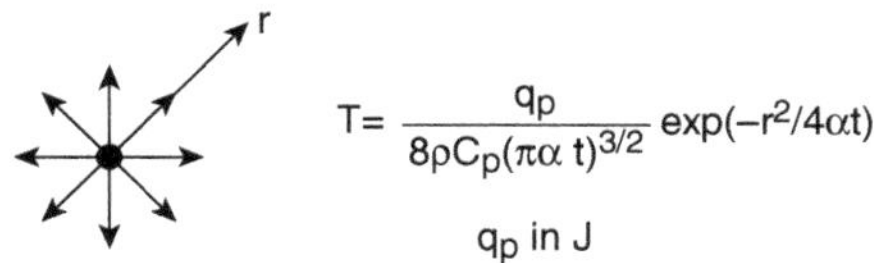

$$T = \frac{q_p}{8\rho C_p(\pi\alpha\, t)^{3/2}} \exp(-r^2/4\alpha t)$$

q_p in J

2. Instantaneous Point Source on an Infinite Plane Emitting into Half-Space

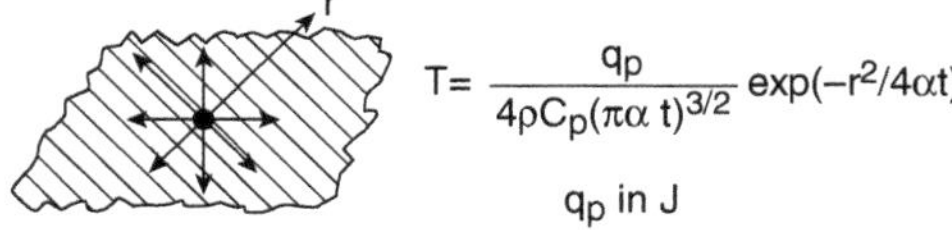

$$T = \frac{q_p}{4\rho C_p(\pi\alpha\, t)^{3/2}} \exp(-r^2/4\alpha t)$$

q_p in J

3. Instantaneous Infinite Plane Source Emitting into Infinite Space

$$T = \frac{q_p}{2\rho C_p(\pi\alpha\, t)^{1/2}} \exp(-x^2/4\alpha t)$$

q_p in J/m^2

4. Instantaneous Infinite Plane Source Emitting into Half Space

$$T = \frac{q_A}{\rho C_p(\pi\alpha\, t)^{1/2}} \exp(-x^2/4\alpha t)$$

q_A in J/m^2

5. Continuous Point Source Emitting into Infinite Space

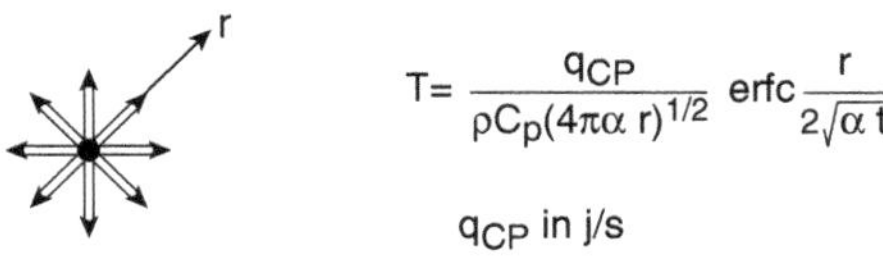

$$T = \frac{q_{CP}}{\rho C_p(4\pi\alpha\, r)^{1/2}} \operatorname{erfc}\frac{r}{2\sqrt{\alpha\, t}}$$

q_{CP} in j/s

6. Continuous Point Source on a Semi-Infinite Plane Emitting into Half Space

$$T = \frac{q_{CP}}{\rho C_p(2\pi\alpha\, r)^{1/2}} \operatorname{erfc}\frac{r}{2\sqrt{\alpha\, t}}$$

q_{CP} in j/s

7. Continuous Infinite Plane Source Emitting into Half Space

$$T = \frac{q_{CA}}{\rho C_p}\left[\left(\frac{t}{\pi\alpha}\right)^{1/2} \exp(-x^2/4\alpha t) - \frac{x}{2\alpha} \operatorname{erfc}\frac{x}{2\sqrt{Dt}}\right]$$

q_{CP} in J/m^2 s

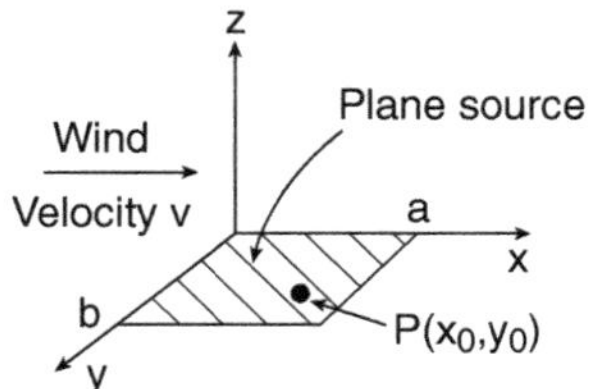

FIGURE 6.7 A finite plane mass source at ground level emitting into the atmosphere.

and obtain the following relation between old and new derivatives using the formulae of Table 5.4:

$$\frac{\partial C}{\partial t} = -v\frac{\partial C}{\partial x} + \frac{\partial C}{\partial \tau}$$

$$\frac{\partial C}{\partial x} = \frac{\partial C}{\partial u} \tag{6.13c}$$

and

$$\frac{\partial^2 C}{\partial x^2} = \frac{\partial^2 C}{\partial u^2}$$

Introduction of these relations into the PDE (Equation 6.13a) yields (after cancellation of terms) the following result:

$$D\left(\frac{\partial^2 C}{\partial u^2} + \frac{\partial^2 C}{\partial y^2}\right) + D_z\frac{\partial^2 C}{\partial z^2} = \frac{\partial C}{\partial \tau} \tag{6.13d}$$

We have thus succeeded in reducing the model to Fick's equation, which we can now proceed to treat by the various superposition techniques established in the previous section.

We start by writing point source solutions for each coordinate direction. We obtain, by adapting Equation 6.7h to the problem:

For the x-direction:

$$C(x,t) = \frac{S}{(4\pi Dt)^{1/2}}\exp[-(x - vt - x_0)^2/4Dt] \tag{6.13e}$$

where we have introduced the new independent variable x – vt into the exponential term.

For the y-direction:

$$C(y,t) = \frac{S}{(4\pi Dt)^{1/2}}\exp[-(y - y_0)^2/4Dt] \tag{6.13f}$$

For the z-direction ($z_0 = 0$):

Here, we must recognize that the domain is a semi-infinite one, with a Type II (no flux) boundary condition at z = 0. The solution to that problem can also be obtained by the method of images, but this time by the addition of two sources. This results in a composite concentration distribution with a minimum at the origin (Figure 6.6C), thus satisfying the no-flux condition. For mass diffusion, this yields:

$$C(z,t) = \frac{2}{(4\pi D_z t)^{1/2}} \exp[-z/4D_z t] \tag{6.13g}$$

Third, with the three unidirectional solutions on hand, we are in a position to solve the three-dimensional case using superposition by multiplication. We obtain:

$$C(x,y,z,t) = \frac{2S}{(4\pi D_z t)^{1/2}} \exp[-z^2/4Dt]$$

$$\frac{1}{(4\pi Dt)} \exp[-(x - vt - x_0)^2/4Dt] \exp[-(y - y_0)^2/4Dt] \tag{6.13h}$$

This is the solution to an instantaneous point source emitting into half-space. The result still needs to be converted to an area source, which is done in the next step.

Fourth, we adapt the method of superposition by integration applied to area sources (Equation 6.9). This leads to the result:

$$C(x, y, z, t) = \frac{2S}{(4\pi D_z t)^{1/2}} \exp[-z_2/4Dt] \times$$

$$\frac{1}{(4\pi Dt)} \int_0^a \int_0^b \exp[-(x - vt - x_0)^2 / 4Dt] \exp[-(y - y_o)^2/4Dt] dx_0 dy_0$$

$$\tag{6.13i}$$

It is left to the practice problems to show that the integrals can be converted into error functions. This yields the final result:

$$C(x,y,z,t) = \frac{S}{(\pi D_z t)^{1/2}} \exp[-z^2/4Dt] \times$$

$$\frac{1}{(4\pi Dt)} \left[erf \frac{x - vt}{(4Dt)^{1/2}} - erf \frac{x - vt - a}{(4Dt)^{1/2}} \right] \left[erf \frac{y}{(4Dt)^{1/2}} - erf \frac{y - b}{(4Dt)^{1/2}} \right] \tag{6.13j}$$

Comments: We have here a fairly impressive success story of the application of various superposition and reduction techniques. The original PDE, although linear, was nevertheless a complex expression in four dimensions (x, y, z, t). A numerical solution would present a fairly formidable task of discretization of four variables. The output would be copious and would lack the cohesiveness of Equation 6.13j.

An additional advantage of the analytical result is its ability to reveal the dependence of the concentration C on various dimensionless parameters. We note in particular the incorporation of the wind velocity into the groups $(x - vt)/(4Dt)^{1/2}$ and $(x - vt - a)/(4Dt)^{1/2}$. Numerical solutions, unless properly nondimensionalized would have to deal with a host of separate parameters.

Further obvious refinements can be added to the model. Continuous sources can be treated by integration over time, as outlined in the previous illustration. Emissions from a smoke stack of height z_0 can be accommodated by the simple change in variable $z \rightarrow z_0$. Variations in diffusivities, particularly D_z, have been expressed in empirical terms described in specialized monographs. All these features and others not covered here have led to a substantial amount of literature on the topic of atmospheric dispersion.

In the illustration that follows, we deviate somewhat from the main theme of this chapter. Although sources are still involved, no superposition methods are applied. The purpose, instead, is to examine the timescale of the diffusional events which accompany the emissions from sources within a practical and realistic context. The numbers we present are startlingly different from what one would expect, and serve as a reminder that models should always be scrutinized for the unusual and unexpected.

Illustration 6.6 Emissions from a Point Source in the Environment and in a Living Cell

To obtain a sense of the speed of mass diffusion, we consider two cases. One involves the instantaneous release of 1 kg of a gaseous substance (say, propane from a small cylinder) and its subsequent diffusion into the surrounding still air for a distance r = 100 m. The second case considers the release of a substance from the center of a living cell into the surrounding cytoplasm. Using a diffusivity in air of $D = 10^{-5}$ m^2/sec, and invoking item 2 of Table 6.4, we obtain for the first case:

$$C = \frac{1}{4(\pi \times 10^{-5} \times 3{,}600 \times 24 \times 10)^{3/2}} \exp\left[-\frac{100^2}{4 \times 10^{-5} \times 3{,}600 \times 24 \times 10}\right] \tag{6.14a}$$

where we have set time t at 10 d. The result, and those for t = 100 d and t = 1000 d, are tabulated in the following:

t (d)	C(kg/m³)
10	0
100	1.4×10^{-24}
1000	6.3×10^{-9}

The data show a very slow, gradual rise in concentration. A subsidiary question that we wish to address before commenting on the results focuses on the time it takes for the concentration to reach its maximum level. That maximum occurs, as was shown in Figure 6.5(2), at some finite point in time, which is independent of the amount released. To show this, we set the time derivative of C equal to zero, obtaining:

$$\exp(-r^2/4Dt^2)t^{3/2} - \exp(-r^2/4Dt)\frac{3}{2}t^{1/2} = 0 \tag{6.14b}$$

and consequently:

$$t_{Max} = \frac{1}{6}(r^2/D) = \frac{1}{6}(100^2/10^{-5}) = 1.67\times 10^8\ s \tag{6.14c}$$

$$C_{Max} = 0.147\ Mp/r^3 = 1.47\times 10^{-7}\ kg/m^3 \tag{6.14d}$$

The striking feature of these results is the extremely slow progress of the diffusion process. It takes almost 6 yr for the concentration to attain its maximum at the modest distance of only 100 m. Evidently these numbers do not present a realistic picture of what would happen in an actual case. Wind and other air currents would intervene to disperse the material at a much higher rate. Turbulent diffusivities tend to be atleast 100 to 1000 times higher than that in still air. It would then still require some 50 to 500 h for the (much attenuated) maximum to reach its target. With wind added, the arrival time is reduced to a few minutes or seconds.

Let us next consider the instantaneous release from a point source at the center of a living cell 100 μm in radius. Although the geometry here is a finite one, the initial stages of the diffusion process can be viewed as taking place into an infinite medium. If we assume D = 10^{-9} m²/sec, typical of a protein, then the time it takes for the maximum to penetrate to a distance of 10 μm is given by:

$$t_{Max} = \frac{1}{12}(r^2/D) = \frac{1}{12}\left(\frac{(10\times 10^{-6})^2}{10^{-9}}\right) = 0.00835s \tag{6.14e}$$

i.e., attainment is almost instantaneous in spite of the much lower diffusivity. Note the vast difference in timescale for the two processes, which is entirely because of the quadratic dependence on r.

Illustration 6.7 Pollutant Diffusion in Water

A semi-infinite column of clear water rests on a long column of solution of pollutant A. The initial concentrations within the water column are: $C_A = C_{A0}$ at $x < 0$ and $C_A = 0$ at $x > 0$. Because of the concentration gradient, pollutant A diffuses along the x-axis into the clear water. The diffusion coefficient is D. We need to obtain a closed-form expression for the concentration of A in water, $C_A(x, t)$.

Our strategy to solve this problem is to use the superposition by integration, i.e., each infinitesimal element in the polluted water column may be considered as a point source, and the overall concentration profile can be obtained by adding the contribution from each of such point sources. We start from the one-dimensional instantaneous point source, Equation 6.7h, for an element dx_o located at x_o away from the interface and integrate this over the overall depth of the polluted column, i.e., $(-\infty, 0)$:

In[1]:= $\int_{-\infty}^{0} \frac{\text{Co}}{2\sqrt{\pi \text{D} t}}$ Exp[- $\frac{(\text{x} - \text{xo})^{\wedge}2}{4 \text{D} t}$] dxD//TraditionalForm

Out[1]//TraditionalForm=

$$\frac{1}{2\sqrt{\pi}\sqrt{Dt}}\left(\text{ColIf}\left[\text{Re}\left(\frac{1}{Dt}\right)>0, \frac{\sqrt{\pi}\,\text{erfc}\left(\frac{1}{2}\sqrt{\frac{1}{Dt}}x\right)}{\sqrt{\frac{1}{Dt}}},\right.\right.$$

$$\left.\left.\text{Integrate}\left[e^{-\frac{(x-xo)^2}{4Dt}}, \{xo, -\infty, 0\}, \text{Assumptions} \to \text{Re}\left(\frac{1}{Dt}\right) \leq 0\right]\right]\right)$$

Apparently, Mathematica encountered difficulty in the integration process. To bypass this issue, let us define a new variable $z = \frac{x - x_o}{2\sqrt{Dt}}$:

In[2]:= **Simplify[$\frac{\text{Co}}{2\sqrt{\pi \text{D} t}}$ Exp[- $\frac{(\text{x} - \text{xo})^{\wedge}2}{4 \text{D} t}$] dxo /.**

{xo → (x - 2$\sqrt{\text{D t}}$ z), dxo -> D[(x - 2$\sqrt{\text{D t}}$ z), z] dz}]//

TraditionalForm

Out[2]//TraditionalForm=

$$\frac{\text{Co}\, e^{-r^2}\,\text{DifferentialD}(e)}{\sqrt{x}}$$

We now try the integration again. Recognizing that the limits are now (∞, x), we obtain the concentration profile in the water column.

In[3]:= $\int_{\infty}^{x} -\frac{\text{Co}}{\sqrt{\pi}} e^{-z^2} dz$

Out[3]= $\frac{\text{Co Erfc}[x]}{2}$

6.4 MORE SUPERPOSITION BY INTEGRATION: DUHAMEL'S INTEGRAL AND THE SUPERPOSITION OF DANCKWERTS

As originally conceived, Duhamel's integral was designed to express the solution of Fourier's equation for a time-varying surface temperature in terms of the solution to the simpler problem with a constant surface temperature. The formula was subsequently extended to Type II and Type III boundary conditions, and to other linear PDEs such as the wave equation, with time-varying forcing function. The general approach for Fourier's equation, which applies equally to Fick's equation, is as follows:

Suppose we wish to solve the following general problem:

$$\frac{\partial v}{\partial t} = \alpha \left[\frac{\partial^2 v}{\partial x^2} + \frac{\partial^2 v}{\partial y^2} + \frac{\partial^2 v}{\partial z^2} \right] \tag{6.15a}$$

or equivalent forms in other coordinate systems, and one of the following boundary conditions applied to the surface S

Type I
$$v(S) = f(t) \tag{6.15b}$$

Type II
$$\frac{\partial v}{\partial n}(S) = g(t) \tag{6.15c}$$

Type III
$$\frac{\partial v}{\partial n}(S) = K[v(S) - h(t)] \tag{6.15d}$$

where n denotes the direction normal to the surface and f, g, and h are arbitrary functions of t.

Suppose further that we have available the solution u to the reduced problem in which f, g, and h are replaced by unity. Then the solution v to the general time-varying problem in the domain V is given by one of the following equivalent integrals:

$$v(V,t) = \frac{\partial}{\partial t}\int_0^t F(t-\tau)u(V,\tau)d\tau \tag{6.15e}$$

$$= \frac{\partial}{\partial t}\int_0^t F(\tau)u(V,t-\tau)d\tau$$

and

$$v(V,t) = \int_0^t \frac{\partial}{\partial t}F(\tau)u(V,t-\tau)d\tau \tag{6.15f}$$

$$= \int_0^t F(t-\tau)\frac{\partial}{\partial t}u(V,\tau)d\tau$$

where we have used the general symbol F to denote any one of the time-varying surface functions f(t), g(t), or h(t).

In practice, the solution to the reduced problem u(V, t) is located in the literature, substituted into either Equation 6.15e or Equation 6.15f, whichever appears more convenient, and the indicated differentiation and integration carried. This certainly requires much less time than the full solution of one of the systems Equation 6.15a to Equation 6.15d, and can be applied to time-varying surface conditions of arbitrary form. If the resulting integral cannot be evaluated analytically, a numerical determination is resorted to.

The following illustration demonstrates the application of Duhamel's integral.

Illustration 6.8 A Problem with the Design of Xerox Machines

The principle of electrostatic copying, or xeroxing, involves charging the copying paper in the image of the original — positive charge for dark areas, negative for light ones — and contacting the paper with negatively charged particles of carbon black encapsulated in a polymer film (Figure 6.8A). The particles adhere to the positive sites and have to be fixed to the paper by raising the temperature to 240°F (115.5°C). Failure to do so results in smudging of the copies.

An early problem that arose in the design of such copiers was the need to bring the paper to the required temperature in the short time stipulated by modern high-speed units. To overcome this difficulty, designers considered exposing the paper to a very intense and short heat flash of 10^{-3}-sec duration produced by a flash-lamp-condenser combination. The power flux from this unit may be regarded as an isosceles

A. Xerox Paper

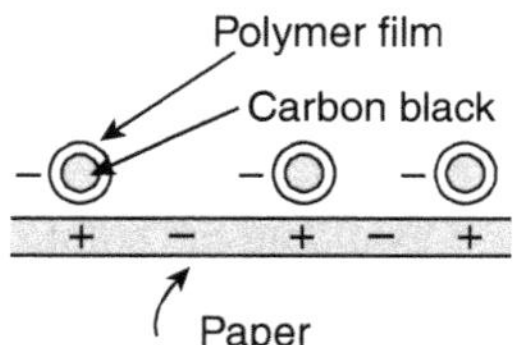

B. Heat Flash and Temperature

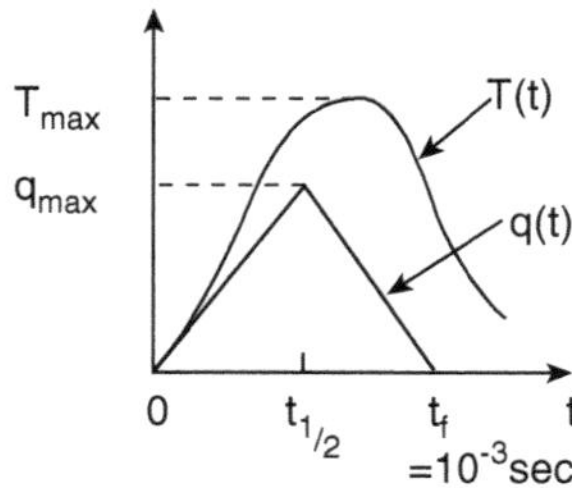

FIGURE 6.8 Heat conduction into xerox paper: (A) Composition of the paper, (B) Surface heat flux and temperature as a function of time.

triangle with base width $\tau_f = 10^{-3}$ sec (see Figure 6.8B). For the design of the flash-lamp-condenser unit, it is required to know the maximum heat flux q_{Max} which will give the required maximum surface temperature of 115.5°C. Note that the maximum lags behind that of the heat flux, and is to be found somewhere between $t_{1/2}$ and t_f.

To solve this rather intricate problem, we propose to proceed as follows. Calculate the temperature profile that results from the time-varying heat flash. This can be done using the known surface temperature that results from a constant heat flux of q = 1, and superposing that solution by means of Duhamel's integral to arrive at a solution for the variable heat flux case. Standard references on heat conduction provide the following temperature distribution $T_c(x, t)$ for constant unit flux q = 1 and a semi-infinite medium.

$$T_c(x,\, t) - T(x,\, 0) = \frac{2}{k}\left[\left(\frac{\alpha t}{\pi}\right)^{1/2} \exp(-x^2/4\alpha t\right] - \frac{x}{2} erfc(x/2\sqrt{\alpha t})$$

The corresponding temperature at the surface is then given by:

$$T_c(0,t) - T(x,0) = \frac{2}{k}\left(\frac{\alpha t}{\pi}\right)^{1/2} \tag{6.16a}$$

where $T_c(0, t)$ denotes the surface temperature resulting from a constant unit heat flux. This is the temperature that is used in the superposition integral to arrive at the solution $T_v(x, t)$ for time-varying flux. We choose Equation 6.15e to superpose and obtain:

$$T_v(0,t) - T(0,0) = \int_0^t \frac{\partial q(\tau)}{\partial t}[T(0,t-\tau) - T(0,0)]d\tau \tag{6.16b}$$

where $T_v(0,t)$ is now the result of variable heat flux.

The next step is to calculate ∂q/∂t, using Figure 6.8B as a guide. Two expressions are obtained, one for the rising portion and one for the falling portion of the isosceles triangle.

Substitute ∂q/∂t and Equation 6.16a into the integral of Equation 6.16b. Integrate in two steps: from 0 to $t_{1/2}$ and from $t_{1/2}$ to t. The result contains q_{Max}, the parameter to be extracted:

$$T_v(0, t) = f(q_{Max}, t) \tag{6.16c}$$

To obtain q_{Max} from this expression, we use the prescribed T_{max} = 115.5 and the corresponding time t_{Max} obtained by differentiating Equation 6.16b and setting the derivation = 0. This is the time at which the maximum temperature shown in Figure 6.8B occurs.

In the final step, we substitute t_{Max} and T_{Max} = 115.5°C into the evaluated Equation 6.16c and solve for q_{Max}. This is the desired result.

Let us see how this works out in practice. Data for xerox paper:

Thermal conductivity h = 0.1295 J/m sec K
Thermal diffusivity $\alpha = 6.65 \times 10^{-8}$ m²/sec
Thickness L = 7.63×10^{-5} m
Initial temperature T(x, 0) = 20.1°C = 203.3 K
Adsorption of incident heat = 30%

We now proceed with a stepwise solution as outlined in the preceding text. Substitution of Equation 6.16b into the Duhamel's integral yields:

$$T_v(0,t) - T(0,0) = \int_0^t \frac{\partial q}{\partial t} \frac{2}{k}\left(\frac{\alpha}{\pi}\right)^{1/2} (t-\tau)^{1/2} d\tau \tag{6.16d}$$

which requires the determination of ∂q/∂t (second step, see the following text) and subsequent integration.

From the slopes of the triangular heat pulse, Figure 6.8B, we obtain for the flux derivative:

$$\frac{\partial q}{\partial t} = \frac{q_{Max}}{t_{1/2}} \quad \textit{for the interval} \quad 0 < \tau < t_{1/2} \tag{6.16e}$$

and

$$\frac{\partial q}{\partial t} = \frac{q_{Max}}{t_{1/2}} \quad \textit{for the interval} \quad t_{1/2} < \tau < t_f$$

Substitution of Equation 6.16e into the integral Equation 6.16d and a two-step integration leads to the sum:

$$T_v(0,t) - T(0,0) = \frac{2}{k}\left(\frac{\alpha}{\pi}\right)^{1/2} \frac{q_{Max}}{t_{1/2}} \left[\int_0^{t_{1/2}} (t-\tau)^{1/2} d\tau - \int_{t_{1/2}}^{t} (t-\tau)^{1/2} d\tau\right] \tag{6.16f}$$

and upon evaluation of the integrals to the expression:

$$T_v(0,t) - T(0,0) = \frac{2}{k}\left(\frac{\alpha}{\pi}\right)^{1/2} \frac{q_{Max}}{t_{1/2}} \left[\frac{2}{3} t^{3/2} - \frac{4}{3}(t - t_{1/2})^{3/2}\right] \tag{6.16g}$$

We now evaluate $\partial T_v/\partial t$, set it equal to zero, and solve for t_{Max}:

$$\frac{\partial T_v}{\partial t} = \frac{2}{k}\left(\frac{\alpha}{\pi}\right)^{1/2} \frac{q_{Max}}{t_{1/2}} \left[t_{Max}^{1/2} - 2\left(t_{Max} - t_{1/2}\right)^{1/2}\right] = 0 \tag{6.16h}$$

This leads to the result:

$$t_{Max} = \frac{4}{3} t_{1/2} \tag{6.16i}$$

i.e., t_{Max} lies between $t_{1/2}$ and t_f, as expected.

t_{Max} and the desired $(T_v)_{Max}$ = 115.5°C = 388.8 K are substituted into Equation 6.16g and the result solved for q_{Max}. We obtain:

$$q_{Max} = \frac{k}{2}\left(\frac{\alpha}{\pi}\right)^{1/2} \frac{[(T_v)_{Max} = T(0,0)]}{\left[\frac{2}{3} t_{Max}^{3/2} - \frac{4}{3}(t_{Max} - t_{1/2})^{3/2}\right]} \tag{6.16j}$$

or:

$$q_{Max} = \frac{0.1295}{2}\left(\frac{\pi}{6.65 \times 10^{-8}}\right)^{1/2} \frac{[388.8 - 293.3]}{\frac{2}{3}\left(\frac{4}{3} 0.5 \times 10^{-3}\right)^{3/2} - \frac{4}{3}\left[\frac{4}{3} 0.5 \times 10^{-3} - 0.5 \times 10^{-3}\right]^{3/2}}$$

$$= 0.5 \times 10^{-3}$$

Hence, $q_{Max} = 2.47 \times 10^6$ J/m² sec at 100% efficiency and $q_{Max} = 8.22 \times 10^6$ J/m² sec at 30% efficiency

This is the desired peak flux required for the design of the flash lamp-condenser unit.

Comments: The reader will have noted that the literature solution for a semi-infinite medium was used in the calculations. This may appear surprising, given the small value of the thickness L of the paper. A quick calculation of the dimensionless time argument that appears in Equation 6.16a, however, shows this to be justified. We obtain:

$$\frac{L^2}{4\alpha t} = \frac{(7.63 \times 10^{-5})^2}{4(6.65 \times 10^{-8})10^{-3}} = 21.8 \tag{6.16k}$$

and

$$\frac{L}{2(\alpha t)^{1/2}} = 4.67$$

For these values of their respective arguments, both the exponential and the erfc in Equation 6.16a become vanishingly small, so that:

$$T_v(L,t) = \mathrm{T}(L,0) \tag{6.16l}$$

i.e., the bottom of the paper x = L, will, for the duration of the heat pulse, remain at the initial of 20.1°C. This justifies the seemingly bizarre description of a sheet of paper as a semi-infinite medium. Had we chosen a finite geometry instead, the resulting solution would have been considerably more complex. We thus made effective use of the suggestions given under Section 5.3.5 in Chapter 5.

Illustration 6.9 Temperature Distribution in Semi-Infinite Solid with Time-Dependent Surface Temperature Using Mathematica

Let us examine the problem of heat transfer in a semi-infinite solid, but this time with a time-varying surface temperature. The initial temperature of the solid is zero, and surface temperature increases linearly with time according to $T(0, t) = kt$. The differential equation and the initial and boundary conditions for this system are given by:

$$\frac{\partial T}{\partial t} = \alpha \frac{\partial^2 T}{\partial x^2} \tag{6.17a}$$

$$T = 0, \quad when \quad t = 0, \quad \& \quad x \geq 0 \tag{6.17b}$$

$$T = 0, \quad when \quad x = \infty \quad \& \quad t > 0 \tag{6.17c}$$

$$T = kt, \quad when \quad x = 0 \quad \& \quad t > 0 \tag{6.17d}$$

We need to find the expression for the temperature distribution within this solid, $T(x, t)$.

The temperature distribution, $u(x, t)$, for a semi-infinite slab with a zero initial temperature subjected to a unit surface temperature at time $t = 0$ can be found readily from literature (see Illustration 8.8 with $T_o = 0$ and $T_s = 1$):

$$u(x,t) = erfc\left(\frac{x}{2\sqrt{\alpha t}}\right) \tag{6.17e}$$

Using Duhamel's integral, Equation 6.15f, the solution to the original problem can be written as:

$$T(x,t) = \int_0^t \frac{\partial F}{\partial x} u(x, t-\tau)\ d\tau \tag{6.17f}$$

where $F(x, t)$ is the time-dependent boundary condition. Now, we continue solving the problem in Mathematica. We start by defining functions u and F:

```
In[1]:= u[x_, t_] := Erfc[x/(2√(α t))]

In[2]:= F[t_] := kt;
```

We now apply these to the Duhamel's integral:

```
In[3]:= DFt = ∂tF[t] /.t → τ
Out[3]= k
```

```
In[4]:= Refine[∫_0^t DFt u[x, t - z]dz, Assumptions → Re[x^2/α] > 0]
        Integrate::idiv:
                                          *
          Integral of Erfc[- - - - - - - - - - - -] does not coverage on (0, t)
                                 Σ Sqrt[α (t  -  τ) ]
          .More...
```

Mathematica has encountered difficulty in converging to the proper result. Here, we try to determine source of the problem and overcome it by finding the indefinite integral and finding the result at the limits:

```
In[5]:= (integral = Refine[∫ DFt u[x, t - τ]d τ])// TraditionalForm
Out[5]//TraditionalForm=
```

$$k\left(\frac{\operatorname{erf}\left\{\frac{x}{2\sqrt{\alpha(t-\tau)}}\right\}x^2}{2\alpha}+\frac{\frac{x^2}{e^{4\alpha\tau-4t\alpha}}\sqrt{\alpha(t-\tau)\,x}}{\sqrt{\pi}\,\alpha}+(\tau-t)\operatorname{erfc}\left(\frac{x}{2\sqrt{\alpha(t-\tau)}}\right)\right)$$

The result of integral at $\tau\to 0$ and $\tau\to t$ will become:

```
In[6]:= (lowerlimit = FullSimplify[PowerExpand[Limit
          [integral, τ → 0]]])// TraditionalForm
Out[6]//TraditionalForm=
```

$$k\left(-t+\frac{e^{-\frac{x^2}{4ta}}x\sqrt{t}}{\sqrt{\pi}\sqrt{\alpha}}+\left(\frac{x^2}{2\alpha}+t\right)\operatorname{erf}\left(\frac{x}{2\sqrt{t}\sqrt{\alpha}}\right)\right)$$

```
In[7]:= upperlimit = FullSimplify[Limit[integral, τ → t]]//
        TraditionalForm
Out[7]//TraditionalForm=
```

$$\lim_{\tau+t} k\left(\frac{\operatorname{erf}\left(\frac{x}{2\sqrt{\alpha(t-\tau)}}\right)x^2}{2\alpha}+\frac{\frac{x^2}{e^{4\alpha\tau-4t\alpha}}\sqrt{\alpha(t-\tau)x}}{\sqrt{\pi}\,\alpha}+(\tau-t)\operatorname{erfc}\left(\frac{x}{2\sqrt{\alpha(t-\tau)}}\right)\right)$$

As you may notice, calculation of the upper limit is the source of our earlier problem in finding the answer to the definite integral. However, we realize that as $\tau \to t$, the first term of the preceding equation approaches $kx^2/2\alpha$, while the remaining terms vanish. Hence:

```
In[8]:= term1 = k x^2/(2 α);
        term2 = 0;
        term3 = 0;
        upperlimit = term1 + term2 + term3
```

Out[11]= $\frac{k\,x^2}{2\,\alpha}$

With this "manual" intervention, we are ready to find the final answer:

```
In[12]:= FullSimplify[upperlimit - lowerlimit]
          //TraditionalForm
```

Out[12]//TraditionalForm=

$$\frac{k\left((x^2+2t\alpha)\operatorname{erfc}\left(\frac{x}{2\sqrt{t}\sqrt{\alpha}}\right)-\frac{2z^{-\frac{x^2}{4t\alpha}}\sqrt{t}\,k\sqrt{\alpha}}{\sqrt{\partial}}\right)}{2\alpha}$$

An alternative way to reach to this expression is using a simple, but not trivial, change of sign of t and τ in the calculation of the Duhamel's integral is:

```
In[13]:= FullSimplify[
           Refine[-∫_0^t DFt u[x, τ - t] dτ,
            Assumptions → (Re[x^2/α] > 0 &&
            Re(x / √α) > 0 && x > 0)] /. t → - t
         ]//TraditionalForm
```

Out[13]//TraditionalForm=

$$\frac{k\left((x^2+2\,t\alpha)\operatorname{erfc}\left(\frac{x}{2\sqrt{\tau}\sqrt{\alpha}}\right)-\frac{2\tau^{-\frac{x^2}{4t\alpha}}\sqrt{t}\,x\sqrt{\alpha}}{\sqrt{\partial}}\right)}{2\alpha}$$

The superposition of Danckwerts — Danckwerts considered the problem of Fickian diffusion with an accompanying first-order irreversible reaction. In one-dimensional Cartesian coordinates, the system is described by the PDE:

$$D\frac{\partial^2 C}{\partial x^2} - k_1 C = \frac{\partial C}{\partial t} \tag{6.18a}$$

His proposal was to express the solution of Equation 6.16 in terms of the solution C of Fick's equation without a reaction term. He succeeded in developing the following superposition formula:

$$C = k_r D\int_0^t e^{-kD\tau} C'(\tau)d\tau + C' e^{-kDt} \tag{6.18b}$$

where C′ is the solution to Fick's equation:

$$D\frac{\partial^2 C'}{\partial x^2} = \frac{\partial C'}{\partial t} \tag{6.18c}$$

Expressions of identical form were derived by Danckwerts for spheres and infinitely long cylinders for boundary conditions of Type I and Type III. For a semi-infinite medium, the following expression results:

$$C/C_0 = \frac{1}{2}\exp(-x\sqrt{k/D})erfc\left[\frac{x}{2\sqrt{Dt}} - \sqrt{k_r t}\right]$$

$$+\frac{1}{2}\exp(x\sqrt{k_r/D})erfc\left[\frac{x}{2\sqrt{Dt}} - \sqrt{k_r t}\right] \tag{6.18d}$$

The derivation of this expression is left to the reader as a practice problems.

PRACTICE PROBLEMS

6.1 Superposition of Functions

Given the two-dimensional function u = f(x, y), which of the following solution pairs can be superposed?

(I)	$u = y\ e^{-x}$	(II)	$\sin y + ku = 0$
	$x^2u + g(y) = 0$		$x \tan u + yu = 0$
(III)	$x/y^2 + k/u = 0$	(IV)	$u + k(xu)^{1/3} = 0$
	$u = g(x, y)$		$u = f(x) + g(y)$

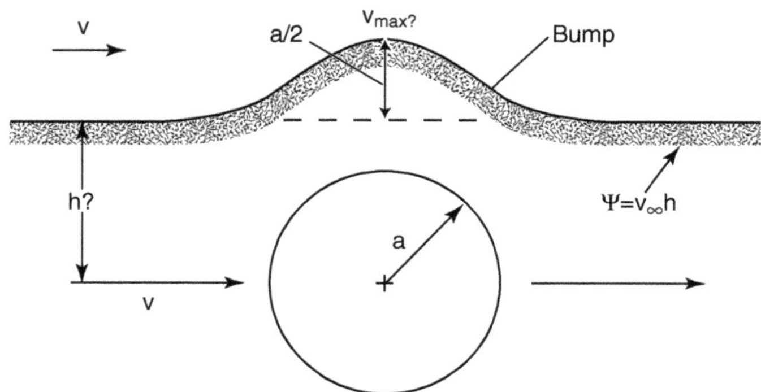

FIGURE 6.9 Flow over a ridge.

6.2 Superposition of Line Sources

A line source is the extension of a point source along an entire coordinate axis $(-\infty, \infty)$. Suppose two line sources of equal strength are placed symmetrically on either side of an infinite plane. Sketch the streamlines that would result.

6.3 Flow over a Ridge

It was pointed out in Illustration 6.1 that the combination of uniform flow and a doublet is also descriptive of flow around bumps or ridges, as well as flow around a circular cylinder.

Consider the ridge shown in Figure 6.9. Its height is defined by the radius a of the associated cylinder and the distance h. What is v_{Max} at the top of the ridge in terms of the free-stream velocity of approach?

(Hint: Apply the equation of the streamline at $\theta = 90°$ to establish the relation between a and h.)

6.4 Wind Velocity and Pressure near the Eye of a Tornado

The flow pattern in tornadoes is well represented by superposition of a vortex and a sink. Given that at a distance of 1 km from the eye, $v_\theta = v_r = -0.5$ m/s and P = 1 atm, calculate the absolute velocity and pressure at a distance of 15 m from the eye. Assume a constant density of air $\rho = 1.25$ kg/m^3.

Answers: v = 27.1 m/s; p = 0.986 atm

Comment: Viscous effects usually start making themselves felt at a distance of 40 m from the eye. These answers are therefore to be regarded as maximum values.

6.5 Wind Force on a Quonset Hut

A Quonset hut, which is a semicylindrical housing structure, is exposed to a wind with transverse velocity of v_∞. Derive an expression for the force exerted on a hut of radius R and length L, given that the internal pressure p_i equals the free-stream pressure.

6.6 More on Product Solutions: Leaching from Solid Particles

Solid particles in the shape of thin disks, R = 2.5 mm and L = 2 mm, are to be leached with solvent to recover valuable material. Estimate the fraction of solute remaining at the center of the particles after 2, 5, and 10 min of leaching. External concentration is taken to be zero, and the internal solute diffusivity is set at 10^{-9} m^2/sec.

Answer: 0.92, 0.52, 0.25

6.7 Embedded Sources

Sources that are embedded in a medium, such as implants in the human body or subsurface deposits, represent a special subclass of emissions. The release of material here is often driven by a prevailing constant concentration or vapor pressure at the external surface. Analyze these systems and state what models or model solutions are appropriate for describing the resulting concentration distributions in the surrounding space and the associated release rates.

6.8 A Moving-Source Problem

A moving point contact is pressed against the plane x = 0 with a constant force F per unit length. It moves with a speed v, producing a constant coefficient of friction μ (Figure 6.10). Obtain the temperature distribution T(x, y, z), assuming an initial temperature of zero and neglecting heat losses to the surroundings. This type of problem arises in milling and lathe operations.

Hint: The problem is equivalent to a stationary continuous source of strength Fμv and a medium moving with velocity v. The PDE is then of the form of Equation 6.13a,

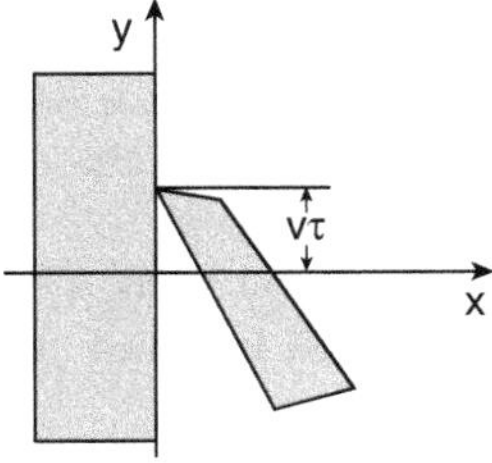

FIGURE 6.10 Configuration of a moving point source.

Illustration 6.5. Reduce it to Fourier's equation in line with what was done there, and follow this up with superposition by multiplication and by integration over time.

Answer: $$T(x,y,z) = \frac{F\mu v}{8\rho\, Cp(\alpha\pi)^{3/2}} \int_0^t \frac{\exp - [x^2 + \{y - v(t-\tau)\}^2 + z^2]/4\alpha(t-\tau)}{(t-\tau)^{3/2}}$$

6.9 Accident Response

Suppose 100 kg of a toxic vapor has been accidentally released at some point in a populated area. Given that an exposure to a concentration level higher than C = 5 ppm is deemed dangerous, should the population at a location 500 m from the accident be evacuated? Assume a very high dispersion coefficient D of 0.1 m^2/sec (D values are subject to a high degree of uncertainty and dependent on atmospheric conditions).

6.10 Temperature Penetration into the Ground

In laying an underground water main, it is desired to establish whether a prolonged severe cold spell would have the potential of causing freezing conditions. Assume the upper part of the pipe to be located 5 m below the surface, and the expected air temperature variables, averaged over a period are as follows:

Midnight to 6 AM	–20°C
6 AM to noon	–10°C
Noon to 6 PM	–7°C
6 PM to midnight	–12°C

Follow the penetration for a period of days and indicate at what point a danger of freezing would arise. Assume T_0 = 20°C, $\alpha = 10^{-6}$ m^2/sec.

(Hint: Use Duhamel's formula.)

6.11 The Danckwerts Superposition

Derive the formula of Equation 6.17d using the Danckwerts superposition integral, Equation 6.17b. The result describes the unsteady concentration distribution for diffusion, accompanied by a first-order irreversible reaction in a semi-infinite medium. Obtain the required concentration distribution C′(x, t) for the case of no reaction from the pertinent literature, and use the error function relations given in Table 6.2.

Hint: (1) Integrate the expression by parts. (2) Differentiate the result from (1) to obtain an error function form. (3) Integrate the result from (2) by parts.

6.12 Method of Images: Semi-infinite Solid with Nonhomogeneous Initial Temperature

A semi-infinite solid has an initial temperature of $T(x,0) = cx^2$ (for $x \geq 0$) , where c is a constant. The surface of this solid is maintained at zero temperature for $t > 0$. Using the method of images, find an expression for the temperature distribution within this solid, $T(x, t)$.

6.13 Product Solution: Diffusion in a Semi-Infinite Slab

Consider a semi-infinite slab ($0 < x < 2L$, $y > 0$) where the initial concentration C_o is constant throughout the slab. At time t = 0, all surfaces of this object (i.e., $x = 0$, $x = 2L$, and $y = 0$) are subjected to the zero surface condition, i.e.,

$$C(0, y, t) = C(2L, y, t) = C(x, 0, t) = 0$$

Using the method of product solution and by utilizing the existing solutions from the literature, obtain an expression for the transient concentration profile in this medium.

6.14 Superposition by Integration: Transient Diffusion of Solute in a Medium

Consider a medium in which solute S is initially confined in the region $-a < x < +a$ with a uniform concentration C_o. At time $t > 0$, this solute starts to diffuse along the x-axis. The diffusion coefficient of solute in the medium is D. Using the method of superposition by integration, find an expression for the transient concentration profile in this system.

6.15 Superposition of Simple Flows: Vortex Around a Moving Cylinder

A cylinder with radius R is submerged in a flowing stream of inviscid liquid with velocity U. The axis of this cylinder is perpendicular to the flow. In addition, there is a clockwise vortex around this cylinder. The angular speed of vortex at the surface of the cylinder is w (rad/sec). By superposition of simple flows, find the velocity components v_r and v_q for the two-dimensional potential flow around this cylinder in terms of U, R, and w.

7 Vector Calculus: Generalized Transport Equations

In the preceding two chapters, we have presented an introduction to multidimensional models in time or space that took the form of partial differential equations (PDEs). These were, for the most part, given in Cartesian coordinates for simplicity and confined to the classical equations of mathematical physics, such as the Laplace, Fourier, and Fick's equations.

In the present chapter, we undertake the derivation of the generalized, multidimensional equations for the transport of mass, energy, and momentum, based on their respective conservation laws. In other words, we will be dealing with mass, energy, and momentum balances in a most general way and at the PDE level. As before, momentum balances may also take the equivalent form of force balances, or extensions of Newton's law.

The generalized approach that we shall be taking calls for the use of vector notation and vector operations principally within the framework of vector calculus. Therefore, we shall start our deliberations with an introduction to this difficult topic, which is often viewed with misgivings by beginners. We shall attempt to demystify the subject matter by placing it within the framework of physical reality and justifying its use by means of illustrations. This is followed by individual sections dealing with mass, energy, and momentum balances, which make use of the tools of vector calculus provided in the lead-in section. Solutions of PDEs, when they do occur, will make use only of the limited methods presented in Chapter 5 and Chapter 6, or those given at the ordinary differential equation (ODE) level. More elaborate and specialized solution methods will be presented and applied to the conservation laws in the concluding chapter, Chapter 8. The present chapter may thus be regarded as a bridge between the introductory treatment of Chapters 5 and 6 and the elaborations that follow in Chapter 8.

7.1 VECTOR NOTATION AND VECTOR CALCULUS

We start this section by listing the principal features of vector operations and the uses to which they can be put.

A first point to note is that vector notation is, in essence, a convenient shorthand for a variety of algebraic, geometric, and differential operations and equations. Thus, the operational form $\nabla^2 u = 0$ is a shorthand version of Laplace's equation, which, in Cartesian coordinates, has the longhand form $\partial^2 u/\partial x^2 + \partial^2 u/\partial y^2 + \partial^2 u/\partial z^2 = 0$.

The symbol ∇^2, the so-called Laplacian, is regarded as an operator, symbolically represented by $\partial^2/\partial x^2 + \partial^2/\partial y^2 + \partial^2/\partial z^2$ in Cartesian coordinates. It resembles, in this respect, the D-operator, which was encountered at the ODE level and symbolically expressed the operation d/dx.

A second important point is that the operational notation or shorthand is independent of geometry. Thus, the expression $\nabla^2 u = 0$ represents Laplace's equation in any arbitrary coordinate system, including Cartesian, spherical, and cylindrical coordinates. This shorthand form can be expanded into conventional notation by means of convenient dictionaries or conversion tables, an example of which appears in Table 7.1.

Any problem that can be treated vectorially can also, in principle, be solved by nonvectorial methods. However, the vectorial approach simplifies the manipulations and renders them less complex and much more elegant. The more complex a system, the greater the need for operational notation. Viscous flow, both Newtonian and non-Newtonian, and mechanical systems involving simultaneous translation and rotation are typical of processes that benefit from vectorial representations. The Maxwell equations of electromagnetics are invariably expressed and manipulated in vectorial form.

Merely putting a differential equation in vector form does not yield a solution. The vector equations must still be converted to scalar form using the appropriate dictionary and solved by scalar methods. However, vector transformations undertaken prior to solution are of considerable aid in simplifying the PDEs.

Some operators, such as the Laplacian mentioned earlier, and the dot product ($\mathbf{A} \bullet \mathbf{B}$) familiar from elementary vector algebra yield, on expansion, single scalar equations; in other words, they are themselves scalar. Others, such as the cross product ($\mathbf{A} \times \mathbf{B}$) and the detested curl $\nabla \times \mathbf{v}$, are themselves vectors and hence decompose into an equivalent set of three scalar equations.

We mention these points as part of our endeavor to make vector notation more palatable to the reader. To further ease the transition, we present by way of an introduction to vector calculus, a brief summary of the principal relations pertaining to vector algebra. Most of these will be familiar from one course or another, although they may have been conveniently forgotten.

A brief synopsis of vector algebra. The principal features and operations of vector calculus bear a distinct relation to those that apply at the more elementary level of vector algebra. Definitions of a vector, and simple operations such as addition, subtraction, and multiplication by a scalar, are identical in both disciplines. Both employ dot and cross products and, for the most part, these operations again lead to identical results. Another common feature is the fact that vector operations can be viewed in a number of different ways, both mathematical and physical. The dot product, for example, can be regarded as the sum of three products, and also has the property of representing the projection of one vector onto another. We shall make use of all these properties, as is indeed the usual case in vector algebra. The only difference, as the names imply, is that vector calculus deals with derivatives and integrals, whereas vector algebra is confined to the use of algebraic (and trigonometric) relations.

TABLE 7.1
Differential Operators in Various Coordinate Systems

A. Gradient ∇u

Cartesian

$$(\nabla u)_x = \frac{\partial u}{\partial x} \qquad (\nabla u)_y = \frac{\partial u}{\partial y} \qquad (\nabla u)_z = \frac{\partial u}{\partial z}$$

Cylindrical

$$(\nabla u)_r = \frac{\partial u}{\partial r} \qquad (\nabla u)_\theta = \frac{1}{r}\frac{\partial u}{\partial \theta} \qquad (\nabla u)_z = \frac{\partial u}{\partial z}$$

Spherical

$$(\nabla u)_r = \frac{\partial u}{\partial r} \qquad (\nabla u)_\theta = \frac{1}{r}\frac{\partial u}{\partial \theta} \qquad (\nabla u)_\varphi = \frac{1}{r\sin\theta}\frac{\partial u}{\partial \varphi}$$

B. Divergence ∇ • v

Cartesian

$$\nabla \bullet \mathbf{v} = \frac{\partial v_x}{\partial x} + \frac{\partial v_y}{\partial y} + \frac{\partial v_z}{\partial z}$$

Cylindrical

$$\nabla \bullet \mathbf{v} = \frac{1}{r}\frac{\partial}{\partial r}(r v_r) + \frac{1}{r}\frac{\partial v_\theta}{\partial \theta} + \frac{\partial v_z}{\partial z}$$

Spherical

$$\nabla \bullet \mathbf{v} = \frac{1}{r^2}\frac{\partial}{\partial r}(r^2 v_r) + \frac{1}{r\sin\theta}\frac{\partial}{\partial \theta}(v\theta \sin\theta\, v_\theta) + \frac{1}{r\sin\theta}\frac{\partial v_\varphi}{\partial \varphi}$$

C. Curl ∇xv

Cartesian

$$(\nabla x\mathbf{v})_x = \frac{\partial v_z}{\partial y} - \frac{\partial v_y}{\partial z} \qquad (\nabla x\mathbf{v})_y = \frac{\partial v_x}{\partial z} - \frac{\partial v_z}{\partial x} \qquad (\nabla x\mathbf{v})_z = \frac{\partial v_y}{\partial x} - \frac{\partial v_x}{\partial y}$$

Cylindrical

$$(\nabla x\mathbf{v})_r = \frac{1}{r}\frac{\partial v_z}{\partial \theta} - \frac{\partial v_\theta}{\partial z} \qquad (\nabla x\mathbf{v})_\theta = \frac{\partial v_r}{\partial z} - \frac{\partial v_z}{\partial r} \qquad (\nabla x\mathbf{v})_\tau = \frac{1}{r}\frac{\partial}{\partial r}(r v_\theta) - \frac{1}{r}\frac{\partial v_r}{\partial \theta}$$

(continued)

TABLE 7.1 (Continued)
Differential Operators in Various Coordinate Systems

Spherical

$$(\nabla x \mathbf{v})_r = \frac{1}{r \sin\theta}\frac{\partial}{\partial\theta}(v_\theta \sin\theta\, v_\varphi) - \frac{1}{r\sin\theta}\frac{\partial v_\theta}{\partial\varphi}$$

$$(\nabla x \mathbf{v})_\theta = \frac{1}{r \sin\theta}\frac{\partial v_r}{\partial\varphi} - \frac{1}{r}\frac{\partial}{\partial r}(r v_\varphi)$$

$$(\nabla x \mathbf{v})_\varphi = \frac{1}{r}\frac{\partial}{\partial r}(r v_\theta) - \frac{1}{r}\frac{\partial v_r}{\partial\theta}$$

D. Laplacian ∇^2u

Cartesian

$$\nabla^2 u = \frac{\partial^2 u}{\partial x^2} + \frac{\partial^2 u}{\partial y^2} + \frac{\partial^2 u}{\partial z^2}$$

Cylindrical

$$\nabla^2 u = \frac{1}{r}\frac{\partial}{\partial r}\left(r\frac{\partial u}{\partial r}\right) + \frac{1}{r^2}\frac{\partial^2 u}{\partial\theta^2} + \frac{\partial^2 u}{\partial z^2}$$

Spherical

$$\nabla^2 u = \frac{1}{r^2}\frac{\partial}{\partial r}\left(r^2\frac{\partial u}{\partial r}\right) + \frac{1}{r^2\sin\theta}\frac{\partial}{\partial\theta}\left(\sin\theta\frac{\partial u}{\partial\theta}\right) + \frac{1}{r^2\sin^2\theta}\frac{\partial^2 u}{\partial\varphi^2}$$

E. Convective Operator (v • ∇)v = A

Cartesian

$$\mathbf{A}_x = v_x\frac{\partial v_x}{\partial x} + v_y\frac{\partial v_x}{\partial y} + v_z\frac{\partial v_x}{\partial z} \qquad \mathbf{A}_y = v_x\frac{\partial v_y}{\partial x} + v_y\frac{\partial v_y}{\partial y} + v_z\frac{\partial v_y}{\partial z}$$

$$\mathbf{A}_z = v_x\frac{\partial v_z}{\partial x} + v_y\frac{\partial v_z}{\partial y} + v_z\frac{\partial v_z}{\partial z}$$

Cylindrical

$$\mathbf{A}_r = v_r\frac{\partial v_r}{\partial r} + \frac{v_\theta}{r}\frac{\partial v_r}{\partial\theta} - \frac{v_\theta^2}{r} + v_z\frac{\partial v_r}{\partial z}$$

$$\mathbf{A}_\theta = v_r\frac{\partial v_\theta}{\partial r} + \frac{v_\theta}{r}\frac{\partial v_\theta}{\partial\theta} - \frac{v_r v_\theta}{r} + v_z\frac{\partial v_\theta}{\partial z}$$

$$\mathbf{A}_z = v_r\frac{\partial v_z}{\partial r} + \frac{v_\theta}{r}\frac{\partial v_z}{\partial\theta} - \frac{v_\theta^2}{r} + v_z\frac{\partial v_z}{\partial z}$$

TABLE 7.1 (Continued)
Differential Operators in Various Coordinate Systems

Spherical

$$\mathbf{A}_r = v_r \frac{\partial v_r}{\partial r} + \frac{v_\theta}{r}\frac{\partial v_r}{\partial \theta} - \frac{v^2{}_\theta + v^2{}_\varphi}{r} + \frac{v_\varphi}{r \sin\theta}\frac{\partial v_r}{\partial \varphi}$$

$$\mathbf{A}_\theta = v_r \frac{\partial r_\theta}{\partial r} + \frac{v_\theta}{r}\frac{\partial v_\theta}{\partial \theta} + \frac{v_r v_\theta}{r} - \frac{v^2{}_\varphi \cot\theta}{r} + \frac{v_\varphi}{r \sin\theta}\frac{\partial v_\theta}{\partial \varphi}$$

$$\mathbf{A}_\varphi = v_\varphi \frac{\partial r_\varphi}{\partial r} + \frac{v_\theta}{r}\frac{\partial v_\varphi}{\partial \theta} + \frac{v_r v_\varphi}{r} - \frac{v_\theta v_\varphi \cot\theta}{r} + \frac{v_\varphi}{r \sin\theta}\frac{\partial v_\varphi}{\partial \varphi}$$

F. Laplacian of a Vector $\nabla^2\mathbf{v}$

Cartesian

$$(\nabla^2\mathbf{v})_x = \frac{\partial^2 v_x}{\partial x^2} + \frac{\partial^2 v_x}{\partial y^2}\frac{\partial^2 v_x}{\partial z^2}$$

$$(\nabla^2\mathbf{v})_y = \frac{\partial^2 v_y}{\partial x^2} + \frac{\partial^2 v_y}{\partial y^2}\frac{\partial^2 v_y}{\partial z^2}$$

$$(\nabla^2\mathbf{v})_z = \frac{\partial^2 v_z}{\partial x^2} + \frac{\partial^2 v_z}{\partial y^2}\frac{\partial^2 v_z}{\partial z^2}$$

Cylindrical

$$(\nabla^2\mathbf{v})_r = \frac{\partial}{\partial r}\left(\frac{1}{r}\frac{\partial}{\partial r}(r v_r)\right) + \frac{1}{r^2}\frac{\partial^2 v_r}{\partial \theta^2} - \frac{2}{r}\frac{\partial v_\theta}{\partial \theta} + \frac{\partial^2 v_r}{\partial z^2}$$

$$(\nabla^2\mathbf{v})_\theta = \frac{\partial}{\partial r}\left(\frac{1}{r}\frac{\partial}{\partial r}(r v_\theta)\right) + \frac{1}{r^2}\frac{\partial^2 v_\theta}{\partial \theta^2} + \frac{2}{r}\frac{\partial v_r}{\partial \theta} + \frac{\partial^2 v_\theta}{\partial z^2}$$

$$(\nabla^2\mathbf{v})_z = \frac{1}{r}\frac{\partial}{\partial r}\left(r \frac{\partial v_z}{\partial r}\right) + \frac{1}{r^2}\frac{\partial^2 v_z}{\partial \theta^2} + \frac{\partial^2 v_z}{\partial z^2}$$

Spherical

$$(\nabla^2\mathbf{v})_r = \nabla^2 v_r - \frac{2}{r^2} v_r - \frac{2}{r^2}\frac{\partial v_\theta}{\partial \theta} - \frac{2}{r^2} v_\theta \cot\theta - \frac{2}{r^2 \sin\theta}\frac{\partial v_y}{\partial \varphi}$$

$$(\nabla^2\mathbf{v})_\theta = \nabla^2 v_\theta + \frac{2}{r^2}\frac{\partial v_r}{\partial \theta} - \frac{v_\theta}{r^2 \sin^2\theta} - \frac{2\cos\theta}{r^2 \sin^2\theta}\frac{\partial v_\varphi}{\partial \varphi}$$

$$(\nabla^2\mathbf{v})_\varphi = \nabla^2 v_\varphi + \frac{v_\varphi}{r^2 \sin^2\theta} + \frac{2}{r^2 \sin\theta}\frac{\partial v_r}{\partial \varphi} + \frac{2\cos\theta}{r^2 \sin^2\theta}\frac{\partial v_\theta}{\partial \varphi}$$

TABLE 7.2
Relations of Vector Algebra

System/Operation	Defining Relation	Other Properties
Free vector **A**	$\mathbf{A} = A_x\mathbf{i} + A_y\mathbf{j} = A_z\mathbf{k}$ $\mathbf{A} = \lvert\mathbf{A}\rvert[\cos(\mathbf{A}, x)\mathbf{i} + \cos(\mathbf{A}, y)\mathbf{j} + \cos(\mathbf{A}, z)\mathbf{k}$ $\mathbf{A} = \lvert\mathbf{A}\rvert\mathbf{a}$ Magnitude $\lvert\mathbf{A}\rvert = (A_x^2 + A_y^2 + A_z^2)^{1/2}$ Unit Vectors **i**, **j**, **k**: Along x, y, z axes **a**: Along vector **A**	Two vectors **A**, **B** are equal if they have identical magnitude and direction, or $A_x = B_x$, $A_y = B_y$, $A_z = B_z$ A = −B implies two parallel vectors of equal magnitude and opposite direction
Addition and subtraction	$(\mathbf{A} \pm \mathbf{B}) = (A_x \pm B_x)\mathbf{i} + (A_y \pm B_y)\mathbf{j} + (A_z \pm B_z)\mathbf{k}$	For graphical construction see Figure 7.1
Multiplication by scalar m	$m\mathbf{A} = mA_x + mA_y + mA_z$ $m\mathbf{A} = \lvert\mathbf{A}\rvert[\cos(\mathbf{A}, x)\mathbf{i} + \cos(\mathbf{A}, y)\mathbf{j} + \cos(\mathbf{A}, z)\mathbf{k}$ $m\mathbf{A} = m\lvert\mathbf{A}\rvert\mathbf{a}$	
Dot product **A** • **B** = a scalar	$\mathbf{A} \bullet \mathbf{B} = \lvert\mathbf{A}\rvert\lvert\mathbf{B}\rvert\cos(\mathbf{A}, \mathbf{B})$ $\mathbf{A} \bullet \mathbf{B} = A_xB_x + A_yB_y + A_zB_z$ $\mathbf{A} \bullet \mathbf{B} = \lvert\mathbf{A}\rvert$ Projection of **B** on **A** $\mathbf{A} \bullet \mathbf{B} = \lvert\mathbf{B}\rvert$ Projection of **A** on **B** **A** • **B** = 0 if **A**, **B** orthogonal $\mathbf{A} \bullet \mathbf{B} = \lvert\mathbf{A}\rvert\ \lvert\mathbf{B}\rvert$ if **A**, **B** parallel $\mathbf{i} \bullet \mathbf{j} = \mathbf{j} \bullet \mathbf{k} = \mathbf{i} \bullet \mathbf{k} = 0$ $\mathbf{i} \bullet \mathbf{i} = \mathbf{j} \bullet \mathbf{j} = \mathbf{k} \bullet \mathbf{k} = 1$	Dot Product is Distributive A • (B + C) = A • B + A • C Commutative **A** • **B** = **B** • **A** Associative (t**A**) • **B** = t(**A** • **B**)
Cross product A × B = a vector	**A** × **B** = **C** = vector normal to A, B $\mathbf{AB} = (A_yB_z - A_zB_y)i + (A_zB_x - A_xB_z)j + (A_xB_y - A_yB_x)k$ $\lvert\mathbf{A} \times \mathbf{B}\rvert = \lvert\mathbf{A}\rvert\mathbf{B}\rvert \sin(\mathbf{A}, \mathbf{B})$ = Area of Parallelogram **A** × **B** = 0 if **A**, **B**, parallel	Cross Product is: Distributive Associative It is not commutative: **A** × **B** = **B** × **A**

We have, in Table 7.2, compiled a list of the basic relations of vector algebra and added to this a graphical representation of elementary vector properties and operations (Figure 7.1). They will serve as a guide in interpreting various aspects of vector calculus. We remind the reader of the following properties, which are shared by both vector algebra and vector calculus.

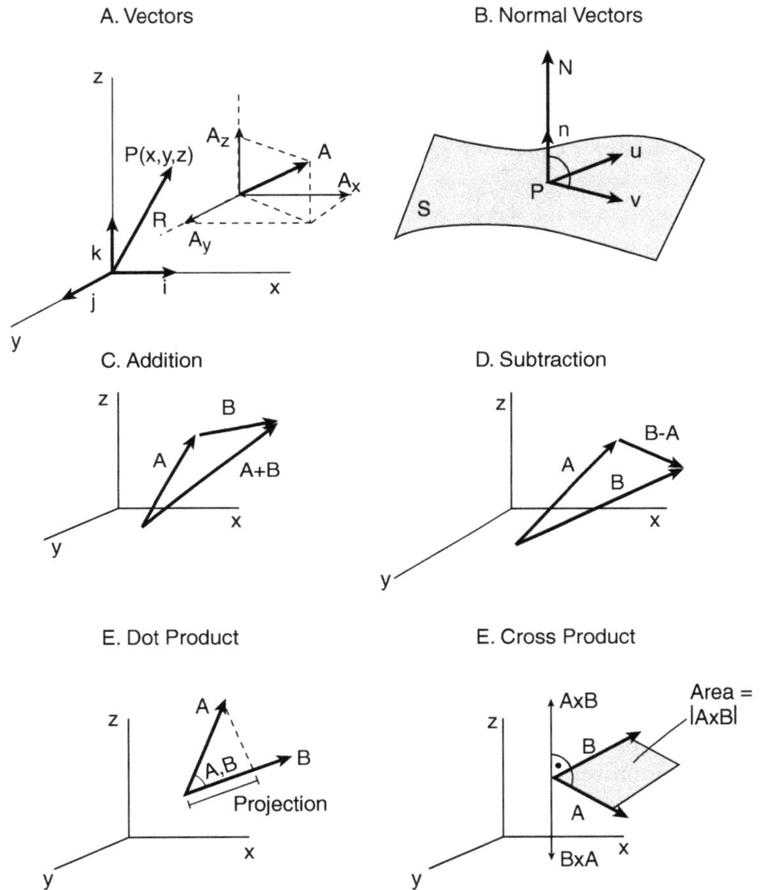

FIGURE 7.1 Vectors and simple vector algebra operations.

- A vector **A** is defined by its magnitude, i.e., length $|\mathbf{A}|$ and its direction, but not its location in space. An exception is the position vector **R** (Figure 7.1A), which always starts at the origin, extends to some point P(x, y, z), and hence has the components x, y, z. Thus,

$$\mathbf{R} = x\mathbf{i} + y\mathbf{j} + z\mathbf{k} \tag{7.1a}$$

Here **i**, **j**, **k** are unit vectors, i.e., have a length of 1 and are directed along the coordinate axes. Unit vectors, premultiplied by the components, provide a complete description of a vector. The vector can also be represented by the product of magnitude and unit vector:

$$\mathbf{A} = |\mathbf{A}|\mathbf{a} \tag{7.1b}$$

or in terms of its directional cosines:

$$\mathbf{A} = |\mathbf{A}|[\cos(\mathbf{A}, x)\mathbf{i} + \cos(\mathbf{A}, y)\mathbf{j} + \cos(\mathbf{A}, z)\mathbf{k}] \tag{7.1c}$$

where length $|\ |$ equals the square root of the sum of component squares:

$$|\mathbf{A}| = (A_x^2 + A_y^2 + A_z^2)^{1/2} \tag{7.1d}$$

- Two vectors **A** and **B** are equal if their corresponding components are equal:

$$A_x = B_x \quad A_y = B_y \quad A_z = B_z \tag{7.1e}$$

- The dot product **A** • **B**, which is a scalar, can be viewed as one or more of the following:
  1. The product of the two magnitudes times the cosine of the angle between them.
  2. The projection of one vector onto the other times its magnitude
  3. The sum of the component products:

$$\mathbf{A} \bullet \mathbf{B} = A_x B_x + A_y B_y + A_z B_z \tag{7.1f}$$

  Note that the dot product vanishes when the two vectors are normal to each other.
- The cross product **A** × **B**, which is itself a vector, is characterized by the following properties.
  1. Its direction is normal to the plane passing through **A** and **B**.
  2. Its magnitude equals the product of the magnitude of the two vectors multiplied by the sine of the angle between them:

$$|\mathbf{A} \times \mathbf{B}| = |\mathbf{A}|\,|\mathbf{B}| \sin(\mathbf{A}, \mathbf{B}) \tag{7.1g}$$

  3. It vanishes when the two vectors are parallel.
- The normal vector **N** (Figure 7.1B) passes through some point P(x, y, z) on the surface S and is orthogonal to two vectors **u** and **v**, which are themselves tangential to the surface at the same point P. Thus,

$$\mathbf{N} = \mathbf{u} \times \mathbf{v} \tag{7.1h}$$

and

$$\mathbf{n} = \frac{\mathbf{N}}{|\mathbf{N}|} = \frac{\mathbf{u} \times \mathbf{v}}{|\mathbf{u} \times \mathbf{v}|} \tag{7.1i}$$

where **n** is the unit normal vector and the vertical bars denote absolute magnitude. We shall encounter **n** again in the definition of certain entities that arise in vector calculus.

TABLE 7.3
The Differential Operators

Operator	Operates On	Is Itself
Gradient grad u = ∇u	Scalar	Vector
Divergence div **v** = $\nabla \bullet \mathbf{v}$	Vector	Scalar
Curl curl **v** = $\nabla \times \mathbf{v}$	Vector	Vector
Laplacian $\nabla^2 u$	Scalar	Scalar

7.1.1 Differential Operators and Vector Calculus

Differential operators represent, as we have noted, a shorthand for certain operations involving partial derivatives. The four principal operators that we wish to consider in more detail here are: the gradient, the divergence, the curl, and the Laplacian. Some operate on vectors or are themselves vectors; others are scalars or operate on scalars. When the result is a vector, there are three equivalent scalar equations to be considered. We summarize these features for the reader in Table 7.3.

Expressions for these operators in Cartesian, cylindrical, and spherical coordinates have previously been presented in Table 7.1, which also lists the convective operator $(\mathbf{v} \bullet \nabla)\mathbf{v}$ and the Laplacian of a vector $\nabla^2\mathbf{v}$. The latter appear principally in momentum transport.

There are several ways in which operators can be viewed. The simplest and perhaps most simplistic is to regard them as a shorthand for certain combinations of partial derivatives that appear in mathematical physics. Thus, the Laplacian $\nabla^2\mathbf{v}$ $(= \partial^2u/\partial x^2 + \partial^2u/\partial y^2 + \partial^2u/\partial z$ in Cartesian coordinates) appears in a host of classical PDEs, including Laplace's and Fourier's equations, as well as those of Poisson and Helmholtz. The divergence of fluid velocity $\nabla \bullet v$ $(= \partial v_x/\partial x + \partial v_y/\partial y + \partial v_z/\partial z)$, makes its appearance in the continuity equation of fluid mechanics. Force or momentum balances that arise in the same field contain the gradient of pressure ∇p, which, being a force, is also a vector, with Cartesian components $\partial p/\partial x$, $\partial p/\partial y$, and $\partial p/\partial z$.

A somewhat more elevated use of operators comes about by exploiting the properties of the "del" or ∇ notation. We have seen a variety of such dels in Table 7.1, a straight "del" that appears in the gradient, a "del dot" that appeared in the divergence, a "del cross" in the curl, and a "del square" associated with the Laplacian. Each of these operators can be expanded into forms that provide a better understanding of the mathematical operation involved. We have done this for Cartesian coordinates and summarize the results in Table 7.4. Some useful applications of this notation will be shown in Illustration 7.3.

A third way of viewing differential operators is to attribute some physical meaning to them. This is not always easy to accomplish, but we shall do our best.

TABLE 7.4
The Del Notation (Cartesian)

Symbol	Expanded Form
∇	$\left(\mathbf{i}\frac{\partial}{\partial x}+\mathbf{j}\frac{\partial}{\partial y}+\mathbf{k}\frac{\partial}{\partial z}\right)$
$\nabla\bullet$	$\left(\mathbf{i}\frac{\partial}{\partial x}+\mathbf{j}\frac{\partial}{\partial y}+\mathbf{k}\frac{\partial}{\partial z}\right)\bullet$
$\nabla\times$	$\left(\mathbf{i}\frac{\partial}{\partial x}+\mathbf{j}\frac{\partial}{\partial y}+\mathbf{k}\frac{\partial}{\partial z}\right)\times$
∇^2	$\mathbf{i}\frac{\partial^2}{\partial x^2}+\mathbf{j}\frac{\partial^2}{\partial y^2}+\mathbf{k}\frac{\partial^2}{\partial z^2}$

Finally, we turn to the most important aspect of operational notation, its role as a powerful and compact tool to express complex physical problems in multi-dimensional space and to facilitate their solution. Once certain rules of the operational "game" have been laid down, as we have attempted to do here, manipulation, simplification and, ultimately, solution of such physical problems becomes much more manageable. A certain sense of uneasiness usually remains because of the unfamiliar nature of the symbolism, but this is no different than other areas of mathematics. In time, and if one persists, the benefits will outweigh the drawbacks.

The gradient, ∇: As seen from Table 7.3, the gradient operates on a scalar, such as pressure or temperature, but is itself a vector. A physical definition of ∇u may be given by considering surfaces in three-dimensional space, for example, isothermal surfaces in a temperature field (Figure 7.2A) that have a constant value of the scalar u. One may then define the gradient as a vector that points in the direction of maximum increase in u (i.e., is normal to the isothermal plane) and has a magnitude equal to the maximum increase per unit distance. In mathematical terms, one obtains:

$$\nabla u \quad = \quad \frac{du}{dn} \quad n \tag{7.2a}$$

Gradient Magnitude Unit Normal Vector

An alternative expression in terms of the vector components of ∇u is obtained by introducing the position vector **R** to the isothermal plane, as well as its differential d**R**, which is tangent to the plane (see Figure 7.3A). We then obtain:

By virtue of orthogonality:

$$\nabla \mathrm{u} \bullet d\mathbf{R} = 0 \tag{7.2b}$$

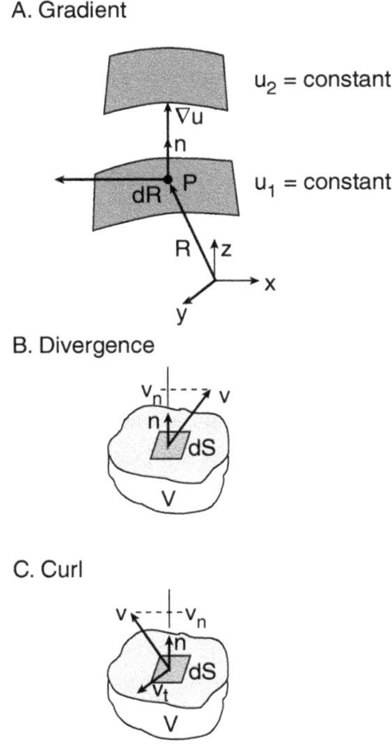

FIGURE 7.2 The three principal operators of differential calculus: (A) The gradient ∇, (B) The divergence ∇, and (C) The curl ∇x.

By virtue of constant u:

$$\frac{\partial u}{\partial x}dx+\frac{\partial u}{\partial y}dy+\frac{\partial u}{\partial z}dz=du=0 \tag{7.2c}$$

As dx, dy, dz are the components of dR, it follows from the definition of the dot product that $\frac{\partial u}{\partial x},\frac{\partial y}{\partial y},\frac{\partial u}{\partial z}$ must be the components of the gradient. We thus obtain the following working equation for the gradient:

$$\nabla u=\mathbf{i}\frac{\partial u}{\partial x}+\mathbf{j}\frac{\partial u}{\partial y}+\mathbf{k}\frac{\partial u}{\partial z} \tag{7.2d}$$

Now, Fick's and Fourier's laws postulate that mass and heat flux, N and q, are proportional to and in the direction of the largest decrease per unit distance of concentration and temperature, i.e., they must be proportional to the negative gradient of C and T. These laws, therefore, take the vectorial form:

$$\mathbf{N}=-\mathrm{D}\nabla\mathrm{C} \tag{7.2e}$$

$$\mathbf{q}=-\mathrm{k}\nabla\mathrm{T} \tag{7.2f}$$

The divergence, $\nabla\bullet$: Here, it is convenient to start with a mathematical definition, which we make more palatable later by giving it some physical meaning. There are several such definitions available, of which we choose the following:

$$\nabla \bullet \mathbf{v} = \lim_{V \to 0} \frac{\int_S \mathbf{n} \bullet \mathbf{v} \, dS}{V} \tag{7.3a}$$

This odd-looking expression, sketched in Figure 7.2B, in essence states that the projection of a vector **v** onto the unit normal vector to the surface dS yields, upon integration over the surface, division by volume V, and allowing V to shrink to a point, the divergence of that vector. Let us attempt to make some physical sense of this expression, and set **v** equal to the velocity of a flowing fluid. Its dot product with the unit normal yields:

$$\mathbf{n} \bullet \mathbf{v} = |\mathbf{n}| \text{ projection of } \mathbf{v} \text{ onto } \mathbf{n} \tag{7.3b}$$

so that $\mathbf{n} \bullet \mathbf{v} = (1)(v_n)$ = normal component of **v**

We can then write the integral in the form

$$\frac{\int_S \mathbf{n} \bullet \mathbf{v} \, dS}{V} = \frac{\rho \int v_n \, dS}{\rho V} \tag{7.3c}$$

which shows that the integral represents the net outflow of fluid from the surface S per unit mass of enclosed fluid. By allowing V to shrink to a point, this becomes the net outflow per unit mass at a point of the flow field, which is the divergence of the velocity.

Two items need to be noted in connection with the divergence.

1. Although we have arrived at some degree of understanding of the physical meaning of $\nabla \bullet \mathbf{v}$, we still lack a convenient mathematical representation for the divergence. This requires evaluation of the Equation 7.3a and will be undertaken in Illustration 7.2. Evidently, the result will have to be a scalar, because the integrand $\mathbf{n} \bullet \mathbf{v}$ is a scalar, and we shall find that the divergence is, in fact, represented by the sum of a set of first-order partial derivatives.
2. A second point concerns the manipulation of the divergence to arrive at a general vectorial representation of the conservation of mass. This can be done by making use of the integral theorems of vector calculus. Specifically, we shall make use of the so-called divergence theorem to arrive at a vectorial representation of the continuity equation. This is taken up in Illustration 7.4.

We now turn to the more difficult task of making sense of the expression referred to as the curl.

The curl, ∇x: The mathematical definition of the curl we have chosen starts innocently enough by replacing the dot product of the divergence in Equation 7.3a by the cross product $\mathbf{n} \times \mathbf{v}$. We obtain (see Figure 7.2C):

$$\nabla \times \mathbf{v} = \lim_{V \to 0} \frac{\int_S \mathbf{n} \times \mathbf{v}\, dS}{V} \tag{7.4a}$$

To make some physical sense of this expression, we note that the magnitude of the cross product is given by:

$$|\mathbf{n} \times \mathbf{v}| = (1)\ |\mathbf{v}|\ \sin(\mathbf{n},\mathbf{v}) = \mathrm{v}_t = \text{tangential velocity component} \tag{7.4b}$$

With some imagination, the curl $\nabla \times \mathbf{v}$ can therefore be regarded as a net tangential velocity over the surface of an infinitesimally small domain. If nonzero, one could consider this to be the result of shear forces, and in fact, almost all viscous flow turns out to be rotational, i.e., obeys $\nabla \times \mathbf{v} \neq 0$, and almost all inviscid flow ($\mu = 0$) is irrotational, characterized by $\nabla \times \mathrm{v} = 0$. There are very few exceptions to these rules, of which we mention three:

1. Gas flow, although essentially inviscid because of its low viscosity, is invariably and strongly rotational in the rear of a submerged bluff body, as evidenced by the eddies formed there.
2. A rotating cylindrical vessel filled with a viscous fluid is nevertheless irrotational. This is a minor case which rarely arises in practice.
3. An inviscid fluid subjected to Cariolis forces due to the rotation of the earth, will be in rotational flow. This is again a highly specialized subject, which will not be dealt with here.

Two further points need to be made:

What precisely is meant, in mathematical terms, by the statement $\nabla \times \mathbf{v} = 0$? The answer is that the components of the curl vector become zero. Inspection of Table 7.1 shows that this results, for Cartesian coordinates, in the following three relations:

$$\frac{\partial v_z}{\partial y} - \frac{\partial v_y}{\partial z} = 0 \qquad \frac{\partial v_x}{\partial z} - \frac{\partial v_z}{\partial x} = 0 \qquad \frac{\partial v_y}{\partial x} - \frac{\partial v_x}{\partial y} = 0 \tag{7.4c}$$

Derivation of these expressions from the integral (Equation 7.4b) is a considerable task, which we relegate to the practice problems. To ease the derivation, some useful hints will be provided.

The second point to be made is, as inviscid and irrotational flow are essentially identical, why not confine ourselves to the former, which is more easily formulated? The answer to this question is provided in detail in Section 7.4, but we summarize the results briefly here. The momentum balances for viscous flow are represented by the three scalar and nonlinear Navier–Stokes equations. By making the assumption of inviscid flow, the equations are shortened, but still remain nonlinear and three in number. Assuming irrotational flow, on the other hand, brings about a drastic reduction from three nonlinear PDEs to a single linear PDE, Laplace's equation. This is a dramatic simplification, made all the more simple by the fact that Laplace's equation has been extensively studied and its numerous solutions, by a variety of methods, are known. Some of these were tabulated in Figure 6.7 of the preceding chapter.

Finally, in our consideration of differential operators, we turn our attention to the Laplacian operator ∇^2:

$$\nabla = \mathbf{i}\frac{\partial}{\partial x} + \mathbf{j}\frac{\partial}{\partial y} + \mathbf{k}\frac{\partial}{\partial z} \tag{7.4d}$$

Also noted in that table was the expanded form of the divergence operator:

$$\nabla \bullet = \left(\mathbf{i}\frac{\partial}{\partial x} + \mathbf{j}\frac{\partial}{\partial y} + \mathbf{k}\frac{\partial}{\partial z}\right)\bullet \tag{7.4e}$$

We now proceed to operate on the gradient operator ∇ with the divergence operator $\nabla\,\bullet$. This leads, in the first instance, to the following dot product:

$$\nabla \bullet \nabla = \left(\mathbf{i}\frac{\partial}{\partial x} + \mathbf{j}\frac{\partial}{\partial y} + \mathbf{k}\frac{\partial}{\partial z}\right)\bullet\left(\mathbf{i}\frac{\partial}{\partial x} + \mathbf{j}\frac{\partial}{\partial y} + \mathbf{k}\frac{\partial}{\partial z}\right) \tag{7.4f}$$

If we further apply the rule given in Table 7.2 that a dot product equals the sum of the products of its components, we obtain formally:

$$\nabla \bullet \nabla = \frac{\partial}{\partial x}\frac{\partial}{\partial x} + \frac{\partial}{\partial y}\frac{\partial}{\partial y} + \frac{\partial}{\partial z}\frac{\partial}{\partial z} \tag{7.4g}$$

The rules of operational notation call for these products of partial derivatives to be equivalent to second-order derivatives. Thus Equation 7.4g becomes:

$$\nabla \bullet \nabla = \frac{\partial^2}{\partial x^2} + \frac{\partial^2}{\partial y^2} + \frac{\partial^2}{\partial z^2} \tag{7.4h}$$

This is no different from what was done with the D-operator at the ODE level where we postulated the equivalence:

$$DD = \frac{d}{dx}\frac{d}{dx} = \frac{d^2}{dx^2} \tag{7.4i}$$

TABLE 7.5
Relations Involving ∇, ∇ •, and ∇ ×

1.	$\nabla(u + w) = \nabla u + \nabla w$
2.	$\nabla(uw) = u\nabla w + w\nabla u$
3.	$\nabla \bullet (\mathbf{A} + \mathbf{B}) = \nabla \bullet \mathbf{A} + \nabla \bullet \mathbf{B}$
4.	$\nabla \bullet (u\mathbf{A}) = \nabla u \bullet \mathbf{A} + u\nabla \bullet \mathbf{A}$
5.	$\nabla u \bullet d\mathbf{R} = du$
6.	$\nabla \bullet (\nabla \times \mathbf{A}) = 0$
7.	$\nabla \bullet (\nabla u \times \nabla w) = 0$
8.	$\nabla \bullet (\mathbf{A} \times \mathbf{B}) = \mathbf{B} \bullet (\nabla \times \mathbf{A}) + (\mathbf{A} \bullet \nabla \times \mathbf{B})$
9.	$\nabla \times (\mathbf{A} + \mathbf{B}) = \nabla \times \mathbf{A} + \nabla \times \mathbf{B}$
10.	$\nabla \times (u\mathbf{A}) = u\nabla \times \mathbf{A} + \nabla u \times \mathbf{A}$
11.	$\nabla \times \nabla u = 0$
12.	$\nabla \times (\mathbf{A} \times \mathbf{B}) = \mathbf{B} \bullet \nabla\mathbf{A} - \mathbf{A} \bullet \nabla\mathbf{B} + \mathbf{A}(\nabla \bullet \mathbf{B}) - \mathbf{B}(\nabla \bullet \mathbf{A})$
13.	$(\mathbf{A} \bullet \nabla)\mathbf{A} = \frac{1}{2}\nabla(\mathbf{A} \bullet \mathbf{A}) - \mathbf{A} \times (\nabla \times \mathbf{A})$
14.	$\nabla^2\mathbf{A} = \nabla(\nabla \bullet \mathbf{A}) - \nabla \times (\nabla \times \mathbf{A})$
15.	$\nabla \times \mathbf{R} = 0$

Readers may at first find these rules arbitrary. We remind them, however, that if these rules are established with consistency and in a manner that yields correct results, operator notation becomes an extremely powerful and compact tool for manipulating differential equations. We repeat that the crux is consistency and a regard for the results obtained.

A number of useful relations have resulted from these procedures, which we summarize for the convenience of the reader in Table 7.5. Proofs for several of these expressions are given in the illustrations that follow. We note that some of these equations are scalar and others are vectorial PDEs. Thus, the equations involving the Divergence ∇ • are all scalar PDEs, and the remainder are vectorial, equivalent to a set of three scalar PDEs.

Illustration 7.2 Derivation of the Divergence

The task to be undertaken here is the conversion of the integral of Equation 7.3a into an equivalent PDE. To accomplish this, we consider a cube of magnitude $\Delta x\Delta y\Delta z$, and focus, in the first instance, on the two faces at the positions x and $x + \Delta x$. We obtain for the integrand:

At position x:

$$\mathbf{n} \bullet \mathbf{v}\ \Delta S = (-1)\text{ projection of }\mathbf{v}\text{ on }\mathbf{n} = -|v|_x\ \Delta y\Delta z$$

At position $x + \Delta x$:

$$\mathbf{n} \bullet \mathbf{v}\ \Delta S = (1)\text{ projection of }\mathbf{v}\text{ on }\mathbf{n} = |v|_{x+\Delta x}\ \Delta y\Delta z$$

Note that at position x, the normal **n** is pointed in the negative x direction, hence its magnitude is 1.

Summing the two components and using the full Expression 7.3a, we obtain:

$$\lim_{\Delta V \to 0} \frac{(v|_{x+\Delta x} - v_x)}{\Delta x} = \frac{\partial v_x}{\partial x} \tag{7.5a}$$

Similar applications of this scheme to the faces in the y and z directions yields, in the limit, the terms $\partial v_y/\partial y$ and $\partial v_z/\partial z$, and the divergence becomes:

$$\nabla \bullet \mathbf{v} = \lim_{V \to 0} \frac{\int_S \mathbf{n} \bullet \mathbf{v}\, dS}{V} = \frac{\partial v_x}{\partial x} + \frac{\partial v_y}{\partial y} + \frac{\partial v_z}{\partial z} \tag{7.5b}$$

This is the expression we had reported in Table 7.1.

Illustration 7.3 Derivation of Some Relations Involving ∇, $\nabla \bullet$, and $\nabla \times \cdot$

In this example, we present the derivation of some of the relations we had previously given in Table 7.5 without proof. To do this, we use the expressions for the del notation shown in Table 7.3, and apply the usual rules of vector dot and cross multiplication.

Item 5 of Table 7.5: $\nabla u \bullet dR = du$ — The expansion of $\nabla \bullet$ and $d\mathbf{R}$ leads in the first instance to the following expression:

$$\nabla u \bullet d\mathbf{R} = \left(\mathbf{i}\frac{\partial u}{\partial x} + \mathbf{j}\frac{\partial u}{\partial y} + \mathbf{k}\frac{\partial u}{\partial z} \right) \bullet (\mathbf{i}\,dx + \mathbf{j}\,dy + \mathbf{k}\,dz) \tag{7.6a}$$

This is

$$\nabla u \bullet d\mathbf{R} = \frac{\partial u}{\partial x}dx + \frac{\partial u}{\partial y}dy + \frac{\partial u}{\partial z}dz \tag{7.6b}$$

where the right-hand side of the equation will be recognized as the total differential du of the scalar u. Hence,

$$\nabla u \bullet d\mathbf{R} = du \qquad Q.E.D.$$

Item 6: $\nabla \bullet (\nabla \times A) = 0$ — Here, the expansion of the component vectors becomes somewhat lengthier. We obtain in the first instance:

$$\nabla \bullet (\nabla \times \mathbf{A}) = \left(\mathbf{i}\frac{\partial}{\partial x} + \mathbf{j}\frac{\partial}{\partial y} + \mathbf{k}\frac{\partial}{\partial z} \right) \bullet \left[\mathbf{i}\left(\frac{\partial A_z}{\partial y} - \frac{\partial A_y}{\partial z} \right) + \mathbf{j}\left(\frac{\partial A_x}{\partial z} - \frac{\partial A_z}{\partial x} \right) + \mathbf{k}\left(\frac{\partial A_y}{\partial x} - \frac{\partial A_x}{\partial y} \right) \right] \tag{7.6c}$$

We now expand the right-hand side by the rules of dot multiplication, noting that the resulting component products of the type $\partial/\partial x(\partial A_z/\partial y)$ are to be regarded as second derivatives. There results the following scalar sum of mixed derivatives:

$$\left(\frac{\partial}{\partial x}\frac{\partial A_z}{\partial y}-\frac{\partial}{\partial x}\frac{\partial A_y}{\partial z}\right)+\left(\frac{\partial}{\partial y}\frac{\partial A_x}{\partial z}-\frac{\partial}{\partial y}\frac{\partial A_z}{\partial x}\right)+\left(\frac{\partial}{\partial z}\frac{\partial A_y}{\partial x}-\frac{\partial}{\partial z}\frac{\partial A_x}{\partial x}\right) \tag{7.6d}$$

Upon inspection, these derivatives are found to exactly cancel each other. We therefore obtain:

$$\nabla \bullet (\nabla \times \mathbf{A}) = 0 \qquad Q.E.D.$$

Item 11: $\nabla \times \nabla u = 0$ — Proceeding as before we first expand the individual operators and obtain:

$$\nabla \times \nabla u = \left(\mathbf{i}\frac{\partial}{\partial x}+\mathbf{j}\frac{\partial}{\partial y}+\mathbf{k}\frac{\partial}{\partial z}\right)\times\left(\mathbf{i}\frac{\partial u}{\partial x}+\mathbf{j}\frac{\partial u}{\partial y}+\mathbf{k}\frac{\partial u}{\partial z}\right)$$

This is followed by cross multiplication of the right-hand side using the rules of vector algebra. We do this for one vector component only, which yields:

$$\mathbf{i}\left(\frac{\partial}{\partial z}\frac{\partial u}{\partial x}-\frac{\partial}{\partial x}\frac{\partial u}{\partial z}\right) \tag{7.6e}$$

Similar expressions result for other vector components. One notes immediately that the mixed derivatives appearing in each component cancel each other so that the vector $\nabla \times \nabla u$ becomes identically zero. This completes the required proof.

7.1.2 Integral Theorems of Vector Calculus

In this section, we present a number of so-called integral theorems of vector calculus that can be proven by fairly elementary applications of the operations of vector algebra and vector calculus. They are somewhat forbidding in appearance, but share the common feature that they all relate surface and volume integrals of certain vector entities to each other. They, therefore, provide a link between the interior of a domain, which could, for example, be represented by the solution space, to surface conditions (for example, the boundary conditions). This is a useful relation to have.

The integral theorem are four in number and take the following form:

1. Divergence theorem:

$$\iint_S (\mathbf{A} \bullet \mathbf{n})dS = \iiint_V (\nabla \bullet \mathbf{A})dV \tag{7.7a}$$

2. First form of Green's theorem (Green's first identity):

$$\iint_S \mathbf{n} \bullet (u\nabla \mathrm{w})dS = -\iiint_V [u\nabla^2 w + \nabla u \bullet \nabla w]dV \tag{7.7b}$$

3. Second form of Green's theorem (Green's second identity):

$$\iint_S \mathbf{n} \bullet (u\nabla \mathrm{w} - \mathrm{w}\nabla \mathrm{u})dS = -\iiint_V [u\nabla^2 w - w\nabla^2 u]dV \tag{7.7c}$$

   Equation 7.7c is derived from the first form of Green's theorem by writing the latter twice, with the scalars u and w interchanged, and subtracting the result.

4. Generalized (Reynolds) transport theorem:

$$\iint_S [X](\mathbf{v} \bullet \mathbf{n})dS \quad = \quad \frac{D}{Dt}\iiint_V [X]dV \quad - \quad \iiint_V \frac{\partial}{\partial t}[X]dV \tag{7.7d}$$

Rate of flux of [X] across the surface	Rate of change of [X] within flowing entity	Rate of change of [X] within volume V

Here D/Dt is the so-called substantial or convective derivative, which records changes within a flowing parcel of constant mass. It is related to the partial derivatives, fixed in space, via the expression:

$$\frac{D}{Dt}[\mathbf{X}] = \frac{\partial[\mathbf{X}]}{\partial \mathbf{t}} + (\mathbf{v} \bullet \nabla)[\mathbf{X}] \tag{7.7e}$$

where [**X**] is an arbitrary entity (scalar vector or tensor) and (**v** • ∇)[**X**] is the convective operator tabulated in Table 7.1. The process of recording changes within a moving entity is referred to as the Lagrangian approach, whereas the analysis in a fixed entity is termed Eulerian. Derivations of these expressions appear in standard texts on transport phenomena. They are best understood by means of applications.

We therefore proceed to use the theorems in the derivation of certain equations in vectorial form and introduce the reader as well to the concept and application of Green's functions, a much-dreaded topic. We shall attempt to ease the task by providing both the physical and mathematical interpretations of these functions and demonstrate their use by means of practical illustrations.

Illustration 7.4 Derivation of the Continuity Equation

Here, we start with the transport theorem and set [X] equal to density ρ. This causes the convective derivative $\frac{D}{Dt}$ to drop out, since ρ dV = dm = 0 for a fluid packet of constant mass. There remains:

$$\iint_S \rho(\mathbf{v} \bullet \mathbf{n})dS = -\iiint_V \frac{\partial}{\partial t}\rho\, dV \tag{7.8a}$$

The surface integral is next converted into a volume integral using the divergence theorem (Equation 7.7a) where we set $\mathbf{A} = \rho\mathbf{v}$. There results:

$$\iiint_V (\nabla \bullet \rho\mathbf{v})dV = -\iiint_V \frac{\partial}{\partial t}\rho\, dV$$

or alternatively,

$$\iiint_V \left[(\nabla \bullet \rho\mathbf{v}) + \frac{\partial}{\partial t}\right] dV = 0 \tag{7.8b}$$

As the volume V can assume any arbitrary value, the integral can only vanish if the integrand itself becomes zero. We obtain:

$$-(\nabla \bullet \rho\mathbf{v}) = \frac{\partial}{\partial t}\rho \tag{7.8c}$$

which is the generalized continuity equation in vector notation. For one-dimensional incompressible flow, the expression reduces to:

$$\rho\frac{dv_x}{\partial x} = 0$$

or

$$v_x\rho = \text{constant} \tag{7.8d}$$

which is equivalent to the classical one-dimensional continuity equation.

Illustration 7.5 Derivation of Fick's Equation

Fick's equation, which was first presented in Chapter 5, Equation 5.18, is to be distinguished from Fick's law, Equation 7.2e. The latter is an empiricism that states that diffusional flux **N** is proportional to the (negative) concentration gradient; whereas the former is an equation to be derived here in general vectorial form. It can also be derived by the traditional "in out = change" approach, as was done for the one-dimensional Fourier equation (Illustration 5.3), but that procedure is less general and has to be repeated for each new geometry. We start with Fick's law (Equation 7.2e) and express it in the following dual form:

$$\mathbf{N}\ (\text{moles/sm}^2) = -D\nabla C = C\mathbf{v} \tag{7.8e}$$

where C is the molar concentration, and **v** represents the velocity vector of the diffusing species. This expression is then operated on by the divergence operator $\nabla \bullet$. We obtain:

$$\nabla \bullet C\mathbf{v} = \nabla \bullet D\nabla C \tag{7.8f}$$

The left-hand side of the equation is recognized from the continuity Equation 7.8c as equaling $-\partial C/\partial t$ so that one obtains:

$$\frac{\partial C}{\partial t} = \nabla \bullet D\nabla C \tag{7.8g}$$

This is the most general form of Fick's equation, which accounts for variable diffusivities. A more easily recognized version is obtained by setting D = constant. This yields:

$$\frac{\partial C}{\partial t} = D\nabla \bullet \nabla C = D\nabla^2 C \tag{7.8h}$$

where $\nabla^2 C$ is seen to be the Laplacian of the concentration C. Note that in this derivation the continuity equation was used, which itself sprang from two integral theorems, and of the divergence operator $\nabla \bullet$, which upon application to the gradient ∇C, yielded the Laplacian $\nabla^2 C$. Thus, considerable use was made of vector notations and operations.

7.2 Superposition Revisited: Green's Functions and the Solution of PDEs by Green's Functions

The purpose of this section is to introduce the reader to the concept of Green's functions and their use in solving the classical linear Laplace, Poisson, Fourier, Fick, and wave equations. We place this topic in this particular location because many manipulations involving Green's functions call for the use of the integral theorem presented in the preceding section.

Let us start by defining these functions: Green's functions, also known as *source*, *influence*, or *response functions*, are solutions to linear homogeneous PDEs or ODEs with homogeneous (i.e., zero) boundary and initial conditions in which the underlying system has been subjected to a *point, line, or plane forcing function*, or *pulse* of unit strength. That function is typically an instantaneous or continuous heat or mass source, an instantaneous or continuous load in mechanical systems, or a point charge in electrical systems. Two applications of Green's functions are sketched in Figure 7.3.

Consider the case of an inantaneous heat source shown in Figure 7.3A. Here, the Green's function is the solution to Fourier's equation in a two-dimensional domain with zero initial conditions (IC) and boundary conditions (BC) (i.e., the BC and IC are homogeneous), and an instantaneous point source at $P(x_0, y_0)$. That Green's function, or solution to the point source problem is given the symbol:

$$\text{Green's function} = G(x_0, y_0, x, y) \tag{7.9a}$$

where x_0, y_0 is the location of the point source in a two-dimensional domain, and x, y are the general coordinates of the domain.

A. Instantaneous Point Source

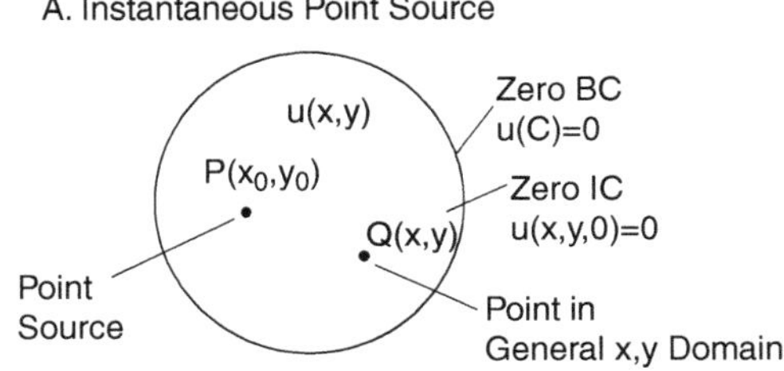

B. Point Load

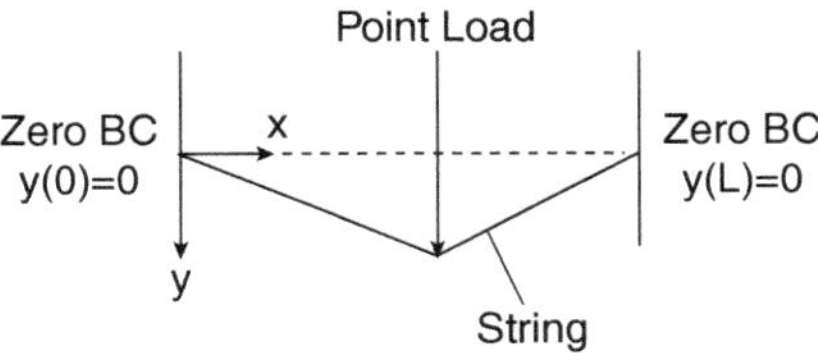

FIGURE 7.3 Physical systems involving: (A) A point source, (B) A point load. Green's functions are the solutions to these problems with homogeneous boundary conditions.

In Chapter 6, Section 6.3, we had given the solution to the instantaneous point heat source problem in infinite space (Equation 6.8a). For the one-dimensional case, this expression reduces to the form:

$$T = \frac{q_P}{2\rho Cp(\pi\alpha t)^{1/2}} \exp(-x^2/4\alpha t) \tag{7.9b}$$

where the source is a plane through $x_0 = 0$, and in Green's function nomenclature, the system is converted to an instantaneous plane source of unit strength, which releases its heat at time $t = \tau$.

The solution to that problem, i.e., the Green's function is then given by:

$$G(0,x,t-\tau) = \frac{1(J/m^2)}{2\rho Cp[\pi\alpha(t-\tau)]^{1/2}} \exp[-x^2/4\alpha(t-\tau)] \tag{7.9c}$$

with the underlying PDE represented by:

$$\alpha \frac{\partial^2 G}{\partial x^2} + \delta(x-x_0)\delta(t-\tau) = \frac{\partial G}{\partial t} \tag{7.9d}$$

Here the Dirac delta functions $\delta(x - x_0)\delta(t - \tau)$ are used as a symbolism for a source of instantaneous unit strength 1 (J/m^2).

We will not delve into the theory underlying the Dirac point function but merely use them as a symbolic way to denote the presence of a point source. To aid in the manipulations that may be required, we summarize its more important properties in Table 7.6.

TABLE 7.6
Properties of the Dirac Delta Function

1. $\int_{-\infty}^{\infty} \delta(x)dx = 1$

2. $\int_{-\infty}^{\infty} \delta(x-x_0)dx = 1$

3. $\int_{x_0-a}^{x_0+a} \delta(x-x_0)dx = 1$

4. $\iiint_V \delta(P,Q)dV_Q = 1$

5. $\iiint_V u(Q)\delta(P,Q)dV_Q = u(P)$

Laplace Transforms

6. $L\{\delta(t)\} = 1$

7. $L(\delta(t - t_0)\} = e^{-st_0}$

TABLE 7.7
Green's Functions for Type I Boundary Conditions

Domain	Green's Function (Unit Strength)
	Fourier's Equation
1. Infinite 1-D Cartesian	$G(x_0,x,t-\tau)=\dfrac{1}{2\rho Cp[\pi\alpha(t-\tau)]^{1/2}}\exp[-(x-x_0)^2/4\alpha(t-\tau)]$
2. Infinite 1-D spherical	$G(0,r,t-\tau)=\dfrac{1}{2\rho Cp[\pi\alpha(t-\tau)]^{1/2}}\exp[-r^2/4\alpha(t-\tau)]$
3. Infinite 3-D Cartesian	$G(P,Q,t-\tau)=\dfrac{1}{8\rho Cp[\pi\alpha(t-\tau)]^{3/2}}\times\exp[-(x-x_0)^2/4\alpha(t-\tau)]$ $\times\exp[-(y-y_0)^2/4\alpha(t-\tau)]\times\exp[-(z-z_0)^2/4\alpha(t-\tau)]$
4. Semi-infinite 1-D Cartesian	$G(x_0,x,t-\tau)=\dfrac{1}{4\rho Cp[\pi\alpha(t-\tau)]}\times$ $\{\exp[-(x-x_0)^2/4\alpha(t-\tau)]-\exp[-(x+x_0)^2/4\alpha(t-\tau)]\}$
	Laplace's equation, Poisson's equation
5. Semi-infinite 2-D	$G(x_0,y_0,x,y)=\dfrac{1}{4\pi}\ln\dfrac{(x+x_0)^2+(y-y_0)^2}{(x-x_0)^2+(y-y_0)^2}$
6. Circle (r,θ), radius R	$G(r_0,\theta_0,r,\theta)=\dfrac{1}{4\pi}\ln\dfrac{R^2+r^2r_0^{\ 2}/R^2-2rr_0\cos(\theta_0-\theta)}{r_0^{\ 2}+r^2-2rr_0\cos(\theta_0-\theta)}$
7. Finite 1-D (0 < x ≤ 1)	$G(x_0,x)=(1-x_0)x \quad for\ x\geq x_0$ $=(1-x)x_0 \quad for\ x\leq x_0$

To further expand on the topic of Green's functions, we list the following important properties:

All solutions to unit source problems with homogeneous (i.e. zero) BCs and ICs are to be regarded as Green's functions. Some of these were presented in Chapter 6. Additional tabulations of Green's functions appear in the accompanying Table 7.7.

Green's functions depend on: (1) the underlying PDE which they solve, (2) the geometry, and (3) the associated boundary conditions. One speaks, for example, of a Green's function for Laplace's equation for a sphere with a Type I (homogeneous) boundary condition; or alternatively, of a Green's function for the Dirichlet problem of the sphere.

In a more general sense, one speaks of Green's function for the heat equation (Fourier), for the diffusion equation (Fick), for the wave equation, etc.

Green's functions can be generated by the superposition procedures outlined in Chapter 6 and by other methods, principally the Laplace transformation. Monographs on heat conduction contain many examples of this type. The reader also is urged to reach into the literature on related topics, such as electrostatics and potential theory for additional source material, even though this may appear to be a forbidding prospect.

The Green's functions carry the immense advantage of allowing us to express solutions of nonhomogeneous systems in closed form and in terms of integrals of the Green's functions and the nonhomogeneous terms of the system. In this, they resemble the Duhamel integral discussed in Chapter 6, Section 6.4.

Illustration 7.6 Solution of Poisson's Equation in Terms of a Green's Function

We use this example to illustrate the use of the integral theorems and other relations by deriving the solution to Poisson's equation for heat conduction in terms of Green's functions. Poisson's equation is itself nonhomogeneous, but we will, in addition, impose nonhomogeneous BCs of a general and arbitrary form. The task is then to solve the following set:

$$\nabla^2 u + \frac{A(Q)}{k} = 0 \quad u(S) = f(S) \tag{7.10a}$$

where Q denotes the coordinates of the domain, S its surface, and k the thermal conductivity. A solution of this model would yield the steady state temperature in a three-dimensional domain containing a distribution of heat sources emitting at a rate A(J/m^{3}sec), and bounded by a surface S maintained at a position-dependent temperature f(S). The corresponding model for a single continuous point source is given by the set

$$\nabla^2 G = \delta(P, Q) \qquad G(S) = 0 \tag{7.10b}$$

where G is the Green's function, i.e., the solution of Poisson's equation for a point source and zero surface temperature.

We start the derivation with Green's second identity, Equation 7.7c, which relates the Laplacian of two scalar functions (here u and G) within a domain, to values of u and G prescribed on the surface:

$$\iiint_V (u\nabla^2 G - G\nabla^2 u)dV = \iint_S (u\nabla G - G\nabla u)\bullet \mathbf{n}dS \tag{7.10c}$$

This is a fairly formidable looking vector equation which does, however, simplify considerably by substitutions from the model equations and certain other relations. Thus, we have:

Within the domain V: (from the model)

$$\nabla^2 u = -A/k \tag{7.10d}$$

and from both the model and item 4 of Table 7.6:

$$\iiint_V u\nabla^2 G\,dV = -\iiint u\delta(P, Q)dV = -u(P) \tag{7.10e}$$

The latter expression is of particular significance, because it gives us directly the solution we are seeking, i.e., u(P).

On the surface S (from the model):

$$u = f(S) \tag{7.10f}$$

and

$$G(S) = 0$$

From vector calculus and algebra,

$$\nabla G = \frac{\partial G}{\partial n}\mathbf{n} \quad \text{(see definition of gradient)}$$

and consequently,

$$\nabla G \bullet \mathbf{n} = \frac{\partial G}{\partial n}\mathbf{n}\bullet\mathbf{n} = \frac{\partial G}{\partial n} \tag{7.10g}$$

With these values in hand, Equation 7.10c reduces to the solution:

$$u(P) = \iiint_V G(A/k)dV - \iint_S \frac{\partial G}{\partial n} f(S)dS \tag{7.10h}$$

Let us illustrate the use of this expression with a simple one-dimensional problem: Consider a thin, insulated wire that generates heat at the rate of $A = 10^3$ W/m sec over the interval $0 < x < 1$. Thermal conductivity k is 400 W/mK (copper), and the ends of the wire where the generated heat is lost are kept at $T = 0$. Drawing on item 7 of Table 7.7 for G and substituting into Equation 7.10h, we obtain the following Green's function integrals:

$$\int_0^{x_0} G dx = \int_0^{x_0}(1 - x_0)x\,dx = \frac{x_0^{\,2}}{2} - \frac{x_0^{\,3}}{2} \tag{7.11a}$$

$$\int_{x_0}^1 G dx = \int_{x_0}^1 (1 - x_0)x_0 dx = x_0 - \frac{x_0}{2} - x_0^{\,2} + \frac{x_0^{\,3}}{2} \tag{7.11b}$$

As there are no variations normal to the wire, the second integral vanishes and for the temperature distribution on the wire:

$$T = (A/k)\int_0^1 G dx = (10^3/400)\frac{1}{2}x_0(1-x_0) = 1.25x_0(1-x_0) \qquad (7.11c)$$

Comments: Equation 7.10h represents an explicit and general solution to Poisson's equation for an arbitrary distribution of sources A(Q) in the interior of the domain, and an equally arbitrary nonhomogeneous surface condition f(S). The latter can be, it will be noted, of a discontinuous nature, a property which is not easily accommodated by other solution methods.

The expression is also a solution for Laplace's equation, as is easily seen by setting the source term A(Q) = 0. We obtain, in this case, the reduced formula:

$$u(P) = -\iint_S \frac{\partial G}{\partial n} f(S)dS \qquad (7.12)$$

where f(S) is, as before, an arbitrary surface distribution of u. We can use this equation, for example, to calculate the steady state temperature distribution in a rectangle whose sides are kept at four different temperatures – T(x, 0), T(x, a), T(0, y), and T(b, y).

The normal derivative of the Green's function, ∂G/∂n, is easily evaluated for simple geometries. Thus, for the aforementioned rectangle, $(\partial G/\partial n)_{y=0} = (\partial G/\partial y)_{y=0}$ as the normal points in the negative y-direction, and $(\partial G/\partial n)_{y=a} = (\partial G/\partial y)_{y=a}$. Similar expressions apply to the other two sides. For radial configurations, these expressions are replaced by the simple relation ∂G/∂n = ∂G/∂r.

There are some disadvantages to the method which need to be noted:

1. The Green's functions vary, as we had noted, with the geometry of the system. This is also evident from the tabulations of Table 7.7. Thus, for a particular problem under consideration, the appropriate Green's function has to be either located in the literature or derived from the basic model.
2. Green's functions also vary with the type of boundary condition assigned to the surface of the domain. Thus, for a Type III BC, the Green's function is the solution to the homogeneous point source problem with radiation or convective transport from the surface to a medium of zero value of the scalar function u. The solution then changes accordingly.
3. The Green's Functions do not always have the simple forms shown in Table 7.7. For finite geometries, in particular, they often consist of an infinite series of complex functions whose manipulation becomes cumbersome.

Having noted some of the stumbling blocks, we must reiterate that the Green's functions remain attractive and powerful tools for obtaining closed-form explicit solutions to linear nonhomogeneous PDEs. They are widely used to particularly obtain solutions to nonhomogeneous forms of Fourier's equations, of which we give an example below.

Illustration 7.7 The Use of Green's Functions in Solving Fourier's Equation

Green's functions can also be applied to obtain closed-form solutions to Fourier's equation with distributed heat sources and with prescribed nonhomogeneous IC and BC. That system is described by the following set of equations:

$$\alpha\nabla^2 T + \frac{1}{\rho Cp}A(Q) = \frac{\partial T}{\partial t} \tag{7.13a}$$

with surface BC:

$$T(S) = g(S, t) \tag{7.13b}$$

and IC:

$$T(Q, 0) = f(Q) \tag{7.13c}$$

where Q, as before, denotes the coordinates of the system, and k, α are the thermal conductivity and diffusivity, respectively. This is clearly the most general form of Fourier's equation and poses a challenging problem even though the underlying equations are all linear.

We present the solution (without proof), which is fairly straightforward and makes use, as before, of Green's identities. It takes the form:

$$T(P, t) = \iiint_V G(P, Q, t)f(Q)dV + \alpha\int_0^t \iiint_V G(P, Q, t-\tau)\frac{A}{k}(Q, \tau)dVd\tau$$

$$-\alpha\int_0^t \iint_S \frac{\partial}{\partial n}G(t-\tau)_S\, g(S, t)dSd\tau \tag{7.13d}$$

Let us try to make some sense of this equation by noting the following points:

G(P, Q, t − τ) is, as usual, a Green's function and, hence, describes the temperature distribution that results from an instantaneous heat source of unit strength releasing heat at t = τ and position P. For our purposes, the Green's functions are obtained from tabulations such as Table 7.7 and other literature sources, i.e., we shall not attempt to derive them ourselves.

The three integrals on the right-hand side of Equation 7.13d contain, in sequence, the nonhomogeneous initial condition T(Q, 0) = f(Q), a general distributed and time-dependent heat source A(Q, τ), and the nonhomogeneous surface BC T(S) = g(s, t). The latter is allowed to vary with position on the surface as well as in time. These are, then, the most general conditions one can expect to encounter in linear versions of Fourier's equation.

To obtain a better grasp of the uses of Equation 7.13d, we consider the simple case of conduction in the semi-infinite one-dimensional domain $x \geq 0$, devoid of heat sources and subject to the following general conditions:

IC $$T(x, 0) = f(x) \tag{7.14a}$$

BC $$T(0, t) = g(t) \tag{7.14b}$$

For these one-dimensional conditions, Equation 7.13d becomes:

$$T(x_0, t) = \int_0^{\infty} G(x_0, x, t) f(x) dx - \alpha \int_0^t \frac{\partial}{\partial n} G(x_0, x, t-\tau)\Big|_{x=0} g(\tau) d\tau \tag{7.14c}$$

where the Green's function, taken from Table 7.7, has the form:

$$G(x_0, x, t-\tau) = \frac{1}{4\rho Cp[\pi\alpha(t-z)]^{1/2}} \{\exp[-(x-x_0)^2/4\alpha(t-\tau)]$$

$$-\exp[-(x+x_0)^2/4\alpha(t-\tau)]\} \tag{7.14d}$$

The only manipulation required here is the evaluation of the normal derivative of the Green's function, $\partial G/\partial n$, which is given by:

$$\left[\frac{\partial G}{\partial n}\right]_{x=0} = -\left[\frac{\partial G}{\partial x}\right]_{x=0}$$

as the normal is taken in the negative x-direction. Hence,

$$\left[\frac{\partial G}{\partial n}\right]_{x=0} = -\frac{1}{4\rho Cp[\pi\alpha(t-z)]^{1/2}} \times \left\{ \exp[-x_0{}^2/4\alpha(t-\tau)]4x_0 \right\} \tag{7.14e}$$

The final solution then has the form:

$$T(x_0, t) = \frac{1}{4\rho Cp[\pi\alpha t]^{1/2}} \int_0^{\infty} f(x) \left\{ \exp[-(x-x_0)^2/4\alpha t] - \exp[-(x+x_0)^2/4\alpha t] \right\} dx$$

$$+ \frac{\alpha}{x_0(\pi\alpha)^{1/2}} \int_0^t g(\tau)(t-\tau)^{-3/2} \exp\left\{-\left[x_0{}^2/4\alpha(t-\tau)\right]\right\} d\tau \tag{7.14f}$$

Comments: The solution is still a fairly intimidating expression, but it should be recalled that it accommodates quite general and arbitrary ICs and BCs. Evaluation of the integrals, which in most cases has to be done numerically, poses no problem as convergence of the first integral is quite rapid.

One notes that the integration in space is with respect to x, not x_0. The latter then becomes the new distance variable in the solution.

Equation 7.14f, seemingly an incomprehensible glob of mathematics, does yield to a physical interpretation. $q_p/\rho Cp$ can be regarded as the heat liberated that would raise a unit volume of the medium to the temperature T, and therefore

$$q_p/\rho Cp = TV \tag{7.14g}$$

It follows from this that an initial distribution of temperature in a no-source problem, here given by f(x), can be regarded as *equivalent to a continuous distribution of instantaneous point sources of strength.*

$$q_p/\rho Cp = TV = [f(x)]avg\ A\Delta x \tag{7.14h}$$

and in the limit:

$$q_p/\rho Cp = f(x)A\ dx \tag{7.14i}$$

The first integral in our solution can thus be seen as representing the temperature at time t of a medium with zero surface temperature resulting from initial distribution of instantaneous heat sources of strength f(x). In other words, we have replaced the initial temperature distribution by an equivalent set of instantaneous heat sources that produce the prescribed initial condition. A physical interpretation of the second integral is a little less straightforward, involving a distribution of sources and sinks, and is described in the literature. This removes to some extent the sense of incomprehension upon being confronted with Equation 7.14b.

7.3 TRANSPORT OF MASS

In Illustration 7.44 we substituted consideration to the transport of total mass, which culminated in the continuity Equation 7.8c. To arrive at this result, we substituted ρ for the variable quantity X. This caused the convective term $\frac{D}{Dt}\iiint_V \rho\, dV$ to drop out. We then introduced the divergence theorem (Equation 7.7d) to convert the surface integral $\iint_S \rho(\mathbf{v} \bullet \mathbf{n})dS$ into a volume integral $\iiint_V (\nabla \bullet \mathbf{v})dV$, which was combined with the remaining volume integral of the transport theorem into a single expression. It was then argued that the integrand must vanish for the integral to be identically zero for any arbitrary volume, as required. This finally led to the continuity Equation 7.8c.

We now consider the transport of a particular species, characterized for example by its molar concentration C_A, and apply the same scheme as before. The difference here is that the convective term no longer vanishes, because the mass of the species A may change due to a chemical reaction. We have in fact:

$$\frac{D}{Dt}\iiint_V C_A \, dV = \pm\iiint_V r_A \, dV \tag{7.15a}$$

beacuse by definition $dC_A/dt = \pm\, r_A$. The transport theorem for the species A then reads:

$$\iint_S C_A(\mathbf{v} \bullet \mathbf{n})dS = \pm\iiint_V r_A \, dV - \iiint_V \frac{\partial}{\partial t} C_A \, dV \tag{7.15b}$$

Introducing the divergence theorem as before and setting the integrand of the resulting volume integral equal to zero thus yields:

$$-\nabla \bullet C_A \, \mathbf{v} = \pm r_A + \frac{\partial C_A}{\partial t} \tag{7.15c}$$

A number of points need to be noted in connection with this expression: Equation 7.15c represents a single (scalar) PDE. This is in contrast to the vector PDE, equivalent to three scalar PDEs, which arises in momentum balances to be taken up in Section 7.5, and which lead to the Navier–Stokes equation. Solution of Equation 7.15c yields the distribution of C_A in three dimensions and in time.

The rate term r_A is to be taken as negative for species consumption and positive for species production. $\nabla \bullet C_A \, \mathbf{v}$ expresses concentration changes due to both convective flow and diffusion. $C_A \, \mathbf{v}$ will be recognized as the molar flux $\mathbf{N}_A$ (moles A per unit time and area) and is represented by the following auxiliary relation

$$\mathbf{N}_A = x_A(\mathbf{N}_A + \mathbf{N}_B) - CD_{AB} \nabla x_A \tag{7.15d}$$

Flux Convection Diffusion

where $\mathbf{N}$ is a molar flux vector. Equation 7.15d thus consists of three scalar PDEs. Upon substitution into the mass balance Equation 7.15c, however, the operation $\nabla \bullet$ converts the expression into a single scalar PDE. We further note that for the special case of diffusion through a stagnant film, $\mathbf{N}_B = 0$, and for equimolar counter diffusion, $\mathbf{N}_A = -\mathbf{N}_B$. For the latter case, as well as for trace diffusion, Equation 7.15d reduces to the simple relation:

$$\mathbf{N}_A = -CD_{AB} \nabla x_A \tag{7.15e}$$

and for constant molar concentration C:

$$\mathbf{N}_A = -D_A \nabla C_A \tag{7.15f}$$

This is the three-dimensional version of Fick's Law.

We now turn to some practical applications of these equations. We consider the following cases:

- Combined convection, reaction, and diffusion (Illustration 7.8)
- Combined reaction and diffusion (Illustration 7.9)
- Combined convection and diffusion (Illustration 7.10)
- Unsteady diffusion (Illustration 7.11 and Illustration 7.12)
- Steady state multidimensional diffusion (Illustration 7.13)

Illustration 7.8 Catalytic Conversion in a Coated Tubular Reactor: Locating Equivalent Solutions in the Literature

The system under consideration here consists of a tubular reactor whose wall is coated with a catalyst, which could for example, be an immobilized enzyme (bioreactor). The purpose of modeling is then usually confined to relating the size of the reactor to conversion for a given flow and set of kinetic parameters (design problem). The device has also been occasionally used to investigate the kinetics of a catalytic reaction without the added complications of heat effects and diffusional resistance within a catalyst pellet (parameter estimation).

We assume laminar flow conditions and plug flow, $v \neq f(r)$, with an irreversible reaction $A \rightarrow B$, $r_A = k_r C_A$ taking place at the wall. This leads to the development of radial and axial concentration profiles described by a partial differential mass balance. The situation is then essentially equivalent to that of the Graetz problem, given in Chapter 5. The mass-transfer equivalent of Equation 5.24d is then given by:

$$v\frac{\partial C_A}{\partial z} = D_{AB}\left[\frac{\partial^2 C_A}{\partial r^2} + \frac{1}{r}\frac{\partial^2 C}{\partial r^2}\right] \tag{7.16a}$$

Three boundary conditions are required, which are as follows:

At the inlet: $C_A(r, 0) = C_A^{\,0}$

At the axis: $\dfrac{\partial C_A}{\partial r}(0, z) = 0$ (7.16b)

At the wall: $-D_{AB}\dfrac{\partial C_A}{\partial r}(R, z) = k_r C_A(R, z)$

These equations can be solved by standard analytical techniques, as will be shown in the next chapter. Our aim here is to avoid this complication and to locate the solution of an equivalent problem in the literature. The reader will already have noted the similarity between Equation 7.16a and that for unsteady radial conduction in a cylinder with convective heat loss to a medium at zero temperature. In fact, by making the following substitutions, one arrives at completely identical models:

C_A	→	Temperature T
z	→	Time t
D_{AB}/v	→	Thermal diffusivity α
D_{AB}	→	Thermal conductivity k
k_r	→	Heat-transfer coefficient h
$(C_A)_0$	→	Initial temperature T_0

The solution to this problem is readily available in the heat-transfer literature. Translated back to the original variables, it assumes the following forbidding form:

$$C_A(r, z) = 2(C_A)_0 \sum_1^{\infty} \frac{\lambda_j}{\lambda_j^2 + \beta^2} \frac{J_1(\lambda_j)}{J_0^2(\lambda_j)} J_0\left(\lambda_j \frac{r}{R}\right) \exp(-\lambda_j^2 z/\gamma^2) \quad (7.16c)$$

where $\beta = k_r R/D_{AB}$, $\gamma = R(v/D_{AB})$, and λ_j = j-th root of the expression

$$\lambda J_1(\lambda) - \beta J_0(\lambda) = 0 \quad (7.16d)$$

J_0 = Bessel function of first kind and order zero
J_1 = Bessel function of first kind and order one

This infinite series represents the concentration profiles as they develop in the axial and radial directions.

The already panic-stricken will be further dismayed to learn that the result has to be converted to the more useful average, or mean integral concentration $(C_A)_{avg}$, i.e., we have to evaluate:

$$(C_A)_{avg} = \int_0^1 C_A 2\pi y\, dy \Big/ \int_0^1 2\pi y\, dy = 2\int C_A\, y\, dy \quad (7.16e)$$

where we have nondimensionalized the radial distance by setting y = r/R. In other words, we have to evaluate the integral of the y-dependent part of Equation 7.16c, $\int_0^1 J_0(\lambda_j y) y\, dy$. This is not as formidable a task as it appears, as we have ready-made formulae for just such cases: We reach back to Table 2.8 of Chapter 2 and extract:

$$\alpha \int x^k J_{k-1}(\alpha x) dx = x^k J_k(\alpha x) + C \quad (7.16f)$$

Some manipulations then result in the following expression:

$$(C_A)_{avg} / (C_A)_0 = 4\sum_1^{\infty} \frac{\beta^2}{\lambda_j^2(\lambda_j^2 + \beta^2)} \exp(-\lambda_j^2 z/\gamma^2) \quad (7.16g)$$

TABLE 7.8
Roots of $\lambda J_1(\lambda) - \beta J_0(\lambda) = 0$ and Their J_0 Values

β	λ_1	$J_0(\lambda_1)$	λ_2	$J_0(\lambda_2)$	λ_3	$J_0(\lambda_3)$	λ_4	λ_5
0	0	1.0	3.8137	−0.403	7.0156	0.300	10.1735	13.3237
0.01	0.1412	0.995	3.8343	−0.403	7.0170	0.300	10.1745	13.3244
0.1	0.4417	0.935	3.8577	−0.403	7.0298	0.300	10.1833	13.3312
1.0	1.2588	0.641	4.0795	−0.391	7.1558	0.295	10.2710	13.3984
10.0	2.1795	0.310	5.0332	−0.168	7.9569	0.181	10.9363	13.9580
100.0	2.3809	0.013	5.4652	−0.017	8.5678	0.023	11.6747	14.7834
∞	2.4048		5.5201		8.6537		11.7915	14.9309

We may at this point be permitted a sigh of relief, because the Bessel functions, at least, have disappeared, although they do lurk in the roots λ_j of Equation 7.16d. *Comments*: The following may also prove soothing. The model we have presented is a rather limited one, as it does not address complex reaction mechanisms or the effect of the parabolic velocity profile which prevails under laminar flow conditions. The main purpose of the exercise is to practice the art of recognizing the equivalence of models for dissimilar processes and reaching into the literature for quick convenient solutions.
The Bessel Functions J_0 and J_1 are of a periodic type (see Figure 2.5) and hence, have an infinite number of roots singly or when they appear in combination, as in Equation 7.16d. That expression is a characteristic transcendental equation, which arises in conduction and diffusion equations with a Type III BC. Its frequent occurrence has led to numerous tabulations of its roots, of which we give an abbreviated version in Table 7.8.

The fact that an infinite series has to be evaluated may at first sight appear discouraging. Closer inspection of Equation 7.16f shows, however, that the axial profile decays rapidly with increasing values of the roots λ_j. Fast convergence of the series may, therefore, be expected in all but very slow reactions. To explore this feature, we consider the following numerical example, taken to apply to a liquid system:

Reactor radius	$R = 5$ cm
Reactor length	$z = 100$ cm
Rate constant	$k_r = 2 \times 10^{-5}$ sec^{-1}
Diffusivity	$D_{AB} = 10^{-3}$ cm^2/sec
Flow velocity	$v = 20$ cm/sec

We obtain, for the parameters,

$$\beta = k_r R / D_{AB} = (2 \times 10^{-5})(5)/10^{-3} = 10^{-1}$$

$$\gamma = Rv/D_{AB} = (5)(20)/10^{-3} = 10^5$$

and from Table 7.8:

$$\lambda_1 = 0.4417,\ \lambda_2 = 3.8577$$

The rate constant here is quite low, and the fluid velocity high, so that conversions are expected to be low.

Substitution of these values into Equation 7.16f yields, for the first two terms of the series

$$(C_A)_{avg} / (C_A)_0 = 4\sum_1^2 \frac{\beta^2}{\lambda_j^2(\lambda_j^2 + \beta^2)} \exp(-\lambda_j^2 z / \gamma^2)$$

$$= 4\frac{0.1^2}{0.4417^2(0.4417^2 + 0.1^2)} \exp(-0.477^2\ 100/10^2)$$

$$+ 4\frac{0.1^2}{3.8577^2(3.8577^2 + 0.1^2)} \exp(-3.8577^2\ 100/10^2)$$

$$(C_A)_{avg}/(C_A)_0 = 0.824 + 6.23 \times 10^{-11}$$

and conversion $X = [1 - (C_A)_{avg}/(C_A)_0]100 = 17.8\%$.

The second term is seen to be negligibly small compared to the first, due primarily to the rapid decay of the exponential term caused by the high value of $\lambda_2^2 = 14.88$. This fast convergence is the norm in many practical applications, and enables us to express the conversion X in terms of the simple expression

$$X = 1 - \frac{4\beta^2}{\lambda_1^2(\lambda_1^2 + \beta^2)} \exp(-\lambda_1^2 z / \gamma^2) \tag{7.16h}$$

The rate constant k_r contained in the parameter β is easily extracted from this expression, and experiments run at different reactor lengths z or feed velocity v contained in γ.

Illustration 7.9 Diffusion and Reaction in a Semi-infinite Medium: Another Literature Solution

We consider in this example the unsteady equimolar or trace diffusion of a species A into a semi-infinite medium initially free of solute and an imposed concentration C_A^0 at the surface x = 0. The solute undergoes an irreversible first-order reaction within the medium, A → B, with the rate given by $r = k_r C$. The partial differential mass balance for this case becomes:

$$\frac{\partial C}{\partial t} = D\frac{\partial^2 C}{\partial x^2} - k_r C \tag{7.17a}$$

Derivation of this expression is given as a practice problem.

The equation can be solved by standard linear techniques (e.g., the Laplace transform) without undue difficulty. We use, instead, the approach applied in the previous illustration and seek a literature solution in the related discipline of heat conduction. This requires some thought and perseverance, as well as a good knowledge of the pertinent literature.

An initial inspection of Equation 7.17a does not appear encouraging: the reaction term k_rC would have to be matched by a corresponding heat sink term hT, which describes heat loss in proportion to the prevailing temperature T at a given point. This is not a realistic physical process, but becomes so if the domain is reduced to that of a thin rod with a uniform temperature T over its cross-section, subject to convective heat loss at a rate hT to a medium at zero temperature. This results in the following two equivalent models.

a) Diffusion in Semi-Infinite Medium with Irreversible Reaction	b) Conduction in Semi-Infinite Thin Rod with Convective Heat Loss	
$\frac{\partial C}{\partial t} = D\frac{\partial^2 C}{\partial x^2} - k_r C$	$\frac{\partial T}{\partial t} = \alpha\frac{\partial^2 T}{\partial x^2} - \beta T$	
$C(x, 0) = 0$	$T(x, 0) = 0$	(7.17b)
$C(0, t) = C_0$	$T(0, t) = T_0$	
$C(\infty, t) = 0$	$T(\infty, t) = 0$	

where

$$\beta = \frac{hP}{\rho Cp A_C}$$

A_C = rod cross-sectional area

P = rod perimeter

Note that the heat loss does not enter the model as a surface BC, but resides instead in the PDE itself. This becomes apparent in the derivation of the PDE, which is given as a practice problem.

The solution to the temperature problem is available in standard texts on heat conduction and takes the form:

$$\frac{T}{T_0} = \frac{1}{2}\exp\left(-x\sqrt{\alpha/\beta}\right) erfc\left[x/2\sqrt{\alpha t} - \sqrt{\beta t}\right] + \frac{1}{2}\exp\left(x\sqrt{\alpha/\beta}\right) erfc\left[x/2\sqrt{\alpha t} + \sqrt{\beta t}\right] \tag{7.17c}$$

Translation into the corresponding diffusion problem is easily accomplished by means of the dictionary provided by Equation 7.17b. Note that the same expression can be derived, in more elaborate fashion, by the Danckwerts superposition (see Practice Problem 6.11).

Illustration 7.10 The Graetz–Lévêque Problem in Mass Transfer: Transport Coefficients in the Entry Region

On several occasions, we have already, referred to the Graetz Problem, and the Lévêque version of it. In Illustration 7.8 we encountered a modified Graetz Problem in which concentration took the place of temperature, and conditions at the wall were described by a BC of Type III. The solution was given in terms of an infinite series, Equation 7.16c, which showed fast convergence for high to moderate diffusivities, and small diameters, long conduits, or low velocities. When this is no longer the case, an alternative method, known as the Lévêque solution, is resorted to. The Lévêque solution focuses on the so-called entry region near the tube inlet, where profile development is still in its initial stage. The following assumption can then be made:

Temperature or concentration changes are confined entirely to a thin boundary layer adjacent to the conduit wall. Within the bulk of the fluid, changes in both the radial and axial directions are negligible.

As the boundary layer typically occupies only a small fraction of the tube diameter, curvature can be neglected. We can unravel the conduit and treat it as a flat plate. A corollary of this approach is that the boundary layer and the neighboring bulk fluid can be regarded as a semi-finite medium, with $C \to C_{bulk}$ as y (or r) $\to \infty$. See in this connection, Chapter 5, Subsection 5.3.5.

The boundary layer lies entirely within the linear portion of the (parabolic) velocity profile. This is justified by its thinness. We encountered a similar situation in Illustration 5.4, where the thermal boundary layer development along a flat plate was seen to lag behind the momentum layer development, with the result that temperature changes along the plate lay entirely within the linear portion of the velocity profile. We can, with some imagination, anticipate that the underlying models and their solutions will therefore be identical in form, if not in detail. This is, in fact, the case as we shall see below.

We start by performing a differential mass balance over the elements Δx and Δy, where x is the coordinate along what is now a flat plate and y the direction perpendicular to it. Diffusion in the x-direction is neglected, and the concentration at the wall is assumed to be zero along the entire length of the plate. This can be brought about by a fast reaction at the wall or rapid permeation through it. We use the "in out = 0" scheme, rather than the generalized mass, and obtain at steady state:

$$\text{Rate of solute in} - \text{Rate of solute out} = 0$$

$$\begin{bmatrix} vC_A\, W\, \Delta y\big|_x \\ -DW\,\Delta x \dfrac{\partial C}{\partial y}\Big|_y \end{bmatrix} - \begin{bmatrix} vC_A\, W\, \Delta y\big|_{x+\Delta x} \\ -DW\,\Delta x \dfrac{\partial C}{\partial y}\Big|_{y+\Delta y} \end{bmatrix} = 0 \tag{7.18a}$$

where W is the width of the plate, equal to the perimeter of the tube.

Dividing by ΔxΔyW and letting the increments go to zero, we obtain the following linear second-order PDE:

$$D\frac{\partial^2 C}{\partial y^2} - v\frac{\partial C}{\partial x} = 0 \tag{7.18b}$$

We note that for the linear portion of the velocity profile we can write:

$$\text{Shear rate } \dot{\gamma} = \frac{dv}{dy} = \frac{v-0}{y-0} = \frac{v}{y} \tag{7.18c}$$

so that the PDE 7.18b becomes:

$$D\frac{\partial^2 C}{\partial y^2} - \dot{\gamma}y\frac{\partial C}{\partial x} = 0 \tag{7.18d}$$

with boundary conditions:

$$\begin{aligned} &\text{At the inlet:} && C(0, y) = C_0 \\ &\text{At the wall:} && C(x, 0) = 0 \\ &\text{In the bulk fluid:} && C(x, \infty) = C_0 \end{aligned} \tag{7.18e}$$

The form of this model is precisely that of the development of the thermal boundary layer along a flat plate (Equations 5.29c and Equation 5.29d). Its solution was obtained by similarity transformation, which can be applied to the present case as well. The details are left to the exercises (see Practice Problem 7.7). The final solution is of the same form as Equation 5.29s and is given by:

$$\frac{C}{C^0} = \frac{\int_0^{\eta} \exp(-\eta^3)d\eta}{\int_0^{\infty} \exp(-\eta^3)d\eta} \tag{7.18f}$$

where the similarity variable η is given by:

$$\eta = y\left(\frac{\dot{\gamma}}{9Dx}\right)^{1/3} \tag{7.18g}$$

Let us plot the concentration profile Equation 7.18f with Mathematica. First we define a function describing the dimensionless concentration C/C^o :

```
In[1]:= c[η_]:= Integrate[Exp[-z^3], {z, 0, η}] / Integrate[Exp[-z^3], {z, 0, ∞}]
```

We always check the definition of the function using the command "?":

```
In[2]:= ?c

        Global `c
```

$$c[\eta\_] := \frac{\int_0^{\eta} e^{-z^3} dz}{\int_0^{\infty} e^{-z^3} dz}$$

That is the same as Equation 7.18f. Now let us plot the profile:

```
In[3]:= Plot[c[η], {η, 0, 2}, PlotRange → All,
         AxesLabel→{"η", "c/c°"}]
```

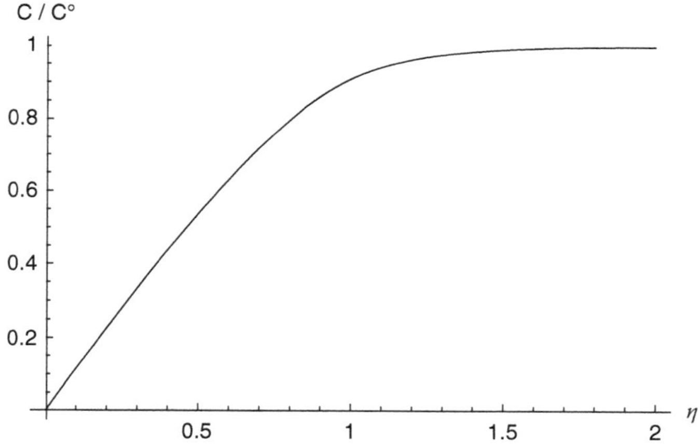


```
Out[3]= -Graphics-
```

We now turn to the task of deriving an effective mass-transfer coefficient k_f from the concentration profile Equation 7.18f. The key to the procedure is the equation:

$$D\frac{\partial C}{\partial y}\bigg|_{y=0} = k_f(C_1^{\ 0} - 0) \tag{7.18h}$$

which merely expresses the rate of arrival of solute in two equivalent forms: one involving Fick's law, the other an effective mass-transfer coefficient k_f and its associated driving force. The crux then is the evaluation of the derivative $(\partial C/\partial y)_{y=0}$.

We start the procedure by first evaluating the integral in the denominator of Equation 7.18f. To do this, the substitution x = η^3 is made, which yields:

$$\int_0^\infty \exp(-\eta^3)d\eta = \frac{1}{3}\int_0^\infty x^{-2/3}e^{-x}dx \tag{7.19a}$$

The integral on the right is known as the gamma function Γ(n) and has the general definition:

$$\Gamma(n) = \int_0^\infty x^{n-1}e^{-x}dx \tag{7.19b}$$

where n is any positive or negative number. It has to be evaluated numerically and is tabulated in most mathematical handbooks for the interval $1 \le n \le 2$. Some selected values are reproduced in Table 7.9. They are seen to be close to unity over the entire range. Other values of n are obtained from the following recursion formula:

$$\Gamma(n) = n\Gamma(n-1) = n(n-1)\Gamma(n-2) \; etc. \tag{7.19c}$$

TABLE 7.9
Values of the Gamma Function

n	Γ(n)
1.000	1.00000
1.050	0.97350
1.100	0.95135
1.150	0.93304
1.200	0.91517
1.250	0.90640
1.300	0.89747
1.350	0.89115
1.400	0.88726
1.450	0.88566
1.500	0.88623
1.550	0.88887
1.600	0.89352
1.650	0.90011
1.700	0.90864
1.750	0.91906
1.800	0.93138
1.850	0.94561
1.900	0.96177
1.950	0.97988
2.000	1.0000

The gamma function can consequently be regarded as a generalized factorial n! applicable to any positive or negative number.

For the case in hand, we obtain from Equation 7.19a and the recursion formula Equation 7.19c:

$$\frac{1}{3}\int_0^\infty x^{-2/3}e^{-x}dx = \frac{1}{3}\int_0^\infty x^{1/3-1}e^{-x}dx$$

$$= \frac{1}{3}\Gamma(1/3) = \frac{1}{3}\frac{\Gamma(4/3)}{1/3} = \Gamma(4/3) = 0.89407 \qquad (7.19d)$$

where the numerical value is interpolated from Table 7.9. Let's check this against Mathematica:

```
In[4]:= Gamma[4/3]//N

Out[4]= 0.89298
```

To evaluate the derivative $(\partial C/\partial y)_{y=0}$ we apply the chain rule of partial differentiation (see Table 5.4) and write, using the profile Equation 7.18f:

$$\left.\frac{\partial C}{\partial y}\right|_{y=0} = \left.\frac{\partial C}{\partial \eta}\frac{\partial \eta}{\partial y}\right|_{y=0} = \left.\frac{C^0}{0.89407}\exp(-\eta^3)\left(\frac{\dot{\gamma}}{9Dx}\right)^{1/3}\right|_{y=0} = \frac{C^0}{0.89407}\left(\frac{\dot{\gamma}}{9Dx}\right)^{1/3}$$

(7.19e)

Substitution of this expression into Equation 7.19h finally yields the effective mass-transfer coefficient:

$$k_f = 0.54\left(\frac{\dot{\gamma}D^2}{x}\right)^{1/3} \qquad (7.19f)$$

For boundary conditions of Type II and Type III, the coefficient on the right is some 10 to 15% higher. This leads us to propose an average coefficient of 0.6, applicable to all three types of BCs. Thus, we have:

$$k_f \cong 0.6\left(\frac{\dot{\gamma}D^2}{x}\right)^{1/3} \qquad (7.19g)$$

Comments: Equation 7.19f has a number of noteworthy features. Foremost among them is the weak, one-third power dependence of k_f on shear rate $\dot{\gamma}$ and distance x. As $\dot{\gamma} = 8v/d$ for laminar flow in a cylindrical tube, k_f will be proportional to $v^{0.33}$. This contrasts with the much stronger dependence, $k_f \propto v^{0.8}$, found in turbulent flow.

The Graetz–Lévêque problem for mass transfer has its origin and counterpart in heat transfer, which will be discussed in some detail in Section 7.4. From the results given there, it can be deduced that the Lévêque (entry) region extends over the following range:

$$0 < \frac{xD}{vd^2} < 10^{-3}$$

This relation serves as a quick means of establishing limits of velocity, tubular diameter, and length beyond which the flow ceases to be in the entry region. In blood flow, for example, with a typical protein diffusivity $D = 10^{-5}$ cm^2/sec, average flow velocity of v = 10 cm/sec, and near-inlet distance x = 0.1 cm, the critical vessel diameter, which arises from the cited condition, is given by:

$$d \geq \left(\frac{xD}{10^{-3}\,v}\right)^{1/2} = \left(\frac{(0.1)(10^{-5})}{(10^{-3})(10)}\right)^{1/2} = 10^{-2}\,cm \tag{7.19h}$$

In other words, for the Lévêque solution to be valid in blood flow, the vessel diameter cannot be less than 0.1 mm.

Illustration 7.11 Unsteady Diffusion in a Sphere: Sorption and Desorption Curves

In this classical problem, which has its exact counterpart in the unsteady conduction of heat, we consider diffusional uptake or release of a solute by a porous sphere. These processes are often referred to as sorption or desorption, although no actual sorptive retention on the solid matrix takes place.

We consider the general case of a Type III BC, descriptive of a film resistance at the surface, and obtain the following model:

$$\frac{\partial C}{\partial t} = D\left(\frac{\partial^2 C}{\partial r^2} + \frac{2}{r}\frac{\partial C}{\partial r}\right) \tag{7.20a}$$

with boundary conditions:

At the surface: $$\pm D\frac{\partial C}{\partial r}\bigg|_{r=R} = k_f(C_{r=R} - C_0)$$

At the center: $$\left.\frac{\partial C}{\partial r}\right|_{r=0} = 0 \tag{7.20b}$$

Initially: $$C(0, r) = C^0$$

The PDE (7.20a) can be obtained by performing a classical mass balance over the spherical increment Δr, or by substitution of the vector flux Equation 7.15d into the generalized mass balance Equation 7.15c and use of the dictionary Table 7.1.

Solution of the model is accomplished by standard linear techniques of which we shall give examples in Chapter 8. For our present purposes, we note that the primary information consists of the unsteady concentration profiles C(r, t), which are usually converted into the more useful fractional uptake or fractional release M_t/M_∞. Here M_t denotes the total amount taken up or released up to time t, M_∞ the same quantity at time infinity. M_t is obtained from the concentration profiles C(r, t) by applying the relation:

$$M_t = -\int_0^t 4\pi R^2 D \left.\frac{\partial C}{\partial r}\right|_{r=R}^{dt} \tag{7.20c}$$

Plots of M_t/M_∞ as a function of Dt/R^2, and the parameter $Bi = k_fR/D$ are displayed in Figure 7.4. Note that the limiting cases of Bi = ∞ and Sh = 0 correspond to the limiting boundary conditions $C\big|_{r=R} = C_0\,(Type\ I)$ and $(\partial C/\partial r)r = R = 0\,(Type\ II)$. The latter case represents an impermeable sphere. Bi is a modified Biot number, a dimensionless group that represents the ratio of internal to external resistance to mass transfer. The reader will recall a similar group, which was used in Illustration 1.7 to characterize the analogous heat-transfer problem.

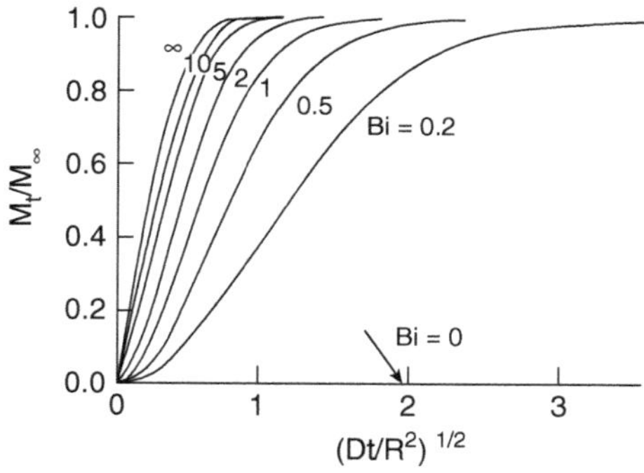

FIGURE 7.4 Fractional uptake and release as a function of dimensionless time for diffusion in and out of a sphere (BC Type III).

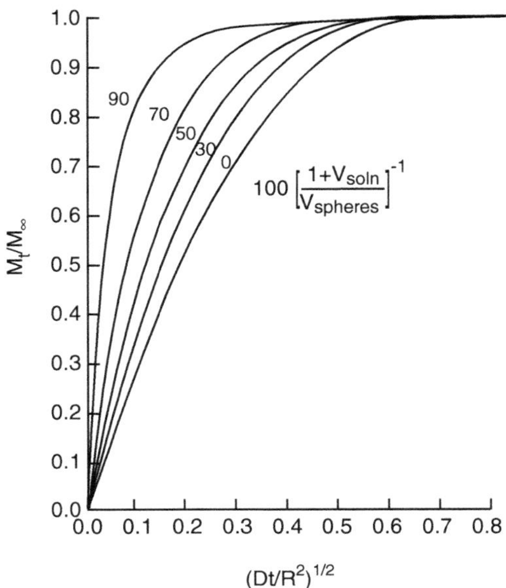

FIGURE 7.5 Fractional diffusional uptake and release as a function of dimensionless time for spheres in a well-stirred solution of limited volume.

Let us briefly demonstrate the use of the plots with a numerical example. We choose:

$$k_f = 10^{-4}\ cm/sec,\ D = 10^{-4}\ cm^2/sec,\ \mathrm{R} = 1\ cm$$

representative of a liquid system and obtain Sh = $(10^{-4})(1)/10^{-4} = 1$. We use this value to calculate the time necessary to deplete a sphere to one half its original solute content. We obtain from Figure 7.5:

$$(Dt/R^2)^{1/2} = 0.5 \text{ at } M_t/M_\infty = 0.5,\ Sh = 1$$

Hence,

$$t = \frac{0.5^2}{10^{-4}} 1^2 = 2500 \text{ sec} \approx 42 \text{ min}$$

Note that the same time is required to achieve a fractional uptake of the same magnitude 0.5.

In this example, the exterior medium was assumed to be infinite in extent, resulting in a constant concentration C_0 of the surroundings. Of more frequent occurrence is the situation in which the surroundings consist of a finite volume of a well-stirred solution. This case is taken up below.

Illustration 7.12 The Sphere in a Well-Stirred Solution: Leaching of a Slurry

We assume in this example that a sphere, or an aggregate of spheres of volume V_{sphere} is suspended in a well-stirred medium of limited volume V_{soln} and concentration C_e. By well-stirred, we wish to imply that the concentration in the fluid is uniform and equal to that at the sphere surface (i.e., film resistance is negligible). The model is similar to that of the previous illustration, differing from it only in the surface BC. We have:

$$\frac{\partial C}{\partial t} = D\left(\frac{\partial^2 C}{\partial r^2} + \frac{2}{r}\frac{\partial C}{\partial r}\right) \tag{7.21a}$$

with boundary conditions:

At the surface: $\pm D4\pi R^2 \left.\frac{\partial C}{\partial r}\right|_{r=R} = V_{soln} \left.\frac{\partial C_e}{\partial t}\right|_{r=R}$

At the center: $\left.\frac{\partial C}{\partial r}\right|_{r=0} = 0$ (7.21b)

Initially: $C(0, r) = C^0$.

where C_e = time varying external concentration.

The concentration profiles initially obtained are again converted into fractional uptake or release M_t/M_∞, with the volume ratio $V_{soln}/V_{spheres}$ now appearing as a parameter. The results are displayed in Figure 7.5. We use the plots to calculate the following numerical example.

Suppose it is desired to calculate the time required to extract 90% of the oil contained in oil-bearing vegetable seeds assumed to be spherical. We set:

$$R = 0.5\ cm,\ D = 10^{-4}\ cm^2/s \text{ and } V_{soln}/V_{spheres} = 1$$

so that $100\,(1 + V_{soln}/V_{spheres})^{-1} = 50$.

From Figure 7.5 we obtain, for $M_t/M_\infty = 0.9$, $(Dt/R^2)^{1/2} = 0.34$, and $t = (0.34^2)(0.5^2)/10^{-4} = 289\ s \approx 5$ min.

Comments: One notes from Figure 7.5 that release time diminishes with decreasing solution volume, which also results in higher extract concentrations. It is advantageous to minimize solvent volume. One has to keep in mind, however, that efficient stirring requires a certain minimum ratio of solvent to solids volume.

A parameter value of zero corresponds to an infinitely large solution volume. This implies, in turn, a constant external concentration, and the plot becomes identical to that of Figure 7.5 at Bi = ∞, i.e., under conditions of no film resistance.

Illustration 7.13 Steady state Diffusion in Several Dimensions

In the absence of convective flow and reactions, steady state diffusional transport is described by Laplace's Equation $\nabla^2 C = 0$. Some solutions to this PDE, principally arrived at by the technique of conformal mapping, have previously been given in Figure 5.7. Additional solutions will be presented in Figure 7.10 in connection with steady state multidimensional conduction, which is also described by Laplace's equation. The latter are not given as distributions, as was the case in Figure 5.7, but rather in terms of so-called shape factors S, which allow the direct calculations of the rate of heat flow q between two bodies maintained at different temperatures. The compilations can, evidently, also be used to calculate the analogous case of steady diffusional mass flow in between two surfaces at different concentrations. The need for this might arise, for example, in connection with emanations from underground deposits, or in connection with controlled release devices.

Let us then reach forward to Table 7.9 and consider the case of a cylinder (or circular hole) of length L and diameter d with a surface temperature of T_1, buried deep in a semi-infinite medium with surface temperature $T_2 < T$. The shape factor for this case, item 8b, is given by:

$$S = 2\pi L \ln(2L/d) \tag{7.22a}$$

and is used directly in Newton's law of cooling to obtain the flux q. Thus,

$$q'\ (J/sec) = kS(T_1 - T_2) \tag{7.22b}$$

or

$$q' = 2\pi L \ln(2L/d) k(T_1 - T_2) \tag{7.22c}$$

The equivalent case of diffusive mass flow N_A is then given by:

$$N'(mole/sec) = 2\pi L \ln(2L/d) D(C_1 - C_2) \tag{7.22d}$$

Suppose the solute is quickly removed from the surface of the semi-infinite medium, for example by a flowing fluid, so that $C_2 \approx 0$. Let us further set L = 100 cm, d = 10 cm, $C_1 = 10^{-4}$ mol/cm^3, and $D = 10^{-4}$ cm^2/sec. Then the total amount diffusing will be given by:

$$N(mole/sec) = 2\pi l \ln(2L/d) D\ C = 2\pi 100 \ln[(2(100)/10)](10^{-4})(10^{-4})$$

$$N_A = 1.88\ 10^{-5}\ mol/s$$

7.4 TRANSPORT OF ENERGY

Derivation of the generalized energy equation proceeds along the lines established for the transport of mass, with one or two extra items added. Briefly, the following steps are involved:

One starts with the general statement of the first law of thermodynamics

$$\begin{matrix}\text{Change in Energy} \\ \text{of the system}\end{matrix} = \begin{matrix}\text{Energy Added} \\ \text{to the System}\end{matrix} + \begin{matrix}\text{Work Done} \\ \text{on System}\end{matrix} \quad (7.23a)$$

Here energy includes transfer by conduction, radiation, or induction heating, as well as heat produced by internal sources (e.g., nuclear reaction). Work encompasses the effects of gravity, buoyancy, electrical and shear forces, as well as conventional piston work. Change in energy consists of two terms: internal and kinetic energy.

The resulting equation is combined with the generalized momentum balance and continuity equation. This results in the cancellation of all work terms because of body forces and kinetic energy.

The surface integrals are converted to volume integrals via the divergence theorem, and the total integrand set equal to zero, as was done in Section 7.3. The following expression, often referred to as the thermal energy balance, is then obtained:

$$\rho\frac{D}{Dt}(H) = \frac{DP}{Dt} + \nabla \bullet k\nabla T + \mu\phi + \rho q_b \quad (7.23b)$$

where $\mu\phi$ is a viscous dissipation term, and q_b represents the rate of heat transfer to the system by radiation, induction, and internal heat sources. This equation is of general validity and applies to compressible and incompressible flow, as well as reacting systems. Heat of reaction ΔH_r for the latter is contained in the enthalpy term DH/Dt but is somewhat cumbersome to extract. It is more convenient, in these cases, to formulate the model by the classical "in–out = change" approach (see Practice Problem 7.13).

In what follows, we confine ourselves to nonreacting systems and address the following topics:

- Conduction with laminar convection (Illustration 7.14)
- Heat transfer in a packed bed (Illustration 7.15 and Illustration 7.16)
- Unsteady conduction (Illustration 7.17)
- Steady state multidimensional conduction (Illustration 7.18)

Illustration 7.14 The Graetz–Lévêque Problem (Yet Again!)

The PDEs pertinent to this problem have previously been derived in Chapter 5, Illustration 5.3 (the Graetz problem) for a cylindrical tube, using the classical "in– out = 0" approach. Here we use the generalized energy balance, Equation 7.24b, as a starting point, neglecting pressure and viscous dissipation terms, which are insignificant compared to enthalpy and conduction, and omitting ρ q_b, which does not apply here. For constant thermal conductivity we then obtain the reduced form:

$$\rho\frac{D}{Dt}H = k\nabla^2 T \quad (7.24a)$$

This equation applies to any arbitrary duct geometry. The convective derivative DH/Dt is next decomposed using the Equation 7.7e and we obtain:

$$\rho\frac{\partial H}{\partial t}+(\rho\mathbf{v}\bullet\nabla)H=k\nabla^2T \tag{7.24b}$$

As a further step, we introduce the auxiliary enthalpy relation:

$$H=C_p\,(T-T_0)+H_0$$

which reduces Equation 7.24a with no loss of generality (except k = const.) to the form

$$\frac{\partial T}{\partial t}+\mathbf{v}\bullet\nabla T=\alpha\nabla^2T \tag{7.24c}$$

To apply this equation to a cylindrical tube with steady flow, we drop the time derivative and draw on our dictionary (Table 7.1) to identify components of the differential operators. The final form obtained, neglecting axial conduction, is identical to that derived by classical means in Illustration 5.3:

$$v(r)\frac{\partial T}{\partial x}=\alpha\left[\frac{\partial^2T}{\partial x^2}+\frac{1}{r}\frac{\partial T}{\partial x}\right] \tag{7.24d}$$

where the velocity will, in general, vary with radial distance.

The Graetz problem was previously alluded to on several occasions in the context of the corresponding mass-transfer problem. Here we address it in its original form, which deals with the steady radial and axial temperature profiles in laminar tubular flow for different BCs. The distributions initially obtained are converted into equivalent Nusselt numbers, Nu = hd/k, using the thermal equivalent of Equation 7.18h. Plots of local Nusselt numbers, i.e., those which prevail at a particular axial position x, are shown in Figure 7.6 for cylindrical and elliptical ducts as a function of dimensionless distance $x^*=x\alpha/vd^2$. Several features can be observed.

The functional form of the Nusselt number shows three distinct domains for ducts of all shapes.

1. The entry region in which Nu varies with the negative 1/3 power of dimensionless distance and which prevails for values $x^* < 10^{-3}$. Here, the principal resistance to heat transfer resides in a thin boundary layer near the wall.
2. The fully developed region that comes about with the penetration of the boundary layer to the center-line of the duct. This domain starts at an approximate value of $x^* \approx 0.03$ and exhibits constant Nusselt numbers.
3. A transition region which falls between the two aforementioned domains and spans the approximate range $10^{-3} < x^* < 0.03$.

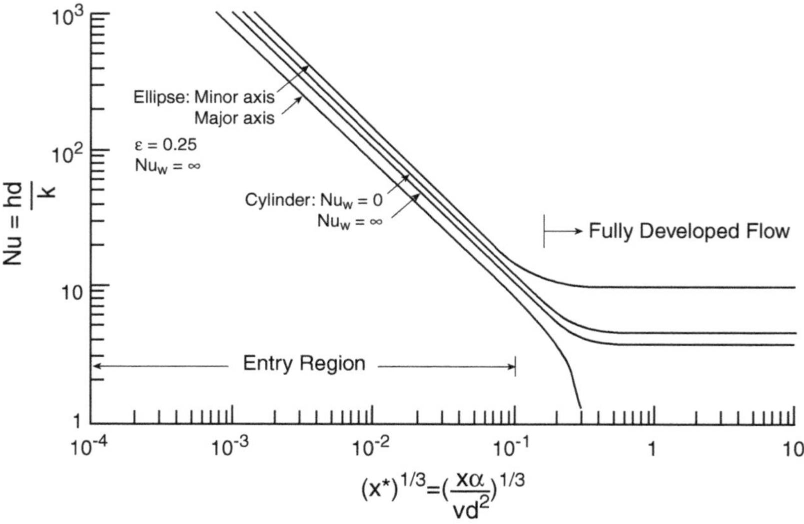

FIGURE 7.6 Local Nusselt numbers as a function of dimensionless distance for steady heat transfer in tubular flow.

A compilation of Nusselt numbers that apply to the entry and fully developed regions are given for some common tubular geometries in Table 7.10.

Variations of Nu with the type of wall BC imposed are slight. The plots shown, which apply to a Type I BC, have the lowest Nusselt numbers, with those for the other extreme of a Type II BC being generally 10 to 15% higher in value, and Type III BCs lying in-between.

TABLE 7.10
Nusselt Numbers for Laminar Tubular Flow and Boundary Conditions of Type I

Duct Geometry	Entry Region	Fully Developed Region
Cylinder	$Nu = 1.08\left(\frac{x\alpha}{vd^2}\right)^{-1/3}$	$Nu = 3.66$
Parallel planes	$Nu = 1.23\left(\frac{x\alpha}{vd^2}\right)^{-1/3}$	$Nu = 7.54$
Annulus, $d_i/d_0 = 0.5$	$Nu = 1.29\left(\frac{x\alpha}{vd^2}\right)^{-1/3}$	—
Square	—	$Nu = 2.98$
Triangular	—	$Nu = 2.47$

A. Through-Flow, Fixed Bed

B. Cross-Flow, Moving Bed

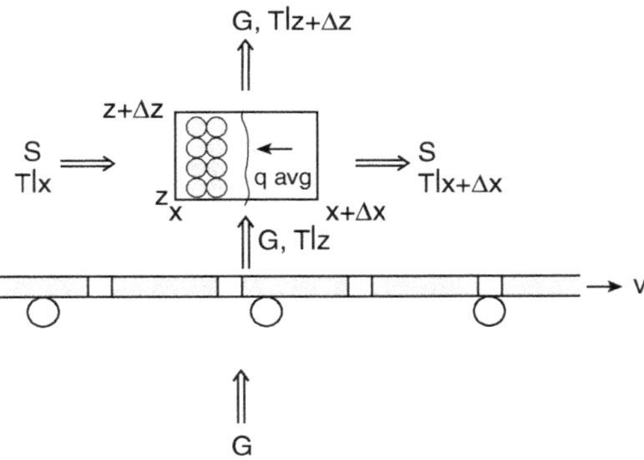

FIGURE 7.7 Heat transfer between a flowing medium and aggregates of solids. The through-flow process (A) is unsteady, and the cross-flow process (B) is at steady state.

For noncircular tubular cross-sections, Nusselt numbers vary along the perimeter as well, and these variations can be quite significant, as shown by the elliptical duct of Figure 7.6.

Illustration 7.15 Heat Transfer in a Packed Bed: Heat Regenerators

In heat regenerators, also termed *heat recuperators*, hot and cold fluids are passed in alternating fashion through a solid matrix, which, in turn, absorbs and then releases the heat, the process being repeated cyclically. Beds packed with solid particles are often employed in this application (see Figure 7.7A), and we use this configuration to derive the equations applicable to the heat uptake step. We consider the heat regenerator to be initially at a uniform temperature T_0, with the inlet gas temperature at T_i. The heat-transfer resistance is assumed to reside principally in the fluid because of its much lower thermal conductivity. Consequently, the solid phase temperature may be considered uniform at a particular position z in the bed, and at time t. Axial conduction is neglected compared to lateral heat transfer.

This description evidently holds only during the initial step of the recuperative process, which in subsequent cycles will have a nonuniform initial temperature. If, however, the heat-transfer front is sharp, as is frequently the case, the degree of nonuniformity will not be severe, and the solution will be a good first approximation of the actual process.

To model the process, we use a classical "in-out = change" approach and obtain for an incremental axial distance Δz.

Gas-phase energy balance:

Rate of energy in – Rate of energy out = Rate of change of energy content

$$v\rho_f A_C Cp_f (T_f - T_{ref})\Big|_z - \begin{bmatrix} v\rho\rho A_C Cp_f (T_f - T_{ref})\Big|_{z+\Delta z} \\ +ha(T_f - T_s)_{avg} \end{bmatrix} = \rho_f Cp_f \varepsilon A_C \rho \Delta z \left(\frac{\partial T_f}{\partial t}\right)_{avg}$$

(7.25a)

where the subscripts f and s refer to fluid and solid, A_C is the cross-sectional area of the regenerator, and a denotes the heat-transfer area in m^2/m^3 bed. Dividing by $A_C\rho_f Cp_f \Delta z$ and letting $\Delta z \rightarrow 0$ yields:

$$-v\frac{\partial T_f}{\partial z} - \frac{ha}{\rho_f Cp_f}(T_f - T_s) = \varepsilon \frac{\partial T_f}{\partial t} \tag{7.25b}$$

A similar procedure for the solid phase yields:

Solid-phase energy balance:

$$\frac{ha}{\rho_f Cp_f}(T_f - T_s) = \frac{\rho_s Cp_s}{\rho_f Cp_f}\frac{\partial T_s}{\partial t} \tag{7.25c}$$

where division by $\rho_f Cp_f$ nondimensionalizes the coefficients.

The reader may have noticed the similarity to the chromatographic process mentioned in Chapter 5 (Equations 5.12e to 5.12g), which has an identical configuration, with convective transport of mass replacing the transport of heat being considered here. We reproduce the pertinent equations for comparison below for the case of a linear phase equilibrium $Y^* = q/H$.

$$-v\frac{\partial Y}{\partial z} - \frac{K_{0Y}a}{\rho_f}(Y - q/H) = \varepsilon \frac{\partial Y}{\partial t} \tag{7.25d}$$

and

$$\frac{K_{0Y}a}{\rho_f}(Y - q/H) = \frac{\rho_s}{\rho_f}\frac{\partial q}{\partial t} \tag{7.25e}$$

Equivalence of terms for the two processes is presented below in the form of a dictionary. We have:

Heat Transfer	**Mass Transfer**
T_f	Y
T_s	q/H
1	H
v	v
$\rho_f Cp_f$	ρ_f
$\rho_s Cp_s$	ρ_s
$ha/\rho_f Cp_f$	$K_0 a$

Both sets of Equation 7.25b and Equation 7.25c for heat transfer and Equation 7.25d and Equation 7.25e for mass transfer, are coupled first-order linear PDEs, which are identical in form and, consequently, have identical solutions. These solutions can be obtained by the method of Laplace transformation (see Chapter 8 and Practice Problem 8.15) and it is customary to express the results in terms of the two dimensionless parameters T (dimensionless time) and Z (dimensionless distance). They take the following form:

	T	**Z**
Mass transfer	$K_0 a \frac{\rho_f}{\rho_s} t$	$K_0 a(z/v)$
Heat transfer	$\frac{ha}{\rho_s C_{ps}} t$	$\frac{ha}{\rho_f C_{pf}}(z/v)$

The numerical values for these solution parameters are shown in Table 7.11 for effluent levels of 1% and 10% of the incoming feed. We refer to the two processes in question, the recuperation of heat and chromatography, as percolation processes and, in general, use the term whenever there is an exchange of mass or heat between a stationary solid and a fluid passing through it.

Let us see how this works out for a particular numerical example. We choose the following parameter values, which are good averages for a typical heat recuperation process:

Inlet temperature	$T_i = 1000°C$
Initial bed temperature	$T_0 = 25°C$
Fluid heat capacity	$Cp_f = 1$ kJ/kg K
Solid heat capacity	$Cp_s = 0.5$ kJ/kg K
Fluid density	$\rho_f = 1.0$ kg/m^3
Solid density	$\rho_s = 3000$ kg/m^3
Heat-transfer coefficient	ha = 50 kJ/sec m^3 K
Height of bed	z = 10 m
Velocity	v = 10 m/sec

TABLE 7.11
Dimensionless Time T and Distance Z for Percolation Processes

Z	T: 1% of feed	10% of feed
1000	900	950
800	700	740
600	520	550
400	330	360
200	150	170
100	70	83
80	52	65
60	37	48
40	22	30
20	7.8	13
10	2.5	5.0
8	1.2	3.5
6	0.38	2.2
5	0.10	1.6

With the use of the dictionary one obtains:

$$Z = \frac{ha}{\rho_f Cp_f}\frac{z}{v} = \frac{50}{(1)(1)}\frac{10}{10} = 50 \text{ (10\% } \textit{breakthrough})$$

The time to 10% breakthrough is then given by:

$$t = \frac{\rho_s Cp_s}{ha} T_{sorption} = \frac{(3000)(0.5)}{50} 39 = 1710 \text{ sec}$$

or approximately twenty minutes. This can also be considered as an estimate of the half cycle time, so that the full cycle would run for about forty minutes.

Cross-flow heating of particulate solids (figure 7.8B) addressed in practice problem 7.16.

Illustration 7.16 Analysis of Heat Recuperators Using the Mathematica Package

Let us now use Mathematica to obtain gas and bed temperature profiles for Illustration 7.15 at various time intervals. We use the parameters given in the earlier illustration with a bed porosity of $\varepsilon = 0.7$.

```
In[1]:= epsilon = 0.7;
        rhof = 1;(*gas density kg/m3*)
        rhos = 3000;(*solid density kg/m3*)
        Cpf = 1000;(*gas heat capacity J/kg/K*)
        Cps = 500;(* solid heat capacity J/kg/k*)
        To = 25+273; (*initial bed temp. C*)
        Ti = 1000+273;(*inlet gas temp. C*)
        ha = 50000;(*heat transfer coeff. J/s/m3/k*)
        v = 10;(*gas velocity m/s*)
        L = 10;(*bed height m*)
```

We use the numerical solver `NDSolve` to find the temperature profiles:

```
In[3]:= sol = {Tf, Ts}/.First[
            NDSolve[
             {-vD[Tf[z, t], z]-ha/rhof/Cpf(Tf[z, t]-Ts[z, t]) ==
               epsilon D[Tf[z, t], t],(Tf[z, t]-Ts[z, t]) ==
               rhos Cps/ha D[Ts[z, t], t],
              Tf[0, t] == 1000, Tf[z, 0] == 25, Ts[z, 0] == 25},
             {Tf, Ts},{z, 0, 10},{t, 0, 2500}]
           ]:
    Print["Temp.@ entrance after 10 sec = ", sol[[1]][0, 10],
    " deg. C"]
    NDSolve::ibcine:
     Warning: Boundary and initial conditions are
     inconsistent. More...
    Temp. @ entrance after 10 sec = 25. deg. C
```

Clearly Mathematica has difficulty evaluating the correct answer: the inlet gas temperature did not increase to 1000°C even after 10 sec. The problem, as stated in the warning message, has to do with the ICs and BCs. To overcome this problem, we assume that the temperature of entering gas increases rapidly but smoothly with time from 25°C to 1000°C according to the following form:

```
In[5]:= Plot[1000 - 975/t + 1), {t, 0, 1000}, Frame → True,
         FrameLabel → {"time, sec", "Entering Gas Temp.,
          deg. C", "", ""},
         PlotRange → {Automatic, {0, 1200}}]
```

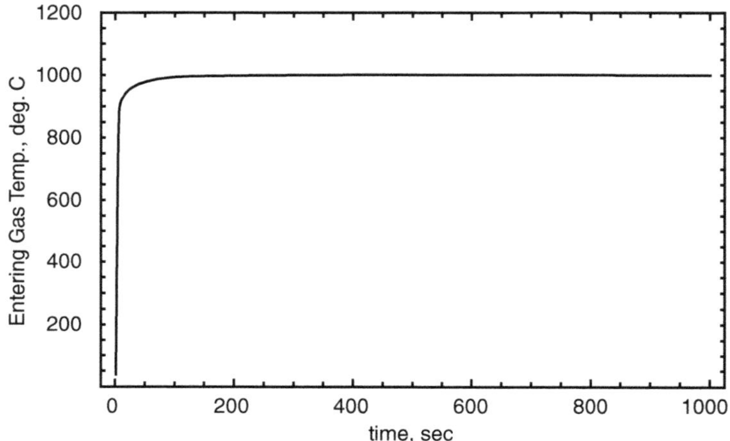

Using this new BC to solve the system and plot the gas and bed temperature profiles after 5, 200, 500, 1000, 1500, and 2000 sec:

```
In[6]:= sol = {Tf, Ts}/.First[
          NDSolve[
          {-vD[Tf[z, t], z] - ha/rhof/Cpf(Tf[z, t] - Ts[z, t]) ==
            epsilon D[Tf[z, t], t], (Tf[z, t] - Ts[z, t])
             == rhos Cps / ha D[Ts[z, t], t],
          Tf[0, t] == 1000 - 975/(t + 1), Tf[z, 0] == 25,
           Ts[z, 0] == 25},
         {Tf, Ts}, {z, 0, 10}, {t, 0, 2500}, MaxSteps → 30000,
          MaxStepSize → .4]
        ] :
```

```
In[7]:= time = {5, 200, 500, 1000, 1500, 2000};
        Show[
         Table[
           {Plot[sol[[1]][z, time[[i]]], {z, 0, 10}, Frame → True,
             Axes → False, TextStyle →{FontFamily → "Times",
            FontSize → 12},
             FrameLabel → {"z(m)", "Tf(deg C)",
              "Temperature Profiles at 5, 200, 500, 1000, 1500
               and 2000 sec", ""},
             DisplayFunction → Identity],
           Plot[sol[[2]][z, time[[i]]], {z, 0, 10},
             DisplayFunction → Identity]},
           {i, 1, Length[time]}],
          DisplayFunction → $DisplayFunction, PlotRange →
           {{0, 10}, {-100, 1200}}]
```

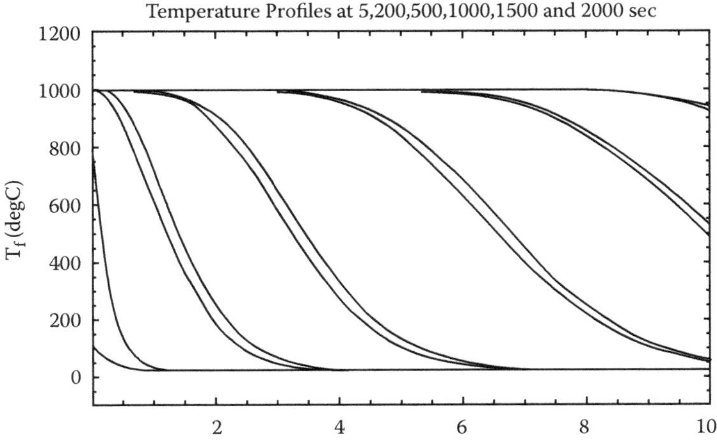

In this graph, the adjacent curves represent the gas and bed profiles, with the bed temperature profile slightly lagging behind.

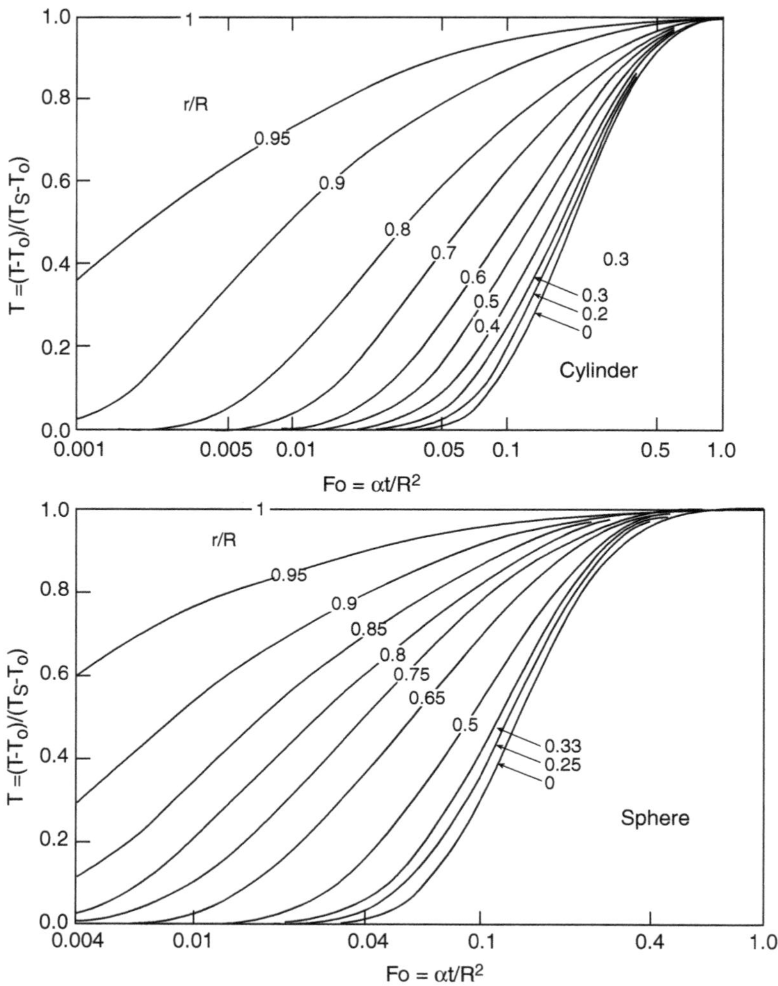

FIGURE 7.8 Dimensionless temperature as a function of the Fourier number in unsteady conduction in a cylinder and a sphere.

Unsteady Conduction:

We briefly allude to this case, described by Fourier's equation, in order to present two typical solutions, those for conduction in a sphere, and radially in a cylinder with an imposed BC of Type I (Figure 7.8). These solutions are usually given as temperature profiles, rather than in terms of cumulative uptake or release, as was the case in mass diffusion (Figure 7.5 and Figure 7.1). A host of other solutions to Fourier's equation in both graphical and analytical form, running literally into the hundreds, if not thousands, can be found in the literature. The plots are given in terms of the fractional approach of the temperature to steady state $(T - T_0)/T_s - T_0)$, where T_0 is the initial temperature, T_s the imposed surface temperature, and the dimensionless Fourier Number $Fo = \alpha t/R_0^2$ characteristic of unsteady conduction. Dimensionless radial distance r/R is a parameter. We note the following points in connection with these plots.

Uptake or release in a sphere is faster than in a cylinder, because the heat-transfer area per unit volume is larger in the former case. Typically, these plots are used to calculate temperatures attained, usually at the surface or center, after a prescribed time interval t. Conversely, one can calculate the time required to attain a prescribed temperature. The solutions are not generally used to extract thermal diffusivities from measured temperature profiles, as simpler steady state methods can be applied.

For BCs of Type II and Type III, additional parameters have to be inserted into the solutions and solution plots. For Type III BCs, the dimensionless Biot number $Bi = \frac{h_f d}{k_s}$ is used, where h_f is the external convective film coefficient and k_s is the conductivity of the solid. That number had previously been alluded to in Illustration 1.7. Note that the Nusselt number $Nu = h_f d/k_f$ is similar in appearance, but contains the fluid rather than the solid thermal conductivity.

7.4.1 Steady State Temperatures and Heat Flux in Multidimensional Geometries: The Shape Factor

The topic of steady state conduction between two isothermal surfaces was briefly taken up in Illustration 7.13 in connection with the corresponding case of steady mass diffusion between surfaces at constant concentration. Both these cases are described by Laplace's equation, and both make use of the so-called shape factor S, which is derived from its solution. S can then be used to calculate heat flux from the convenient linear relation:

$$q'(J/s) = kS(T_1 - T_2) \tag{7.26}$$

where S has the dimension of length. Figure 7.9 gives a short tabulation of the more common geometries and their associated shape factors. Note that for items 1, 2, 5, 9, and 10, heat flux is given as q″(J/ms), i.e., per unit length of the "buried" objects, which extend into the paper to infinity.

The simple equation Equation 7.26 hides the fact that the underlying process is one of considerable complexity. To obtain the shape factor S, one must first solve Laplace's equation, which for the complex geometries in question is often done by conformal mapping. This results in temperature distributions of the type we had listed in Figure 5.7. To obtain the heat flux, the temperature gradient at the buried object has to be derived from the primary T-distribution, and that gradient integrated over the surface of the buried object. The result is then matched to Equation 7.26 to yield the shape factor S, which is tabulated in Figure 7.9.

In Practice Problem 7.18, the reader is asked to apply the tabulated shape factors to a problem of practical significance.

	1. Thin vertical strip in semi-infinite solid	
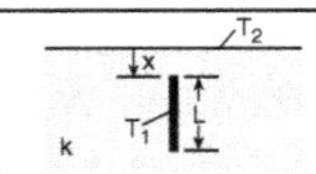	$S=2.38/(x/L)^{0.24}$	$\frac{1}{2}<\frac{x}{L}<12$
	2. Thin horizontal strip in simi-infinite solid	
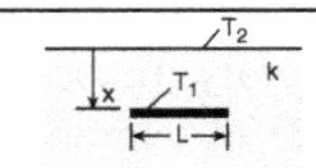	$2.94/(x/L)^{0.32}$	$\frac{1}{2}<\frac{x}{L}<12$
	3. Thin rectangular plate in semi-infinite solid	
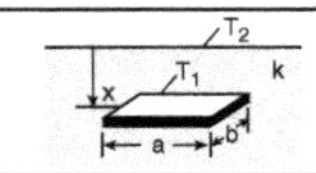	a) $\pi a \ln(4a/b)$ b) $2\pi a \ln(2\pi x/b)$ c) $2\pi a \ln(4a/b)$	$x=0$ $a \gg b,\ x>2b$ $x \gg a$
	4. Thin circular disk in semi-infinite solid	
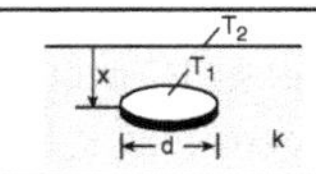	$2d$ $4d$	$x=0$ $x \gg d$
	5. Rectangular hole in semi-infinite solid	
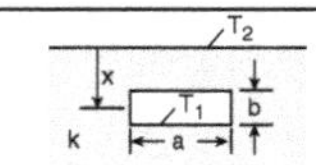	$(a/2b+5.7)/\ in\ (3.5x/a^{1/4}b^{3/4})$	$a>b$
	6. Spherical hole in semi-infinite solid	
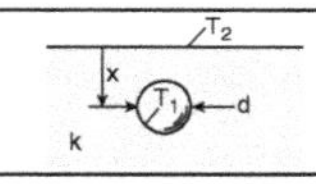	$2\pi d/(1-d/4x)$	$x>d/2$ $x \gg d$
	7. Vertical circular surface hole in semi-infinite solid	
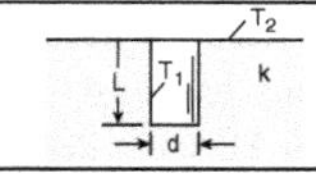	$2\pi L/(4L/d)$	$L>d$
	8. Horizontal circular hole in semi-infinite solid	
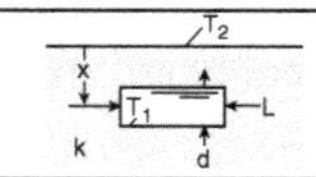	a) $2\pi L/\ln(4x/d)$ b) $2\pi L/\ln(2L/d)$	$d<x<L$ $x \gg d$
	9. Circular hole in square solid	
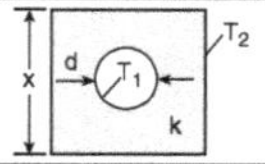	$2\pi/\ln(1.08x/d)$	$x>d$
	10. Eccentric circular hole in cylindrical solid	
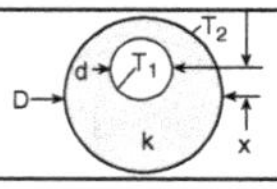	$2\pi/\cosh^{-1}[(D^2+d^2-4x^2)/2Dd]$	$D>d$

FIGURE 7.9 Shape factors.

7.5 TRANSPORT OF MOMENTUM

We turn, in this last section of the chapter, to the task of establishing a generalized momentum or force balance to complement the generalized mass and energy balances of the preceding sections. Here, it is convenient to dispense with the generalized transport theorem and start directly with a general statement of Newton's law, applied to a moving packet of fluid of fixed mass. The convective derivative then appears in a natural way and we obtain:

$$\rho\frac{D\vec{v}}{Dt} = \sum F_b + \sum F_s \tag{7.27a}$$

Body forces per unit volume — Surface forces per unit volume

Here the body forces include gravitational, thermal, electrical, and magnetic forces, whereas the term ΣF_s deals mainly with forces arising from viscous stresses on the fluid element. Those stresses depend on the nature of the fluid, i.e., whether Newtonian or non-Newtonian (Bingham, Viscoelastic, fluids, etc.). The stresses for these various cases consist of nine component tensors, which can, in turn, be related to the more convenient velocity gradients. One such relation based on linear or Hookean deformation, i.e., for so-called Newtonian fluids, was derived by Stokes and can be used directly in the force balance previously derived by Navier. We dispense with the details of this part of the development, which can be found in standard texts on transport phenomena, and present instead a first version of the generalized momentum balance, applicable to Newtonian flow of constant density and expressed in terms of velocity components. We have:

$$\rho\frac{D\mathbf{v}}{Dt} = -\nabla p + \rho\mathbf{g} + \mu\nabla^2\mathbf{v} \tag{7.27b}$$

or alternatively,

$$\frac{\partial}{\partial t}\boldsymbol{v} + (\mathbf{v}\bullet\nabla)v = -\frac{1}{\rho}\nabla p + \boldsymbol{g} + \nu\nabla^2\boldsymbol{v} \tag{7.27c}$$

Unsteady term — Convective term — Pressure term — Gravitation — Viscous term

where $(\mathbf{v}\bullet\nabla)\mathbf{v}$ is the convective operator, and $\nabla^2\mathbf{v}$ the Laplacian of the velocity vector, both of which we had tabulated in the dictionary, Table 7.1. These expressions represent the vectorial formulations of the Navier–Stokes equations. The following points need to be noted:

Equation 7.27b and Equation 7.27c represent three scalar PDEs that, together with the continuity equation, describe the velocity and pressure distributions in a viscous, Newtonian flow field. To obtain the scalar components, use is made of the dictionary shown in Table 7.1. Thus, for the x-component in Cartesian coordinates, the dictionary yields:

$$\frac{\partial v_x}{\partial t} + v_x\frac{\partial v_x}{\partial x} + v_y\frac{\partial v_x}{\partial y} + v_z\frac{\partial v_x}{\partial z} = -\frac{1}{\rho}\frac{\partial P}{\partial x} + \nu\left[\frac{\partial^2 v_x}{\partial x^2} + \frac{\partial^2 v_x}{\partial y^2} + \frac{\partial^2 v_s}{\partial z^2}\right] \tag{7.27d}$$

Unsteady — Convective — Pressure — Viscous term

Note that the convective term is nonlinear.

The Navier–Stokes equations hold for viscous Newtonian flow with constant density and viscosity. For compressible and non-Newtonian flow, one has to return to the tensorial form of the shear stresses and convert them to velocity gradients through use of appropriate relations.

Consideration of body forces was limited to gravity only. Other body forces, such as the buoyancy forces that arise in free convention, have to be added to the Navier–Stokes equations. The gravity force itself is usually taken in the vertical direction z only and can be expressed by the equivalent terms:

$$g = g\nabla z = g\mathbf{k} \tag{7.27e}$$

where we have made use of the fact that $\nabla z = (\partial z/\partial x)\mathbf{i} + (\partial z/\partial y)\mathbf{j} + (\partial z/\partial z)\mathbf{k} = \mathbf{k}$.

In what follows, we present a number of Illustrations, each one of which will deal with a particular reduced form of the Navier–Stokes equation. We take up in turn the following topics:

- Duct flow (Illustration 7.17)
- Creeping flow (Illustration 7.18)
- Boundary layer flow (Illustration 7.19)
- Inviscid flow (Illustration 7.20)
- Irrotational flow (Illustration 7.21).

Illustration 7.17 Steady, Fully Developed Incompressible Duct Flow

We consider flow in a duct of arbitrary cross section, and place it in a Cartesian framework. The aim here will be to develop working expressions like the Poiseuille or Fanning equations, which will allow us to calculate pressure drop as a function of volumetric flow rate Q or average velocity v. The following simplifications apply:

$$v_y = v_z = 0 \text{ (Impermeable walls)} \tag{7.28a}$$

$$\frac{\partial \mathbf{v}}{\partial t} = 0 \quad \text{(Steady flow)} \tag{7.28b}$$

$$\frac{\partial v_x}{\partial x} = 0 \quad \text{(Continuity)} \tag{7.28c}$$

$$\frac{\partial p}{\partial x} = \Delta p/L \quad \text{(Constant pressure drop)} \tag{7.28d}$$

Equation 7.28a implies that only one velocity component, v_x, needs to be considered; hence, one component momentum balance, that in the direction of flow x, suffices. Introducing the simplifications, that equation becomes:

$$0 = \frac{\Delta P}{L} + \mu\left[\frac{\partial^2 v_x}{\partial y^2} + \frac{\partial^2 v_x}{\partial z^2}\right] \tag{7.28f}$$

We note that Equation 7.28f is of the same form as the classical Poisson equation:

$$\nabla^2 u + f(x, y, z) = 0 \tag{5.14}$$

which is linear and can be solved by a variety of standard methods. The fact that ΔP/L is a constant, simplifies the solution, which takes the form:

$$v_x = f(y,\ z,\ \Delta p/L) \tag{7.28g}$$

In order to solve Equation 7.28f, Δp/L (as well as the viscosity μ) have to be specified. The flow rate Q produced by the stipulated pressure drop is then obtained by integrating the velocity (Equation 7.28g) over the duct cross section. The reverse problem of calculating Δp/L for a specified flow rate Q cannot, unfortunately, be addressed in the same direct manner. One has to solve Poisson's equation repeatedly for arbitrary values of the pressure drop until the resulting flow rate matches the specified value Q. To avoid this complication and to provide a working tool for the practicing engineer, it is customary to translate the results into an equivalent friction factor f, which can then be used in the empirical Fanning equation found in standard fluid mechanics texts:

$$\frac{\Delta p}{L} = 4f\frac{\rho v^2}{2}\frac{1}{d_h} \tag{7.29a}$$

where $v = Q/A_C$ = average velocity and $d_h = 4 \times A_C/P$ = hydraulic diameter. f itself is related to the wall shear stress τ_w via the relation:

$$f = \frac{2}{\rho v^2 P}\int_P \tau_w dP \tag{7.29b}$$

where P is the perimeter of the duct in question, and τ_w is obtained from the velocity gradients using Newton's viscosity laws.

Table 7.12 provides a listing of these computed friction factors for two noncircular duct geometries. Note that for a circular duct, 4f Re equals 64, close to the numbers shown in Table 7.12. To obtain a sense of the magnitude of the pressure losses involved, consider the rather academic example of water flow, v = 1 cm/sec, through a square duct of width b = 1 cm and length L = 1 m. With a kinematic viscosity for water at 25° of $\nu = 10^6$ m²/sec, we obtain:

$$d_h = 4\ b^2/4b = b = 10^2\ m$$

$$\text{Re} = \frac{vd_h}{\nu} = \frac{10^{-2}\ 10^{-2}}{10^{-6}} = 100$$

$$f = 56.9/\ \text{Re}_{D_h} = 56.9/100 = 0.569$$

TABLE 7.12
Friction Factor for Laminar Flow in Noncircular Ducts

Rectangular		Isosceles Triangular	
b/a	4f $Re_{D/h}$	Apex angle α, deg	4f $Re_{D/h}$
0.0	96.0	0	48.0
0.1	84.7	10	51.6
0.25	72.9	30	53.3
0.50	62.2	50	52.0
0.75	57.9	70	49.5
1.0	56.9	90	48.0

The Fanning Equation then yields the following pressure drop:

$$\Delta p = 4f\rho\frac{v^2}{2}\frac{L}{d_h} = (0.569)(1000)\frac{(10^{-2})^2}{2}\frac{1}{10^{-2}}$$

$$\underline{\Delta p = 2.85 \text{ Pa}}$$

To confirm this result, we introduce the calculated pressure drop Δp/L = 2.85 Pa/m into the Poisson equation, which is solved by the Mathematica package and the resulting velocity profile integrated over the duct to obtain the average velocity v. However, before solving the equation, let us transform Equation 7.28f into dimensionless form:

$$\left[\frac{\partial^2 u}{\partial y^{*2}} + \frac{\partial^2 u}{\partial z^{*2}}\right] = -1 \tag{7.29c}$$

with $u = \frac{\mu L}{b^2 \mu P} v_x$, $y^* = \frac{y}{b}$, and $z^* = \frac{z}{b}$.

Let us first discretize Equation 7.29c using a central difference scheme with step size k:

```
In[1]:= FanningEqn=
        u[-1+i, j]/k^2+u[i,-1+j]/k^2-(2*u[i, j])/
         k^2-(2*u[i, j])/k^2+
        u[i, 1+j]/k^2+u[1+i, j]/k^2== -1
```

$$\text{Out[1]=}\ \frac{u[-1+i,j]}{k^2} + \frac{u[i,-1+j]}{k^2} - \frac{4u[i,j]}{k^2} + \frac{u[i,1+j]}{k^2} + \frac{u[1+i,j]}{k^2} == -1$$

Using n = 40 steps in each of the y- and z-directions, a system of linear algebraic equations (eqns) can be formed:

```
In[2]:= n = 40;
        k = 1/n;
        eqns = Flatten[Simplify[
            Table[FanningEqn, {i, 1, n - 1}, {j, 1, n - 1}]]];
```

The BCs for this problem are:

```
In[5]:= BCs =
         Union[
          Flatten[
           Append[
            Table[{u[0, j] -> 0, u[n, j] → 0}, {j, 0, n}],
            Table[{u[i, 0] → 0, u[i, n] → 0}, {i, 1, n}]
           ]
          ]
         ];
```

By inserting the above boundary in eqns, the system of equations is solved and dimensionless local velocities u[i, j] are obtained:

```
In[6]:= uapprox := Flatten[Table[u[i, j], {i, 1, n - 1},
        {j, 1, n - 1}]];
        Uprofile = Flatten[Solve[eqns /. BCs, uapprox]];
        Uprofile = Join[BCs, Uprofile];
```

The average velocity can be determined using the Mean function:

The approximate numerical result, $v_{x,avg}$ = 0.95 cm/sec is close to the exact solution of 1 cm/sec.

```
In[9]:= Uavg = Mean[
          Table[Uprofile[[i, 2]], {i, 1, Length[Uprofile]}]
         ]// N;
        Dp = 2.85;  (*Pa*)
        L = 1; (* m *)
        mu = 10^-3; (*kg/m.s *)
        a = 0.01; (*m*)
        Vx = Uavg * Dp * a^2 / L / mu* 100
Out[10]= 0.951415
```

Illustration 7.18 Creeping Flow

The principal difficulty in solving the Navier–Stokes Equation, apart from its dimensionality, resides in the nonlinear convective term $(\mathbf{v} \bullet \nabla)\mathbf{v}$. That term was neatly eliminated in the previous Illustration by virtue of Equation 7.28, leaving us with a single linear PDE of the Poisson type. A similar result can be achieved by assuming the flow to be very slow, so that the convective term becomes small in comparison to the viscous term. We then refer to the system as being in creeping flow. This approximation is used in a number of applications of practical importance.

Low Reynolds number external flow around submerged bodies ($Re_p < 1$): The celebrated Stokes law for the drag on a submerged sphere is directly derived from the velocity and pressure distributions arrived at by the solution of the linearized Navier–Stokes equation.

Slow viscous flow in narrow passages of varying cross section: This topic is usually treated under the heading, lubrication theory.

Incompressible flow through porous media: A vast subject with a number of important subtopics: filtration, reservoir and petroleum engineering, groundwater seepage, geohydrology, etc.

In most of these cases, except flow through porous media (see Practice Problem 7.21), gravity forces are neglected and one winds up with the following operative equations:

Continuity equation $$\nabla \bullet v = 0 \tag{7.30a}$$

Reduced Navier–Stokes equation $$\nabla p = \nu \nabla^2 \mathbf{v} \tag{7.30b}$$

This is still a fairly formidable system of four PDEs, albeit linear ones, with dependent variables pressure p and the three velocity components. We now proceed to simplify this system, and, in doing so, make fruitful use of certain operations of vector calculus presented in Section 7.1.

We start by noting that one can quickly decouple pressure and velocity by forming the divergence of Equation 7.30b, i.e., by dot multiplying both sides with $\nabla\bullet$. There results, by virtue of Formula 14 of Table 7.5,

$$\nabla \bullet \nabla\, p = \mu[\nabla \bullet \nabla(\nabla \bullet \mathbf{v}) - \nabla \bullet \nabla x(\nabla x \mathbf{v})] \tag{7.30c}$$

The left-hand side is immediately seen to be the Laplacian of p. The first term on the right vanishes because of the continuity equation, Equation 7.30a, the second term by virtue of Relation 6 in Table 7.5. We thus obtain:

$$\nabla^2 p = 0 \tag{7.30d}$$

i.e., the pressure distribution in creeping flow is described by Laplace's equation.

We still have the three velocity components of Equation 7.30b to contend with, but for the two-dimensional case at least, a reduction to a single PDE is possible. To achieve this for the Cartesian case, we first differentiate the equation with the y-component with respect to x and the x-component with respect to y, and subtract the result. This gives us a third order PDE in v_x and v_y. These two dependent variables are then coalesced into a single variable, the stream function ψ, using the relations given in Chapter 6, Table 6.1. A single fourth order PDE in ψ results, which is the biharmonic equation:

$$\nabla^4\psi = \frac{\partial^4\psi}{\partial x^4} + 2\frac{\partial^4\psi}{\partial x^2\,\partial y^2} + \frac{\partial^4\psi}{\partial y^4} = 0 \tag{7.30e}$$

The full creeping-flow solution can thus be obtained by solving, independently, Laplace's equation in pressure p, and the biharmonic equation in ψ. The latter is also the governing equation for two-dimensional elasticity problems and solutions found in that discipline are often applicable to creeping-flow problems.

The reader should note that not all creeping-flow problems require the full solution of Equations 7.30d and 7.30e. Simple geometries often reduce the problem to the ODE level or revert to familiar duct flow problems. In other cases, one may exploit the linearity of Equation 7.30b and apply superposition techniques to arrive at a solution of problems of some complexity. The following illustrates the application of these concepts.

Consider the velocity distributions shown in Figure 7.10 that arise when one wall of a parallel plane channel is moved at a velocity u while the other wall remains stationary. The distribution, which will be a linear one is obtained by simple inspection and is given by:

$$v = \frac{u}{2}\left(1 + \frac{y}{h}\right) \tag{7.31a}$$

The same result may be obtained, although slightly more circuitously, from Equation 7.30b, which here becomes

$$\nabla^2 \boldsymbol{v} = 0 \tag{7.31b}$$

Imagine, now, that a forced viscous flow is induced in the system by imposing a pressure gradient Δp/L on the system. We know the solution to a reduced form of this case when both channel walls are stationary. It was taken up in Illustration 4.7. Its solution, translated into the coordinates of Figure 7.10, is given by:

$$v = \frac{h^2}{2\mu}\left(1 - \frac{y^2}{h^2}\right)\frac{\Delta p}{L} \tag{7.31c}$$

Here again, the same result can be obtained from the reduced Navier–Stokes Equation 7.30b, opening the door to a superposition of the two solutions, Equations 7.31a and 7.31c.

We obtain:

$$v = \frac{u}{2}\left(1 + \frac{y}{h}\right) + \frac{h^2}{2\mu}\left(1 - \frac{y^2}{h^2}\right)\frac{\Delta p}{L} \tag{7.31d}$$

This is the velocity distribution for forced viscous flow in a parallel plane channel with one wall moving at a velocity u. A plot of this profile is shown in Figure 7.9B, and reveals that the maximum velocity no longer resides at the center line, which

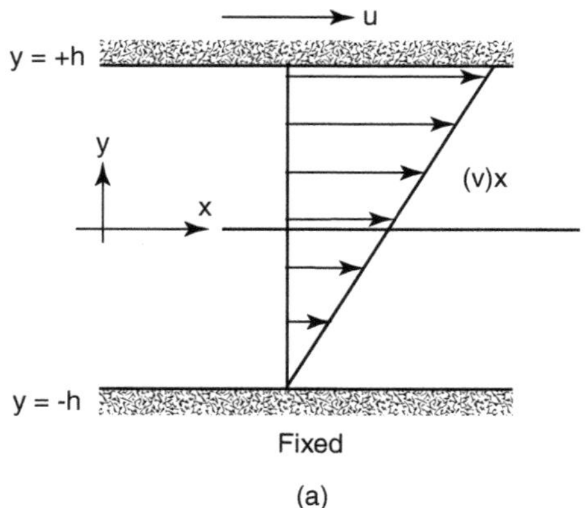

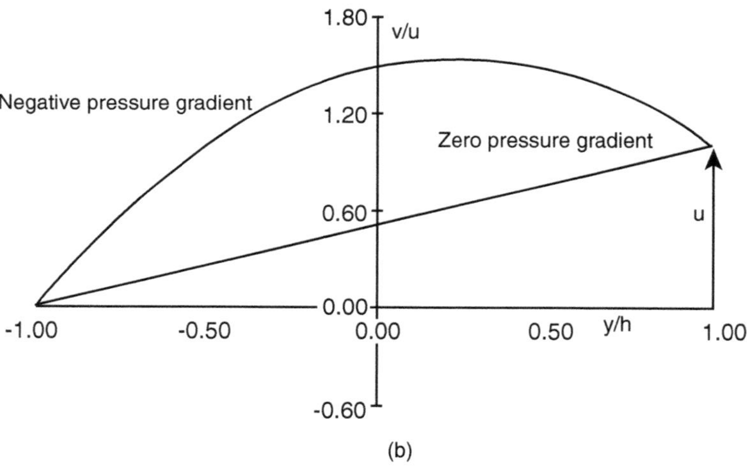

FIGURE 7.10 Flow in a parallel plane channel with a moving wall. (a) Free flow, (b) Free and forced flow.

was the case in Poiseuille flow, but is shifted instead toward the moving plane. It is an easy matter to integrate the profile Equation 7.31d into a Q-Δp relation, which is the preferred form for engineering use.

Illustration 7.19 The Prandtl Boundary Layer Equations

We have already, on a number of occasions, referred to boundary layer theory, which divides the external flow field around submerged bodies into a thin boundary layer adjacent to the body surface in which viscous effects predominate, while the bulk fluid further away was essentially in inviscid or irrotational flow (Chapter 5, Figure 5.3). This concept, due to Prandtl, has been particularly successful in gas flow and aerodynamics. We reexamine the underlying equations briefly in order to place them on a firmer foundation, limiting ourselves to steady, incompressible two-dimensional flow. The governing equations for this case are given by:

Continuity:

$$\frac{\partial v_x}{\partial x}+\frac{\partial v_y}{\partial y}=0 \tag{7.32a}$$

x-Momentum (Navier–Stokes):

$$v_x\frac{\partial v_x}{\partial x}+v_y\frac{\partial v_x}{\partial y}=-\frac{1}{\rho}\frac{\partial P}{\partial x}+\nu\left(\frac{\partial^2 v_x}{\partial x^2}+\frac{\partial^2 v_y}{\partial y^2}\right) \tag{7.32b}$$

y-Momentum (Navier–Stokes):

$$v_x\frac{\partial v_y}{\partial x}+v_y\frac{\partial v_y}{\partial y}=-\frac{1}{\rho}\frac{\partial P}{\partial y}+\nu\left(\frac{\partial^2 v_y}{\partial x^2}+\frac{\partial^2 v_y}{\partial y^2}\right) \tag{7.32c}$$

where x is the direction along the wall and y is normal to it.

Prandtl made two crucial assumptions:

1. The velocity normal to the wall is much smaller than that along it:

$$v_y \ll v_x \tag{7.32d}$$

2. Changes in the x-direction are much smaller than those in the y-direction, with the result that:

$$\frac{\partial}{\partial x}\ll\frac{\partial}{\partial y} \tag{7.32e}$$

Applying these relations to the y-momentum equation we find that all velocity terms disappear and, hence,

$$\frac{\partial p}{\partial y}\approx 0 \tag{7.32f}$$

In other words, Equation 7.32c disappears entirely, and pressure varies along the boundary layer only, not through it. This means that the pressure distribution can be recovered from the solution of the inviscid outer field via Bernoulli's equation and Laplace's equation. The governing model has thus been reduced to the following four relations:

Continuity:

$$\frac{\partial v_x}{\partial x} + \frac{\partial v_y}{\partial y} = 0 \tag{7.33a}$$

x-Momentum (boundary layer):

$$v_x \frac{\partial v_x}{\partial x} + v_y \frac{\partial v_x}{\partial y} = -\frac{1}{\rho}\frac{\partial p}{\partial x} + \nu \frac{\partial^2 v_x}{\partial y^2} \tag{7.33b}$$

Bernoulli equation (inviscid field):

$$U\frac{dU}{dx} = -\frac{1}{\rho}\frac{dp}{dx} \tag{7.33c}$$

Laplace's equation:

$$\frac{\partial^2 \psi}{\partial x^2} + \frac{\partial^2 \psi}{\partial y^2} = 0 \tag{7.33d}$$

where the stream function ψ is related to the velocity components v_x, v_y via the expressions given in Chapter 5, Table 5.6, and U is the outer field velocity.

The normal solution procedure then consists of the following steps:

1. Solve Laplace's equation for the outer field, and from the stream function ψ, recover the corresponding velocity components v_x and v_y.
2. Compute the outer field velocity U from its components: $U = (v_x^2 + v_y^2)^{1/2}$.
3. Introduce the stream function ψ into Equation 7.33a and Equation 7.33b. This has the effect of collapsing the two variables and, hence, the two equations into a single entity. The resulting third-order PDE is then reduced to an ODE by similarity transformation and the ODE usually solved numerically. For a flat plate, this ODE is the Blasius equation, Equation 5.32a.
4. The solutions for the inner and outer regions are "patched together" or matched at the edge of the boundary layer where we typically specify:

$$v/U = 0.99 \tag{7.33e}$$

This admittedly brief sketch is meant to convey the main steps of the solution procedure, which has led to a host of successful solutions. One will note, in particular, the ingenious simplifications of Prandtl and the use of various classical tools, such as the use of the stream function and Laplace's equation, to describe the velocity distributions. Obviously a good deal of detail was left out, for which the reader is referred to specialized texts.

Illustration 7.20 Inviscid Flow: Euler's Equation of Motion

We describe here briefly the inviscid version of the momentum balance, usually attributed to Euler. It takes the form:

$$\rho \frac{D\mathbf{v}}{Dt} = -\nabla p - \rho \mathbf{g} \tag{7.34a}$$

or

$$\frac{\partial \mathbf{v}}{\partial t} + (\mathbf{v} \bullet \nabla)\mathbf{v} = -\frac{1}{\rho} - \mathbf{g} \tag{7.34b}$$

and is introduced here mainly to provide a stepping stone to the more powerful Bernoulli equation. It also has important applications in the study of waves and open channel flows when one can neglect viscous forces, including wind-induced stresses.

Although Equations 7.34a and 7.34b seem to have arisen from the Navier–Stokes Equations 7.27b and 7.27c valid for incompressible flow, a direct derivation from the force balance Equation 7.27a shows that it applies to compressible flow as well. Thus Bernoulli's equation, which we derive below from Euler's equation of motion, is quite generally applicable to both compressible and incompressible flow.

Illustration 7.21 Irrotational (Potential) Flow: Bernoulli's Equation

To derive Bernoulli's equation from Euler's equation of motion, we start by expanding the convective term $(\mathbf{v} \bullet \nabla)\mathbf{v}$ using the vector relation 13 given in Table 7.5. We obtain, for steady flow:

$$\nabla \frac{1}{2}|v|^2 - \mathbf{v}x(\nabla x\mathbf{v}) + \frac{1}{\rho}\nabla p + g\nabla z = 0 \tag{7.35a}$$

where we have also made use of Equation 7.27e to convert the gravity vector $\mathbf{g}$ to the gradient of the vertical distance z. We now introduce the irrotational flow condition $\nabla x\mathbf{v} = 0$, which, as was pointed out in Section 7.1, is in most instances equivalent to the inviscid condition. This relatively modest increase in the stringency of the conditions imposed leads to an enormous simplification of Euler's equation. We obtain, in the first instance:

$$\nabla\left(\frac{1}{2}|\mathbf{v}|^2\right) + \frac{1}{\rho}\nabla p + g\nabla z = 0 \tag{7.35b}$$

where $|\mathbf{v}|$ represents the magnitude of the velocity vector.

This is beginning to look very much like Bernoulli's equation, and a final operation will lead us to that goal. We dot multiply each term of the equation by the differential position vector $d\mathbf{R}$ and invoke relation 5 of Table 7.5: $\nabla u \bullet d\mathbf{R} = du$. This yields:

$$v\,dv + \frac{1}{\rho}dp + g\,dz = 0 \tag{7.35c}$$

which is recognized as the differential form of Bernoulli's equation, applicable to both compressible and incompressible flow.

Comments: One notices here, as in the case of creeping flow, Illustration 7.18, the power of vector notation and the simplicity of the operations, once certain basic tenets have been accepted. Their absence would make the transformation of Euler's equation to Bernoulli's equation a fairly cumbersome task, which would have to be repeated for each new coordinate system. The result obtained is thus quite general and not dependent on the geometry of the system.

A careful reading of the present derivation of Bernoulli's equation will reveal that it differs from that given in standard fluid mechanics texts. The latter relies on a force balance performed for inviscid flow on a section of a stream tube and, consequently, requires any two positions to which the equation was applied to lie on the same streamline. No such restriction applies here. One can integrate Equation 7.35c between any two points of the flow field without violating its validity. The present version is, for that reason, often referred to as the *strong form of Bernoulli's equation*. It must be remembered, however, that this greater freedom was bought at the cost of imposing the somewhat more stringent condition of irrotational flow. In either its weak or strong form, Bernoulli's equation has become (perhaps apart from the continuity equation) the most frequently used relation in fluid mechanics.

PRACTICE PROBLEMS

7.1 Differential Operators

Prove the relations given by item 7, item 8, and item 15 of the Table 7.5.

(Hint: If proof can be provided for one coordinate system, the relation is valid for all coordinate systems.)

7.2 Volume of a Cone

Use the divergence theorem to show that the volume of a cone is $V = \frac{1}{3}\pi r^2 h$.

(Hint: Use a position vector R with the apex of the cone as the origin and start by showing that $V = \frac{1}{3}\iint_S \mathbf{n} \bullet \mathbf{R}\, dS$. Note: Volumes of other shapes can be determined in similar but less easy fashion.)

7.3 Maxwell's Equations of Electromagnetic Theory

Maxwell's equations relate electric and magnetic fields to the charges and currents which produce them. They are four in number and are expressed as follows:

$$\nabla \bullet \mathbf{E} = 4\pi\rho \tag{1}$$

$$\nabla \bullet \mathbf{B} = 0 \tag{2}$$

$$\nabla \times \mathbf{E} = -\frac{1}{C}\frac{\partial \mathbf{B}}{\partial t} \tag{3}$$

$$\nabla \times \mathbf{B} = -\frac{1}{C}\frac{\partial \mathbf{E}}{\partial t} + \frac{4\pi}{C}\mathbf{J} \tag{4}$$

where **E** denotes the electric field, **B** the magnetic field, ρ charge density (charge q per volume), **J** the current density (current I per unit cross-sectional area), and C the speed of light. Recall that by field is meant the force exerted on a unit charge q_0.

Equation 3 and Equation 4 express the fact that time-varying magnetic fields produce electric fields (i.e., an electric current), whereas time-varying electric fields, in turn, produce magnetic fields. Equation 2 states that the total flux of magnetic forces over a closed surface is zero, i.e., what goes in must come out (recall the definition of divergence, Equation 7.4a). Vector fields with a zero divergence are therefore often referred to as solenoidal. Equation 1, finally, relates the flux of electric forces to the charge density ρ. The answers sought are as follows.

1. Which of Maxwell's equations are scalar, which are vectorial? Choose one of the latter category and expand it into an equivalent set of three scalar equations in Cartesian coordinates.
2. Classify the PDEs, which result from Maxwell's equations, as to order, linearity, and homogeneity.

7.4 Conservation of Charge: The Continuity Equation of Electricity

Use the divergence and generalized transport theorems, Equations 7.7a and 7.7d and to derive the following continuity equation of electricity:

$$\frac{\partial \rho}{\partial t} = -\nabla \bullet \mathbf{J}$$

where, as before, ρ and **J** are charge and current densities. Note the similarity to the continuity equation of fluid mechanics.

7.5 The Interior Dirichlet Problem for a Circle

When an arbitrary potential u(R,θ) is imposed on the circumference of a circle or infinitely long cylinder of radius R, with u held at zero as $r \to \infty$, a steady state distribution of the potential, u(r,θ), results both within and outside the circle. Derivation of the former is referred to as the interior Dirichlet problem for a circle, the latter as the exterior Dirichlet problem for a circle. The same language is applied to other geometries. Thus, one speaks of the Dirichlet problem for a half-plane (see Problem 7.6) and the Dirichlet problem for a rectangle.

Using the Green's functions of Table 7.7 as a guide, show that the steady state temperature within a circle which results from a prescribed boundary temperature distribution T(R,θ) is given by:

$$T(r,\theta) = \frac{R^2 - r^2}{2\pi} \int_0^{2\pi} \frac{T(R,\theta')d\theta'}{R^2 - 2Rr\cos(\theta' - \theta) + r^2}$$

7.6 Dirichlet's Problem for a Half-Plane

Derive the potential distribution which results from a distribution u = f(y) imposed along the positive x-axis x > 0, with u → 0 as x → ∞.

(Hint: Use the Green's Function of Table 7.7.)

Answer: $$u(x,y)=\frac{x}{\pi}\int_{-\infty}^{\infty}\frac{f(y')dy'}{x^2+(y-y')^2}$$

Note that in view of the arbitrary form of the imposed boundary condition, the solutions to Practice Problems 7.5 and 7.6 can be regarded as solutions to an infinite set of different problems. The method of Green's function allows us to derive these solutions in a terse, closed form.

7.7 A Lévêque Problem

Consider steady laminar flow in a tubular reactor, L = 100 cm, which releases solute from the wall at an unknown constant rate N (mol/cm^{2}sec). Average velocity v of the fluid is 10 cm/sec, diffusivity D of the solute 10^{-5} cm^2/sec.

a. What is the concentration boundary layer thickness δ at the exit?
   (Hint: Recall the definition of the mass-transfer coefficient k_f and that $\dot{\gamma} = 8$ v/d.)
b. If the average outlet concentration is found to be 10^{-5} mol/l, what is the value of N?
   (Hint: Use the ratio d/δ to find the wall concentration C_w, then apply $N = k_f C_w$.)
   Answer: (a) 0.33 mm

7.8 The Electrochemical Method

Local or spot wall shear rates $\dot{\gamma}$ in complex tubular geometries can be determined experimentally by measuring the current that results from an induced electrochemical reaction at electrodes installed in the wall of the conduit. Typically, in this method, an upstream section of metallic tube serves as an anode. Tiny electrodes, 0.1 to 1 mm in diameter, embedded in the wall at various locations of the downstream test section act as cathodes. The fluid carries a dissolved ionic solute, typically a ferric cyanide. A voltage is applied to the electrodes, and the current from the resulting redox reaction (e.g., $Fe(CN)_6^3 \rightarrow Fe(CN)_6^2$) measured.

Show that the shear rate $\dot{\gamma}$ at a given anode is related to the measured current by the expression:

$$\dot{\gamma}=\frac{1.9\,L}{(LWFC^0)^3D^2}i^3$$

where

L and W = length and width of a cathode
F = Faraday number

C^0 = solute concentration in the bulk fluid
D = solute diffusivity
i = Measured current

(Hint: Obtain the ion flow N (mole/sec) by integrating Equation 7.19e from 0 to L, then substitute the result into the electrochemical relation N = i/F.)

Note that because of their cubic dependence, high-precision measurements of cathode dimensions, solute concentration, and current are required.

7.9 Derivation of the Lévêque Relation

Apply a similarity transformation to Equations 7.18d and 7.18e to derive Equation 7.18f.

7.10 Batch Adsorption of a Trace Substance

When a diffusing solute partitions or adsorbs onto the solid matrix, one can often use standard solutions for nonsorbing solids to follow the course of adsorption by suitably modifying one of the solution parameters. For the case of adsorption by a sphere from a well-stirred solution, for example, $V_{soln}/V_{spheres}$ in Figure 7.6 is replaced by $V_{soln}/(KV_{spheres})$, where K is the partition coefficient or Henry's constant.

Assume the following parameter values: K = 10, $V_{soln}/V_{spheres}$ = 10, D = 10^{-5} cm^2/sec, R = 0.5 cm.

a. By making a cumulative mass balance, show that the modified parameter $100/(1 + V_{soln}/K\ V_{spheres})$ also represents the percentage of solute in the solution which is ultimately taken up by the solids at $t \to \infty$ (50% here).
b. What is the fraction of solute content of the solution taken up from the solution after 1 h?
(Hint: Multiply the ordinate value by the parameter value.)
Answer: 37%

7.11 Diffusion with a Type III Boundary Condition: The Modified Sherwood Number (Biot Number for Mass Transfer)

Figure 7.5 shows the fractional uptake and release for diffusion in and out of a sphere with a BC of Type III.

Why are there no solution curves available for Sh 0 to 0.2? How would you fill this gap?

(Hint: Consult Illustration 1.7.)

7.12 Solutions from Solutions

Suppose all you have available is a standard heat-transfer text, which only gives solutions to Fourier's equation for a plane sheet and a sphere with Type I BCs. You need to find a solution to Fick's equation for a plane sheet with one face impermeable, the other exposed to a sinusoidally varying concentration.

How would you proceed?

7.13 Temperature Transients in a Tubular Reactor

Consider a nonisothermal tubular reactor with an irreversible reaction A → products taking place in it. Show that for constant pressure operation, the temperature transients, occasioned for example by fluctuations in the feed, is given by:

$$Cp_v v\frac{\partial T}{\partial z} - \Delta H_r r_A - \frac{UP}{A}(t - T_{ext}) = Cp_v \frac{\partial T}{\partial t}$$

where Cp_v represents the volumetric heat capacity of the reaction mixture. Use the classical "in-out = change" approach.

7.14 Entry Length for Laminar Flow Heat Transfer

Entry length refers to the length of conduit necessary to establish fully developed temperature profiles, i.e., the distance from the inlet required to allow the thermal boundary layer to penetrate to the tubular centerline. When that point is reached, the Nusselt number becomes constant (see Figure 7.7).

a. Show that for laminar flow heat transfer in a cylindrical tube with a Type I BC, the thermal entry length L is given by the approximate expression:

$$L/d \cong 0.03 \text{ Re Pr}$$

where the product Re Pr is also known as the Peclet Number, $Pe = vd/\alpha$.

b. Derive the corresponding expression for laminar flow mass transfer.

7.15 Steam-Heated Tube in the Entry Region

Viscous fluids such as glycerol, which has a Prandtl number of the order 10^3 at room temperature, can have entry lengths of several meters (cf. Problem 7.13). Heating of such fluids in conventional heat exchangers will consequently lie entirely in the entry or Lévêque region. Consider the case of a viscous fluid being heated by isothermally condensing steam in a single-pass shell and tube heat exchanger. Show that the fluid temperature profile T(x) is given by the expression

$$T_s - T = (T_s - T_i)\exp\left[-6,48\left(\frac{\alpha x}{vd^2}\right)^{2/3}\right]$$

(Hint: Consult Table 7.10.)

7.16 Cross-Flow Heating of Solids

Loose granular solids can be heated by passing a hot gas through a perforated moving belt carrying the solids in cross-flow to the incoming gas (Figure 7.8B). Temperatures at any particular position x, z are then time-invariant, in contrast to the situation found in fixed-bed through-flow heating (Figure 7.8A and Illustration 7.15), which is an unsteady process.

Consider the differential element shown in Figure 7.8B and perform steady state, "in-out = 0" energy balances in the gas and solid phases, with S kg/sec of solids moving in the x-direction, and G kg/sec of gas in the z-direction. Show that the resulting PDEs are of exactly the same form as the through-flow Equations 7.25b and 7.25c, when the unsteady term in Equation 7.25b, which is of minor significance, is dropped. Solutions to the cross-flow problem can then be drawn from the same source as was used in Illustration 7.15, with the horizontal distance x replacing time t, and other parameters suitably changed. Make a small dictionary of the substitutions that have to be made.

7.17 Heat Sealing of Plastic Sheets

Sheets of plastic and other materials are frequently bonded together by applying heated elements (platens) to the surfaces of the sheets. In order to establish the proper heating cycle, it is necessary to know the time required for the interface to reach a specified sealing temperature. Using literature solutions, calculate this quantity for the following three cases:

1. Both upper and lower platens are kept at a constant temperature T_p.
2. The upper platen is at T_p, the lower is neutral, i.e., neither heated nor cooled. The thermal diffusivity of the platen is assumed to be approximately that of the plastic sheets, and their dimensions are large compared to the thickness of the sheet. This is the most common situation.
3. The upper platen is at T_p, the lower platen is cooled to maintain its temperature at the initial value T_i.

Data:

Initial temperature $T_i = 25°C$
Upper platen temperature $T_p = 200°C$
Required sealing temperature $T_s = 100°C$
Thickness of each sheet $L = 5$ mm
Thermal diffusivity $\alpha = 1.6\ 10^7\ m^2/sec$

Contact resistance is to be neglected. Calculated values are consequently minimum times.

(Hint: Consult both the conduction and diffusion literature.)

Answer: (1) 51 sec; (2) 124 sec; (3) 141 sec

7.18 Use of Shape Factors

Devise a meaningful problem, which would make use of the shape factors shown in Figure 7.9.

7.19 Radial Velocity Profiles for Small Leakages through a Tubular Wall

Consider a fluid in steady creeping flow through a cylindrical tube with a permeable wall. The leakage velocity v_w is taken to be small, so that we may assume $\frac{\partial p}{\partial r} \approx 0$ and $\partial^2 v_x / \partial x^2 \approx 0$, i.e., axial velocity v_x diminishes linearly with distance x.

a. Show that the profile of axial velocity $v_x(r)$ remains parabolic, but is associated with the derivative of pressure dp/dx rather than a constant pressure drop Δp/L. (Hint: Integrate the x-momentum balance, having first dropped the convective term.)
b. Show that the profile of the radial velocity $v_r(r)$ is given by the following expression:

$$\frac{v_r}{v_w} = 2\frac{r}{R} - \left(\frac{r}{R}\right)^3$$

(Hint: Integrate the continuity equation for cylindrical coordinates from 0 to r and from 0 to R to eliminate the pressure gradient.)

Comments: A situation of this type arises in urine flow through the kidney where tubular Reynolds numbers are of the order 10^{-2}. Water and various solutes pass from the blood and through the permeable wall of the tubules into the urine for ultimate elimination from the body.

The profile given above has been used to determine radial solute concentration profiles and the magnitude of the internal mass-transfer resistance $1/k_f$. What equations would you draw upon for this task?

7.20 Velocity Profiles Near a Moving Boundary

Suppose that one of the retaining walls of a parallel plane channel is suddenly set in motion with a constant velocity v_0.

a. Show that the relation describing the resultant time-dependent velocity profiles is given by the equation,

$$\frac{\partial \mathbf{v}}{\partial t} = \nu \nabla^2 \mathbf{v}$$

b. Indicate what disciplines might provide ready-made solutions to this problem, where $\nabla^2 \mathbf{v}$ is the Laplacian of the velocity vector v, and is tabulated in the dictionary Table 7.1.

7.21 Flow in Porous Media

Flow in a porous medium is customarily described by the empirical D'Arcy's law which, in three dimensions and with gravity included, becomes:

$$\mathbf{v}_s = -\frac{K}{\mu}(\nabla p - \rho \mathbf{g})$$

where $\mathbf{v}_s$ is the superficial velocity vector and K is the permeability of the medium. Show that for incompressible flow with constant K/μ, D'Arcy's law leads to Laplace's equation in the pressure p. Back-substitution then yields the velocity profiles which can be integrated to yield total volumetric flow rates. Note that the corresponding case of compressible flow was taken up in Illustration 5.11.

7.22 Tangential Velocity Distributions Between Rotating Cylinders

a. Show that for two concentric cylinders of radius R_i and R_0, with the latter rotating with an angular velocity ω, the tangential velocity v_θ is given by:

$$v_\theta = \omega R_0 \left(\frac{R_i}{r} - \frac{r}{R_i} \right) \Big/ \left(\frac{R_i}{R_0} - \frac{R_0}{R_i} \right)$$

(Hint: Integrate the equation for θ momentum, using the BCs $v_\theta(R_i) = 0$, $v_\theta(R_0) = \omega R_0$.)

b. Derive the corresponding expression for the case of the inner cylinder rotating, with the outer cylinder held stationary.

7.23 Solutions from Solutions

The steady state temperature distribution in a rectangle with constant internal heat generation A (J/m³s) and the surfaces x = ±a and y = ±b kept at zero temperature is given by the formidable expression

$$T = \frac{A(a^2 - x^2)}{2k} - \frac{16 A a^2}{k\pi^3} \sum_{n=a}^{\infty} \frac{(-1)^n \cos[(n+1)\pi x/2a] \cosh(2n+1)\pi yk/2a}{(2n+1)^3 \cosh(2n+1)\pi bk/2a}$$

What, if any, fluid mechanics problem can be solved by this equation?

8 Analytical Solutions of Partial Differential Equations

In this last chapter we present outlines of three important classical methods for the solution of partial differential equations (PDEs). We start with the method of separation of variables, which dates back to the 18th century and finds its principal application in the solution of second-order homogeneous and linear PDEs. A host of solutions to Fourier's and Fick's equations are arrived at by this method, and we present several illustrations to explain and expand on the procedure. An opening preamble is devoted to the twin topics of Fourier series and orthogonal functions, which play a key role in the application of the method of separation of variables.

The second section deals with the Laplace and other integral transform methods, this time in the context of solving PDEs. These methods again apply to linear PDEs only, but are capable of handling nonhomogeneous systems as well. They can thus be used to reach a wider range of the classical PDEs of mathematical physics. Although the main focus is on the Laplace transformation, some time is spent in explaining and illustrating the use of the less common integral transforms.

The final section introduces the reader to the method of characteristics. This elegant and powerful procedure extends our reach considerably, and enables us to address first-order quasilinear, as well as linear PDEs. We limit ourselves to the treatment of single equations, but provide sufficient detail for the reader to grasp the general methodology of the procedure. The solution of systems of quasilinear PDEs, which is a much wider topic, can be pursued in one of the many excellent monographs available.

8.1 SEPARATION OF VARIABLES

8.1.1 Orthogonal Functions and Fourier Series

We open this segment of the chapter with a preamble to introduce the reader to the concepts of orthogonal functions and Fourier series expansion. Both of these topics make their appearance in the course of solving PDEs by separation of variables, and it is therefore natural to introduce them within the framework of this important solution method. Both orthogonal functions and Fourier series have other important applications, such as the representation of arbitrary functions, but our focus here will be their role in the solution of PDEs. In the course of our narrative, we will present some of their general properties, thus preparing the ground for their use in other areas as well.

To demonstrate the genesis of these two concepts — or tools, as they become here — we outline, in a step-by-step fashion, the application of the method of separation of variables. We shall see that near the end of the procedure we reach a seeming impasse that cannot be overcome by conventional means. It is at this point that we introduce the notion of orthogonal functions and Fourier series expansion. In what follows, we develop the various steps that lead us to that point.

Step 1 — The essence of the method of separation of variables lies in the assumption that the solution can be expressed as the product of functions of a single variable. Thus, for a PDE in the dependent variable u, and in Cartesian coordinates, the solution is assumed to have the general form

$$u = f(x)\, g(y)\, h(z)\, k(t) \tag{8.1a}$$

This assumption, the validity of which has to be ultimately proven, can be applied to any arbitrary PDE, but is usually successful only in the case of linear second-order homogeneous PDEs with constant or variable coefficients. The reasons for the restrictions will become apparent in the course of the development of the procedure.

At times it is possible to provide a physical rationale for the assumed solution form. If the reader will cast a glance back at Chapter 1, Figure 1.3, describing the quenching of a steel billet, it will become apparent that the temperature profiles that arise in this case can be viewed as sine half waves with a time-dependent amplitude. It seems reasonable to assume, therefore, that the solution might have the form:

$$T = \sum A(t) \sin bx \tag{8.1b}$$

where A(t) is the time-varying amplitude. We have craftily included a summation sign, because we surmise that a single sine function will not suffice to represent all profiles, in particular the discontinuous ones that appear at the start of the operation. This concept of using a sum of functions to represent another function will, as we shall see, ultimately lead to the notion of a Fourier series expansion.

Step 2 — We introduce the assumed form of Equation 8.1a into the PDE. We had previously indicated in Section 5.3 that when this is done for the case of the one-dimensional conduction (Fourier) equation, with an assumed solution form u = T(t)X(x), slight rearrangement of the result led to the expression:

$$\frac{1}{\alpha}\frac{g'(t)}{g(t)} = \frac{f''(x)}{f(x)} \tag{5.26c}$$

It was then argued that the two sides, which are functions of different independent variables, can only be equal if they are constant. We, therefore, wrote:

$$\frac{1}{\alpha}\frac{g'(t)}{g(t)} = \frac{f''(x)}{f(x)} = constant = -\lambda^2 \tag{5.26d}$$

where the constant is set $= -\lambda^2$ rather than λ to avoid redundant solutions and square roots.

Note that by this procedure we have reduced the PDE to a set of equivalent ODEs. For PDEs in more than two independent variables, similar results are obtained. This is a major simplification that does, however, hinge on the validity of the assumed solution form. If the PDE had been nonhomogeneous, the product solution would have yielded:

$$\frac{1}{\alpha}g'(t)f(x) = g(t)f''(x) + h(x,t) \tag{8.1c}$$

and the felicitous form Equation 5.26d would not have been obtained. This is the principal reason for restricting the method of separation of variables to homogeneous systems.

Step 3 — After this promising start, we turn to the relatively mundane task of solving the ODEs, which can usually be accomplished by standard methods. It is to be noted, however, that because λ is arbitrary and yet to be defined, we must accommodate the possibility of λ being zero as well as nonzero. This will, in general, give rise to two different sets of solutions that we accommodate by invoking the superposition principle, i.e., we argue that because each set is presumed to be a solution to the PDE, their sum must also be a solution. Note that if the PDE had been nonlinear, and thus ineligible for superposition procedures, the proceedings would have come to a halt at this point. This explains our restriction of the method of separation of variables to linear PDEs.

Step 4 — Solution of the ODEs is followed by an evaluation of the integration constants and of the eigenvalues λ. We use for this purpose the available boundary conditions, leaving the initial condition to the last for reasons that will become apparent later. After most of the integration constants have been evaluated, and with one BC left over, we may have the following situation:
Solution to this point:

$$u = C \sin(\lambda x)\exp(-K\lambda^2 t) \tag{8.1d}$$

Remaining BC:

$$u(a,t) = 0 \tag{8.1e}$$

where the latter might represent a normalized temperature or concentration at $x = a$. One sees immediately that this condition can be used to evaluate λ, for with $u = 0$ at $x = a$, we must have $\sin(a\lambda) = 0$, and λ takes on the infinite set of values $\lambda = n\pi/a$ (the so-called eigenvalues of the PDE), with $n = 1,2,3, \ldots$. This neat result would not have come about if Equation 8.1e had been nonhomogeneous. Thus, we have a second good reason to require homogeneity of both the PDE and the boundary condition for a successful application of the method.

To accommodate this infinite set of solutions, we invoke the superposition principle as before so that the solution now becomes:

$$u = \sum_{n=1}^{\infty} C \sin(n\pi x/a) \exp(-kn^2\pi^2 t/a^2) \tag{8.1f}$$

We have here our second glimpse of the dreaded infinite series that crop up regularly in solutions of PDEs. At least now we know the culprit — it is superposition.

Step 5 — At this stage we are seemingly left with only one integration constant, C, to be determined, for which we have the initial condition available. This turns out to be the most difficult step and leads to an impasse. For suppose the initial condition were given as:

$$u(x, 0) = f(x) = 1 \tag{8.1g}$$

representing, for example, a uniform and normalized initial temperature or concentration. Then substitution into the solution (Equation 8.1d) yields:

$$C = \left[\sum_{n=1}^{\infty} \sin(n\pi x/a)\right]^{-1} \tag{8.1h}$$

which is clearly unacceptable, because the right side is a function of x and not a constant.

Suppose that we assigned a different constant to each sine term, hoping that by properly weighting them, the sine would converge to unity. This is legitimate as long as the initial condition is satisfied. We would then have:

$$f(x) = 1 = C_1 \sin(\pi x/a) + C_2 \sin(2\pi x/a) + \ldots C_n \sin(n\pi x/a) + \ldots \tag{8.1i}$$

This may lead to the desired result, but compounds our difficulties because an infinite set of constants will now have to be evaluated. It speaks for the genius of the early workers in this field that they not only persisted in this line of attack, but ultimately devised a way of evaluating the coefficients. To do this, they drew on two seemingly unrelated and innocuous expressions, which are nowadays routinely found in all tables of integrals:

$$\int_0^a \sin(n\pi x/a) \sin(n\pi x/a) dx = 0 \tag{8.1j}$$

$$\int_0^a \sin^2(n\pi x/a) dx = a/2 \tag{8.1k}$$

They then multiplied each term of Equation 8.1i by sin(nπx/a)dx and integrated from 0 to a. This causes all terms on the right side to vanish except the n-th coefficient that by virtue of Equation 8.1k becomes:

$$C_n = \frac{\int_0^a f(x)\sin(n\pi x/a)dx}{\int_0^a \sin^2(n\pi x/a)dx} = \frac{2}{a}\int_0^a 1\ \sin(n\pi x/a)dx \tag{8.1l}$$

or, upon evaluation of the last integral

$$C_n = \frac{2}{n\pi}[1-(-1)^n] \tag{8.1m}$$

Thus, we have obtained a general expression for the n-th integration constant and are now in a position to write the infinite series (Equation 8.1f) in explicit form

$$u = \frac{2}{\pi}\sum_{n=1}^{\infty}\frac{1}{n}[1-(-1)^n]\sin(n\pi x/a)\exp(-Kn^2\pi^2 t/a^2) \tag{8.1n}$$

Comments: We note that the sequence in which we used the boundary and initial conditions is now justified, for it was the earlier evaluation of λ that led to the infinite series, and ultimately the determination of its coefficients. Had the initial condition been introduced prior to that point, we would have obtained:

$$C = [\sin(\lambda x)]^{-1} \tag{8.1p}$$

a self-contradictory result with no means of resolution.

The need to have a homogeneous PDE as well as homogeneous BCs was justified on several occasions. That requirement, however, does not extend to the initial condition, because any nonhomogeneity f(x) can be accommodated easily in the first integral of Equation 8.1h. This is in agreement with what was stated in Section 5.3.3 of Chapter 5, dealing with the elimination of nonhomogeneous terms. The approach taken there was to accept even severe nonhomogeneities in the initial condition, provided the PDE and remaining BC could be rendered homogeneous. This has now been justified.

We use this occasion to introduce the reader to the terminology associated with these proceedings. We have already referred to λ as the *eigenvalues* of the PDE. Associated with them are the so-called *eigenfunctions* f(λ), here sin λx. The infinite series (Equation 8.1i) is referred to as the *Fourier series expansion* of f(x) = 1, and the associated constants C_i as *Fourier coefficients*. Sequences of functions such as sin(nπx/a) that obey the type of relations expressed by Equations 8.1j and 8.1k are referred to as *orthogonal functions*. The latter will be taken up in greater detail in the following.

The reader may have wondered whether there are other functions with the felicitous properties of sin(nπx/a), and if so, whether they can be used in the solution of PDEs. The answer to both questions is yes. There are, in fact, a broad range of such orthogonal functions and, what is more, they arise as solutions of certain linear second-order ODEs that in turn are generated in the course of applying the method of separation of variables. The conditions that the ODEs have to satisfy so as to yield an orthogonal set of functions are enshrined in the Sturm–Liouville theorem. We shall take a brief leave from PDEs and separation of variables to address these important concepts, as well as Fourier series in general, after which we shall return with further examples of PDE solutions.

8.1.1.1 Orthogonal and Orthonormal Functions. The Sturm-Liouville Theorem

We start with a definition. A sequence of continuous functions $y_1(x)$, $y_2(x)$, ..., $y_n(x)$, ... are said to be *orthogonal* with respect to the weight function p(x) in the interval $a \le x \le b$ if the following two conditions are met:

$$\int_a^b p(x)\ y_m(x)\ y_n(x)dx = 0 \tag{8.2a}$$

$$\int_a^b p(x)\ y_n^2(x)dx = C^2 \tag{8.2b}$$

where $C \neq 0$ is the so-called norm or normalizing factor. By dividing each y_j by the norm C, we obtain a new set of functions ϕ termed *orthonormal*:

$$\phi_j = y_j/C \tag{8.2c}$$

and with it a modification of the condition given in Equation 8.2b:

$$\int_a^b p(x)\phi_m(x)\phi_n(x)dx = 0 \tag{8.2d}$$

$$\int_a^b p(x)\phi_n^2(x)dx = 1 \tag{8.2e}$$

These new terms are related to the functions of our previous discussion as follows:

Orthogonal functions:	$y_m(x) \rightarrow \sin(m\pi x/a)$
	$y_n(x) \rightarrow \sin(n\pi x/a)$
Weight function:	$p(x) \rightarrow 1$
(Norm)^2:	$C^2 \rightarrow a/2$
Orthonormal function:	$\phi_j \rightarrow (2/a)^{1/2} \sin(j\pi x/a)$

A number of sets of orthogonal functions that are of frequent occurrence in the solution of PDEs by separation of variables are listed in Table 8.1. We note some of the implications of this table, and follow this up with an illustration.

TABLE 8.1
Sets of Orthogonal Functions

Set	Interval	Normalizing Factor C	Weight Function
1. sin nx n = 0, 1, 2,...	$(-\pi,\pi)$	$\pi^{1/2}$	1
2. cos nx n = 0, 1, 2,...	$(-\pi,\pi)$	$\pi^{1/2}$, $(2\pi)^{1/2}$	1
3. sin $(n\pi x/a)$ n = 0, 1, 2,...	(0,a)	$(a/2)^{1/2}$	1
4. cos $(n\pi x/a)$ n = 0, 1, 2	(0,a)	$(a/2)^{1/2}$, $a^{1/2}$	1
5. $(\sin \lambda_j x)/x$ λ_j = roots of $\tan a\lambda = h(a\lambda)$	(0,a)	$\dfrac{\sin a\lambda}{2^{1/2}}$	x^2
6. $\cos(\lambda_j x)$ λ_j = roots of $\tan \lambda = h/\lambda$	(0,1)	$\left(\dfrac{h+\sin^2\lambda_j}{2h}\right)^{1/2}$	1
7. $J_k(\lambda_j x)$ λ_j = roots of $J_k(\lambda a) = 0$	(0,a)	$\dfrac{a}{\sqrt{2}}\left[J_{k+1}(\lambda_j a)\right]$	x
8. $J_k(\lambda_j x)$ λ_j = roots of $\lambda a J_{k+1}(\lambda a) - hJ_k(\lambda a) = 0$	(0,a)	$\left[\dfrac{\lambda_j^2 a^2 + h^2 - k^2}{2}\right]^{1/2} \dfrac{J_k(\lambda_j a)}{\lambda_j}$	x

We start by pointing out that the weight function p(x) of Equation 8.2b is unity in some instances, and a simple function of x in others. This does not unduly complicate the evaluation of the integrals (Equations 8.2a and 8.2b). A second noticeable feature is that the running index of the sequence, n or λ, is not necessarily composed of positive integers. This is so for certain trigonometric functions, items 1 to 4; but in other cases, items 5 to 8, λ_j are the roots of certain transcendental equations. We have previously, in Table 7.8, Chapter 7, given a short list of the roots of item 8. Roots of the equations contained in items 5 to 7 can be found in standard texts on diffusion and conduction, and in mathematical handbooks. Using the table entries for item 7, for example, the orthogonality relations (Equations 8.2a and 8.2b) now become:

$$\int_0^a xJ_k(\lambda_m x)J_k(\lambda_n x)dx = 0 \tag{8.2f}$$

$$\int_0^a xJ_k^{\,2}(\lambda_n x)dx = C^2 = \frac{a^2}{2}J_{k+1}^2(\lambda_n a) \tag{8.2g}$$

These somewhat abstract-looking relations in fact have their uses. One may anticipate that Bessel functions of this type will arise in solving PDEs in cylindrical coordinates, and that the relations (9.1.16c) and (9.1.16d) would then be very handy in evaluating Fourier coefficients in much the same way as the simpler relations, such as Equations 8.1j and 8.1k.

Illustration 8.1 The Cosine Set

It will be of further comfort to the uninitiated to have a particular set of entries in Table 8.1 derived in detail. We choose for this purpose the cosine sequence, item 2, and set out to prove orthogonality and to derive the norm C.

$$m \neq n \neq 1$$

We use the trigonometric formula

$$\cos mx \cos nx = \frac{1}{2}[\cos(n-m)x + \cos(m+n)x] \tag{8.3a}$$

and obtain from integral tables, using a weight function p(x) = 1 in Equation 8.2a:

$$\int_{-\pi}^{\pi} (1)(\cos mx)(\cos nx)\,dx = \frac{1}{2}\left[\frac{\sin(n-m)x}{m-m} + \frac{\sin(m+n)x}{m+n}\right]_{-\pi}^{\pi} = 0 \tag{8.3b}$$

We repeat this for the first term and obtain:

$m = 1$:

$$\int_{-\pi}^{\pi} (1)(1)\cos nx\ dx = -\frac{1}{2}\sin y\Big|_{-n\pi}^{n\pi} = 0 \tag{8.3c}$$

To fulfill the second condition (Equation 8.2b) and obtain the norm, we show, using the appropriate formula from integral tables:

$m = n \neq 0$:

$$\int_{-\pi}^{\pi} (1)\cos^2 nx\ dx = \frac{1}{2}\left[x + \frac{\sin 2nx}{2n}\right]_{-\pi}^{\pi} = \pi = C_1^2 \tag{8.3d}$$

$m = n = 0$:

$$\int_{-\pi}^{\pi} (1)1^2\,dx = 2\pi = C_2^2 \tag{8.3e}$$

This shows, as was indicated in Table 8.1, that for the cosine sequence only, we have two norms, a general one C_1, for the case $n \neq 0$, and a special one, C_2, for $n = 0$. This does not, however, invalidate the condition (9.1.16b), and the set can still be considered orthogonal,

The Sturm–Liouville Theorem

We have seen that orthogonal functions are highly useful tools, not only in the solution of PDEs by the method of separation of variables, but also in the representation of both continuous and discontinuous functions. Finding these functions by choosing them at random and guessing both the weight function p(x) and the interval of orthogonality (a,b) are obviously unrewarding tasks. Fortunately, a theorem is available that generates these functions and the associated weight functions and intervals from a general linear second-order ODE with variable coefficients of the form:

$$[r(x)y']^1 + [q(x) + \lambda p(x)]y = 0 \tag{8.4a}$$

with the following boundary conditions of Type III

$$A_1 y(a) + A_2 y'(b) = 0 \tag{8.4b}$$

$$B_1 y(a) + B_2 y'(b) = 0$$

Such second-order systems arise, as we have seen, in the course of applying the method of separation of variables. They are referred to as Sturm–Liouville systems. The theorem may be phrased as follows.

Provided that over the interval $a \le x \le b$, p(x), q(x), r(x) and r′(x) are real and continuous, the solutions of the Sturm–Liouville system will be a set of functions $y_n(x)$ that are orthogonal with respect to the weight function p(x) over the interval $a \le x \le b$. Furthermore, any function f(x) that is sectionally continuous in this interval (translation: $f(x) \ne \infty$) can be expanded in terms of the orthogonal set:

$$f(x) = \sum_{n=0}^{\infty} C_n y_n(x) \tag{8.4c}$$

The following are to be noted. If p(x) or r(x) should vanish at the end points, the theorem will still hold, provided y(x) remains finite at those points. The theorem also applies if the interval is replaced by an unbounded one.

This is a powerful statement, which will ease our task considerably. It merely requires an inspection of an ODE and its boundary conditions to establish orthogonality and the validity of a series expansion of arbitrary functions. We shall use it in subsequent illustrations to solve PDEs by the method of separation of variables.

Fourier series — We have seen in Table 8.1 that the sequence of both sine and cosine functions form an orthogonal set. They can be combined into a more general form known as a *Fourier series*:

$$\frac{1}{2}a_0 + \sum_{n=1}^{\infty} (a_n \cos nx + b_n \sin nx) \tag{8.5a}$$

where the Fourier coefficients are given by:

$$a_n = \frac{1}{\pi}\int_{-\pi}^{\pi} f(x)\cos nx\, dx \tag{8.5b}$$

$$b_n = \frac{1}{\pi}\int_{-\pi}^{\pi} f(x)\sin nx\, dx \tag{8.5c}$$

f(x) is an arbitrary function over (−π, π), and π in the denominator is recognized as the square of the norm C^2 (see Table 8.1).

We note the following properties of this series. Any function defined arbitrarily over the interval (−π,π) and outside it by the equation f(x + 2π) = f(x), and which has a finite number of discontinuities and extrema over that interval, can be represented by a Fourier series. Thus, the Fourier series can be used to represent both a function f(x) over (−π,π) for values of x in that interval, or a periodic function with period 2π for all values of x.

The Fourier series can be extended to an arbitrary interval (−a, a) and then becomes:

$$\frac{a_0}{2} + \sum_{n=1}^{\infty}\left[a_n \cos(n\pi x/a) + b_n \sin(n\pi x/a)\right] \tag{8.5d}$$

with the Fourier coefficients given by:

$$a_n = \frac{1}{a}\int_{-a}^{a} f(x)\cos(n\pi x/a)dx \tag{8.5e}$$

$$b_n = \frac{1}{a}\int_{-a}^{a} f(x)\sin(n\pi x/a)dx \tag{8.5f}$$

If f(x) is an even function f(x) = f(−x), all the sine terms vanish. For odd functions f(x) = −f(−x), on the other hand, the cosine terms drop out. The coefficient a_0(n = 0) carries a factor $\frac{1}{2}$ designed to accommodate the differences in norms for cos nx and cos 0 (see Equations 8.3d and 8.3e). With this factor in place, a single norm C = $(\pi)^{1/2}$ can be used for the entire series. Let us illustrate the use of Fourier series with a particular example.

Illustration 8.2 Fourier Series Expansion of a Function f(x)

We set out to represent the function f(x) = x, i.e., a straight line of slope 1 in the interval (−π,π) by means of a Fourier series expansion. The form of the function, and its periodic extension, are shown in Figure 8.1A. Using Equations 8.5b and 8.5c

A. The Function

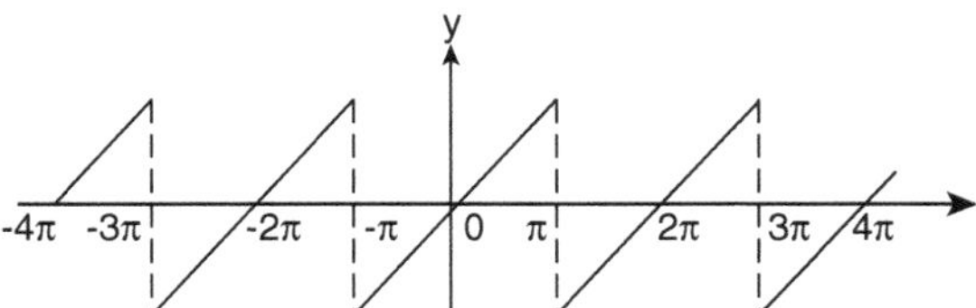

B. Convergence

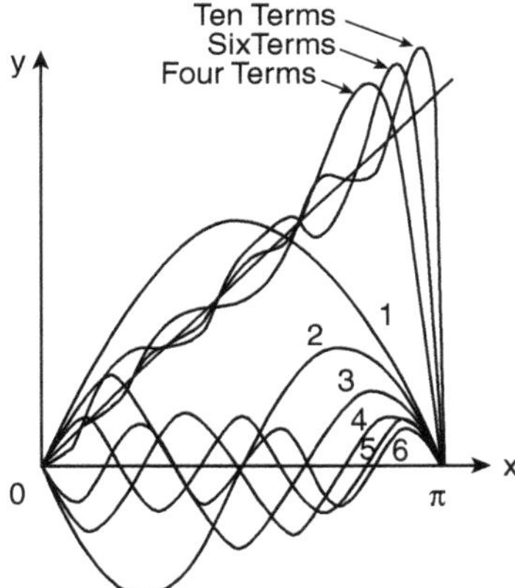

FIGURE 8.1 Expansion of saw-tooth function (A) by means of a Fourier sine series. (B) Lower curves represent individual sine terms, the upper curves their sum.

to evaluate the Fourier coefficients, and making appropriate use of tables of integrals, we find:

$$a_n = \frac{1}{\pi}\int_{-\pi}^{\pi} x \cos nx \, dx = 0 \tag{8.6a}$$

$$b_n = \frac{1}{\pi}\int_{-\pi}^{\pi} x \sin nx \, dx = -\frac{2}{n}\cos n\pi \tag{8.6b}$$

Only the sine terms remain, and we obtain:

$$x = \sum_{n=1}^{\infty} b_n \sin nx$$

that is

$$x = 2\left[\left(-\frac{1}{1}\cos\pi\right)\sin x + \left(-\frac{1}{2}\cos 2\pi\right)\sin 2x + \ldots\right] \tag{8.6c}$$

or

$$f(x) = x = 2\left[\sin x - \frac{\sin 2x}{2} + \frac{\sin 3x}{3} - \ldots\right]$$

This is the desired expansion of x over the interval $(-\pi, \pi)$.

Comments: The disappearance of the cosine terms was to be expected because the function in question is an odd one, with $f(x) = -f(-x)$. Because the series (Equation 8.6c) has the period 2π, it also represents the periodic extension of $f(x) = x(-\pi,\pi)$, shown as a graph of discontinuous parallel lines in Figure 8.1A. Note that at the points of discontinuity, the series converges to zero, which is one-half of the sum of the right-hand and left-hand limits.

Convergence of the Fourier series to $f(x) = x$ over the interval $(0,\pi)$ is graphically depicted in Figure 8.1B. The lower terms represent the individual sine terms, the upper curves their sum. The degree of convergence is marked, but will evidently require a considerable number of terms to achieve acceptable agreement.

This completes our intermezzo on orthogonal functions and Fourier series. We return to the topic of separation of variables and illustrate its use in the solution of linear second-order PDEs with a number of examples.

Illustration 8.3 The Quenched Steel Billet Revisited

This example considers the temperature transients that arise when a hot steel billet is exposed, at time $t = 0$, to an external temperature T_s. We assume a uniform initial temperature T_0 and neglect external heat transfer resistance. The system and its boundary and initial conditions are depicted in Figure 8.2.

Prior to proceeding to a solution, we nondimensionalize and normalize the temperature variable by setting:

$$\theta = \frac{T - T_s}{T_1 - T_s} \tag{8.7a}$$

The following set of equations and conditions is obtained:

Fourier's equation: $$\frac{\partial\theta}{\partial t} = \alpha\frac{\partial^2\theta}{\partial x^2} \tag{8.7b}$$

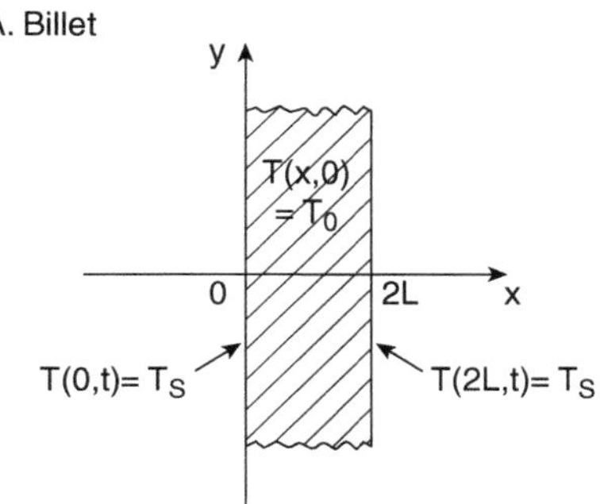

FIGURE 8.2 Geometry, BCs, and IC for a quenched steel billet.

This eases the representation of the results and, as a bonus, yields homogeneous boundary conditions.

At the two surfaces $$\theta(0, t) = 0 \tag{8.8a}$$

$$\theta(2L, t) = 0 \tag{8.8b}$$

Initially $$\theta(x, 0) = 1 \tag{8.8c}$$

Although these three conditions are sufficient in principle, two additional conditions are given which are helpful in evaluating integration constants. They must, at any rate, also be satisfied by the solution.

Steady state $$\theta(x, \infty) = 0 \tag{8.8d}$$

Symmetry $$\frac{\partial \theta}{\partial x}(L, t) = 0 \tag{8.8e}$$

We now introduce the assumption of separation of variables and substitute the resulting expression into Fourier's equation (Equation 8.7b) (Steps 1 and 2 of our preamble). After some rearrangement, there results:

$$\frac{1}{a}\frac{T'(t)}{T(t)} = \frac{X''(x)}{X(x)} = -\lambda^2 \tag{8.9a}$$

The constant value on the right, which can be positive, negative, or zero, has already been justified, but the reason for squaring it and assigning it a negative sign needs to be explained. Both moves are taken on anticipatory grounds. We expect the temperature $\theta = T(t)X(x)$ to decay exponentially with time, hence the negative sign. To ensure that this feature is not wiped out by negative values of λ, the latter is squared. This also eliminates square roots that we expect will arise in the solution of the ODEs in X(x).

Step 3: Solution of ODEs — We apply separation of variables to the ODE in t, and the standard D-operator method to the ODE in x. There results:

For $\lambda^2 \neq 0$
$$T_1 = C_1 \exp(-\alpha\lambda^2 t) \tag{8.9b}$$

$$X_1 = C_2 \cos \lambda x + C_3 \sin \lambda x$$

For $\lambda^2 = 0$
$$T_2 = C_4 \tag{8.9c}$$

$$X_2 = C_5 + C_6\, x$$

We now invoke the superposition principle by adding the two solutions, and obtain:

$$\theta = T_1X_1 + T_2X_2 = C_7 + C_8\,\mathrm{x} + (C_9 \cos \lambda \mathbf{x} + C_{10} \sin \lambda x)\exp(-\alpha\lambda^2 t) \tag{8.9d}$$

Step 4: Evaluation of constants — We note that there are five constants to be evaluated, hence it becomes convenient to draw on the auxiliary boundary condition (Equation 8.8d). The reader is also reminded of the necessity to leave the initial condition unused for the time being. We obtain the following results:

BC (Equation 8.8d): $\theta(x,\infty) = 0$ — This results in the exponential term vanishing. The remaining terms must also vanish and we obtain $C_7 = C_8 = 0$.

BC (Equation 8.8a): $\theta(0,t) = 0$ — Here, the sine terms vanish; hence, to obtain $\theta = 0$, the cosine terms must also drop out. This requires $C_9 = 0$.

BC (Equation 8.8b): $\theta(2L,t) = 0$ — This condition is used to determine the eigenvalues λ. At this stage the solution is composed of:

$$\theta = C_{10} \sin \lambda x \exp(-\alpha\lambda^2 t) \tag{8.9e}$$

To satisfy the BC, the sine terms must vanish, which is accomplished by setting $\lambda = n\pi/2L$, n = 1, 2, 3, … . We obtain:

$$\lambda = n\pi/2L \qquad \text{eigenvalues}$$

$$f(x) = \sin \lambda x \quad \text{eigenfunctions}$$

Linear superposition of this result now gives us the following infinite series:

$$\theta = \sum_{n=1}^{\infty} C_{10} \sin(n\pi x/2L)\exp(-\alpha n^2\pi^2 t/4L^2) \tag{8.9f}$$

Step 5: Evaluation of Fourier coefficients — The problem that faces us at this stage has already been noted. To satisfy the initial condition (Equation 8.8c), we must decompose C_{10} into an infinite set of coefficients C_n and hope for a valid Fourier expansion in terms of orthogonal functions. We have previously shown orthogonality of the sine sequence by making use of certain integrals of that function. We will not repeat that near-miraculous solution, but make use instead of the Sturm–Liouville theorem, which proves orthogonality and guarantees a Fourier expansion. Scrutiny of our ODEs (Equation 8.9a) shows that the Sturm–Liouville system associated with our model is given by the ODE in x:

$$X'' + \lambda^2 X = 0 \tag{8.9g}$$

with associated boundary conditions:

$$X(0) = 0 \tag{8.9h}$$

$$X(2L) = 0 \tag{8.9i}$$

Comparison of these equations with the general Sturm–Liouville system (Equations 8.4a and 8.4b) shows that the weighting function here is p(x) = 1, and that the interval of orthogonality is given by the physical boundaries of the system, (0, 2L). Expansion of some function f(x) (here f(x) = 1) in terms of the eigenfunction that results from Equation 8.9g, i.e., sin(nπx/2L), is also guaranteed. We thus have a valid representation of the initial condition:

$$1 = C_1 \sin(\pi x/2\mathrm{L}) + \mathrm{C}_2 \sin 2\pi x/2\mathrm{L} + \ldots + \mathrm{C}_n \sin(n\pi x/L) + \ldots \tag{8.9j}$$

The Fourier coefficients are now evaluated by multiplying each term by p(x)sin(nπx/2L)dx and integrating over the interval (0, 2L), where p(x) = 1. We obtain

$$\int_0^{2L} (1)(1)\sin(n\pi x/2L0dx = \int_0^{2L} C_1(1)\sin(\pi x/2L)\sin(n\pi x)/2Ldx = 0$$

$$+\int_0^{2:} C_2(1)\sin(2\pi x/2L)\sin(n\pi x/2L)dx + \ldots + \int_0^{2L} C_n(1)\sin^2(n\pi x/2L)dx + \ldots = 0 \tag{8.9k}$$

All terms on the right side except the n-th one drop out by virtue of orthogonality, and we obtain the following result for the n-th Fourier coefficient:

$$C_n = \frac{\int_0^{2L} \sin(n\pi x/2L)dx}{\int_0^{2L} \sin^2(n\pi x/2L)dx} \tag{8.9l}$$

The integral in the denominator is the square of the norm, and is tabulated in Table 8.1: $C^2 = 2L/2 = L$. The integral in the numerator yields:

$$\int_0^{2L} \sin(n\pi x/2L)dx = -\frac{2L}{n\pi}\cos(n\pi x/2L)\Big|_0^{2L} = \frac{2}{n\pi}[1-(-1)^n] \tag{8.9m}$$

Our solution then takes the final form of the following infinite series:

$$\theta(x,t) = \frac{2}{\pi}\sum_{n=1}^{\infty}\frac{1}{n}[1-(-1)^n]\sin(n\pi x/2L)\exp(-\alpha n^2\pi^2 t/4L^2) \tag{8.9n}$$

Comments: One notes that for even values of n, $[1-(-1)^n] = 0$ so that only odd terms remain in Equation 8.9n. This has led to an alternative formulation, frequently found in the literature, which takes the form:

$$\theta(x,t) = \frac{4}{\pi}\sum_{n=1}^{\infty}\frac{1}{2n-1}\sin\left(\frac{2n-1}{2L}x\right)\exp(-\alpha n^2\pi^2 t/4L^2) \tag{8.10a}$$

In this formulation, none of the terms vanish.

It is of some interest to verify whether the unused symmetry condition $\partial\theta/\partial x\big|_{x+L} = 0$ is indeed satisfied. This is done by a term-by-term differentiation of the Fourier series and yields the result:

$$\left.\frac{\partial\theta}{\partial x}\right|_{x=L} = \sum_{n=1}^{\infty}\frac{1}{L}\underbrace{[1-(-1)^n]}_{\substack{0\text{ for}\\\text{even terms}}}\;\underbrace{\cos(n\pi/2)}_{\substack{0\text{ for}\\\text{odd terms}}}\;\exp(-\alpha n^2\pi^2 t/4L^2) \tag{8.10b}$$

Thus, all terms of the series vanish, and the BC is consequently satisfied.

For small times t and low values of thermal diffusivity α, convergence of the series Equation 8.9n is quite slow, because representation close to the initial rectangular temperature distribution (Figure 8.2B) requires a considerable number of terms. In these cases, use is made of an alternative solution, arrived at by

Laplace transformation, which takes the form of an infinite series of complementary error functions. This solution has fast convergence for low values of α and t. Its derivation is taken up in Section 8.2, Illustration 8.9.

Illustration 8.4 Conduction in a Cylinder with External Resistance: Arbitrary Initial Distribution

In Illustration 7.8 of the preceding chapter, the unsuspecting reader was confronted with an infinite series of Bessel functions, which were used to describe concentration profiles in a tubular reactor undergoing a first-order irreversible reaction at the wall. It was shown there that, provided radial velocity gradients were neglected, the underlying model was identical to that describing unsteady conduction in an infinitely long cylinder with an external heat transfer resistance and constant initial temperature.

In the present illustration, we consider the same problem under the somewhat broader condition of an arbitrary initial temperature distribution $T(r,0) = f(r)$, otherwise retaining the same features as before. The model is then comprised of the following equations and boundary or initial conditions in terms of the dimensionless radial distance $y = r/R$:

Fourier equation:
$$\frac{\partial T}{\partial t} = \alpha/R^2 \left[\frac{\partial^2 T}{\partial y^2} + \frac{1}{y}\frac{\partial T}{\partial y} \right] \tag{8.11a}$$

At steady state:
$$T(y,\infty) = T_e \tag{8.11b}$$

Symmetry:
$$\left.\frac{\partial T}{\partial y}\right|_{y=0} = 0 \tag{8.11c}$$

or
$$T\Big|_{y'=0} = \mathit{finite}$$

At the surface:
$$-(k/R)\left.\frac{\partial T}{\partial y}\right|_{y=1} = h(T|_{y=1} - T_e) \tag{8.11d}$$

Initially:
$$T(y,0) = f(y) \tag{8.11e}$$

where T_e represents the temperature of external medium.

The solution proceeds over the same steps as those of the previous illustration.

Step 1: Separate variables — This is done by assuming the following form of solution:

$$T = F(y)G(t) \tag{8.12a}$$

Step 2: Substitution into the PDE — This step, followed by some rearrangement, yields the separated expression:

$$\frac{R^2}{\alpha}\frac{G'(t)}{G(t)} = \frac{1}{F}\left[F'' + \frac{1}{y'}F'\right] = -\lambda^2 \tag{8.12b}$$

Justification of the constant term $-\lambda^2$ follows along the lines given in the previous illustration.

Step 3: Solution of ODEs — We apply separation of variables to the ODE in t, p-substitution for the ODE in y when $\lambda^2 = 0$, and for $\lambda^2 \neq 0$, the generalized solution for second-order ODEs with variable coefficients, Equation 2.22a of Chapter 2. This yields:

For $\lambda^2 \neq 0$
$$F_1 = C_1 J_0(\lambda y) + C_2 Y_0(\lambda y) \tag{8.12c}$$
$$G_1 = C_3 \exp(\alpha\lambda^2 t)$$

For $\lambda^2 = 0$
$$F_2 = C_4 + C_5 \ln y \tag{8.12d}$$
$$G_2 = C_6$$

Superposition of these two solutions leads to the expression:

$$T = F_1 G_1 + F_2 G_2 = [C_1 J_0(\lambda y) + C_8 Y_0(\lambda y)]\exp(-\alpha\lambda^2 t) + C_9 + C_{10} \ln y \tag{8.12e}$$

Step 4: Evaluation of constants — It is best to start here with the symmetry condition (Equation 8.11c), and follow this up with the conditions (b) and (d). The initial condition is as usual left to the last. We obtain the following results:

BC (9.1.52b) $T|_{y=0} =$ *finite (i.e., bounded)* — Here, we note that the Bessel function $Y_0(\lambda y)$ goes to minus infinity for zero values of the argument (see Figure 2.5), as does the logarithmic term. For the solution to remain finite we must therefore set $C_8 = C_{10} = 0$.

BC (9.1.52a) $T(y, \infty) = T_e$ — This condition causes the exponential term in Equation 8.12e to vanish so that $C_9 = T_e$. At this point the solution has been reduced to the form:

$$T - T_e = C_7 J_0(\lambda y)\exp(-\alpha\lambda^2 t) \tag{8.12f}$$

BC (9.1.52*c*) $-(k/R)(\partial T/\partial y)_{y=1} = h(T|_{y=R} - T_e)$ — This relation is used to determine the eigenvalues λ, and requires evaluation of the derivative of the Bessel function $J_0(\lambda y)$. This can be done using the formulas given in Table 2.8 of Chapter 2 and yields:

$$\left.\frac{\partial T}{\partial y}\right|_{y=1} = C_7[-\lambda J_1(\lambda)]\exp(-\alpha\lambda^2 t) \tag{8.12g}$$

Back-substitution into the boundary condition yields the transcendental equation:

$$\lambda J_1(\lambda) = \beta K\, J_0(\lambda) \tag{8.12h}$$

where $\beta = hR/k$ (in other words, the Biot number; see Illustration 1.7). This is seen to be identical in form to the expression given in item 8 of Table 8.1. We can consequently expect $J_0(\lambda;y)$ to be an orthogonal set over the interval (0,1) with weight function $p(y) = y$ and λ_j given by the roots of Equation 8.12h, which we have previously tabulated in Table 7.8. We note that the solution has now become, by virtue of superposition:

$$T - T_1 = \sum_{n=1}^{\infty} C_n J_0(\lambda_n y)\exp(-\alpha\lambda_n^2 t/R^2) \tag{8.12i}$$

The same result can also be obtained by a Strum–Liouville analysis.

Step 5: Evaluation of Fourier coefficients — We follow the procedure outlined in the previous illustration, i.e., after introducing the initial condition, we multiply each term of the infinite series by $[yJ_0(\lambda_m y)dy]$ and integrate from 0 to 1. Dropping all terms but the n-th, we arrive at the following formula for C_n:

$$C_n = \frac{\int_0^1 y[f(y) - T_e]J_0(\lambda_n y)dy}{(Norm)^2} \tag{8.12j}$$

Using the expression for the norm given in Table 8.1, with k set = 0, we finally obtain:

$$T - T_e = 2\sum_{n=1}^{\infty} \frac{\lambda_n \int_0^R y[f(y) - T_e]J_0(\lambda_n y)dy}{(\lambda_n^2 R^2 + \beta^2)J_0^2(\lambda_n R)} J_0(\lambda_n y)\exp(-\alpha\lambda_n^2 t) \tag{8.12k}$$

Comments: Before being overwhelmed by the complexity of this expression, the reader is reminded that such series often converge quite fast, as was seen in Illustration 7.8. The integral will, in general, have to be evaluated numerically, but this poses no great problem because J_0 is a well-behaved periodic function.

Equation 8.12k is of the same form as Equation 7.16c of Illustration 7.8, but has slightly different boundary and initial conditions. In Illustration 8.6, we shall instruct the reader how to quickly extract useful information from this equation in spite of its formidable appearance.

ILLUSTRATION 8.5 STEADY STATE CONDUCTION IN A HOLLOW CYLINDER

So far in our Illustrations we have been able to apply the method of separation of variables in a rather mechanical way once the principles of orthogonality have been established. This is not always the case. Even modest changes in the underlying model may require rather substantial changes in our approach, although the basic stepwise procedure we established previously still applies. We show this with the following example.

Consider an infinitely long hollow cylinder with the interior and exterior surfaces maintained at the angle-dependent temperature $f_i(\theta)$ and $f_0(\theta)$. We wish to establish the resultant steady state temperature distribution $T(r,\theta)$ in the interior of the cylinder. The underlying model is Laplace's equation:

$$\nabla^2 T = 0 \tag{8.13a}$$

which we translate with the help of our dictionary Table 7.1 into radial and angular coordinates. There results:

$$\frac{\partial^2 T}{\partial r^2} + \frac{1}{r}\frac{\partial T}{\partial r} + \frac{1}{r^2}\frac{\partial^2 T}{\partial \theta^2} = 0 \tag{8.13b}$$

So far we have merely traded the angular derivative $(1/r^2)(\partial^2\theta)/\partial r^2)$ for the time derivative $\partial T/\partial t$ of the previous illustration. The first difficulty arises when we attempt to formulate boundary conditions. We appear to require at least four BCs, two for each second derivative, but seem to have only two available:

At the outer surface $$T(R_0,\theta) = f_0(\theta) \tag{8.13c}$$

At the inner surface $$T(R_i,\theta) = f_i(\theta) \tag{8.13d}$$

With some imagination one might add a third condition expressing periodicity of the temperature:

Periodicity $$T(r,\theta) = T(r,\theta + 2n\pi) \tag{8.13e}$$

or $$T(r,\theta - \pi) = T(r,\theta + \pi)$$

but this exhausts the possible condition. One hopes that a single BC will be capable of evaluating more than one constant, as was the case with the Fourier coefficients. Let us therefore proceed with the solution.

Steps 1 and 2 — We assume the solution to be of the form:

$$T = R(r)S(\theta) \tag{8.14a}$$

and obtain, after substitution into the PDE:

$$r^2\frac{R''}{R} + r\frac{R'}{R} = -\frac{S''}{S} = +\lambda^2 \tag{8.14b}$$

Note that we have postulated positive real eigenvalues because we require the solution of $S(\theta)$ to be periodic, not exponential. This is the first departure from convention.

Step 3 — We proceed with the solution of the ODEs for the two cases $\lambda^2 \neq 0$ and $\lambda^2 = 0$, which is accomplished by standard D-operator and p-substitution methods.

A second departure from the routine occurs in the solution of the ODE $r^2R'' + rR' - \lambda^2R = 0$, which is of the Euler Cauchy type (see Table 2.3), and requires the substitution $r = e^z$ to reduce to D-operator form. We obtain:

For $\lambda^2 = 0$:
$$S_1 = C_1' + C_2'\,\theta \tag{8.14c}$$

$$R_1 = C_3' + C_4' \ln r$$

For $\lambda^2 \neq 0$:
$$S_2 = C_5' \cos \lambda\theta + C_6 \sin'\lambda\theta \tag{8.14d}$$

$$R_2 = C_7'\, r^{\lambda} + C_8'\, r^{-\lambda}$$

Adding the solutions by superposition yields:

$$T = C_1 + C_2 \ln r + C_3\theta + C_4\theta \ln r$$

$$+ (C_5r^{\lambda} + C_6r^{-\lambda})\cos \lambda\theta + (C_7r^{\lambda} + C_8r^{-\lambda})\sin \lambda\theta \tag{8.14e}$$

Step 4 — To evaluate the integration constants, we start by utilizing the periodicity condition Equation 8.13f, which in the first instance leads to the condition $C_3 = C_4 = 0$ because $\theta \neq \theta \pm \pi$. We then use the same condition to argue that because the solution, including the surface condition, is periodic and has a finite number of discontinuities and extrema, it can be represented in terms of the Fourier series (Equation 8.5a). In other words, we have equivalence of the following expressions:

$$f_{i,0}(R_{i,0}) = \frac{a_0}{2} + \sum_{n=1}^{\infty}[a_n \cos n\theta + b_n \sin n\theta] \tag{8.14f}$$

$$f_{i,0}(R_{i,0}) = C_1 + C_2 \ln R_{i,0} + \sum_{\lambda=1}^{\infty}[C_5R_{i,0}^{\lambda} + C_6R_{i,0}^{-\lambda}] \cos \lambda\theta$$

$$+\sum_{\lambda=1}^{\infty}[C_7{R_{i,0}}^{\lambda} + C_8R_{i,0}^{-\lambda}] \sin \lambda\theta \tag{8.14g}$$

where we use $R_{i,0}$ to denote either the inner or outer radius. The Fourier coefficients a_n, b_n are equivalent to four infinite sets of coefficients C_5 to C_8, and are evaluated from Equations, 8.5b and 8.5c. The anomaly here is that each term in the series has two Fourier coefficients associated with it due to the appearance of the sets of two constants (C_5, C_6) and (C_7, C_8) in Equation 8.14g. This causes no difficulty, however, because we have two boundary conditions available in Equations 8.13c and 8.13d. Application of Equation 8.5b then yields the following set of relations:

For $a_0/2$:

$$C_1 + C_2 \ln R_0 = \frac{1}{2\pi}\int_{-\pi}^{\pi} f_0(\theta)d\theta \tag{8.14h}$$

$$C_1 + C_2 \ln R_i = \frac{1}{2\pi}\int_{-\pi}^{\pi} f_i(\theta)d\theta \tag{8.14i}$$

For a_n:

$$C_5 R_0^{\lambda} + C_6 R_0^{-\lambda} = \frac{1}{\pi}\int_{-\pi}^{\pi} f_0(\theta)\cos\lambda\theta \, d\theta \tag{8.14j}$$

$$C_5 R_i^{\lambda} + C_6 R_i^{-\lambda} = \frac{1}{\pi}\int_{-\pi}^{\pi} f_i(\theta)\cos\lambda\theta \, d\theta \tag{8.14k}$$

For b_n:

$$C_7 R_o^{\lambda} + C_8 R_o^{-\lambda} = \frac{1}{\pi}\int_{-\pi}^{\pi} f_o(\theta)\sin\lambda\theta \, d\theta \tag{8.14l}$$

$$C_7 R_i^{\lambda} + C_8 R_i^{-\lambda} = \frac{1}{\pi}\int_{-\pi}^{\pi} f_i(\theta)\sin\lambda\theta \, d\theta \tag{8.14m}$$

with $\lambda = 1, 2, 3, \ldots$.

These linear algebraic relations in the constants can be solved to yield explicit relations for $C_1 \ldots C_8$.

Comments: This is clearly a fairly complex problem of mainly academic interest. Its principal purpose was to induce the reader to "stretch" known principles and theorems to accommodate unusual circumstances. Mathematicians would want to provide more formal proof of the validity of the solution, but we prefer to content ourselves with the somewhat intuitive procedure used here.

Evaluation of the solution does not present overwhelming difficulties. The integrals in Equation 8.14f can be determined numerically, and the result substituted into the final solution, which now has the form:

$$T(r,\theta) = C_1 + C_2 \ln r + \sum_{\lambda=1}^{\infty}(C_5 r^{\lambda} + C_6 r^{-\lambda})\cos\lambda\theta + \sum_{\lambda=1}^{\infty}(C_7 r^{\lambda} + C_8 r^{-\lambda})\sin\lambda\theta \tag{8.14n}$$

This form can be further compressed (see Practice Problem 8.5).

ILLUSTRATION 8.6 QUICK RESULTS FROM COMPLEX SOLUTIONS

Although the infinite series solutions that inevitably result in applications of the separation of variables methods are nowadays easily evaluated numerically, the reader should not be discouraged from extracting some useful results by more mundane means. The calculations need not occupy more than a few minutes and carry the advantage of retaining the analytical solution form.

We draw on Equation 8.12k for our illustration and simplify it by setting $f(y) = T_i$ (constant initial temperature). Noting from Equation 8.12h that $J_1(\lambda_j)/J_0(\lambda_j) = \beta/\lambda_j$, the solution simplifies to the following form:

$$\frac{T - T_e}{T_i - T_e} = 2\sum \frac{\beta}{(\lambda_j^2 + \beta^2)} \frac{J_0(\lambda_j y)}{J_0(\lambda_j)} \exp(-\lambda_j^2 \alpha t/R^2) \tag{8.15a}$$

The solution of the analogous mass diffusion case is identical in form and reads:

$$\frac{C - C_e}{C_i - C_e} = 2\sum \frac{\beta}{(\lambda_j^2 + \beta^2)} \frac{J_0(\lambda_j y)}{J_0(\lambda_j)} \exp(-\lambda_j^2 D_e t/R^2) \tag{8.15b}$$

where β is now the mass transfer Biot number $Bi = k_C R/D_e$ and D_e, is the effective diffusivity in the medium, assumed to be a porous solid.

Much of the success of a rapid calculation depends on the behavior of the exponential term. If convergence is fast and all but the first exponential vanishes, calculations become extremely straightforward. To estimate when this will happen, we have compiled the following table of typical (order of magnitude) thermal and mass diffusivities:

	Metals	Nonmetals
α	10^{-5} m²/sec	10^{-7} m²/sec
	Gases	Liquids
D_e	10^{-6} m²/sec	10^{-10} m²/sec

A useful feature of the eigenvalues λ_j is that they increase approximately in increments of 3 (or π), starting with λ_1, which is of the order of unity for Biot numbers of interest (see Table 7.8). This completes the list of tools we require for our calculations.

The items we deem to be of particular interest are three in number: temperature at the central axis ($y = 0$), temperature at the surface ($y = 1$), and surface heat flux ($-k(\partial T/\partial y)\ y = 1$). Let us extract this information for a metal bar of radius $R = 10$ cm and a Biot number of $\beta = 10$. For these values, convergence is fast after $t = 100$ sec, and we use this time interval for our computations.

1. *Surface temperature after 100 sec.* For this case, the irksome Bessel function drops out, and Equation 8.15a reduces to:

$$\frac{T_s - T_e}{T_i - T_e} = 2\sum \frac{\beta}{(\lambda_j{}^2 + \beta^2)} \exp(-\lambda_j{}^2 \alpha t/R^2) \tag{8.15c}$$

$$= 2\left[\begin{array}{l} \dfrac{10}{2,17^2 + 10^2}\exp(-2.18^2 \times 10^{-5} 100/0.1^2 \\ + \dfrac{10}{5.03^2 + 10^2}\exp(-5.03^2 \times 10^{-5} \times 100 / 0.1^2 + \ldots \end{array}\right]$$

$$\frac{T_s - T_e}{T_i - T_e} = 2[0.059 + 0.0064 + \ldots] = 0.13$$

Convergence is essentially complete after two terms, and the surface temperature has dropped to approximately one-tenth of its initial value.

2. *Surface flux after 100 sec.* Here, we reach back to Table 2.8 of Chapter 2 to obtain an expression for the derivative of $J_0(\lambda_j y)$, which is required for the flux calculation. We have:

$$\left.\frac{d}{dy} J_0(\lambda_j y)\right|_{y=1} = -\lambda_j J_1(\lambda_j) \tag{8.15d}$$

which by virtue of Equation 8.12h becomes:

$$\left.\frac{d}{dy} J_0(\lambda_j y)\right|_{y=1} = -\beta J_0(\lambda_j) \tag{8.15e}$$

Introducing this relation into Equation 8.15a, we obtain for the flux:

$$q/A = k(T_i - T_e) 2\sum \frac{\beta}{\lambda_j{}^2 + \beta^2} \exp(-\lambda_j{}^2 \alpha t/R^2) \tag{8.15f}$$

where the summation is identical to that of Equation 8.15c. Consequently:

$$q/A = k(T_i - T_e) \times 0.13 \tag{8.15g}$$

For example, for a typical metal thermal conductivity k = 100 W/m²K, and assuming a difference between initial and external temperature $T_i - T_e = 1000$ K, we obtain for the flux:

$$q/A = 100 \times 1000 \times 0.13 = 13 \text{ kW/m}^2 \quad (8.15h)$$

3. *Temperature at central axis after 100 sec.* Here, the variable Bessel function in Equation 8.15a becomes unity, $J_0(\lambda_j y)_{y=0} = J_0(0) = 1$ (see Figure 2.5), and we obtain from Equation 8.15a:

$$\frac{T - T_e}{T_i - T_e} = 2\sum \frac{\beta}{\lambda_j^2 + \beta^2} \frac{1}{J_0(\lambda_j)} \exp(-\lambda_j^2 \alpha t/R^2) \quad (8.16a)$$

To evaluate this expression, we again draw on Table 7.8 but this time also utilize the values of the Bessel function $J_0(\lambda_j)$, and obtain:

$$\frac{T - T_e}{T_i - T_e} = 2\left[\frac{10}{2.18^2 + 10^2} \frac{1}{0.310} \exp(-2.18^2 \times 10^{-5} \times 100/0.1^2)\right.$$

$$+ - \frac{10}{5.93^2 + 10^2} \frac{1}{0.168} \exp(-5.03^2 \times 10^{-5} \times 100/0.1^2)$$

$$\left. + \frac{10}{7.96^2 + 10^2} \frac{1}{0.181} \exp(-7.96^2 \times 10^{-5} \times 100/0.1^2 - \cdots\right] \quad (8.16b)$$

$$\frac{T - T_e}{T_i - T_e} = 2[0.19 - 0.038 + 0.00065 - \ldots] \quad (8.16c)$$

$$\frac{T - T_e}{T_i - T_e} = 0.306 \quad (8.16d)$$

Convergence is again essentially complete after two terms and the central temperature is seen to be still at approximately 30% of its initial value.

4. *Recovery of a valuable substance by leaching.* Here we turn to the mass diffusion counterpart of our previous thermal examples and consider the extraction of a substance from a fibrous material of fiber radius R = 1 mm. Because of the low values of $D_e \approx 10^{-10}$ m²/sec, fast convergence of the series (Equation 8.15b) can only be expected after about 1000 sec.

Suppose, however, that the substance is a valuable one, and we stipulate near complete recovery so that at the central axis the remaining concentration is no more than 0.01% of its original value. We ask: What is the time required to achieve this goal?

Here convergence within one term is assured because of the long time interval involved, and we can write:

$$\frac{C_{y=0}-C_e}{C_i-C_e}=2\frac{\beta}{\lambda_1{}^2+\beta^2}\frac{1}{J_0(\lambda_1)}\exp(-\lambda_1{}^2D_e t/R^2) \tag{8.16e}$$

$$10^{-4}=2\frac{10}{2.18^2+10^2}\frac{1}{0.310}\exp(-2.18^2\times10^{-10}\,t/(10^{-3})^2) \tag{8.16f}$$

and hence

$$t = 1.73 \times 10^4 \text{ s} = 4.82 \text{ h}$$

5. *Numerical evaluation using the Mathematica package.* We conclude this illustration by demonstrating the use of the Mathematica package in evaluating the series (Equation 8.16a). Let us first determine λ_j s for j =1 to 10:

```
In[1]:= b  = 10; (* Biot number *)
        λ  = Union[
             Sort[
              Table[
               FromDigits[
                RealDigits[
                 x/.
                  FindRoot[x BesselJ[1, x]-b BesselJ[0, x] ==0,
                     {x, a}]
                     , 10, 5]
             ],
             {a, 1, 30}]]] // N
Out[1]= {2.1795, 5.0332, 7.9569, 10.936, 13.958, 17.01,
         20.083, 23.171, 26.27, 29.377}
```

Next, we define the dimensionless temperature function within the cylinder:

```
In[2]:= ft[n_, y_, t_]:=
```

$$\sum_{i=1}^{n} 2 \frac{b}{\lambda[[i]]^{\wedge}2 + b^{\wedge}2} \frac{\text{BesselJ}[0, \lambda[[i]]y]}{\text{BesselJ}[0, \lambda[[i]]]}$$

```
        Exp[-λ[[i]]^2] at/R^2];
```

Now, we plot the surface temperature (y = 1) for the initial 100 sec:

```
In[3]:= a = 10^-5;(* Thermal Diffusivity *)
        R = 0.1;(* Radius of Cylinder *)
        Plot[ft[10, 1, t], {t, 0, 100},
          TextStyle → {FontFamily → "Arial - Bond",
            FontSize → 14},
          Frame → True,
          FrameLabel → {"t", "(TS-Te)/(Ti-Te)",
            "Surface Temperature, Ts", ""}]
```

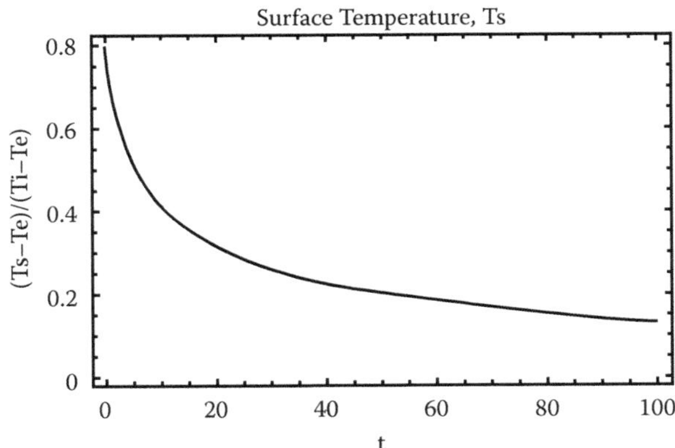


```
Out[5]= -Graphics-
```

8.1.2 Historical Note

Early steps in the development of the method of separation of variables were taken in the 18th century and are associated with the names of the English mathematician Brook Taylor of Taylor series fame (1685–1731), Daniel Bernoulli (1700–1782) and

Leonhard Euler (1707–1783) of Switzerland, and Jean d'Alembert (1717–1783) of France. Given the preeminence of music in the social life of the times, these scientists were drawn to the study of the mathematical theory of musical vibrations. By the 1750s, the wave equation was known and a solution of the boundary value problem of a vibrating string had been found. The twin notions of superposition and representation of arbitrary functions by trigonometric series made their first appearance around this time, and somewhat later Euler gave the formulas for the constants, which are now popularly known as *Fourier coefficients*. It was left to Jean Baptiste Joseph Fourier (1768–1830) to illustrate the basic procedure of separation of variables and superposition and help popularize trigonometric series representation. His book *Théorie Analytique de la Chaleur*, published in 1822, provides many examples of expansions in trigonometric series that arise in the conduction of heat. Although a relative latecomer to the field, his contributions were such that many of the tools used in separation of variables are now associated with his name.

Illustration 8.7 Diffusion in a Sphere Using Mathematica

Consider unsteady radial diffusion in a sphere with radius R:

$$\frac{\partial u}{\partial t} = D\left(\frac{\partial^2 u}{\partial r^2} + \frac{2}{r}\frac{\partial u}{\partial r}\right) \tag{8.17a}$$

where u is the dimensionless concentration within the sphere defined as $(C - C_0)/(C_1 - C_0)$. C_1 is the initial concentration in the sphere, and C_0 is the concentration required to maintain equilibrium with the surrounding atmosphere (which is a constant). At t = 0, the surface of sphere is exposed to a Type III boundary condition:

$$-D\,\partial u/\partial r\big|_{r=R} = k\,u(R,t) \tag{8.17b}$$

where D is the diffusion coefficient, and k is the mass transfer coefficient. Using the method of separation of variables, we find an expression for the dimensionless concentration inside the sphere, i.e., as a function of radial location r and time t:

```
In[1]: = eqn = D (∂r,r u[r, t] + 2/r ∂r u[r, t]) - ∂t u[r, t];

Out[1] = -u^(0,1)[r, t] + D(2 u^(1,0)[r, t]/r + u^(2,0)[r, t])
```

Let us perform a change of variable v = ru.

```
In[2]:= Simplify[
          eqn/.{∂_{r,r}u[r, t] → D[V[r, t]/r, {r, 2}],
             ∂_r u[r, t] → D[v[r, t]/r, r],
             ∂_t u[r, t] → D[v[r, t]/r, t]}
        ]
```

Out[2]= $\dfrac{-v^{(0,1)}[r, t] + D\, v^{(2,0)}[r, t]}{r}$

The transformed PDE becomes:

```
In[3]: = (eqn=Numerator[%]) == 0
```

Out[3]= $-v^{(0,1)}[r, t] + D\, v^{(2,0)}[r, t] == 0$

Assuming the solution can be represented by T(t) R(r), the preceding equations become:

```
In [4]:= eqn = eqn/.(∂_t v[r, t] → T'[t]R[r],
             ∂_{r,r} v[r, t] → T[t]R''[r]}
```

Out [4] = -(R[r] T ' [t]) + D T[t] R ' ' [r]

```
In[5]: = (eqn = Simplify[Expand[eqn/(D R[r]T[t])==0]) == -λ²
```

Out[5] = $(\dfrac{R''[r]}{R[r]} == \dfrac{T'[t]}{D\,T[t]}) == -\lambda^2$

The ODEs can now be solved independently.

```
In[6]:= R[r] =
         R[r]/.
          First[DSolve[eqn[1]] == -λ², R[r], r,
            GeneratedParameters → C1]]
```

Out[6] = C1[1] Cos[r λ] + C1[2] Sin[r λ]

```
In[7]:= T[t]=
        T[t]/.
         First[DSolve[eqn[[2]] == -λ², T[t], t,
          GeneratedParameters → C2]]
```

Out[7] = $\dfrac{C2[1]}{E^{D t \lambda^2}}$

Note that the sign of λ^2 is decided such that the solution to the ODE in the r-direction with homogeneous boundary conditions resulted in a set of orthogonal functions; i.e., sine and cosine. Now, the function u(r,t) can be constructed by multiplying T(t) and R(r):

```
In[8]:= u[r, t] = Factor[1/r
          Expand[R[r]T[t]]/.{C1[1] C2[1]→C1, C1[2] C2[1]→C2}
        ]
```

$$\text{Out[8]} = \frac{\text{C1 Cos}(r\,\lambda) + \text{C2 Sin}[r\,\lambda]}{E^{D\,t\,\lambda^2}\,r}$$

By applying the boundary condition, we determine C_1 to be zero:

```
In[9]:= C1 = C1/.First[Solve[% == 0, C1]/.r → 0]
Out[9] = 0
```

The second boundary conditions will lead us to the eigenvalues:

```
In[10]:= Simplify[(D D[u[r, t], r]/.r→R)==(k u[r, t]/.r→R)]
```

$$\text{Out[10]}= \frac{\text{C2}\,(D\,R\,\lambda\,\text{Cos}[R\,\lambda] - (D + k\,R)\text{Sin}[R\,\lambda])}{E^{D\,t\,\lambda^2}R} = 0$$

```
In[11]:= eigen = Numerator[%[[1]]]/C2 == 0
Out[11]= D R λ Cos[R λ] - (D + k R) Sin[R λ] == 0
```

The preceding equation will result in an infinite number of solutions for λ_is. However, to determine their values, the numerical values for D, k, and R are required. For a given λ_i, the solution will be:

$$\text{In[12]:= } u_i = \text{u[r, t]/C2/. } \lambda \to \lambda_i$$

$$\text{Out[12]}= \frac{\text{Sin}[r\,\lambda_i]}{E^{D\,t\,\lambda_i^2}\,r}$$

The overall solution to the PDE will be the linear superposition of all u_is obtained for all of the eigenvalues:

$$\text{In[13]: = } u[r, t] = \sum_{i=1}^{\infty} C_i u_i;$$

The constants C_is in this function can be now obtained from the initial condition:

In[14]:= **(u[r, t]/.t → 0) == 1 // TraditionalForm**

Out[14]//TraditionalForm =

$$\sum_{i=1}^{\infty} \frac{\sin(r\,\lambda_i)C_i}{r} == 1$$

We multiply the preceding equation by r Sin(r λ_j) and integrate the result over (0,R). Recognizing that the $\int_0^R r\,Sin(\lambda_j r)\,Sin(\lambda_i r)\,dr = 0$ for i ≠ j, we find:

In[15]:= $\int_0^2 \mathbf{C_i Sin[\lambda_i\ r]\,Sin[\lambda_i\ r]dr} == \int_0^2 \mathbf{rSin[\lambda_i r]dr}$

$$\text{Out[15]} = C_i\left(\frac{R}{2} - \frac{Sin[2\,R\,\lambda_i]}{4\,\lambda_i}\right) == \frac{Sin[R\,\lambda_i] - R\,Cos[R\,\lambda_i]\lambda_i}{\lambda_i^2}$$

Therefore, the coefficient C_i becomes:

In[16]:= **C_i=C_i/. First[Solve[%, C_i]]**

$$\text{Out[16]}= \frac{4\,(Sin[R\,\lambda_i] - R\,Cos[R\,\lambda_i]\lambda_i)}{\lambda_i(-Sin[2\,R\,\lambda_i] + 2\,R\,\lambda_i)}$$

All that is left is to insert this into our earlier result for u(r, t), and we get:

In[17]:= **u[r, t] // TraditionalForm**

Out[17]//TraditionalForm =

$$\sum_{i=1}^{\infty} \frac{4e^{-Dta_i^2}\,\sin(r\lambda_i)\,(\sin(R\lambda_i) - R\cos(R\lambda_i)\lambda_i)}{r\lambda_i(2R\lambda_i - \sin(2R\lambda_i))}$$

Now, let us calculate the concentration profile after different time intervals. We start by finding the eigenvalues. But, first we replace λ_i/R by $\lambda_{o,i}$ and assign $\beta = kR/D$:

In[18]:= **eigen = eigen / .R * λ → λ_0/. R * λ → λ_0**

Out[18] = D λ0 Cos[λ0] – (D + k R) Sin[λ0] == 0

In [19]:= **eigen = Simplify[Expand[eigen[[1]]/D] /.kR/D → β] == 0**

Out[19]= λ_0 Cos[λ_0] – (1 + β) Sin[λ_0] == 0

It is a good idea to plot the preceding function before attempting to solve it. Here we plot it for $\beta = 10$:

```
In[20]:= eqn = eigen[[1]] /.β → 10;
         Plot[eqn, {λ0, 0, 30}, PlotRange → All]
```

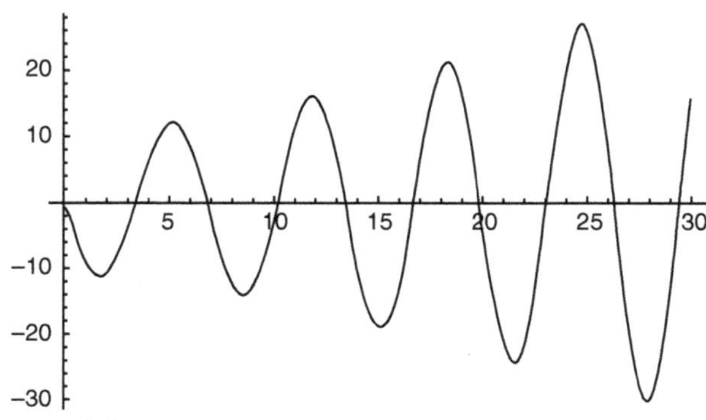


```
Out[21]= -Graphics-
```

Let us determine all the eigenvalues smaller than $\lambda_{o,i} < 1000$ and store them in table "λtab":

```
In[22]:=  β = 10;
          eqn = eigen[[1]];
          astep = 0.5;
          λtab = Drop[
            Union[
              Sort[
               Table[
                 FromDigits[
                  RealDigits[
                   λ0/.
                    Chop[
                     FindRoot[eqn == 0, {λ0, a, astep}]
                     ]
            , 10, 5]
          ],
          {a, 1, 1000, astep}]]] // N
       , 1]
```

Next, we refine the function u(r,t) by transforming the independent variables to dimensionless forms $\zeta = r/R$ and $\tau = Dt/R^2$:

```
In[26]:= Sum[c_i u_i /. λ_i -> λo_i/R /. r -> ξR /. Dt/R^2 -> τ, {i, 1, n}]
```

$$\text{Out[26]} = \text{Sum}\left[\frac{4\,\text{Sin}[\xi\,\lambda o_i]\,[\text{Sin}[\lambda o_i] - \text{Cos}[\lambda o_i]\,\lambda o_i]}{E^{t\,\lambda o_i^2}\,\xi\,\lambda o_i(-\text{Sin}[2\,\lambda o_i] + 2\,\lambda o_i)}, \{i, 1, n\}\right]$$

Therefore, the dimensionless form of the solution becomes:

```
In[27] := un[n_,ξ_, z_]:=
        Sum[(4*(-(Cos[λ0[[i]]]*λ0[[i]])+Sin[λ0[[i]]])*
    Sin[ξ*λ0[[i]]])/(E^(z*λ0[[i]]^2)*ξ*λ0[[i]]*
    (2*λ0[[i]]-Sin[2*λ0[[i]]])),{i,1,n}]
```

We are now ready to plot the solution at four different time intervals:

```
In[28] := λ0 = λtab;
        tau = {0.01,0.05,0.1,0.5};
        Show[
 Table[
  Plot[un[Length[λ0], ξ, tau[[i]]],{ξ, 0.01, 1},
   PlotRange -> All, DisplayFunction -> Identity],
   {i, 1, Length[tau]}],
DisplayFunction -> $DisplayFunction, PlotRange -> All,
Axes -> False,
Frame -> True, TextStyle -> {FontFamily -> "Arial - Bold",
FontSize -> 12},
FrameLabel -> {"ξ = r / R","u(ξ, τ)","Conc. profiles at
at τ = 0.01,0.05,0.1 & 0.5",
""}
]
```

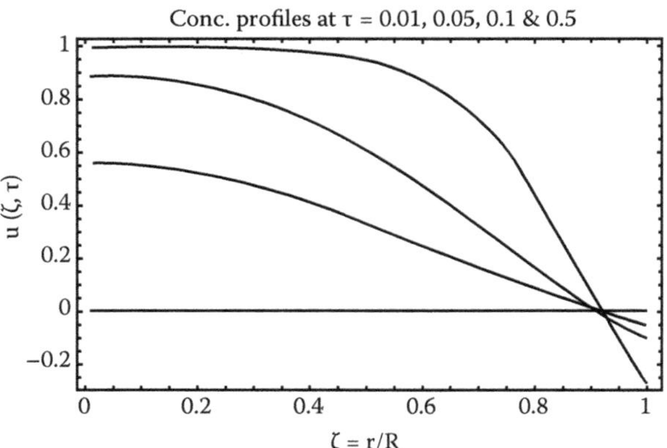

The preceding profiles are for $\tau = 0.01$, 0.05, 0.1, and 0.5; respectively, from top to bottom.

8.2 LAPLACE TRANSFORMATION AND OTHER INTEGRAL TRANSFORMS

8.2.1 General Properties

We have already pointed out in Chapter 2 that the Laplace transform is an important but special case of a larger class of operations termed *integral transformations*. We formalize this operation in the following expression:

$$T\{F(x_j)\} = \int_a^b f(x_j)K(s, x_j)dx_j = f(s) \tag{8.18a}$$

where T is the operational symbol for the transformation, $F(x_j)$ is the function to be transformed, $K(s, x_j)$ is the so-called kernel, f(s) is the transformed function, and s is a free parameter. Let us note a number of features of this operation, some of which have already been mentioned in Chapter 2.

$F(x_j)$ is a general function of the independent variable x_j, and can take the form of an explicit function such as sin x_j, $\exp(-x_j)$, 1, or can consist of implicit functions, such as the dependent variable itself, $y(x_j)$ and its derivatives $y'(x_j)$, $y''(x_j)$, etc. $F(x_j)$ can also depend on several independent variables, only one of which, x_j, is "transformed," i.e., eliminated by integration. One can then write Equation 8.18a in the form:

$$T\{F(x_1 \ldots x_n)\} = \int_0^a F(x_1 \ldots x_n)K(s, x_1 \ldots x_n)dx_j = f(s, x_1 \ldots x_{j-1}\, x_{j+1} \ldots x_n) \tag{8.18b}$$

Transformation of explicit functions of x_j results in explicit functions of s. Thus, the Laplace transform of 1 is 1/s, and of exp(x_j), 1/(s–1). When implicit functions are transformed, a different result is obtained. The dependent variable y(x_j) becomes y(s), often written as $\overline{y}(s)$, i.e., it remains an unknown. Ordinary derivatives become algebraic expressions that incorporate boundary or initial conditions. Thus, the Laplace transform of a first derivative becomes, as we have seen:

$$L\{dy/dt\} = \int_0^\infty (dy/dt)e^{-st}\,dt = sy(s) - y(0) \tag{8.18c}$$

where y(t) is the initial condition. Transforms of partial derivatives in two independent variables can yield two different results. Let us demonstrate this by applying the Laplace transform to the derivatives ∂y/∂t and ∂y/∂x of the variable y(x,t). For the former case, we obtain:

$$L\{\partial y/\partial t\} = \int_0^\infty (\partial y/\partial t)e^{-st}\,dt = sy(x,s) - y(x,0) \tag{8.18d}$$

i.e., the result is the same as for an ordinary derivative, except that the transformed variable and the initial condition are still functions of x. For ∂y/∂x on the other hand, we obtain:

$$L\{\partial y/\partial t\} = \int_0^\infty (\partial y/\partial x)e^{-st}\,dt = \frac{d}{dx}\int_0^\infty y\;e^{-st}\;dt = \frac{dy(x,s)}{dx} \tag{8.18e}$$

i.e., by reversing the order of differentiation and integration we have shown that the partial derivative with respect to the untransformed variable x becomes an ordinary derivative in x. These results apply to integral transforms in general and can be summarized as follows:

a. The transform of a derivative with respect to the independent variable being eliminated yields an algebraic expression that contains boundary and initial conditions.
b. The transform of a derivative with respect to the other independent variable that is not being eliminated yields an ordinary derivative. When there are more than two independent variables, the result (A) still holds, but the ordinary derivative of (B) is replaced by a new partial derivative:

$$T\{\partial y/\partial t\} = \int_a^b \partial y/\partial x_j\;\; K(s,x_i,x_j,x_n)dx_i = \frac{\partial y}{\partial x_j}(x_1 \ldots s \ldots x_n) \tag{8.18f}$$

Integral transforms are thus, in essence, a tool to reduce the number of independent variables. In the case of an ODE, the result is an algebraic equation; in the case of a PDE in two independent variables, the result is an ODE. Finally, when there are n independent variables, a new PDE in (n − 1) variables is obtained.

8.2.2 The Role of the Kernel

The kernel can assume various functional forms, the more common ones being exponential, trigonometric, and Bessel functions. Each of these kernels is capable of transforming a particular derivative, or set of derivatives, to algebraic expressions that incorporate specific boundary and initial conditions. The Laplace transform, for example, transforms partial derivatives of any order to algebraic equations in the initial values of the dependent variables and their derivatives. It is thus the preferred tool for initial value problems. Trigonometric kernels preferentially transform second derivatives to algebraic form, but incorporate boundary values of the dependent variable and its derivative. They are consequently used in boundary value situations. Kernels in various types of Bessel functions transform the group of derivatives $[\partial^2 y/\partial r^2 + (1/r)\partial y/\partial r]$ into algebraic forms containing boundary values. This group is associated with radial diffusion or conduction problems and its transform, termed a *Hankel transform*, finds its principal application in such processes.

We have summarized these and other properties of some common integral transforms in Table 8.2. To help in deciphering the various expressions, we note that defining equations for the sine and cosine transforms are identical in form to the Fourier coefficients given by Equations 8.5e and 8.5f. The defining equations for the Hankel transforms are likewise identical in form to the Fourier coefficients that arise in the corresponding cylinder problems. One may surmise, therefore, that classes of integral transforms can be generated by examining Fourier coefficients and the underlying Sturm–Liouville systems from which they arise. This can in fact be done, as shown in the specialized literature.

The inversion formulae for both trigonometric and Hankel transforms are given in explicit form in terms of an infinite series. These series look suspiciously like the Fourier series we obtained by separation of variables and are, in fact, identical to them in many cases (see Illustrations 8.3 and 8.4). In the case of the Laplace transformation, no such simple inversion formula exists, and the general procedure requires evaluation of the contour integral in the complex plane shown in Table 8.2. This procedure is rarely followed, however, and use is made instead of the extensive tabulations of Laplace transforms available in the literature. A condensed version applicable to PDE problems is shown in Table 8.3, to be used in conjunction with the previous listings of Table 2.9.

The transforms of derivatives given in Table 8.2 dictate the type of transform appropriate to a particular problem. Suppose, for example, that one wishes to solve the unsteady conduction problem for an infinitely long cylinder with a prescribed surface temperature. This is described by the Fourier equation:

$$\frac{\partial T}{\partial t} = \frac{\alpha}{R^2}\left[\frac{\partial^2 T}{\partial y^2} + \frac{1}{y}\frac{\partial T}{\partial y}\right] \tag{8.19a}$$

where the radial variable has been normalized to y = r/R.

TABLE 8.2
Integral Transforms

	1. Laplace Transform
Defining equation	$L\{F(t)\} = \int_0^\infty F(t)e^{-st}\,dt = f(s)$
Inversion formula	$L^{-1}\{f(s)\} = \frac{1}{2\pi}\int_c f(s)e^{st}\,ds$
Transform of derivative	L{F′(t)} = sf − F(0) L{F″(t)} = s^2f − sF(0) − F′(0)
Application	Initial value problems
	2. Finite Fourier Sine Transform
Defining equation	$S\{F(x)\} = \int_0 F(x)\sin(n\pi x/a)dx = f_s(n)$
Inversion formula	$S^{-1}\{f(n)\} = \frac{2}{a}\sum_{n=1}^{\infty} f_s(n)\sin nx$
Transform of derivative	$S\{F''(x)\} = -\frac{n^2\pi^2}{a^2}f_s(n) + \frac{n\pi}{a}[F(0) + (-1)^{u+1}F(1)]$
Application	Type I BCs on slab surfaces
	3. Finite Fourier Cosine Transform
Defining equation	$C\{F(x)\} = \int_0^\pi f(x)\cos(n\pi x/a)dx = f_c(n)$
Inversion formula	$C^{-1}\{f_c(n)\}_c = \frac{1}{a}f_s(0) + \frac{2}{a}\sum_{n=1}^{\infty} f_{c(n)c}(n)\cos(n\pi x/a)$
Transform of derivative	$C\{F''(x)\} = -\frac{n^2\pi^2}{a^2}f_c(n) + (-1)^n F'(a) - F'(0)$
Application	Type II BCs on slab surfaces
	4. Finite Hankel Transform
Defining equation	$H\{F(y)\} = \int_0^1 F(y)yJ_0(\lambda_j y)dy = f_\pi(\lambda_j)$ where λ_j = roots of $J_0(\lambda) = 0$
Inversion formula	$H^{-1}\{f_H(\lambda_j)\} = 2\sum_{j=1}^{\infty}\frac{f_H(\lambda_j)J_0(\lambda_j y)}{J_1^2(\lambda_j)}$
Transform of derivative	$H\left\{F'' + \frac{1}{y}F'\right\} = -\lambda_j^2 f_H(\lambda_j) + \lambda_j J_1(\lambda_j)F(1)$
Application	Type I BCs on cylinder surface

(*continued*)

TABLE 8.2 (Continued)
Integral Transforms

	5. Modified Hankel Transform
Defining equation	$H_M\{F(y)\} = \int_0^1 F(y) y J_0(\lambda_j y) dy = f_{MH}(\lambda_j)$ where λ_j = roots of $\lambda J_1(\lambda) - \beta J_0(\lambda) = 0$
Inversion formula	$H_M^{-1}\{f_{MH}(\lambda_j)\} = 2\sum_{j=1}^{\infty} \frac{f_{MH}(\lambda_j)\lambda_j^2 J_0(\lambda_j y)}{(\lambda_j^2 + \beta^2) J_0^2(\lambda_j)}$
Transform of derivatives	$H\left\{F'' + \frac{1}{y}F'\right\} = -\lambda_j^2 f_{MH}(\lambda_j) + J_0(\lambda_j)[\beta F(1) + F'(1)]$
Application	Type III BCs on cylinder surface

If we use separation of variables, a second-order ODE with variable coefficients results, which leads to a rather lengthy solution procedure. The use of Hankel transforms, on the other hand, leads to a simple first-order ODE, as can be shown by using item 4 of Table 8.2 and the rules for transform of derivatives previously described. We obtain the following ODE in the transformed temperature f_H ($= \overline{T}$):

$$\frac{R^2}{\alpha}\frac{df_H}{dt} = -\lambda_j f_H + \lambda_j J_1(\lambda_j) T(1) \tag{8.19b}$$

where T(1) is the prescribed surface temperature. Equation 8.19b, although somewhat complex in appearance, can be immediately integrated by separation of variables to give a simple exponential expression for $f_H(t)$. That result is then directly substituted into the inversion formula of Table 8.2 to arrive at the solution of the PDE, i.e., T(y, t), where y is, as mentioned, the normalized radial variable. The integral transform is thus quite rapid and straightforward in its application. It has other advantages as well as disadvantages, which are taken up in the following.

8.2.3 Pros and Cons of Integral Transforms

8.2.3.1 Advantages

The method is applicable to a large class of linear, nonhomogeneous first- and second-order PDEs, in particular (using a Cartesian description):

$$\nabla^2 u + ku + f(x,y,z,t) = \frac{\partial u}{\partial t} \tag{8.20a}$$

which includes the Helmholtz, Poisson, Laplace, Fourier, and Fick's equations as subcases, as well as the following

$$\nabla^2 u + ku + f(x,y,z,t) = \frac{\partial^2 u}{\partial t^2} \tag{8.20b}$$

TABLE 8.3
Laplace Transforms for PDEs

f(s)	F(t)
1. p(s)/q(s), where a_i = distinct real roots of q(s) and $\varphi_I = (s - a_i)[p(s)/q(s)]$	$\sum_{j=1}^{\infty} \lim_{s \to a_i} \varphi_i(s) e^{st}$ (a) or $\sum_{j=1}^{\infty} \lim_{s \to a_i} [p(s)/q'(s)] e^{st}$ (b)
2. $\exp(-k\sqrt{s})$	$\frac{k}{2\sqrt{\pi t^3}} \exp[-k^2/4t]$
3. $[\exp(-k\sqrt{s})]/s$	$erfc\left[k/2\sqrt{t}\right]$
4. $\left[\exp\left(-k\sqrt{s}\right)\right]/\sqrt{s}$	$\frac{1}{\sqrt{\pi t}} \exp[-k^2/4t]$
5. $\left[\exp\left(-k\sqrt{s}\right)\right]/s\sqrt{s}$	$2\sqrt{t/\pi}\ \exp[-k^2/4t] - k\ erfc(k/2\sqrt{t})$
6. $[\exp(-k/s)]/s$	$J_0\left(2\sqrt{kt}\right)$
7. $[\exp(-k/s)]/s^n$	$\left(\frac{t}{k}\right)^{(n-1)/2} I_{n-1}\left(2\sqrt{kt}\right)$

which includes the wave equation as a subcase, and

$$\frac{\partial u}{\partial x} + k_1 u + k_2 v = k_3 \frac{\partial u}{\partial t} \tag{8.20c}$$

and

$$\frac{\partial v}{\partial x} + k_3 u + k_4 v = k_5 \frac{\partial v}{\partial t}$$

which includes the linear chromatographic equations as a subcase. Integral transforms, thus, are in principle capable of handling nonhomogeneous terms of any description; in contrast to the method of separation of variables that required homogeneity in both the PDEs as well as the boundary conditions. The restriction to linear systems in both cases arises from the need to apply superposition, and can be traced to their common foundation of Strum–Liouville systems.

Application of the method is quite mechanical in many instances, made so by the use of tables of transforms and explicit inversion formulae. Most boundary and initial conditions are automatically included in the transforms, thus reducing the necessity to evaluate integration constants.

8.2.3.2 Disadvantages

The Laplace transform is generally restricted to initial value situations, and is mainly applied to the transformation of first-order derivatives. The transformed PDE is then often a second-order ODE, which has to then be solved and the solution inverted. This can lead to cumbersome procedures. The transforms based on trigonometric and Bessel functions, on the other hand, are limited to very specific geometries and boundary conditions. Note that item 2 in Table 8.2, for example, is restricted to a slab configuration with Type I BCs. To handle Type III BCs for the same geometry, a completely new transform has to be developed with the help of the Sturm–Liouville theorem.

Nonhomogeneous terms in the PDE cannot be processed unless the relevant integrals have analytical forms. The Hankel transform of a sine forcing function, for example, would require evaluation of the integral:

$$H\{\sin ay\} = \int_0^1 (\sin ay)y\,J_0(\lambda_j y)dy \tag{8.20e}$$

which may not easily yield an analytical form. Because of this difficulty, tabulated expressions of the more unusual transforms and their uses are rather limited. The Laplace transform has remained the most frequently used integral transformation, exactly because of the extensive tabulations available (about 2000). The reader, nevertheless, is encouraged to explore the use of other transforms when confronted with boundary value situations involving nonhomogeneous linear PDEs.

8.2.4 The Laplace Transformation of PDEs

Our main attention in Section 8.2 will be on applications of the Laplace transform, although the use of other transforms will also be illustrated. To aid in this task, we present a short compilation of Laplace transforms which find frequent use in the solution of PDEs (Table 8.3). Some brief comments will be of help:

- Item 1 may be viewed as an extension of the Heaviside expansion given in Chapter 2, Table 2.9. p(s) and q(s) are now arbitrary functions, in lieu of the more restricted polynomials used in the original definition, with q(s) having an infinite number of real or imaginary roots. Typically, q(s) is composed of trigonometric, hyperbolic, or Bessel functions. An application of this inversion formula will be given in Illustration 8.7.
- Items 2 to 5 make their appearance in solutions of Fick's and Fourier's equations, among others, as does item 1.
- Item 7 is typically encountered in the solution of the first-order PDEs, in which they lead to convolution integrals.
- Much more extensive tabulations are, of course, available, for which the reader is referred to the literature.

Illustration 8.7 Inversion of a Ratio of Hyperbolic Functions

We illustrate here the inversion of the ratio $\sinh(x\sqrt{s})/\sinh(\sqrt{s})$ by the extended Heaviside expansion, item 1b of Table 8.3. This requirement arises in the solution of certain diffusion and conduction problems.

We start by noting that a direct application of the inversion formula is not possible, because the roots of the denominator are all imaginary. We craftily circumvent this difficulty by converting the hyperbolic function to a trigonometric one. Making use of the relations we had given in Table 2.6, we obtain:

$$\sinh\sqrt{s} = \frac{1}{i}\sin i\sqrt{s} \tag{8.21a}$$

which has an infinite number of roots at:

$$i\sqrt{s} = \pm n\pi$$

or

$$s = -n^2\pi^2 \tag{8.21b}$$

$$n = 1, 2, 3, \ldots$$

Having obtained these equivalent real roots, we are in a position to apply the inversion formula. We write:

$$L^{-1}\left\{\frac{(s)}{q(s)}\right\} = \sum_{n=1}^{\infty} \lim_{s\to n^2\pi^2} \frac{\sinh x\sqrt{s}}{(\sinh\sqrt{s})'} e^{st} \tag{8.21c}$$

$$= \sum_{n=1}^{\infty} \frac{\frac{1}{i}\sin i x(-i)n\pi \exp(-n^2\pi^2 t)}{-\frac{n\pi i}{2} - \cos i(-1)n\pi} \tag{8.21d}$$

$$= \sum_{n=1}^{\infty} \frac{2}{n\pi}\frac{\sin n\pi x}{\cos n\pi}\exp(-n^2\pi^2 t) \tag{8.21e}$$

and hence,

$$L^{-1}\left\{\frac{\sinh x\sqrt{s}}{\sinh\sqrt{s}}\right\} = \frac{2}{\pi}\sum_{n=1}^{\infty}\frac{(-1)^n}{n}\sin n\pi x \exp(-n^2\pi^2 t) \tag{8.21f}$$

One notes the similarity to the Fourier series solution (Equation 8.9n), which confirms that such inversions arise in conduction and diffusion problems.

Illustration 8.8 Conduction in a Semi-Infinite Medium

To give a simple illustration of the use of the Laplace transform and its short Table 8.3, we consider a semi-infinite medium initially at T_0 with its surface at x = 0 subjected to a lower temperature T_s. The model consists of the following equation:

$$\alpha \frac{\partial^2 T}{\partial x^2} = \frac{\partial T}{\partial t} \tag{8.22a}$$

with boundary and initial temperatures given by:

At the surface $$T(0,t) = T_s \tag{8.22b}$$

At infinity $$T(\infty,t) = \text{bounded} \tag{8.22c}$$

Initially $$T(x,0) = T_0 \tag{8.22d}$$

We normalize the temperature using the new variable:

$$\Theta = \frac{T - T_s}{T_0 - T_s} \tag{8.22e}$$

and obtain the revised model:

$$\alpha \frac{\partial^2 \Theta}{\partial x^2} = \frac{\partial \Theta}{\partial t} \tag{8.22f}$$

$$\Theta(0,t) = 0 \tag{8.22g}$$

$$\Theta(\infty,t) = \text{bounded} \tag{8.22h}$$

$$\Theta(x,0) = 1 \tag{8.22i}$$

We now use the Laplace transform with respect to time t, noting that the left side of Equation 8.22f becomes an ordinary derivative in the transformed variable $\bar{\theta}(x,s)$, while the right side reduces to an algebraic expression. We obtain:

$$\alpha\bar{\theta}''(x,s) = s\bar{\theta}(x,s) - 1$$

or equivalently,

$$\bar{\theta}'' - (5/\alpha)\bar{\theta} = -1/\alpha \tag{8.23a}$$

This is a second-order linear, nonhomogeneous ODE with constant coefficients and can be solved by the standard D-operator method and the use of a particular integral for the nonhomogeneous term. There results:

$$\bar{\theta}(x,s) = A\ \exp\left(-x\sqrt{s/\alpha}\right) + B\ \exp\left(x\sqrt{s/\alpha}\right) + \frac{1}{s} \tag{8.23b}$$

Because $\bar{\theta}(x,s)$ has to remain finite for $x \to \infty$, we obtain $B = 0$, and from the transformed surface condition Equation 8.22g, $A = -1/s$. Hence,

$$\bar{\theta}(x,s) = \frac{1}{s} - \frac{\exp\left(-x\sqrt{s/\alpha}\right)}{s} \tag{8.23c}$$

The first term is inverted by item 3 of Table 2.9, the second term by item 3 of Table 8.3. We obtain:

$$\theta = \frac{T - T_s}{T_0 - T_s} = 1 - erfc\left(x/2\sqrt{\alpha t}\right) = erf\left(x/2\sqrt{\alpha t}\right) \tag{8.23d}$$

We have here a particularly simple use of the transform tables, as well as a simple, terse solution, which is a characteristic of conduction or diffusion problems in a semi-infinite medium. The proceedings are somewhat more complex for finite geometries, as shown in the next illustration.

Illustration 8.9 Conduction in a Slab: Solution for Small Time Constants

It has previously been pointed out that the classical Fourier series solution of the type represented by Equation 8.9n shows slow convergence for low values of the time constant appearing in the exponential term. The Laplace transform provides a means of expressing the solution in terms of an equivalent series of error functions, which shows fast convergence under these same conditions, but is conversely less suitable for large values of time. This equivalent series is obtained by means of an ingenious expansion of the transformed variable, undertaken midway through the solution procedure and prior to the inversion. We illustrate this with the following example.

Consider a slab $-a < x < +a$ initially at $T = 0$, subjected to a surface temperature T_s at time zero. The model for the transients is given by:

$$\alpha \frac{\partial^2 T}{\partial x^2} = \frac{\partial T}{\partial t} \tag{8.24a}$$

$$T(a,t) = T_s \tag{8.24b}$$

$$\frac{\partial \bar{T}}{\partial x}(0,t) = 0 \tag{8.24c}$$

$$T(x,0) = 0 \tag{8.24d}$$

and in its Laplace transformed form by:

$$\frac{d^2\bar{T}}{dx^2} - q^2\,\bar{T} = 0 \tag{8.24e}$$

$$\bar{T}(a,s) = T_s/s \tag{8.24f}$$

$$\frac{\partial \bar{T}}{\partial x}(0,s) = 0 \tag{8.24g}$$

where $q^2 = s/\alpha$, and α denotes thermal diffusivity.

Note that the transform of T_s is T_s/s, not T_s, as is often erroneously assumed. Note also that the symmetry condition Equation 8.28c is used in place of the second surface condition $T(-a,t) = T_s$ for greater convenience. This is not a requirement, but avoids functions with negative arguments.

Solution of Equation 8.28e by the D-operator method, and evaluation of the integration constants yields the transformed temperature:

$$\bar{T}(x,s) = \frac{T_s}{s}\frac{\cosh q\,x}{\cosh qa} \tag{8.25a}$$

We could at this stage proceed to invert, using the technique of expressing cosh in terms of equivalent cosines, outlined in the previous illustration. This would yield a Fourier series similar to Equation 8.21f that converged well for large values of the exponential arguments. To obtain a form suitable for small values, we first rewrite Equation 8.25a in exponential form:

$$\bar{T} = \frac{T_s}{s\,\exp(qa)}\frac{\exp(qx) + \exp(-qx)}{[1 + \exp(-2qa)]} \tag{8.25b}$$

and expand the bracketed term by means of the binomial theorem. We obtain:

$$\bar{T}(x,s) = \frac{T_s}{s}\{\exp[-q(a-x)] + \exp[-q(a-x)]\}\sum_{n=0}^{\infty}(-1)^n\,\exp(-2naq) \tag{8.25c}$$

or alternatively,

$$\bar{T}(x,s) = \frac{T_s}{s}\sum_{n=0}^{\infty}(-1)^n \exp\left\{-q[(2n+1)a - x]\right\} + \frac{T_s}{x}\sum_{n=0}^{\infty}(-1)^n \left\{-q[(2n+a)+x\right\}$$

(8.25d)

Because q = $(s/\alpha)^{1/2}$, each term in the two series has the form $[-\exp(-k\sqrt{s})]/s$, and can therefore be inverted by item 3 of Table 8.3. We obtain upon inversion:

$$\frac{T(x,t)}{T_s} = \sum_{n=0}^{\infty}(-1)^n \, erfc\frac{(2n+1)a - x}{2\sqrt{\alpha t}} + \sum_{n=0}^{\infty}(-1)^n \, erfc\frac{(2n+1)a + x}{2\sqrt{\alpha t}} \quad (8.25e)$$

Had we proceeded directly with the inversion of Equation 8.25a, the result would have been:

$$\frac{T(x,t)}{T_s} = 1 - \frac{4}{\pi}\sum_{n=0}^{\infty}\frac{(-1)^n}{2n+1}\cos\frac{(2n+1)\pi x}{2a}\exp\{-\alpha(2n+1)^2\pi^2 t/4a^2\} \quad (8.25f)$$

Proof of this is left to the Practice Problems.

The reader may at this stage conclude that we have merely replaced an already complex expression with an even more complex one, Equation 8.25e. A numerical example may help dispel this notion.

We choose a low value for the time constant $\alpha t/a^2 = 10^{-2}$, and evaluate T/T_s from both Equation 8.25d and Equation 8.25e for the midpoint of the slab, x = 0. We obtain in the first case:

$$\frac{T}{T_s} = 2\{erfc\ 5 - erfc\ 15 + erfc\ 25 - \cdots] \quad (8.26a)$$

Because these small values of erfc are not tabulated, we use the expansion for large values of the argument given in Table 6.2:

$$erfc\ x \cong \pi^{-1/2}e^{-x^2}\left[\frac{1}{x} - \frac{1}{2x^3} + \cdots\right] \quad (8.26b)$$

This yields:

$$\frac{T}{T_s} = 3.1\times10^{-12} - 1.5\times10^{-98} + \cdots \quad (8.26c)$$

i.e., the temperature at the midpoint is for all practical purposes still at the initial temperature of zero. Convergence of the series is very rapid and practically complete after the first term.

Let us next look at the predictions of Equation 8.25f. Here, the result obtained is:

$$\frac{T}{T_s} = 1 - \frac{4}{\pi}\left[\exp(-10^{-2}\pi^2/4) - \frac{1}{3}\exp(-9\times 10^{-2}\pi^2/4) + \cdots\right]$$

or

$$\frac{T}{T_s} = 1 - [1.2422 + 0.3390 - 0.1374 + 0.0426 - \cdots] \tag{8.26d}$$

One sees that convergence is painfully slow. In addition, the terms would have to be evaluated at least fourteen decimal places to match the result given by the error function series. This demonstrates the power and convenience of the latter, and the total inadequacy of the exponential series for small values of the time constant.

Illustration 8.10 Conduction in a Cylinder Revisited: Use of Hankel Transforms

We have previously (in Illustration 7.8) used the solution for unsteady radial conduction in a cylinder with surface resistance and zero external temperature to mimic radial diffusion in a tubular reactor with a first-order reaction at the wall. The solution given there was taken from the literature and given without proof. We undertake that solution now, using the modified Hankel transform listed in Table 8.3, and examine its advantages compared to other methods.

Using a normalized radial variable y = r/R, the model is given by the following equations:

PDE:

$$\frac{\alpha}{R^2}\left[\frac{\partial^2 T}{\partial y^2} + \frac{1}{y}\frac{\partial T}{\partial y}\right] = \frac{\partial T}{\partial t} \tag{8.27a}$$

At the centerline:

$$T(0,t) = \text{bounded} \tag{8.27b}$$

or

$$\frac{\partial T}{\partial y}(0,t) = 0$$

At the surface:

$$-\left.\frac{\partial T}{\partial y}\right|_{y=1} = \beta T\Big|_{y=1} \tag{8.27c}$$

where

$$\beta = hR/k$$

Initially

$$T(y,0) = T_0 \tag{8.27e}$$

Application of the modified Hankel transform implies that each term in the model is transformed into the integral:

$$H_{M\pi}\{Term\} = \int_0^1 (Term) y J_0(\lambda_j y) dy \tag{8.28a}$$

where λ_j are the roots of:

$$\lambda J_1(\lambda) - \beta J_0(\lambda) = 0 \tag{8.28b}$$

tabulated in Table 7.8. In particular, the left side of the PDE becomes an algebraic expression:

$$H_M \left\{ \frac{\partial^2 T}{\partial y^2} + \frac{1}{y}\frac{\partial T}{\partial y} \right\} = -\lambda_j^2 \bar{T}(\lambda_j) + J_0(\lambda_j) \left[\beta T + \frac{\partial T}{\partial y} \right]_{y=1} \tag{8.28c}$$

where the bracketed term vanishes because of the boundary condition Equation 8.27e. The right side of the PDE (Equation 8.27a) is transformed into the total derivative $d\bar{T}/dt$, where we denote the transformed temperature by $\bar{T}$. The entire PDE is thus reduced to the simple first-order ODE in $\bar{T}$.

$$-\lambda_j^2 (\alpha/R^2)\bar{T} \doteq \frac{d\bar{T}}{dt} \tag{8.28d}$$

with the solution:

$$\bar{T} - \bar{T}_0 \ \exp(-\alpha\lambda_j^2 t/R^2) \tag{8.28e}$$

Here, $\bar{T}_0$ is the transformed initial condition:

$$\bar{T}_0 = \int_0^1 T_0 \, y J_0(\lambda_j) dy = (T_0/\lambda_j) J_1(\lambda_j) \tag{8.28f}$$

in which we have used the tabulations of Bessel function integrals given in Table 4.8 of Chapter 4 to evaluate the integral. The transformed temperature $\bar{T}$ is therefore given by:

$$\bar{T}(\lambda_j, t) = (T_0/\lambda_j) J_1(\lambda_j) \exp(-\alpha\lambda_j t/R^2) \tag{8.28g}$$

We are now ready to invert, and introduce $\bar{T}$ = f_{MH} into the inversion formula given in Table 8.2. The result is:

$$T(y,t) = 2\ T_0 \sum_{j=1}^{\infty} \frac{\lambda_j}{\lambda_j^2 + \beta^2} \frac{J_1(\lambda_j)}{J_0^2(\lambda_j)} J_0(\lambda_j y) \exp(-\lambda_j^2 \alpha t/R^2) \tag{8.28h}$$

Because $J_1(\lambda_j)/J_0(\lambda_j) = \beta/\lambda_j$ (cf. Equation 8.28b), we can also write:

$$\frac{T}{T_0} = 2\sum \frac{\beta}{\lambda_j^{\,2} + \beta^2} \frac{J_1(\lambda_j g)}{J_0(\lambda_j)} \exp(-\lambda_j^{\,2}\alpha t/R^2) \tag{8.28i}$$

which is identical to Equation 8.15a for $T_e = 0$.

Comments: The method obviously requires some getting used to, given the wealth of new symbols. Once these are accepted and understood, however, the transform reveals itself as a compact tool for solving problems in cylindrical coordinates with a Type III boundary condition. Application becomes quite mechanical, but one still has to keep a watchful eye on potential pitfalls. There is a temptation, for example, to set the transform of the initial condition equal to the condition itself: $\overline{T}_o = T_0$, instead of evaluating the full integral (Equation 8.28f). With some care, such errors can be avoided.

The Hankel transform was shown capable of reducing a second-order PDE with variable coefficients into a simple, separable first-order ODE (Equation 8.28d). The Laplace transform, by contrast, would only succeed in reducing the PDE to a second-order ODE with variable coefficients, which has to be solved and, more importantly, inverted. This is a cumbersome process. The method of separation of variables is similarly unwieldy, as shown in Illustration 8.4. Both methods lead to valid results, however, and are attractive because of their greater range of applications.

Illustration 8.11 Analysis in the Laplace Domain: The Method of Moments

We introduce the reader here to an analysis applied to a PDE that is designed specifically for use in parameter estimation from experimental data. The procedure does not require an inversion to be carried out, but relies instead entirely on information obtained from the transformed PDEs. We term this *Laplace domain analysis*, a designation that will be familiar from studies of process control in which it occupies an important place.

The example we use to illustrate the procedure is that of linear chromatography, previously described by the fluid and solid mass balances, Equations 5.12e and 5.12f. For our present purposes, we combine the two equations into a single two-phase mass balance and retain the solid-phase balance as a second relation. Thus:
Two-phase mass balance:

$$G_s \frac{\partial Y}{\partial z} + \rho_G \frac{\partial Y}{\partial t} + \rho_b \frac{\partial q}{\partial t} = 0 \tag{8.29a}$$

Solid-phase mass balance:

$$K_{0S}a(q^* - q) = \rho_b \frac{\partial q}{\partial t} \tag{8.29b}$$

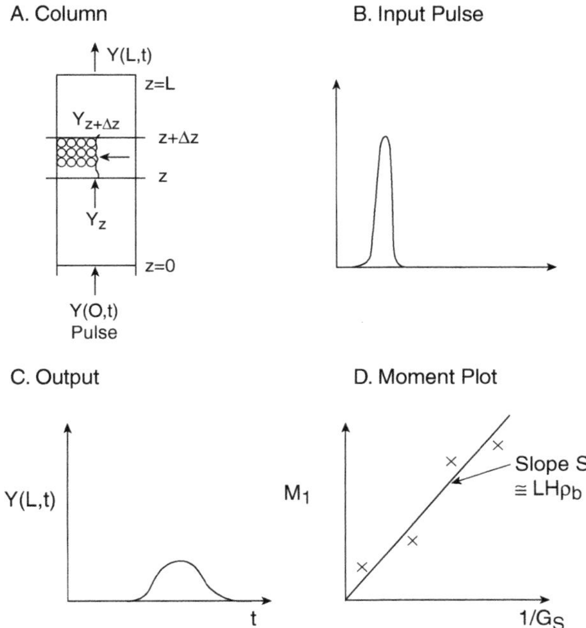

FIGURE 8.3 Parameter estimation from chromatographic data by the method of moments.

Equilibrium relation:

$$q^* = HY \tag{8.29c}$$

We note that the fluid-phase driving force (Y – Y*) of Equation 5.12f has been replaced by a solid driving force (q* – q) that is better suited for the extraction of parameters of interest.

Let us look at both the experimental and mathematical sides of the procedure. The experiment consists of introducing a solute pulse Y(0,t) into a chromatographic column (Figure 8.3A and Figure 8.3B). The column is initially clean, i.e., q(z,0) = Y(z,0) = 0. The solute at first becomes adsorbed near the inlet of the column and is subsequently eluted by continued purge with carrier gas, making its appearance at the outlet as an attenuated pulse Y(L,t), Figure 8.3C. Attenuation is because of the interphase mass transfer resistance residing in the transfer coefficient K_{0S}.

On the mathematical side, we introduce the equilibrium relation (Equation 8.28c) into the solid-phase mass balance, and then the Laplace transform Equations, 8.28a and 8.28b with respect to time t. We obtain

$$G_s \frac{d\overline{Y}}{dz} + \rho_G s\overline{Y}\rho_b s\overline{q} = 0 \tag{8.29d}$$

$$K_{0S}a(H\overline{Y} - \overline{q}) = \rho_b s\overline{q} \tag{8.29e}$$

where $\bar{Y}$ and $\bar{q}$ denote the transformed fluid and solid-phase concentration respectively. Note that no initial conditions appear in these expressions because the column is clean at time t = 0.

Eliminating $\bar{q}$ algebraically, we obtain the first-order ODE in $\bar{Y}$:

$$G_s \frac{d\bar{Y}}{dz} + P(s)\,\bar{Y} = 0 \tag{8.29f}$$

which is integrated by separation of variables to yield:

$$\bar{Y} = \bar{Y}(0,s)\exp(-Pz/G_s) \tag{8.29g}$$

Here, $P(s) = s[(\varepsilon\rho_G + HK^a_{0S})/(s + K^a_{0S}\rho_b)]$ and $\bar{Y}(0,s)$ is the transformed input pulse at z = 0, i.e., the boundary condition for the ODE (Equation 8.29f). When that pulse is instantaneous, that is a Dirac function $\delta(0)$, its Laplace transform is 1 (see item 6 in Table 7.6), i.e., we can set $\bar{Y}(0,s) = 1$.

The question now arises as to how we can relate the measured output Y(L,t) to the parameters contained in our transformed PDE, Equation 8.29g. An ingenious method has been developed to achieve this: We reach back to Table 2.9 of Laplace transforms and extract item 9, which relates the derivative of the transformed variable to the time integral of the variable itself. We have:

$$\frac{d^n\bar{Y}}{ds^n} = (-1)^n \int_0^\infty t^n Y(t) e^{-st}\,dt \tag{8.29h}$$

and in particular:

$$\left.\frac{d^n\bar{Y}}{ds^n}\right|_{s=0} = (-1)^n \int_1^\infty t^n Y(t)\,dt \tag{8.29i}$$

where the integral on the right is referred to as the *n-th moment of Y(t)*.

The reader will note that this expression establishes the desired link between experiment and model. The left side can be established by differentiating the transformed PDE (Equation 8.29g) with respect to s, and contains the physical parameters of the system, such as the partition coefficient or Henry's constant H and the mass transfer coefficient K_{0C} a. The right side represents various time integrals of the measured outlet concentration Y(L,t). For n = 0, for example, the integral represents the area under the curve of Figure 8.3C. For higher-order moments, the output concentration is multiplied by t^n and then integrated. The existence of n such moments gives us the ability, in principle, to extract n physical parameters. The precision of the moments, however, quickly diminishes with increasing n, and in practice one is therefore confined to two or three such parameter determinations.

Suppose we wish to extract the Henry's constant H from experimental output data. We proceed as follows: We normalize the moments by dividing them by the zero-th moment. This leads to simpler expressions and a cancellation of errors. We obtain, for the first normalized moment M_1:

$$M_1 = -\frac{\int_0^\infty tY(t)dt}{\int_0^\infty Y(t)dt} = \left[\frac{d\bar{Y}}{ds} / \bar{Y}\right]_{s=0} = \frac{L}{G_s}(\varepsilon\rho_G + H\rho_b) \approx \frac{L}{G_s} H\rho_b \tag{8.29j}$$

Similar expressions containing additional parameters can be obtained from the higher moments. Note again that the left side of Equation 8.29j represents the experiment and the right side represents the model. The first term of the latter is usually small compared to the product $H\rho_b$ and can be neglected.

In practice, it will be desirable to perform a series of experiments to obtain the best parameter value by a least square fit. This can be done by running the column at different carrier flow rates G_s, or by using columns of different lengths. In the former case, a plot of the first moment M_1 against $1/G_s$ should yield a straight line with a slope $S \cong LH\rho_b$ from which the Henry constant H can be extracted (Figure 8.3D).

Comments: We have here an example of the intertwining of imaginative modeling and experimentation. The accurate determination of Henry's constant by conventional equilibration involves a considerable expenditure in time and equipment. The procedure described here uses standard equipment available in most laboratories, and can be carried out rapidly with relatively little effort. Electronic integrators are available for the evaluation of moments from the experimental output.

Extraction of the parameters from the data and model can, in principle, also be achieved by equating the theoretical transform $\bar{Y}$ to the Laplace transform integral of the experimental output, i.e., we write:

$$\bar{Y} = \bar{Y}(0,s)\exp(-P_z/G_s) = \int_0^\infty Y(L,t)e^{-st}dt \tag{8.29k}$$

and evaluate the integral for various values of the Laplace parameters, using the results to evaluate the parameters contained in P(s). This is evidently much more cumbersome than the method of moments, which ultimately leads to a simple linear plot from which the parameters are extracted with ease.

Historical Note

Early work on matters related to the Laplace transform was undertaken by the French mathematician Pierre Simon de Laplace (1749–1827), who investigated the properties of integrals of the form $\int_0^\infty F(t)e^{-st}$. Its application to the solution of differential equations had to await the work of the English electrical engineer Oliver Heaviside (1850–1925). He invented for that purpose what is now known as the *Heaviside operational calculus* and used it to solve a host of practical problems related particularly to electrical systems. His work was not immediately accepted, partly because of the difficult nature of his operational calculus, and also because of his lack of rigor, for which he was much derided by the mathematical community. It was left to other workers to replace his calculus by the simple procedures which are now everyday tools used in the solution of differential equations. It was felt appropriate to name the method after Laplace for his early investigations of the relevant integral. Heaviside himself died a bitter recluse.

Illustration 8.12 Conduction in a Semi-Infinite Solid Using Mathematica

We intend to find a solution to the one dimensional Fourier equation using the Laplace transformation technique with the aid of Mathematica:

$$\alpha \frac{\partial^2 T}{\partial x^2} = \frac{\partial T}{\partial t} \tag{8.30a}$$

The boundary and initial conditions are:

$$T(x,0) = 0 \tag{8.30b}$$

$$T(\infty,t) = 0 \tag{8.30c}$$

$$\left.\frac{\partial T}{\partial x}\right|_{x=0} = -1 \tag{8.30d}$$

Let us begin by taking the Laplace transformation of the PDE:

```
In[1]:= eqn = LaplaceTransform[a∂x,x T[x, t]-∂t T[x, t]==0, t, s]

Out[1] = -(s LaplaceTransform [T [x, t] , t, s] ) +
        α LaplaceTransform [T^(2,0) [x, t] , t, s] + T [x, 0] == 0
```

Now we denote the L{T} by q, and note that $L\left\{\frac{\partial^2 T}{\partial x^2}\right\} = \frac{\partial^2}{\partial x^2} L\{T\}$:

```
In[2]:= eqn/. {T[x, 0]→0, LaplaceTransform[T[x, t], t, s]→θ[x],
  LaplaceTransform[Derivative[2, 0][T][x, t], t, s] → θ''[x]}

Out[2] = -(s θ[x]) + α θ''[x] == 0
```

The above ODE can be solved readily using the `Solve` function:

```
In[3]: = ans = θ[x]/.First[DSove[%, θ[x], x]]
```

$$\text{Out[3]} = E^{(\text{Sqrt[s]x})/\text{Sqrt}[\alpha]}\ \text{C[1]} + \frac{\text{C[2]}}{E^{(\text{Sqrt[s] x})/\text{Sqrt}[\alpha]}}$$

Integration constants `C[1]` and `C[2]` can be now determined from boundary conditions:

```
In[4]:= Solve[(ans/.x → ∞) == 0, C[1]]//TraditionalForm
```

$$\text{Out[4]//TraditionalForm} = \left\{\left\{c_1 \to -e^{\frac{\sqrt[-\infty]{\text{sgn}(s)}}{\sqrt{\text{sgn}(\alpha)}}}\ c_2\right\}\right\}$$

As the reader may have already noticed, because s and a are unknown to Mathematica, the numerical value for `C[1]` is not determined. We can help Mathematica by assigning any finite positive real values to these parameters (this does not affect the generality of the solution):

```
In[5]:= C1 = C[1]/. First[Solve[(ans/.
            {x → ∞, α → 1, s → 1}) == 0, C[1]]]
Out[5] = 0
```

and for `C[2]`:

```
In[6] := C2 = C[2]/.First[Solve[(D[ans, x]/.
            {x → 0, C[1] → 0}) == -1/s, C[2]]]
```

$$\text{Out[6]} = \frac{\text{Sqrt}[\alpha]}{S^{3/2}}$$

The solution in the Laplace space then becomes:

```
In[7]: = ans/.{C[1] → C1, C[2] → C2}
```

$$\text{Out[7]} = \frac{\text{Sqrt}[\alpha]}{E^{(\text{Sqrt[s] x})/\text{Sqrt}[\alpha]}S^{3/2}}$$

Now we use the `InverseLaplaceTransform` function to obtain the desired solution:

```
In[8]:= InverseLaplaceTransform[%, s, t] // TraditionalForm
```

Out[8]//TraditionalForm =

$$\frac{\sqrt{\frac{1}{t}}\sqrt{t}x\left(\frac{4e^{-\frac{\pi^2}{4t\alpha}}}{\sqrt{\frac{\pi^2}{t\alpha}}}-2\sqrt{\pi}\,\mathrm{erfc}\left(\frac{1}{2}\sqrt{\frac{\pi^2}{t\alpha}}\right)\right)}{2\sqrt{\pi}}$$

Let us tidy up the answer:

```
In[9]: = Simplify[PowerExpand[Simplify[Refine[%, t > 0 && x
        0]]], α > 0 && t   0] // TraditionalForm
```

Out [9]//TraditionalForm =

$$\frac{2e^{-\frac{\pi^2}{4t\alpha}}\sqrt{t\alpha}}{\sqrt{\pi}}-x\,\mathrm{erfc}\left(\frac{x}{2\sqrt{t\alpha}}\right)$$

Let us now plot this function; but before doing this, we define $z = x/\sqrt{\alpha t}$. Therefore, the solution becomes:

```
In[10] := Refine[Simplify[%/Sqrt[tα]] /.(x → z Sqrt[tα])]
```

Out[10] =

$$\frac{2}{E^{z^{2}/4}\,\mathrm{Sqrt[Pi]}}-z\mathrm{Erfc}\left[\frac{z}{2}\right]$$

```
In[11]: = Plot[%, {z, 0, 10}, PlotRange → All, Frame → True,
        FrameLabel → {" x/√αt ", "T/√αt", "", ""}]
```

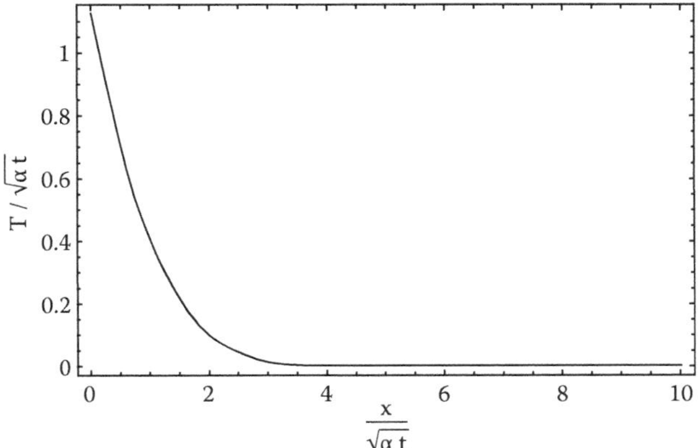

8.3 THE METHOD OF CHARACTERISTICS

8.3.1 General Properties

Similar to the two preceding methods of separation of variables and integral transforms, the method of characteristics achieves its goal of simplification by reducing the PDE or system of PDEs to an equivalent set of ODEs. This is the only common feature of the three methods. In all other respects, they differ both in concept as well as in details of applications. In the following, we summarize the principal features and properties of the method.

The principal area of application of the method of characteristics is the solution of:

- Single first-order PDEs of arbitrary form.
- Systems of hyperbolic PDEs of otherwise arbitrary form.
- Single second-order hyperbolic PDEs, e.g., the wave equation.

The term arbitrary form encompasses both linear, quasilinear, and fully nonlinear PDEs. We note, however, that full theories have been developed only for systems of quasilinear PDEs and, to a lesser extent, single nonlinear PDEs. Sets of fully nonlinear PDEs still elude complete treatment, as they do at the ODE level.

A particularly rich area of application is that of certain homogeneous quasilinear first-order PDEs, in which one or more PDEs are combined with one or more auxiliary algebraic equations, i.e., those of the form:

$$\frac{\partial u}{\partial x}+\frac{\partial v}{\partial y}=0 \tag{8.31a}$$

$$v=f(u) \tag{8.31b}$$

The PDE is in this instance termed *reducible* because it lacks algebraic terms in the dependent variables. It is homogeneous because it lacks isolated terms f(x,y), and quasilinear by virtue of the linear appearance of its highest derivative.

We know from previous examples that the combination (Equation 8.31a) arises in a natural way in all unsteady, convective transport processes, i.e., those lacking second-order diffusive terms. We noted in Chapter 5 its appearance in equilibrium chromatography:

$$G_s \frac{\partial Y}{\partial x} + [\varepsilon\rho_g + \rho_b f'(Y)]\frac{\partial Y}{\partial t} = 0 \tag{8.31c}$$

$$q = f(Y) \tag{8.31d}$$

in traffic problems:

$$\frac{\partial q'}{\partial x} + \frac{\partial C}{\partial t} = 0 \tag{8.31e}$$

$$q' = vC = g(C) \tag{8.31f}$$

and in sedimentation:

$$\frac{\partial (vC)}{\partial z} + \frac{\partial C}{\partial t} = 0 \tag{8.31g}$$

$$v = h(C) \tag{8.31h}$$

Note that when a mass transfer resistance is included in Equation 8.31a, the equation is no longer reducible, and its analysis becomes correspondingly more complex. The treatment given in this introductory chapter will principally deal with single quasilinear first-order PDEs of the reducible form.

The reduction of the PDEs to ODE form is achieved by adopting a Lagrangian approach, i.e., instead of using a fixed Eulerian reference framework, we move with the physical entity such as concentration, temperature, or a vehicle, and establish its trajectory or path of propagation. In other words, we replace the previous variables, e.g., x, y, z, t, by a single independent variable s, taken along the path of propagation. This concept, and the consequences that flow from it, require some getting used to, but provides us with rich benefits.

To provide the reader with a tangible example of the application of the method, we consider the unidirectional movement of vehicular traffic, depicted in the diagrams of Figure 8.4. Both normal traffic as well as conditions leading to rear-end collisions are examined. The Eulerian representation is shown in Figure IA and Figure IIA as the "velocity profiles" of the cars, i.e., the velocity as a function of distance. The Lagrangian representation on the other hand, Figure 8.4 (IB, IIB), utilizes the z-t plane to trace the trajectory of each car, i.e., its position z at a particular time t. The top figures describe the movement of three cars, all traveling at constant but different velocities, and the car with the highest velocity being the furthest advanced. The bottom diagrams depict the reverse situation, in which the slower car ② is ahead of the faster-moving vehicle ①. If no evasive action is taken by either car, i.e., if we limit ourselves to a single coordinate direction z, the slower car will undergo a rear-end collision with the approaching faster car, and the velocity will drop essentially to zero. There will consequently be a discontinuity in the velocity, a condition that we shall refer to as a *shock*. The Eulerian representation depicts this situation, as well as that for normal traffic, in terms we are accustomed to, i.e., profiles of a state variable, here the vehicle velocity v. The Lagrangian representation, on the other hand, utilizes the velocity as a parameter that equals the inverse of the slope of each trajectory, termed a *characteristic*. Slow cars have steep characteristics, fast cars have shallower trajectories. If the vehicle velocity is constant, these trajectories will be straight lines of slope v^{-1}.

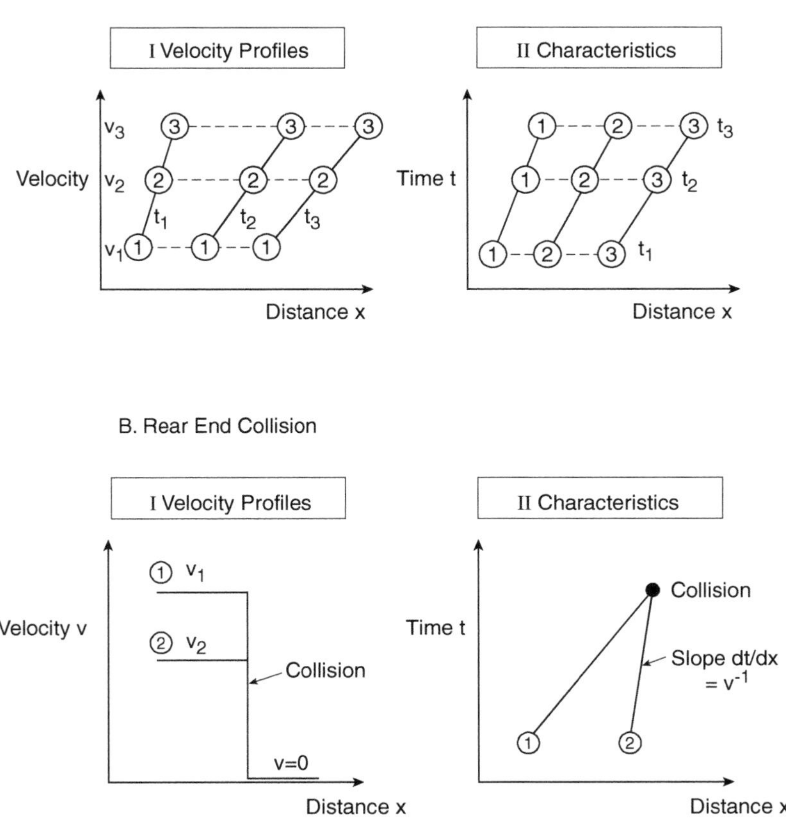

FIGURE 8.4 Vehicle movement represented in the v-x and the t-x (characteristic) planes.

Let us now examine the mathematical formulation of these characteristics and present a summary of the various types encountered in practice.

8.3.2 The Characteristics

If we consider two independent variables only, for example, distance z and time t, and limit ourselves to a single state variable u, a first-order quaslinear PDE takes on the general form:

$$A(x,t,u)\frac{\partial u}{\partial x} + B(x,t,u)\frac{\partial u}{\partial t} = C(x,t,u) \tag{8.32a}$$

Because quasilinearity is assumed here, this requires that A, B, and C be free of derivatives of u, but does permit the presence of nonlinear terms in u in these coefficients.

The transformation of this PDE into an equivalent set of ODEs is often based on the notion of a directional derivative, encountered in basic calculus. To arrive at an expression for this derivative, we start with the total differential of u:

$$dx\frac{\partial u}{\partial x} + dt\frac{\partial u}{\partial s} = du \tag{8.32b}$$

and after dividing by ds, we obtain:

$$\frac{dx}{ds}\frac{\partial u}{\partial x} + \frac{dt}{ds}\frac{\partial u}{\partial t} = \frac{du}{ds} \tag{8.32c}$$

where ds may be viewed as the differential arc along a characteristic. Comparing Equation 8.32a and Equation 8.32c, we see that the two expressions will be equivalent, provided the following ODEs are satisfied:

$$\frac{dx}{ds} = A(x,t,u) \tag{8.32d}$$

$$\frac{dt}{ds} = B(x,t,u) \tag{8.32e}$$

$$\frac{du}{ds} = C(x,t,u) \tag{8.32f}$$

Many textbooks use more sophisticated arguments to arrive at these expressions, but the net result in each case is that the original PDE (Equation 8.32a) has been transformed into an equivalent system of three ODEs in what are now the dependent variables x, t, and u. Because the arc length s is ultimately redundant to the solution, we may, as an alternative, eliminate ds by division, reducing the system to two simultaneous ODEs in the dependent variables x and u

Velocity of propagation
$$\left(\frac{dx}{dt}\right)_c = \frac{A(x,t,u)}{B(x,t,u)} \tag{8.32g}$$

State variable
$$\left(\frac{du}{dt}\right)_c = \frac{C(x,t,u)}{B(x,t,u)} \tag{8.32h}$$

The subscript c is used as a reminder that the derivatives are taken along a characteristic. Either set of equations may be integrated by standard ODE solution methods. The numerical procedure used in these cases is referred to as the *method of lines.*

When C = 0, the PDE (Equation 8.32a) becomes reducible. If, in addition, the coefficients A and B are functions of the state variable u only, we obtain as a special case:

Velocity of propagation
$$\left(\frac{dx}{dt}\right)_c = f(n) \tag{8.32c}$$

This case arises in many applications of equilibrium chromatography, traffic theory, sedimentation, and other processes, and can be analyzed in a particularly fruitful manner. This will be shown in several of the illustrations that follow.

We end this section by summarizing certain important categories of characteristics that arise in practice. They are displayed in Figure 8.5.

Figure 8.5A consists of parallel characteristics of equal slope, i.e., of equal velocities of propagation. This is representative of all vehicles in a traffic problem moving at the same speed, or a fixed constant concentration being fed to a chromatographic column, and is termed a *constant state.*

In Figure 8.5B, the velocity along a characteristic is constant but varies among different entities. This state is referred to as a *simple wave*. An important subcategory is the so-called centered simple wave, shown in Figure 8.5C. It may be thought of, for example, as representing a range of concentrations or temperatures emanating from a particular point in time and space.

Figure 8.5D, finally, represents the most general case of velocities of propagation that vary among physical entities or with time. This state is referred to as a *complex wave* and leads to curved characteristics.

Let us start our illustrations with a simple example that admits a closed-form solution of the state variable.

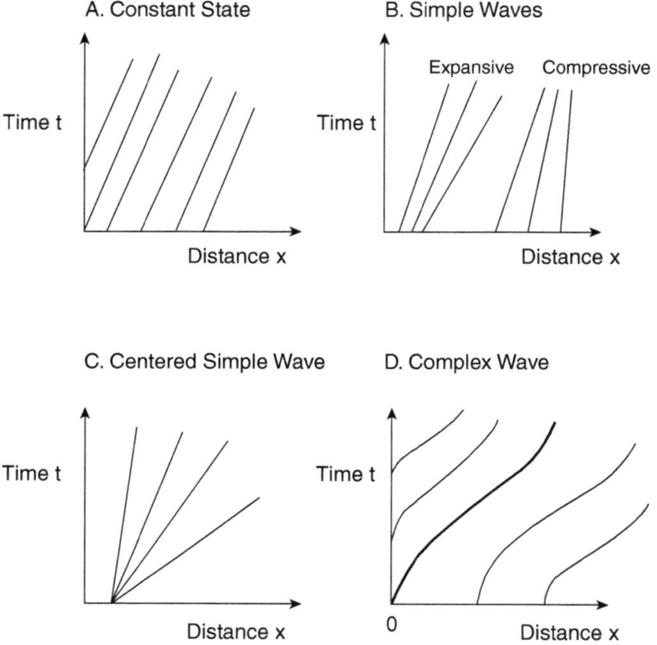

FIGURE 8.5 Types of characteristics.

Illustration 8.13 The Heat Exchanger with a Time-Varying Fluid Velocity

The case considered here is that of a single-pass steam-heated shell-and-tube heat exchanger. The fluid being heated, assumed to be on the tube side, has a time-varying inlet velocity v(t), which will also affect the heat transfer coefficient h(t), as the latter generally depends on Reynolds number.

The relevant model is given by the following first-order linear PDE:

$$v(t)\rho(\pi D^2/4)\frac{\partial T}{\partial x} + U(t)(\pi d)[T_s - T] = \rho Cp\frac{\pi d^2}{4}\frac{\partial T}{\partial t} \tag{8.33a}$$

or alternatively:

$$-v(t)\frac{\partial T}{\partial t} + K(t)[T_s - T] = \frac{\partial T}{\partial t} \tag{8.33b}$$

where $K(t) = 4/d\rho Cp.$

By casting the PDE in the form of Equation 8.32a and applying the relations (Equations, 8.32d, 8.32e, and 8.32f), we obtain the following characteristic equations:

$$\frac{dt}{ds} = 1 \tag{8.33c}$$

$$\frac{dx}{ds} = v(t) \tag{8.33d}$$

$$\frac{dT}{ds} = K(t)[T_s - T] \tag{8.33e}$$

These ODEs can be solved numerically to obtain a relation between T, x, and t at a particular point s along the characteristic. Alternatively, we can arrive at analytical forms by eliminating ds by division and integrating the result. We obtain in the first instance:

Velocity of propagation $$\frac{dx}{dt} = v(t) \tag{8.33f}$$

State variable $$\frac{dT}{dt} = K(t)[T_s - T] \tag{8.33g}$$

The first equation can be formally integrated by separation of variables. The solution to the second ODE is given by item 6 of our listing of ODE solutions, Table 2.4. The result is given by the two expressions:

$$x = x_0 + \int_{t_0}^{t} v(t')dt' \tag{8.33h}$$

and

$$T(t,t_0) = T(t_0)\exp(-\bar{K}) + \int_{t_0}^{t} T_s K(t')\exp[-\bar{K}(t')]dt'' \tag{8.33i}$$

where $\bar{K} = \int_{t_0}^{t} K(t')dt'$.

Note that the resultant characteristics form a complex wave, shown in Figure 8.5D, because the slope dt/dx = 1/v(t), i.e., it varies with time. A distinction is now made between the characteristics emanating from the x-axis, and those originating on the t-axis. The former describe the propagation of the temperature distribution T(x,0) = f(x) initially present in the heat exchanger, whereas the latter represent the pathways of the incoming feed temperature. For the characteristics emanating from the abscissa, Equation 8.33g and Equation 8.33h become:

$$x_0 = x - \int_0^t v(t')dt' \tag{8.33j}$$

$$T(x,t) = f\left[x - \int_0^t v(t')dt'\right]\exp(-\bar{K}) + \int_0^t T_s K(t')\exp[-\bar{K}(t')]dt \tag{8.33k}$$

It is left to the exercises to derive the corresponding expressions for the characteristics emanating from the ordinate.

Comments: Equation 8.33k, although somewhat cumbersome, represents a closed-form expression for the unsteady temperature distribution of the fluid initially present in the exchanger. These solutions are therefore valid only during the initial period of displacement, $t_d = L/\bar{v}$, where L represents the heat exchanger length, and $\bar{v}$ is the mean integral inlet velocity over the period t_d.

The reader should note that the deviations from the usual steady state profiles produced by this model are to be viewed as maximum values. In actual practice, the temperature peaks and valleys produced by the velocity fluctuations will be attenuated because of the heat capacity of the tubular wall. To take account of this effect, however, would require a second energy balance, thus complicating the model considerably.

Illustration 8.14 Saturation of a Chromatographic Column

The present illustration and that which follows deal with the two simplest and most common chromatographic or sorption operations. We consider, in the first instance, the saturation of a clean bed with a feed of constant solute concentration, and follow this up with the purge of a uniformly loaded column with pure carrier fluid or solvent. The latter process is alternatively termed *elution* or *desorption*.

The saturation step that appears to be the simpler of the two does, in fact, require special treatment when one applies the method of characteristics. We have already introduced the reader to the intuitive notion that, in the absence of an interphase transport resistance, instantaneous equilibrium is established between the fluid and solid phases, and the solute penetrates the bed in the form of a rectangular discontinuity. We now reexamine this phenomenon in a more thorough fashion within the framework of the method of characteristics. The operative model is represented by Equations 8.31c and 8.31d, which upon elimination of the solid-phase concentration q, lead to the single expression:

$$\frac{\partial Y}{\partial x} + \left[\varepsilon\rho_G/G_s + (\rho_b/G_s)f'(Y)\right]\frac{\partial Y}{\partial t} = 0 \tag{8.34a}$$

with boundary and initial conditions:

$$\text{Feed } Y(0,t) = Y_F \tag{8.34b}$$

$$\text{Clean bed } Y(x,0) = 0 \tag{8.34c}$$

Here, f′(Y) is the derivative or slope of the equilibrium relation q = f(Y). Comparison of this expression with Equation 8.32g shows that the bracketed term equals the inverse of the propagation velocity, i.e.,

$$[\text{Propagation Velocity}]^{-1} = \left(\frac{dt}{dx}\right)_c = \varepsilon\rho_G/G_x + (\rho_b/G_s)f'(Y) \tag{8.34d}$$

We note that in practice, the fluid-phase accumulation term ρ_G can be neglected compared to its solid-phase counterpart, so that:

$$\left(\frac{dt}{dx}\right)_c \doteq \frac{\rho_b}{G_s} f'(Y) \tag{8.34e}$$

where $(dt/dx)_c$ is the slope of the characteristics, shown in Figure 8.6. For Langmuir-type equilibria, also termed *Type I isotherms*, the slope of the equilibrium curve f′(Y) decreases with increasing values of Y. Consequently, the slopes of the characteristics themselves will be high for low solute concentrations, and decrease with an increase in concentration. This is reflected in the plots shown in Figure 8.6. Note that all characteristics are straight lines for a given solute concentration Y, i.e., for constant values of f′(Y). We term this situation a *constant state*.

Let us now examine these diagrams in more detail. Figure 8.6 (AI) shows straight lines emanating from the abscissa, which describe the propagation of the initial (clean) bed condition. Similarly, the characteristics starting from the ordinate represent the pathways of the incoming feed concentration Y_F. The latter has the lower slope because of the higher value of $Y_F > 0$.

A special situation arises at the origin representing the inlet at time t = 0. Here, characteristics for both $Y = Y_F$ and Y = 0 must perforce emanate, and because the space between them cannot be left void, we must have a continuous spectrum of concentrations between those two limits, propagating at different but constant velocities. This produces a simple wave centered at the origin.

The structure of these three sets of characteristics leads to anomalies, which are depicted in the initial model of Figure 8.6A. Because the higher concentrations of the simple centered waves have the lower slope, they propagate faster then their lower-concentration cousins. This leads to an "overhanging profile" of the type shown in Figure 8.6 (IB), and is unacceptable on physical grounds. A second anomaly arises from the intersection of three characteristics at a single point P in the single wave region. This, in turn, implies the coexistence of three distinct concentrations carried by these characteristics at the same point in time and space. We have represented this situation by the three concentration levels PO, PP′, and PP″ in Figure 8.6 (IB). Clearly, such a multiplicity of solutions is as unacceptable on physical grounds as the overhanging profile. The only way to overcome these twin anomalies

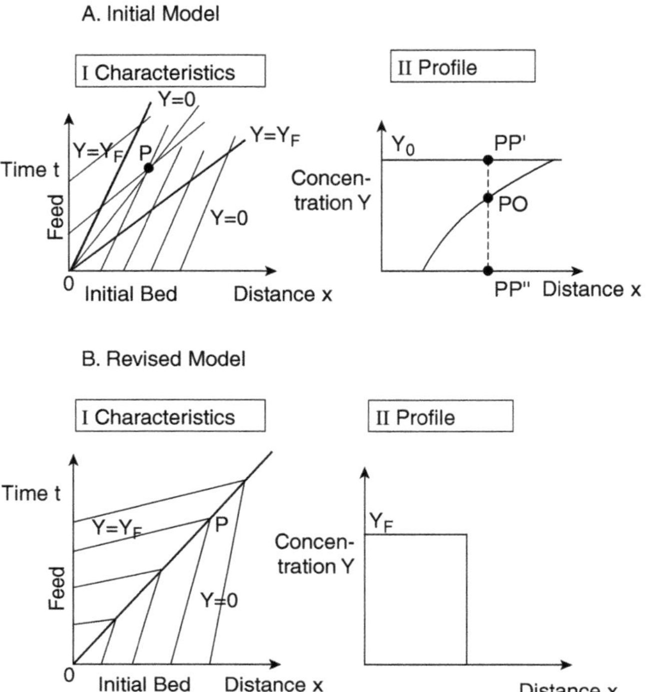

FIGURE 8.6 Characteristic diagrams and the resulting concentration profiles for the saturation of a clean chromatographic column or adsorber.

is to introduce the notion of a discontinuous front that disposes of the overhang and eliminates the multiplicities at one and the same time. The characteristics through the origin are then reduced to a single pathway OP termed the *shock path*.

A consequence of the model revision is that the movement of the discontinuity is no longer described by the PDE (Equation 8.34a). We must abandon that equation and replace it instead by a cumulative algebraic mass balance. This has already been done in Chapter 4, and we repeat the result using appropriately revised symbols:

$$\frac{x}{t} = \frac{\rho_g v}{\rho_b (q/Y)_F} \tag{8.35a}$$

where q_F is the solid-phase concentration in equilibrium with the feed Y_F.

Comments: We start by noting that the development given here benefits considerably from the fact that the mass transfer resistance was neglected. This enables us to combine the two differential balances that would otherwise have arisen into the single PDE (Equation 8.34a). That equation, furthermore, is of the reducible type, which leads to the immediate conversion into a single ODE, Equation 8.34d.

The fact that the original PDE had to be abandoned in favor of an algebraic balance merely confirms that in modeling, as in other endeavors, dogma often has to yield to physical reality. Acceptance of this fact is part of the art of modeling.

Equation 8.35a can be applied in a variety of ways. In its most frequently used application, it allows us to calculate the minimum bed requirement per unit of feed treatment. In the present case, this becomes:

$$W_m \text{ [kg bed/kg carrier]} = Y_F/q_F \tag{8.35b}$$

Conversely, one can use Equation 8.35a to calculate the time a column can remain on stream before breakthrough occurs. That value is perforce a maximum one, because mass transfer resistance will inevitably erode the discontinuity into an S-shaped front (see Figure 4.10), resulting in shorter breakthrough times.

Illustration 8.15 Elution of a Chromatographic Column

We turn here to the counterpart of the previous illustration and consider the elution or desorption of a uniformly loaded column with a clean purge. The same PDE as before, Equation 8.34a, applies, and it reduces to the same characteristic, Equation 8.34b and Equation 8.34c, or Equation 8.34d. What have changed are the boundary and initial conditions that are now reversed, i.e., we have:

Clean purge $$Y(0,t) = 0 \tag{8.36a}$$

Uniform initial bed $$Y(x,0) = Y_0 \tag{8.36b}$$

Both of these conditions are again represented by straight-line characteristics emanating from the ordinate and abscissa, respectively. The special case of t = x = 0 likewise leads to the same simple wave centered on the origin that we had seen before. There is, however, an important difference. None of the characteristics intersect, because they are either parallel or far away from each other. This is shown in Figure 8.7A. As a consequence, no shocks arise and the PDE and its characteristics are retained as the underlying model.

To derive the corresponding profiles, we intersect the characteristics with constant time lines, for example, t_1 and t_2. Note that the slope of the characteristics increases and the propagation velocity of a particular concentration diminishes as we move from right to left. Low concentrations will consequently lag behind higher ones, leading to the type of expanding profiles shown in Figure 8.7B.

Suppose now, that we wish to establish the time required to purge a loaded column completely from adsorbed solute. We apply the characteristic (Equation 8.34e) to the final concentration of the desorption process, i.e., Y = 0. Noting that the characteristics have a constant slope, we obtain:

$$\frac{dt}{dx} = \frac{t}{x} = \frac{\rho_b f'(Y)}{G_s} = \frac{\rho_b H}{\rho_g v} \tag{8.36c}$$

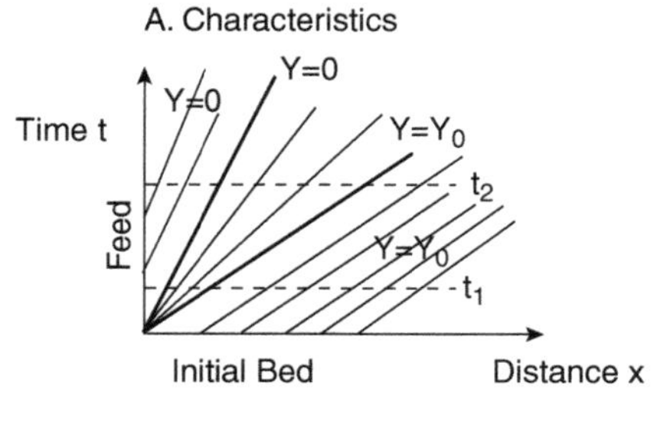

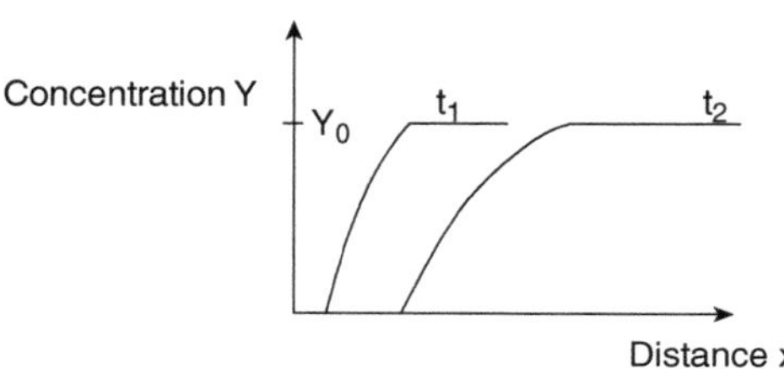

FIGURE 8.7 Characteristic diagram and concentration profiles for the desorption of a uniformly loaded chromatographic column or adsorber.

or

$$t_{des} = \frac{\rho_b H}{\rho_g v} L \tag{8.36d}$$

where H is Henry's constant and L is the length of the column.

Comments: One notes that Equations 8.36c and 8.36d are identical in form to that describing the saturation step, Equation 8.35a, with Henry's constant H taking the place of the ratio q_F/Y_F. A comparison of the two expressions also reveals that desorption is a slow, drawn out process compared to saturation, because the slope at the origin of the equilibrium, the Henry constant H, is always larger than the ratio q_F/Y_F. This fact, long known to practitioners in the field, has led to the use of a hot purge to speed up the desorption process and bring it in line with the saturation step. This becomes necessary when operating a dual-bed system, with one bed being on stream, while the other being regenerated.

The reader is reminded that the purge time calculated from Equation 8.36d is a minimum value, because the presence of transport resistance, which was neglected here, will slow down the desorption process.

Illustration 8.16 Development of a Chromatographic Pulse

Hitherto in our illustrations of chromatographic processes, we have confined ourselves to uniform boundary and initial conditions. We now consider a slightly different situation in which the initial concentration is still uniform (Y = 0), but the feed is introduced as a rectangular solute pulse of duration t_0, followed by elution with clean purge. We have, for the BC and IC:

$$Y(x,0) = 0 \tag{8.37a}$$

$$Y(0,t) = \begin{matrix} Y_F & 0 \le t \le t_0 \\ 0 & t > t_0 \end{matrix} \tag{8.37b}$$

The characteristic diagram, shown in Figure 8.8A, now consists of four sets of linear characteristics, some of which intersect, and others which do not. Let us examine each set in turn.

The initial bed concentration Y = 0 emanates, as usual, from the abscissa. An identical set of characteristics also originates from the t-axis for $t > t_0$, because the concentration in the clean purge is also Y = 0. Between these two sets lies the region of pulse introduction during $0 \le t \le t_0$, for which the characteristics are also linear

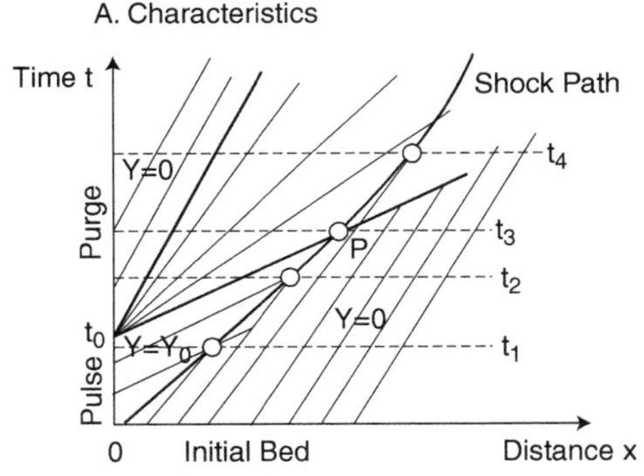

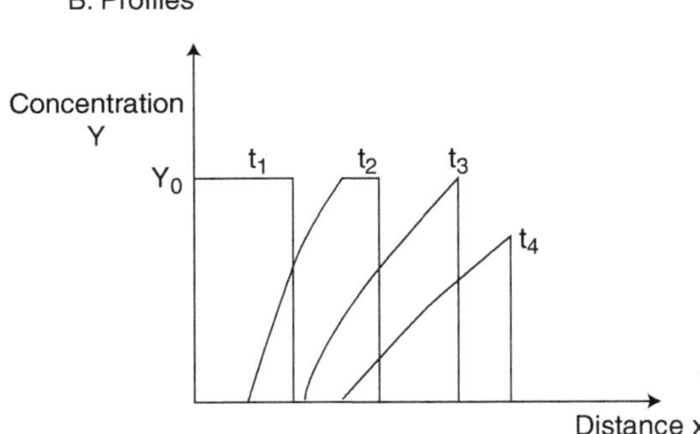

FIGURE 8.8 Deposition of a chromatographic pulse and subsequent elution with clean carrier gas. Note erosion of the plateau and diminishing shock strength with the passage of time.

but of a lower slope, because $Y_F > 0$. The fourth set, comprising a simple wave centered at $t = t_0$, was anticipated, because our previous deliberations have shown that two constant states of different velocities will always be separated by a simple wave.

Let us now examine the interactions of these characteristics. The initial bed characteristics interact with those of the pulse in much the same way as was seen in bed saturation (Illustration 8.14). The two sets intersect and, in fact, give rise to a fifth set (not shown here for clarity), consisting of a simple wave centered on the origin. The arguments we use in the saturation case lead us to the conclusion that the three sets merge into a single straight shock path OB, identical to the shock path OP seen in Figure 8.6 (BI). These shocks propagate, for the time being with a constant velocity given by the inverse of the slope of OP. At t_1, this gives the rectangular profile shown in Figure 8.8B.

When $t > t_0$, for example, $t = t_2$, and the simple wave centered on t_0 comes into play. As we move horizontally, we enter a region of diminishing solute concentrations, with ever-decreasing propagation velocities. This leads to a slow, expanding rear zone desorption whose concentrations increasingly lag behind the movement of the shock front. At $t = t_3$, this phase of profile development comes to an end. The plateau of $Y = Y_0$ has been completely eaten away, and the expanding rear joins up directly with the shock front.

What happens beyond $t = t_3$? Here, we see an intersection of the initial bed characteristics with those of the centered simple wave. Concentrations in that wave diminish with increasing values of t and result in a decrease of the height, or strength of the shock, as shown by the profile for $t = t_4$. Note that the shock path now curves upward, resulting in a lower propagation velocity of the shock front.

We do not derive quantitative relations here that require the use of an actual equilibrium relation $q = f(Y)$, but note that the construction of the characteristic diagram is, by itself, capable of revealing all the qualitative features of a chromatographic process.

Mass transfer resistance will have the effect of spreading the leading shock, so that the eluted profile takes on the shape of a skewed Gaussian distribution that flattens out with distance. When the equilibrium is linear, the profiles become symmetrical and undiminished. This is what one sees in chromatographic analysis.

Illustration 8.17 A Traffic Problem

We turn here to the application of the method of characteristics to traffic movement as described by Equations 8.31g and 8.31h. We had previously noted in Chapter 5 that the relation between vehicle velocity v and concentration C in its simplest form is described by the expression:

$$v = v_m\left(1 - \frac{C}{C_m}\right) \tag{8.38a}$$

Equation 8.38a satisfies the elementary conditions that velocity is at its maximum v_m on an empty highway where $C = 0$, and in turn drops to zero when vehicle density reaches its own maximum value of C_m. That maximum is representative of stalled, bumper-to-bumper traffic.

Substitution of Equation 8.38a into Equation 8.31h and introduction of the result into Equation 8.31g yields the characteristic:

$$\frac{dt}{dx} = \frac{1}{1 - 2\bar{C}} \tag{8.38b}$$

where $\bar{C}$ is the normalized vehicle concentration C/C_m.

Let us consider the situation where traffic has temporarily come to a halt at a red light, represented by the origin of the characteristic diagram shown in Figure 8.9A. Vehicle density to the left of the light is $\bar{C} = 1$ (bumper-to-bumper). To the right of it, $\bar{C} = 0$, representative of a road devoid of traffic. We wish to trace the vehicle movement when the light turns green.

We start by noting that the characteristics for $\bar{C} = 1$ all emanate from the negative x-axis and have a slope of −1, deduced from Equation 8.38b. Those bearing the density $\bar{C} = 0$ (no traffic), originate on the positive x-axis and all have a slope of +1.

A. Characteristics

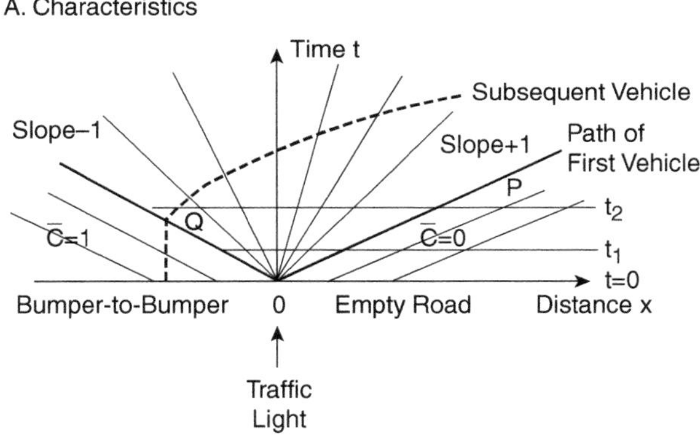

B. Profiles

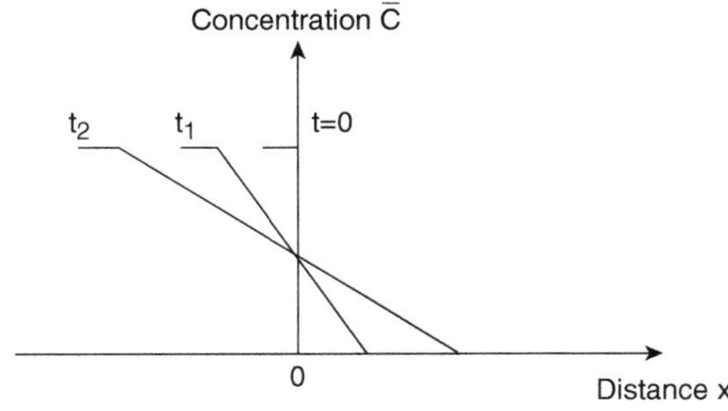

FIGURE 8.9 Traffic concentration at a red traffic light and subsequent vehicle movement when the light turns green.

These two constant states must be separated by a simple wave, which is here centered on the origin and encompasses all vehicle concentrations, between the two lines $\bar{C} = 1$ and $\bar{C} = 0$.

Movement starts when the light turns green. The resulting vehicle density "profiles" can be sketched by intersecting various horizontal lines for t = constant with the characteristics. The results are shown in Figure 8.9B and indicate that the initial discontinuous distribution at t = 0 quickly converts into a continuous profile that becomes increasingly drawn out with the passage of time.

It is important to note that the characteristics shown in Figure 8.9A describe the propagation pathways of various concentration levels, not those of the vehicles themselves. An exception occurs in the case of the first vehicle, which, facing an empty road, immediately accelerates to the maximum velocity v_m and continues its trajectory along the characteristic OP. To trace the movement of subsequent vehicles in the t-x plane, one must eliminate vehicle density C between Equation 8.38a and Equation 8.38b and integrate the result. An example of a typical pathway that is obtained in this fashion is shown in Figure 8.9A. It is left to the Practice Problems to work out the details of the solution (see Practice Problem 8.21). Note that the vehicle remains stalled until it reaches the line 0Q and gradually accelerates thereafter.

Comments: Once the application of the method of characteristics has been demonstrated by example, one is inclined to regard its relevance to traffic problems as self-evident. This is certainly not the attitude one has on first being confronted with such problems. Here is a system composed of discrete entities (the vehicles), each entity subject to the whims of the driver. Traffic lights and other control mechanisms introduce some order into the proceedings, but the flow is still intermittent and has an air of unpredictability. To describe a system so at odds with our usual transport processes by a PDE certainly required a leap of imagination, and attests to the genius of the early workers in the field. As we have noted before, such departures from conventional thinking are one of the ingredients of successful modeling.

PRACTICE PROBLEMS

8.1 A Fourier Series Expansion of a Square Wave

Develop f(x) in Fourier series in the interval (–2,2) if f(x) = 0 for –2 < x < 0 and f(x) = 1 for 0 < x < 2.

$$\text{Answer:}\quad f(x) = \frac{1}{2} + \frac{2}{\pi}\left[\sin(\pi x(2) + \frac{1}{3}\sin(3\pi x/2) + \frac{1}{5}\sin(5\pi x/2) + \cdots\right]$$

8.2 Cosine Expansion of a Sine

Write the cosine series for f(x) = sin x in the interval (0,π), and sketch the periodic extension of the result.

$$\text{Answer:}\quad \sin x = \frac{2}{\pi} + \frac{2}{\pi}\sum_{n=2}^{\infty}\frac{1+(-1)^n}{1-n^2}\cos nx$$

8.3 Drying a Porous Slab

A slab of porous material is to be dried from both faces at a constant rate W(kg/sm^2), (dielectric, microwave, or induction heating), i.e., under conditions of constant heat supply. The initial moisture content is C_0 (kg/m^3), and the movement of moisture is described by Fick's law $N = -D_{eff}\, A(\partial C/\partial x)$. If the slab is of thickness L, determine the moisture concentration profile C(x,t).

(Hint: Make the boundary conditions homogeneous.)

Answer:

$$C(x,t) - C_0 = \frac{W}{L\, D_{eff}}(Lx - x^2 - L^2/6 - 2\, D_{eff}\, t)$$

$$+\frac{2W}{L\, D_{eff}} \sum_{n=1}^{\infty} \frac{(-1)^n + 1}{n^2} \cos(n\pi x/L) \exp(-D_{eff}\, n^2\pi^2 t/L^2)$$

Comments: The solution yields $C = -\infty$ for $t \rightarrow \infty$. It is thus valid only for small finite values of time. This stands to reason, because the surface gradient cannot stay constant indefinitely but must eventually diminish as the slab dries out. The homogeneous problem also represents a slab with sealed or insulated faces and an initial parabolic concentration or temperature distribution that spreads out through the solid with time.

8.4 The Vibrating String

Consider a string fixed at both ends and subjected to an initial arbitrary displacement $u = f(x)$. The model consists of the following set of equations:

$$C^2 \frac{\partial^2 u}{\partial x^2} = \frac{\partial^2 u}{\partial t^2} \quad \text{(P8.4a)}$$

$$u(0,t) = 0 \quad \text{(P8.4b)}$$

$$u(L,t) = 0 \quad \text{(P8.4c)}$$

$$u_t(x,0) = 0 \quad \text{(P8.4d)}$$

$$u(x,0) = f(x) \quad \text{(P8.4e)}$$

Use the method of separation of variables to show that the variations in amplitude u(x,t) are given by the expression:

$$u(\alpha,t) = \frac{2}{L} \sum_{n=1}^{\infty} \left[\int_0^L f(x) \sin(n\pi x/L)\, dx \right] \sin(n\pi x/L) \cos(n\pi x/L)$$

8.5 Conduction in a Hollow Cylinder Revisited

Show that for $f_0 = 0$ upon solving for the constants C_1 to C_8 in Equations 8.14h, 8.14i, 8.14j, 8.14k, 8.14l, and 8.14m, Equation 8.14n can be reduced to the form:

$$T(r,\theta) = \frac{1}{2}\frac{\ln(R_0/r)}{\ln(R_0/R_i)}a_0 + \sum_{\lambda=1}^{\infty}\frac{(R_0/r)^{\lambda} - (r/R_0)^{\lambda}}{(R_0/R_i)^{\lambda} - (R_i/R_0)^{\lambda}}[a_n \cos\lambda\theta + b_n \sin\lambda\theta] \qquad \text{(P8.5)}$$

where the coefficients a_n, b_n are given by the integrals of Equations 8.14h, 8.14i, 8.14j, 8.14k, 8.14l, and 8.14m.

8.6 Cooling of a Solid Sphere

Consider a solid insulated sphere of unit radius and an initial temperature distribution $T(r,0) = f(r)$. Show that this distribution spreads out with time according to the expression:

$$T(r,t) = 3\int_0^1 s^2 f(s)ds + \sum_{n=1}^{\infty} C_n \frac{\sin(\lambda r)}{r}\exp(-\lambda^2 kt)$$

Identify the eigenvalues, Fourier coefficients, and the constant k. Verify that the steady state condition is satisfied.

8.7 Double Fourier Series: Vibrations of a Membrane

Solve the following boundary value problem describing the vibrations of a square membrane of length π subjected to an initial displacement $u(x,y,0) = f(x)$.

$$\frac{\partial^2 u}{\partial t^2} = c^2\left(\frac{\partial^2 u}{\partial x^2} + \frac{\partial^2 u}{\partial y^2}\right)$$

$$u(0,y,t) = u(T,y,t) = u(x,0,t) = u(x,\pi) = \frac{\partial u}{\partial t}(x,y,0) = 0$$

$$u(x,y,0) = f(x,y)$$

Procedure: Start by showing that separation of variables leads to three Sturm–Liouville systems with solutions:

$$X = \sin mx \qquad m = 1, 2, 3,\ldots$$

$$Y = \sin ny \qquad n = 1, 2, 3,\ldots$$

$$T = \cos\left[ct\sqrt{m^2 + n^2}\right]$$

Combine these results to arrive at the formal solution:

$$u(x,y,t) = \sum_{m=1}^{\infty}\sum_{n=1}^{\infty} C_{mn} \sin mx \sin nx \cos\left[ct\sqrt{m^2+n^2}\right]$$

and show that a Fourier expansion of the initial condition f(x,y) yields the following expression for the Fourier coefficients.:

$$C_{mn} = \frac{4}{\pi^2}\int_0^{\pi} \sin ny \int_0^{\pi} f(x,y)\sin mx \, dx \, dy$$

Note that this problem requires an imaginative use of the principles learned in this section.

8.8 Inversions from Other Inversions

Once certain inversion formulas are established, they can often be used to invert similarly structured functions by means of the convolution integral, item 7 of Table 2.9. Consider the two related transforms, items 2 and 3 of Table 8.3:

$$\exp\left(-k\sqrt{s}\right) \quad and \quad \left[\exp\left(-k\sqrt{s}\right)/s\right]$$

Derive the inverse of the latter by applying the convolution integral and using the known inverse of $\exp(-k\sqrt{s})$.

(Hint: Make the substitution $\lambda = k/2\sqrt{t}$.)

8.9 Inversion to Fourier Series

Invert Equation 8.25a using the extended Heaviside expansion, item 1 of Table 8.3, and converting the hyperbolic functions to trigonometric functions as shown in Illustration 8.7. Show that this leads to the solution (Equation 8.25f) that is the alternative to the error function series (Equation 8.25e).

8.10 Use of the Finite Cosine Transform

Consider a slab with an initial temperature distribution T(x,0) = f(x) and both faces at x = 0 and x = a insulated. This is a problem particularly well-suited for the finite Fourier cosine transform (item 3, Table 8.2), which has the property of transforming second derivatives to boundary conditions of Type II. Show that its application yields the solution:

$$T = \frac{1}{a}\int_0^a f(x)dx + \frac{2}{a}\sum_{n=1}^{\infty}\left[\int_0^a f(x)\cos\frac{n\pi x}{a}dx\right]\cos\frac{n\pi x}{a}\exp(-\alpha n^2\pi^2 t/a^2)$$

The expression confirms that at steady state ($t \rightarrow \infty$), the temperature in the slab becomes uniform and equal to the mean integral of the initial distribution f(x).

8.11 Fourier Transforms in Infinite Media

Integral transforms also exist for semi-infinite domains. The definitions and inversions are similar in form to those for finite domains, with the exception that the summations in the inversion formulas are replaced by integrals with a continuous integrating variable r taking the place of the integers n. The Fourier sine transform, for example, is described by the following relations:

$$\text{Definition: } S_r\{F(x)\} = \int_0^{\infty} F(x)\sin rx\,dx = f(r) \tag{P8.11a}$$

$$\text{Inversion: } S_r^{-1}\{f(r)\} = \frac{2}{\pi}\int_0^{\infty} f(r)\sin rx\,dx \tag{P8.11b}$$

$$\text{Transform or derivative: } S_r\{F''(x)\} = -r^2 f(r) + rF(0) \tag{P8.11c}$$

Devise a problem suitable for the application of this transform, and by making use of Fourier sine transform tables in the literature, arrive at a solution of the problem. Compare the result with existing solutions.

8.12 Diffusion and Reaction

Consider the system of equations:

$$K_1\frac{\partial^2 u}{\partial x^2} - K_2 u = \frac{\partial u}{\partial t} \tag{P8.12a}$$

$$u(0,t) = u_0 \tag{P8.12b}$$

$$u(\infty,t) = 0 \tag{P8.12c}$$

$$u(x,0) = 0 \tag{P8.12d}$$

The model describes unsteady diffusion and reaction in a semi-infinite medium, as well as conduction in a thin rod with convective heat exchange with the surroundings (see Illustration 7.4 in this connection). Using Laplace transformation, show that the solution is given by:

$$u(x,t) = \frac{2u_0}{\sqrt{\pi}}\int_{x/2\sqrt{K_1 t}}^{\infty} \exp\left[-\lambda^2 - \frac{K_2 x^2}{4K_1\lambda^2}\right] d\lambda$$

Use of special formulas reduces this expression to error function form.

8.13 Collision Frequency in the Coagulation of Aerosols

The theory of coagulation by Brownian motion postulates that any collision between two particles of radius R constitutes a coagulation event, and that the movement of the particles itself is described by Fick's equation.

a. Show that the model for this process in a field with an initially uniform concentration C_0 is given by:

$$D\frac{1}{r^2}\frac{\partial}{\partial r}\left(r^2\frac{\partial C}{\partial r}\right)=\frac{\partial C}{\partial t}$$

$$C(R,t) = 0$$

$$C(\infty,t) = C_0$$

$$C(r,0) = C_0$$

b. Solve the system by Laplace transformation to obtain the concentration distribution.

$$\text{Answer:}\quad C(r,t)=C_0\left[1-\frac{R}{r}erfc\left(\frac{r-R}{2\sqrt{Dt}}\right)\right]$$

c. Show that the flux at r = R, i.e., the collision frequency, is given by:

$$N=4\pi R^2 D C_0\left[\frac{1}{R}+\frac{1}{\sqrt{\pi Dt}}\right]$$

8.14 Determination of Liquid Diffusivities

Liquid diffusivities can be determined by passing a solvent over the open end of a capillary sealed at the bottom and containing a solution of the solute whose diffusivity is to be measured (see Illustration 5.2 and the accompanying Figure 5.1B). The average solute concentration in the capillary is determined at various time intervals, and the results are used to extract values of D from the solution of an appropriate model. Show that the model leads to the following expression for the percentage of solute E_t remaining in the capillary:

$$E\%=\frac{800}{\pi^2}\sum_{n=1}^{\infty}\frac{1}{(2n-1)^2}\exp[-(2n-1)^2\pi^2Dt/4L^2]$$

8.15 Integration of the Chromatographic Equation

A particularly simple form of the linear chromatographic equations is given by their nondimensionalized version, Equation 5.38m. Consider the situation where the column is initially clean, and is subjected at time t = 0 to a step change in concentration $Y(0,t) = Y_0$.

a. Apply the Laplace transformation to both equations, and eliminate the transformed concentration Y*.
b. Solve the resulting first-order ODE in the transformed fluid-phase concentration. Repeat for Y*.
c. Invert by using the convolution theorem and item 7 of Table 8.3.
d. Convert to Y by using the relation $\partial Y/\partial N = \partial Y^*/\partial T$.

The result is given by:

$$\frac{Y}{Y_0} = 1 - \int_0^N \exp(-T - N')I_0\left(2\sqrt{TN'}\right)dN'$$

where $\alpha = N, \ \beta = T.$

The right side is known as the J-Function J(N,T) and is tabulated in the literature.

8.16 The Unsteady Heat Exchanger

Show that the characteristics emanating from a point t_0 of the τ axis of the heat exchanger model given in Illustration 8.13 are described by the relations:

$$x = \int_{t_0}^{t} v(t')dt$$

$$T(x,t) = T(0,t_0)\exp(-\bar{K}) + \int_{t_0}^{t} T_s K(t')\exp(-\bar{K})dt' \qquad \text{(P8.16)}$$

8.17 Linear Chromatography

Show that for systems with linear equilibria, q = HY, adsorption and desorption times are identical.

8.18 Linear Chromatography Again

Apply the problem discussed in Illustration 8.16 to a system with a linear isotherm, q = HY, and show that:

a. The rectangular pulse moves through the column unchanged and undiminished.
b. Its velocity of propagation v_p of the pulse is given by:

$$v_p = \frac{\rho_g v}{\rho_b H}$$

8.19 The Type III Isotherm

Adsorption equilibria are often classified according to the shape of the equilibrium isotherm. The classical Langmuir equilibrium curve, for example, which is concave to the Y-axis, is termed a *Type I isotherm*. Its inverse, i.e., a curve that is convex to the Y-axis, is referred to as a *Type III isotherm*. Type II, Type IV, and Type IV have inflection points and are generally known as BET isotherms. Show that the saturation step for a Type III Isotherm yields an elongated adsorption profile, whereas desorption leads to a shock front. Sketch the resulting profiles.

8.20 The Freundlich Isotherm

Freundlich isotherms are described by the relation:

$$q = kY^{1/n}$$

where n is a positive integer ≠ 1.

Consider the equilibrium elution of a column uniformly saturated with a solute obeying the above equation. Show that the concentration at the outlet of the column (x = L) is given by the relation:

$$Y(L,t) = \left[\frac{1}{k}\left(\frac{vt}{L} - 1\right)\right]^{(1-n)/n}$$

8.21 Vehicle Pathway

Analyze the general vehicle pathway for Illustration 8.17. Show that the vehicle is at first stationary over a time interval $0 < t < t_0$, and subsequently follows a parabolic path, as shown in Figure 8.9A.

8.22 Steel Billet with Asymmetrical Boundary Conditions

A hot steel billet with a uniform initial temperature, T(x,0)= T0, is exposed to a cold environment at time t > 0. The left face of this slab is subjected to a constant temperature, i.e., T(0,t) = 0, whereas the right face is insulated, i.e., $\partial T/\partial x$ (L,t) = 0. Using the method of separation of variables, find the transient temperature profile inside this billet, T(x,t).

8.23 Flat-Plate Bioreactor

A flat-plate bioreactor is developed to study the role of oxygen in modulating cellular functions. In this system, a mixture of oxygen and nitrogen flows over a cell layer, and the concentration of oxygen is monitored at various locations within the reactor chamber. Assuming the steady state transport in a uniform flow field along the reactor (x-direction) and across the reactor (y-direction), the dimensionless differential equation for the concentration of oxygen can be written as:

$$\frac{\partial c}{\partial x} = \frac{\alpha}{Pe}\frac{\partial^2 c}{\partial y^2}$$

$$\frac{\partial c}{\partial x}(x,0) = 0$$

$$\frac{\partial c}{\partial x}(x,1) = -D_a$$

$$c(0, y) = 0$$

with $x = X/L$, $y = Y/H$, and $D_a = \rho V_{max} H/D C_{in}$. Here, H and L are the depth and length of the reactor chamber, and c is the dimensionless oxygen concentration in the gas stream defined in terms of the local concentration of oxygen, C, and the inlet concentration of oxygen, C_{in}, according to: $c = (C_{in} - C)/C_{in}$. Using the method of separation of variables, find an expression for the concentration of oxygen in the reactor chamber, c(x,y).

8.24 Semi-Infinite Solid with Convective Boundary Condition

A semi-infinite solid (x > 0) has a uniform initial temperature of zero. At time t > 0, this solid is heated by the convection of heat from the surroundings:

$$\left[k\frac{\partial T}{\partial x} - h\,(T - T_a)\right]_{x=0} = 0$$

Using the method of Laplace transformation, find an expression for the transient temperature distribution, T(x,t).

8.25 One-Dimensional Wave Propagation

Consider the one-dimensional wave propagation in an elastic string of length L vibrating as a result of an initial perturbation. We assume that the string is initially at rest with a nonzero displacement having the form of a parabola.

The differential equation and the boundary and initial conditions for this system are given as:

$$\frac{\partial^2 z}{\partial t^2} - c^2 \frac{\partial^2 z}{\partial x^2} = 0$$

$$z = 0, \quad \text{when} \quad x = 0, \quad \& \quad t > 0$$

$$z = 0, \quad when \quad x = L \quad \& \quad t > 0$$

$$z = x(L - x), \quad when \quad t = 0 \quad \& \quad 0 \le x \le L$$

$$\frac{\partial z}{\partial t} = 0, \quad when \quad t = 0 \quad \& \quad 0 \le x \le L$$

Using the Laplace transformation, find an expression for the instantaneous displacement of string, z(x,t).

(Hint: $\dfrac{1}{\mathrm{Sinh}(a)} = 2e^{-a} \sum_{n=0}^{\infty} e^{-2na} \quad for \; |e^{-2a}| < 1$).

Selected References

A. MATHEMATICAL TOPICS

1. There exists a plethora of texts and monographs dealing with ordinary differential equations (ODEs). They range from the profound and abstract to more applied and pedagogical treatments. Among the former, the following stand out:

Ince, E.I., *Ordinary Differential Equations*, Dover, New York, 1956.
Arnold, V.I., *Ordinary Differential Equations*, MIT Press, Cambridge, MA, 1980.

Textbooks dealing with the same subject addressed to students cover a fairly wide range of topics, often including treatments of nonlinear phenomena such as chaos, Fourier series with glimpses of their use in solving PDEs, and some practice in the modeling of physical systems. Two excellent and representative texts are by:

Rainville, E.D., Bedient, P.E., and Bedient, R.E., *Elementary Differential Equations*, 8th ed., Prentice-Hall, Upper Saddle River, NJ, 1997.
Edwards, C.H. and Penney, D.E., *Differential Equations and Boundary Value Problems*, 3rd ed., Pearson-Prentice Hall, Upper Saddle River, NJ, 2004.

2. The monograph:

Churchill, R.V., *Operational Mathematics*, 3rd ed., McGraw-Hill, New York, 1972,

now dating back some three decades, remains unsurpassed in its treatment of the Laplace transformation and various other integral transforms. The contents range from the profound to the applied and include numerous detailed solutions of both ordinary and partial differential equations.

A highly readable compendium of theorems and solved problems, both in ODEs and PDEs, appears in:

Spiegel, M.R., *Theory and Problems of Laplace Transforms*, Schaum, New York, 1965.

Extensive use of the Laplace transformation is also made in the area of process control. The reader will find the following texts on the topic of interest:

Coughanowr, D.R., *Process Systems Analysis and Control*, 2nd ed., McGraw-Hill, New York, 1991.
Luyben, W.L., *Process Modeling, Simulation and Control for Chemical Engineers*, 2nd ed., McGraw-Hill, New York, 1990.
Bequette, B.W., *Process Control*, Prentice-Hall, Upper Saddle River, NJ, 2004.

Tables of Laplace transforms appear in all of these texts. Churchill's and Spiegel's books carry, in addition, short tales of other integral transforms. More extensive tabulations can be found in:

Erdelyi, A. (Ed.), *Tables of Integral Transforms*, Vol. I and II, McGraw-Hill, New York, 1954.

3. The topic of partial differential equations is taken up in a host of monographs, some of which provide general treatments of the major analytical methods, whereas others are more narrowly focused on a specific solution technique. Of the former, the following is a good representative example:

Haberman, A., *Applied Partial Differential Equations with Fourier Series and Boundary Value Problems*, 4th ed., Pearson-Prentice Hall, Upper Saddle River, NJ, 2004.

For a lucid and thorough treatment of the solution of PDEs by integral transforms, the reader is referred to the monograph by R.V. Churchill, cited earlier. A companion volume by the same author provides a similarly outstanding treatment of the method of separation of variables. See:

Churchill, R.V., *Fourier Series and Boundary Value Problems*, 2nd ed., McGraw-Hill, New York, 1962, and
Churchill, R.V. and Brown, J.W., *Fourier Series and Boundary Value Problems*, 3rd ed., McGraw-Hill, New York, 1978.

of which the first is somewhat more explicit in its applications. Both texts devote separate chapters to the twin topics of Fourier series and Orthogonal functions.

The method of characteristics, which we describe briefly in our last chapter, is most often subjected to elaborate treatments that make fairly heavy reading. See for example:

Jeffrey, A., *Quasilinear Hyperbolic Systems and Waves*, John Wiley & Sons, New York, 1974.

A more readable account displaying both mathematical rigor and numerous interesting applications appears in the two-volume treatise:

Rhee, H.K., Aris, R., and Amundson, N.R., *First Order Partial Differential Equations, Vol. 1: Theory and Applications of Single Equations, Vol. 2: Coupled Systems of Equations*, Prentice-Hall, Upper Saddle River, NJ, 1986, 1989.

4. Entire monographs devoted to the topic of vector calculate are now becoming more common. See for example:

Marsden, J.E. and Tromba, A.J., *Vector Calculus*, 5th ed., W.H. Freeman, 2003.

Texts in this category generally include detailed discussions of Green's functions and Green's theorems. A more light-hearted treatment of vector calculus appears in:

Schey, H.M., *Div, Grad, Curl and All That*, Norton, New York, 1973.

B. TRANSPORT PHENOMENA. FLUID MECHANICS, HEAT TRANSFER, AND MASS TRANSFER

1. Since the introduction in 1986 of the concept of transport phenomena by Bird, Stewart, and Lightfoot, its use as a unifying theme for mass, energy, and momentum transfer has continued and grown. The original seminal work is now in its second edition. See:

Bird, R.B., Stewart, W.R., and Lightfoot, E.N., *Transport Phenomena*, 2nd ed., John Wiley & Sons, New York, 2000.

A somewhat more relaxed version of the topic, containing many examples drawn from the environmental and biological disciplines, appears in:

Welty, J.R., Wicks, C.R., Wilson, R.E., and Rorrer, G., *Fundamentals of Momentum, Heat and Mass Transfer*, 4th ed., John Wiley & Sons, New York, 2001.

2. The concept of transport phenomena has found fertile ground in a number of important subdisciplines, particularly in the biosciences. See:

Fournier, R.L., *Transport Phenomena in Biomedical Engineering*, Taylor and Francis, London, 1996.
Treiskey, F.E., Yuan, F., and Katz, D.F., *Transport Phenomena in Biological Systems*, Pearson-Prentice Hall, Upper Saddle River, NJ, 2004.
Poirier, D.R. and Geiger, G.H., *Transport Phenomena in Materials Processing, Minerals*, Metals and Materials Society, Warrendale, PA, 1994.

3. A recent trend is to limit the consideration of transport to the twin topics of heat and mass transfer. See:

Middleman, S., *Introduction to Heat and Mass Transfer*, John Wiley & Sons, New York, 1997.
Baehr, H.D., *Heat and Mass Transfer* (translated from the German), Springer-Verlag, New York, 1998.

4. Separate treatments of the three transport modes appear in a number of excellent texts dealing with fluid mechanics, heat transfer, and mass transfer. The following have been found particularly useful:

White, F., *Fluid Mechanics*, 5th ed., McGraw-Hill, New York, 2003,

which includes an extensive discussion of potential flow and the superposition of simple flows.

Kreith, F. and Bohn, R.S., *Principles of Heat Transfer*, 6th ed., Brooks/Cote, Pacific Grove, CA, 2001.
Cussler, E.L., *Diffusion: Mass Transfer in Fluids Systems*, 2nd ed., Cambridge University Press, New York, 1997.

For purely diffusive processes, and their mathematical description, the reader will wish to consult the definitive monograph by Crank:

Crank, J., *Mathematics of Diffusion*, 2nd ed., Oxford University Press, Oxford, 1978, p. 171.

The accompanying topic of heat conduction is treated in outstanding fashion in the massive compendium of solutions to Fourier's equation by Carslaw and Jaeger:

Carslaw, H.S. and Jaeger, J.C., *Conduction of Heat in Solids*, Oxford University Press, Oxford, 1959, p.266.

The work also contains numerous solutions to important heat source problems and some enlightening sections on various superposition methods.

Index

A

B

C

D

G

H

I

K

L

Q

R

S

T

V

W

X

Milton Keynes UK
Ingram Content Group UK Ltd.
UKHW021919071024
449327UK00022B/1684